Reliability-Based Design

Reliability-Based Design

Singiresu S. Rao
School of Mechanical Engineering
Purdue University

McGraw-Hill, Inc.
New York St. Louis San Francisco Auckland Bogotá
Caracas Lisbon London Madrid Mexico Milan
Montreal New Delhi Paris San Juan São Paulo
Singapore Sydney Tokyo Toronto

Library of Congress Cataloging-in-Publication Data

Rao, S. S.
Reliability-based design / S. S. Rao
p. cm.
Includes bibliographical references and index.
ISBN 0-07-051192-6
1. Reliability (Engineering) 2. Engineering design. I. Title
TA169.R37 1992
620'.00452—dc20 91-45937
CIP

1 2 3 4 5 6 7 8 9 0 DOC/DOC 9 8 7 6 5 4 3 2

ISBN 0-07-051192-6

The sponsoring editor for this book was Robert Hauserman, and the production supervisor was Pamela A. Pelton. It was set in Century Schoolbook by ETP Services Company.

Printed and bound by R. R. Donnelley & Sons Company.

To my parents
Narasimham and Manikyamma

Contents

Preface

In today's competitive world, all engineering systems are expected to perform satisfactorily over the duration of their expected life span. No system will perform reliably unless it is designed specifically for "reliability." The field of reliability-based design has come into greater prominence after the recent system failures such as the ones experienced in the space shuttle Challenger disaster and the Chernobyl (USSR) nuclear accident. This book is intended for use as an introduction to the subject of reliability-based design and reliability engineering. The theory and applications of reliability-based design are presented in a simple manner. The mathematical background necessary in the area of probability and statistics is covered briefly so as to make the presentation complete and self-contained. Expanded explanations of the fundamentals are given throughout the book, with emphasis on the physical significance of the concepts. Numerous examples and problems are used to illustrate the principles and concepts.

Features

All the concepts are explained fully and the derivations are presented with complete details for the benefit of the reader. The computational aspects of reliability-based design are adequately emphasized. Specific features include:

Nearly 130 illustrative examples that follow the presentation of most of the topics

More than 240 review questions to help readers in reviewing and testing their understanding of the text material

Over 320 references to lead the reader to specialized and advanced literature

More than 350 practice problems that aid in the practical implementation of the theory

Biographical information about the mathematicians and scientists who contributed to the development of the theories of probability and reliability is given on the opening page of each chapter.

Contents

Reliability-Based Design is organized into 15 chapters and 4 appendices. All the aspects of reliability engineering and reliability-based design are discussed comprehensively. The material of the book provides flexible options for different types of reliability courses such as: reliability based design, mechanical reliability, reliability engineering, and probabilistic methods in engineering. The relative simplicity with which the various topics are presented makes the book useful to students as well as to practicing engineers for purposes of self study and as a reference source. The mathematical background of engineering undergraduates is sufficient to understand and apply the techniques discussed in the book.

Chapter 1 gives an introduction to reliability analysis and design. The patterns of failure of electrical/electronic and mechanical/structural systems and the relationship between the factor of safety and reliability are discussed while a brief outline of the historical development of the subject of reliability is presented. Chapters 2–5 deal with the basic concepts of the theory of probability, Bayes' rule, discrete and continuous random variables, probability distributions, moments of random variables, extremal distributions, and functions of random variables. The time-dependent reliability of components and systems is presented in Chapter 6. The construction of the failure rate curve from empirical data and the analysis of series, parallel, and complex multicomponent systems are also discussed. The probabilistic modeling of component geometry, material strength and loads is the topic of Chapter 7. The statistical treatment of dimensions, including tolerances and clearances, is considered in the modeling of component geometry. Brittle, plastic, and fiber-reinforced materials are considered in the modeling of material strength. The modeling of loads includes the static, wind, and earthquake loads. Strength-based reliability and the computation of reliability using interference theory is discussed in Chapter 8. The use of experimentally determined distributions for the reliability analysis of components is also considered. Chapter 9 deals with the design of mechanical components and systems. The analysis and design of different types of machine components, including a connecting rod, pressure vessel, helical spring and fatigue design of shafts, are considered. The reliability analysis and design of multicomponent systems is illustrated by considering a gear train, cam-follower system, and four-bar mechanism as examples. Chapter 10 presents the methods of analyzing structural reliability. Both weakest link and fail safe systems are considered. Truss and frame examples are included for illustration.

The application of optimization techniques such as graphical procedures, calculus principles, and nonlinear and dynamic programming approaches are outlined along with illustrative examples in Chapter 11. The reliability allocation problem involving the minimization of total cost or maximization of system reliability for series-parallel systems is considered. The optimum designs of a two-bar truss, shoe brake, and gear train are presented as illustrative examples. The reliability analysis of maintainable and repairable systems is outlined in Chapter 12. The influence of imperfect maintenance is also examined. The optimal replacement strategy for minimum cost and the aspect of spare parts requirement are discussed. The availability analysis, based on the Markovian approach, is presented for single components as well as series and parallel systems with one or more repair persons. Chapter 13 discusses several types of design and safety review techniques, including the failure modes, event tree, and fault tree analyses. These methods aid in understanding how accidents occur, how their probabilities can be estimated, and how to reduce the probability of their occurrence. The concept of minimal cut sets and its use in the reliability analysis of systems, with the aid of fault trees, is also introduced. The design of a machine tool gear train is presented to illustrate the various concepts. Chapter 14 presents the Monte Carlo simulation method for the reliability analysis of engineering systems. The analyses of a contact stress problem, plate-clutch mechanism and a function-generating mechanism are presented as illustrative examples.

Reliability testing plays a crucial role in the successful development and operation of any system. The analysis of failure-time data and the aspects of accelerated-life testing are discussed in Chapter 15. The concepts of magnified load, sudden-death testing and sequential life testing are introduced. The roles of statistical inference and parameter estimation methods in reliability tests are also discussed in detail. Appendices A, B, and C present the critical values of standard normal-, t- and χ^2-distributions, respectively. Finally, Appendix D provides a brief introduction to product liability and its relationship to reliability.

Acknowledgments

I would like to express my appreciation to the many students who took my course, "Reliability Based Design," and whose queries and comments helped me improve the presentation of the material. The comments and suggestions made by the reviewers have been of great help in presenting some of the topics. Finally, I would like to thank my wife Kamala and daughters Sridevi and Shobha for their patience, encouragement, and support in completing the book.

Singiresu S. Rao

Reliability-Based Design

Chapter

1

Introduction

Biographical Note

Alfred Martin Freudenthal

Alfred Martin Freudenthal, born in 1906 in Austria, was the son of a civil engineer. In 1930, he was awarded the degree of Doctor of Technical Services by the German Technical University in Prague for his dissertation on the theory of plasticity. His professional career as a structural designer began in Prague (Czechoslovakia) in 1930. In 1935 he emigrated to Palestine (Israel) and later became a professor of bridge engineering at the Hebrew Institute of Technology in Haifa. In 1948, Freudenthal moved to the United States and taught first at the University of Illinois at Urbana and then at Columbia University, New York (1949–1969). He moved to the George Washington University, Washington, D.C., in 1969 where he served as professor of civil and materials engineering and director of the institute for the study of fatigue and structural reliability. He was awarded the Norman and the von Karman medals from the ASCE, nominated to the National Academy of Engineering, and received several other honors. He is considered one of the pioneers in structural reliability. His work on reliability started with a paper on the statistical aspects of fatigue that was published in the Proceedings of the Royal Society, London in 1947. His subsequent work included random fatigue failure of multiple-load-path redundant structures, structural reliability under earthquake loads, probabilistic design of maritime structures, reliability analysis of reactor components and systems, reliability assessment of aircraft structures and reliability based optimization of structures. He died in 1980 [1.117].

1.1 Definition

Due to the increasing complexity of modern engineering systems, the concept of reliability has become a very important factor in the overall system design. Reliability is important because a designer must design machinery and equipment that will work both in theory and in practice. In order to express the reliability of a system in quantitative terms, it is necessary to develop a mathematical model of the overall system and analyze its performance under realistic operating conditions. The generally accepted definition of reliability is as follows:

> Reliability is the probability of a device performing its function over a specified period of time and under specified operating conditions.

Thus, reliability can be viewed as a measure of successful performance of a system. The foregoing definition includes such diverse cases as that of a television that is switched on and off several times and is expected to function for years without failure, and that of a missile that must be in standby condition for long periods of time and then must respond just once, but it must do so without failure. In all the cases, the success in meeting reliability requirements is critically dependent on the extent to which reliability is designed and built into the system.

Reliability is an analytical problem that involves both statistical and engineering aspects. It must be given critical attention throughout the life of a system—including its development, design, production, quality control, shipping, installation, operation, and maintenance. It requires the integrated application of many disciplines: statistics, probability, materials engineering, circuit analysis, mechanical design, structural analysis, production engineering and several other engineering disciplines. Reliability also requires sound organization for the discovery and correction of the causes of failure.

1.2 Importance of Reliability

Throughout the history of modern engineering, failures of systems have been observed in every field of engineering. For example, in 1940, the Tacoma Narrows bridge collapsed after just four months of its existence because of torsional oscillations induced by a wind velocity of 42 mph. In 1943, the Schenectady, the first welded tanker built at the shipyard of the Kaiser Company, Portland, Oregon, broke in two (due to welded structural failure) while lying afloat in the calm waters of a fitting-out dock. In recent times, the space shuttle Challenger exploded in midair in January 1986. In 1985, the worst industrial accident in history occurred at the Union Carbide plant in Bhopal, India, in which thousands of people died. In 1986, history's worst nuclear-power reactor accident occurred in Chernobyl, USSR, which resulted in the leakage of radioactivity into the atmosphere. The importance of reliability of components and systems is also recognized at every stage of daily life, ranging from consumer products such as TV sets, clothes washers and dryers, lawn mowers, and automobiles, to larger systems such as trains and airlines.

In reliability studies, the distinction between a component and a system should be clearly understood because the same physical item can be treated as either a component or a system depending on the viewpoint. For example, an automobile engineer may view the engine, gear box, rear axle, and so on, as components and the complete vehicle as a drivetrain system. However, the engine manufacturer treats the engine as a system having pistons, piston rings, cylinders, crankshaft, and so on as its components. Thus, the same item, in this case the engine, can be regarded as a component or a system depending on the context. For the purpose of reliability analysis, we shall use the word *component* to denote an integral item which is nonmaintained, and the word *system* to indicate an assembly of components, which may be maintained or nonmaintained. By a *nonmaintained* component we mean that if it fails, it is removed and discarded from the aggregation of components. According to this terminology, if a failed component is repaired to original standards and returned to service, it will be called a *new component*. Similarly if a system fails due to the failure of some of its components, it is removed and discarded from the population if it is a nonmaintained system. The failed components are replaced, and the system is put into service if it is a *maintained system*.

In the case of consumer products such as electric fans, clothes washers and dryers, TV sets, and automobiles, the simplest concept of reliability (in terms of length of service) is assessed against specified values at the product inspection stage. The products passing the inspection stage are then delivered to the customers. Usually the reliability of the product is related to the price of the product. The customer buys the product with the understanding that it might fail at a future time. Although the manufacturer gives a warranty to cover the failures of the product during its early stages of life, too many failures during the warranty period causes inconvenience to the customer and high cost of repair to the manufacturer. In addition, too many product failures, either during or after the warranty period, will mean a loss of reputation which might adversely affect the future business for the manufacturer.

In the case of large mission-oriented systems, such as missiles and space shuttles, reliability (program) plays a crucial role. These systems are composed of several subsystems and components. All the individual components must be designed properly to achieve the specified overall reliability of the system. Often, it is not the complex elements that cause troubles but the simple ones; a 10-cent resistor can and has aborted the flight of a 300,000-part, million-dollar missile. In fact, much of the common loss of reliability is not mainly due to the failure of complex parts, but due to the malfunction of simple parts such as faulty joints—electrical, hydraulic, mechanical. It is well-known that it is the failure of O-rings that caused the destruction of the space shuttle Challenger, which cost the lives of seven astronauts, including the life of a school teacher.

1.3 Pattern of Failures

Before studying the methods of improving the reliability of a component or system, we need to understand the patterns and modes of failure associated with the particular component or system. In practice, a major difference can be noted between the patterns of failure of electronic/mechanical systems and structural systems. Most electronic and mechanical components and systems deteriorate during use as a result of elevated temperatures, chemical changes, mechanical wear, fatigue, and a number of other reasons. Thus the reliability of an electronic/mechanical system is closely associated with the life of the system. The time at which failure occurs will be the primary random variable in this case. On the other hand, many structural systems tend not to deteriorate with time, except by the mechanisms of fatigue and corrosion. In some cases, the structure may even get stronger with time; for example, the strength of concrete increases with time. Usually, a structural component fails when extreme loads are encountered. Hence, the pattern of failures of structural components depends on the pattern of loads exceeding the strengths of the components.

1.3.1 Component failures

For electronic, mechanical, and electromechanical components, there are three basic types in which the rate of failure varies with time. The failure rate may be decreasing, constant, or increasing. In fact, the variation over time of the failure rate for most mechanical, electronic, and electromechanical components can be represented as shown in Fig. 1.1. This failure-rate curve is commonly known as the bathtub curve because of its shape. In most cases, we can tell much about the cause of failure and the reliability of the product by knowing the shape of its failure-rate curve. In general, a decreasing failure rate is observed in the early stages of the product life. This indicates the presence of substandard components due to manufacturing defects and poor quality control procedures. As these defective products fail and are repaired or replaced, the failure-rate decreases as time progresses. The constant failure-rate indicates failures due to random causes. For example, mechanical failures caused by accidental overloads and electrical failures caused by transient circuit overloads occur randomly, usually, at a constant rate. The constant failure-rates observed during the useful life of some mechanical and electrical components are given in Table 1.1. An increasing rate of failures is observed after the useful life of the product and the increase represents, in mechanical components, fatigue brought about by strength deterioration due to cyclic loading. In engineering product design, it is important to know the underlying failure patterns so that their causes can be understood and corrective actions can be taken to improve the product reliability. The experimental data on the failure of a component are usually

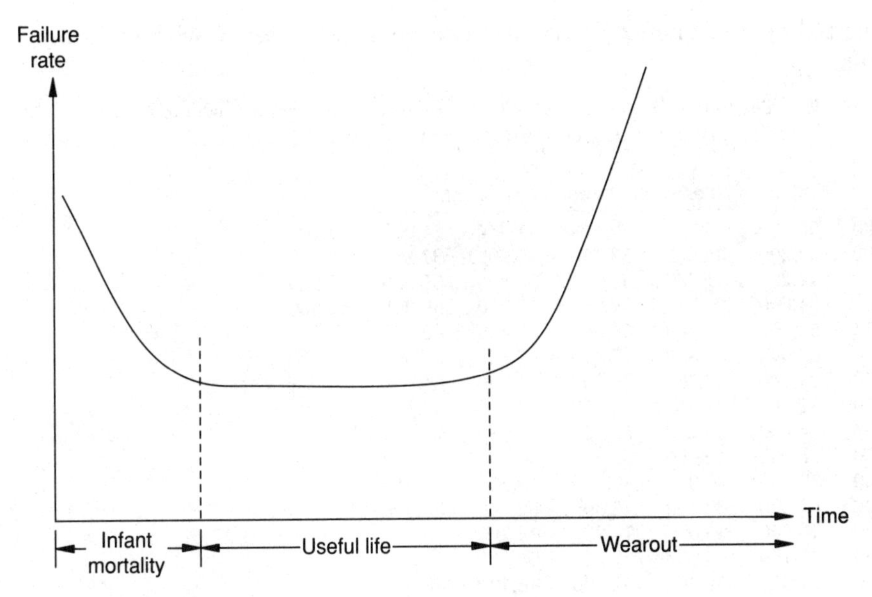

Figure 1.1 The bathtub curve.

TABLE 1.1 Representative Failure Rates of Some Mechanical and Electrical Components [1.111, 1.114]

Mechanical components	Failures per million hours	Electrical components	Failures per million hours
Accelerometer	35.1	AC generator	0.81
Actuator	50.5	Ammeter/Voltmeter	26.0
Air compressor	6.0	Circuit breaker	1.2
Air pressure gauge	2.6	Coaxial connector	0.19
Ball bearing	1.1	DC generator	36.8
Boiler feed pump	0.42	Electric heater	2.3
Brake	4.3	Emergency light	2.0
Clutch	0.6	Fractional horse power motor	1.5
Differential	15.0	Incandescent lamp	18.6
Fan	2.8	Indicator	3.9
Gasket and seal	1.3	Large electric motor	0.9
Gear	0.17	Lead acid battery	0.44
Gear shaft	6.7	Neon lamp	0.49
Gyroscope	513.9	Nickel cadmium battery	0.25
Heat exchanger	1.1	Printed circuit board	0.24
Hydraulic valve	9.3	Rechargeable battery	1.5
O-ring	2.4	Solder joint	0.001
Roller bearing	0.28	Solenoid	2.4
Shock absorber	0.81	Switch	107.3
Spring	5.0	Tachometer	10.7
Storage tank	1.6	Turbine/Generator	626.2
Thermostat	17.4	Voltage regulator	3.0

represented by a frequency distribution curve, as illustrated by the following example.

Example 1.1 The data collected on the lives of brakes of a particular model of an automobile (measured in miles of successful operation) are shown in the following table.

Lives of Brakes in Miles of Successful Operation

75,900	55,100	64,900	60,100	51,350	74,050	54,700
70,600	50,200	34,900	58,150	45,900	67,200	83,250
68,300	42,200	41,300	51,000	53,800	57,950	46,250
56,500	61,000	84,750	79,250	37,200	59,300	48,500
50,050	50,350	50,500	66,950	51,450	45,750	
26,100	33,000	56,850	47,400	47,650	59,950	
61,200	86,300	66,450	35,100	57,550	64,050	
55,450	71,450	77,450	55,850	67,850	53,350	
39,950	74,850	55,700	50,850	80,000	57,850	
64,550	66,000	45,450	62,250	60,750	67,500	
40,050	72,500	46,950	57,150	49,600	60,950	
52,800	63,800	69,600	73,300	62,800	59,600	
69,100	45,250	35,050	74,200	44,700	55,900	
31,050	47,150	56,050	69,950	53,050	38,800	
56,350	61,650	52,200	57,200	46,700	45,900	
52,650	43,500	64,750	44,200	51,700	63,450	

Show the distribution of the life of brakes on a *relative frequency diagram*. Also find the mean and standard deviation of the life of brakes.

Solution Let X denote the life of brakes. To show the distribution of the life of brakes on a relative frequency diagram, we proceed as follows:

1. The lowest- and the highest-values of life (X) observed are identified as 26,100 and 86,300 miles, respectively. The nearest rounded-off numbers are chosen for plotting purpose as 25,000 and 90,000 miles. This gives the range of life of brakes as $90{,}000 - 25{,}000 = 65{,}000$ miles.
2. The range of life of brakes is divided into thirteen equally spaced intervals of 5000 miles each.
3. The number of lives observed in each interval is counted and the following *frequency table* is set-up as:

Interval of life (in miles of operation)	Number of failures observed in the interval	Interval of life (in miles of operation)	Number of failures observed in the interval
25,001–30,000	1	60,001–65,000	14
30,001–35,000	3	65,001–70,000	10
35,001–40,000	5	70,001–75,000	7
40,001–45,000	8	75,001–80,000	4
45,001–50,000	11	80,001–85,000	2
50,001–55,000	16	85,001–90,000	1
55,001–60,000	18		

4. The number of lives observed in each interval is represented as a bar as shown in Fig. 1.2*a*. This diagram is known as a *histogram*, or *bar chart*.
5. If the ordinate in Fig. 1.2*a* is nondimensionalized by dividing the number of lives observed in each interval by the total number of lives (100 in this case), we get Fig. 1.2*b*. This diagram is known as the relative frequency diagram and shows the desired distribution of the life of brakes.
6. Often, the relative frequency diagram is approximated by a continuous curve as shown by the dotted line in Fig. 1.2*c*. This diagram can be used to obtain what is known as the probability density function of the life of brakes by rescaling the ordinate such that the area under the curve is unity.

To determine the mean value of brake life we compute the arithmetic average of the observed lives, $X_i, i=1,2,\ldots,100$.

$$\overline{X} = \text{mean life of brakes} = \frac{1}{100}\sum_{i=1}^{100} X_i$$

$$= (75{,}900 + 70{,}600 + \cdots + 48{,}500)/100 = 56{,}669.5 \text{ miles}$$

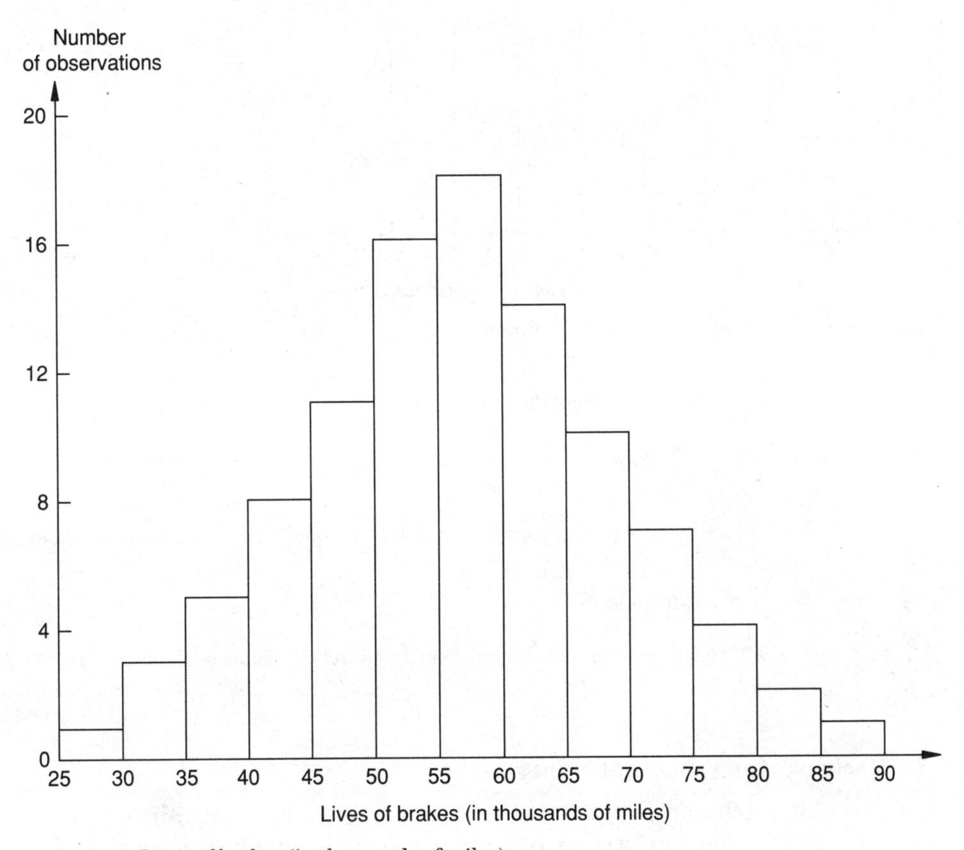

Figure 1.2*a* Lives of brakes (in thousands of miles).

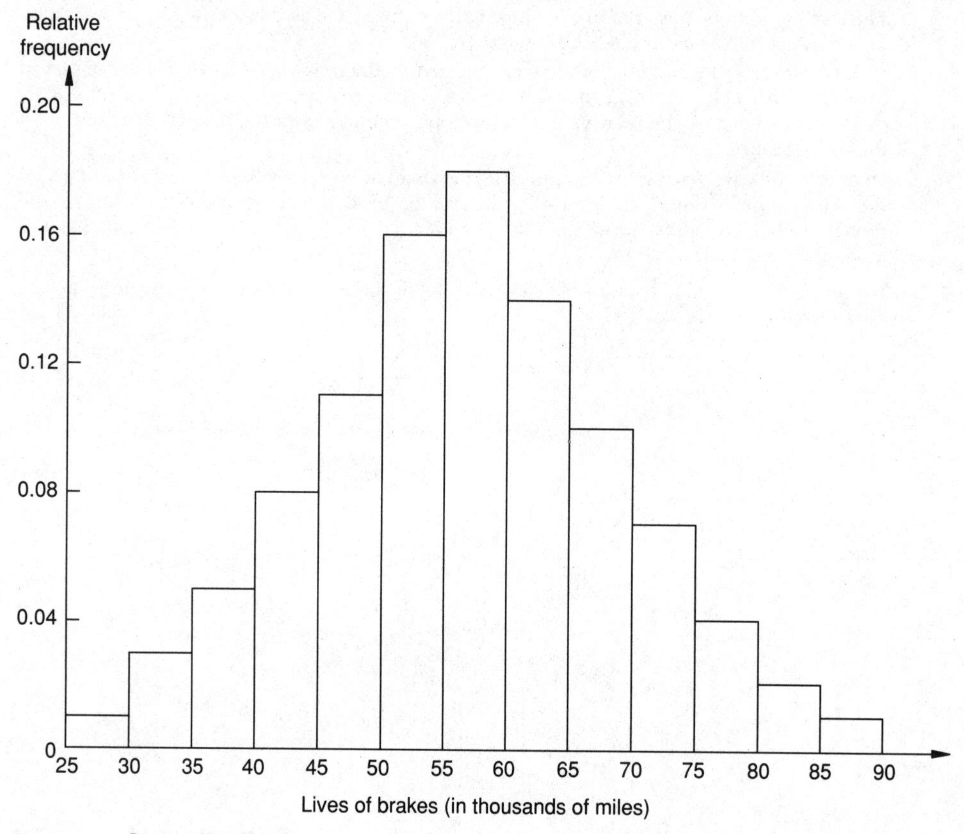

Figure 1.2b Lives of brakes (in thousands of miles).

The standard deviation of the life of brakes, σ_X, can be determined as

$$\sigma_X = \left\{\frac{1}{100}\left\{\sum_{i=1}^{100}(X_i-\bar{X})^2\right\}\right\}^{1/2}$$

$$= \left\{\frac{1}{100}\left[(75{,}900-56{,}669.5)^2+(70{,}600-56{,}669.5)^2+\cdots+(48{,}500-56{,}669.5)^2\right]\right\}^{1/2}$$

$$= 12{,}393.64 \text{ miles}$$

As will be shown later, the standard deviation indicates a measure of the *spread out* of the life of brakes.

1.3.2 Mechanical and structural failures

All mechanical and structural components have certain useful lives within which they perform satisfactorily, but beyond which they can not be used. A mechanical or structural component is considered to have failed when it

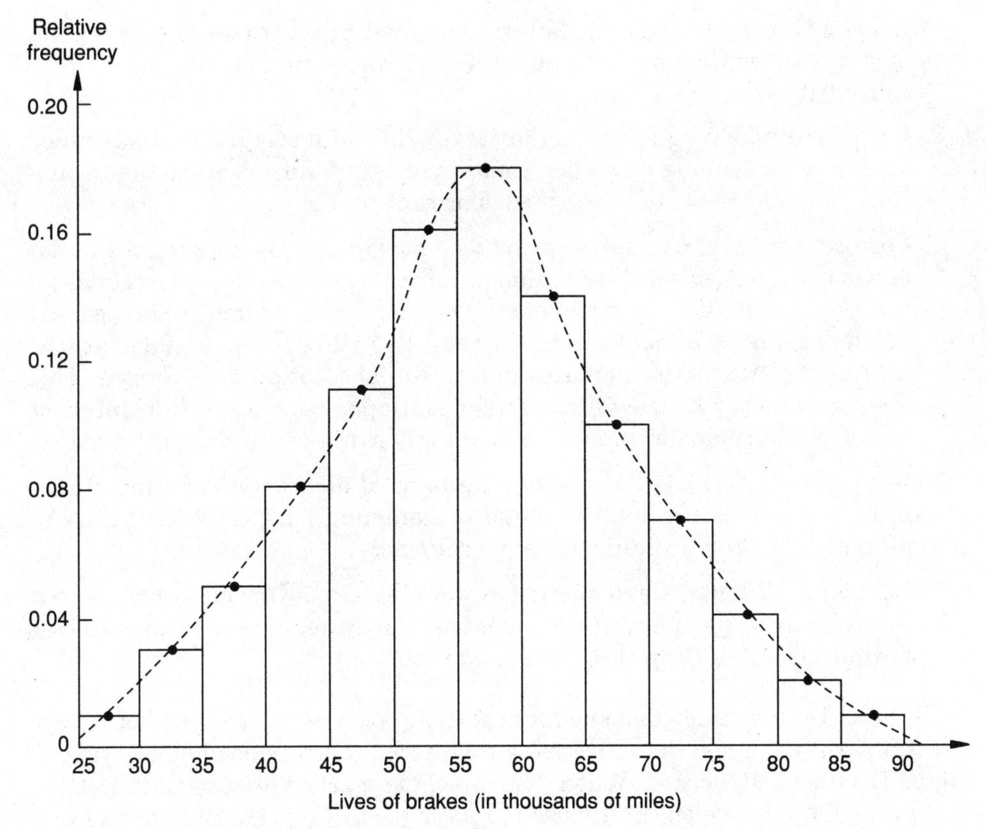

Figure 1.2c Lives of brakes (in thousands of miles).

ceases to function properly for its intended use. One reason for the failure of a component is the magnitude and type of load that it is required to withstand. There are three basic types of loads: static, dynamic, and cyclic. Buckling, creep, fracture, crushing, tearing, spalling, abrading, and wear are some types of failures that result from different types of loads.

Another cause of failure is corrosion or chemical attack by the environment. General thinning, pitting, cavitation, hydrogen embrittlement, and intergranular corrosion are some types of damage caused by corrosion. Often, several modes of failure occur simultaneously.

Some of the important failure modes and their characteristics can be summarized as follows [1.116].

1. *Static failure.* When excessive static loads, beyond the ultimate strength, are applied, most materials fail by fracture.
2. *Fatigue failure.* When cyclic loads are applied, materials can fail by fracture even when the magnitude of the maximum cyclic stress is far below

the yield strength. Initially, failures usually begin with a crack at a point of stress concentration and end with the propagation of the crack, which causes the final fracture.

3. *Creep failure.* Creep refers to the steady flow of metal under a sustained load. The continuing creep deformation causes failure when either a limiting tolerable distortion is exceeded or rupture occurs.

4. *Corrosion failure.* Corrosion is the deterioration of metal surfaces under service or storage conditions due to direct chemical or electrochemical reaction with its environment. The corrosion damage is usually accelerated by the application of stress. In hydrogen embrittlement, the ductility of the metal increases due to the absorption of hydrogen. This results in either brittle failure under static loads at low-strain rates, or fracture failure under impact loads at high-strain rates.

5. *Wear failure.* Wear is the gradual mechanical destruction of a metal surface by contact with another metal or nonmetal surface. Wear occurs in all types of contact: sliding, rolling, or impact.

6. *Instability.* When a small change in work done by external forces exceeds the strain energy stored in the member, the system becomes unstable as in the case of buckling of columns and plates.

In practice, machine components experience several other types of failure as well. For example, manufacturing tolerances can produce higher stresses than the ideal structure. When two machine parts rub together fretting fatigue will occur, which can reduce the strength by 50 percent or more. Casting and welding defects, poor surface finishes, and overtightening or faulty bolting or riveting cause larger stresses. Improper grinding may embrittle a surface. Hammering a part to straighten it can introduce residual stresses. Each type of failure requires a different analysis, inspection, and test procedure.

1.4 Factor of Safety and Reliability

The quantity that is traditionally used to maintain a proper degree of safety in structural and mechanical design is the *factor of safety*. Generally the factor of safety is understood to be the ratio of the expected strength to the expected load. The strength of the component and the acting load are assumed to be unique in conventional design. However, in practice both the strength and load are variables, the values of which are scattered about their respective expected (or mean) values. This results in an overlap in the distributed values of strength and load that might lead to the failure of the system. In fact, the interference area between the strength and load distributions can be used to compute the probability of failure of the component. The conventional design methods based on the factor of safety are not rational in the sense that the same factor of safety might imply different values of reliability in different situations.

For illustration, consider a bolted joint in a machine. If experiments are conducted to find the tensile strength S and the load (stress) applied on the bolt L in several seemingly identical machines, the results would be as indicated in Figs. 1.3*a* and *b*. These distributions (histograms) can be approximated by the curves shown by dotted lines. The average or expected values of the strength and load are indicated as $\overline{S} = 150\text{k}\,(\text{lb/in}^2)$ and $\overline{L} = 75\text{k}\,(\text{lb/in}^2)$.

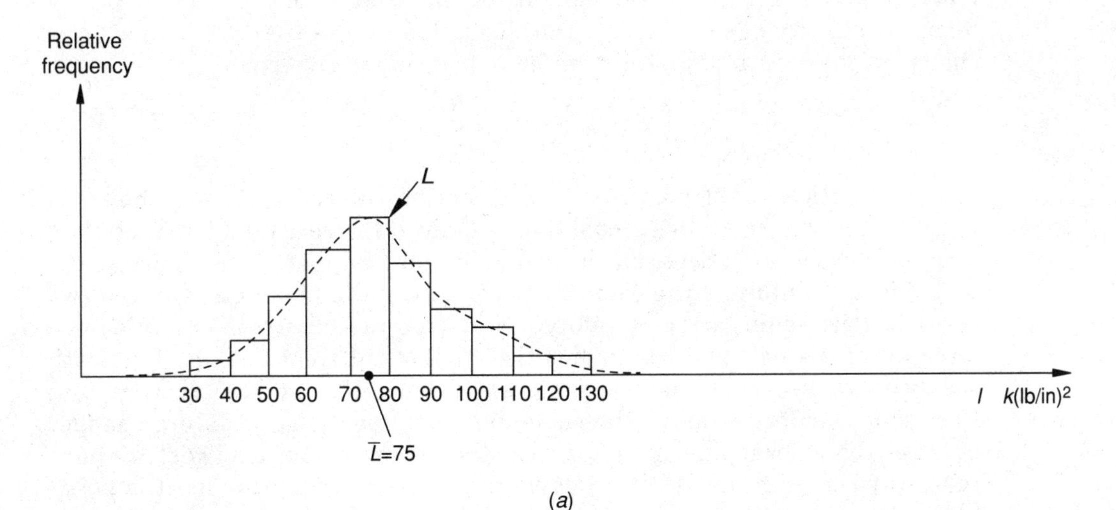

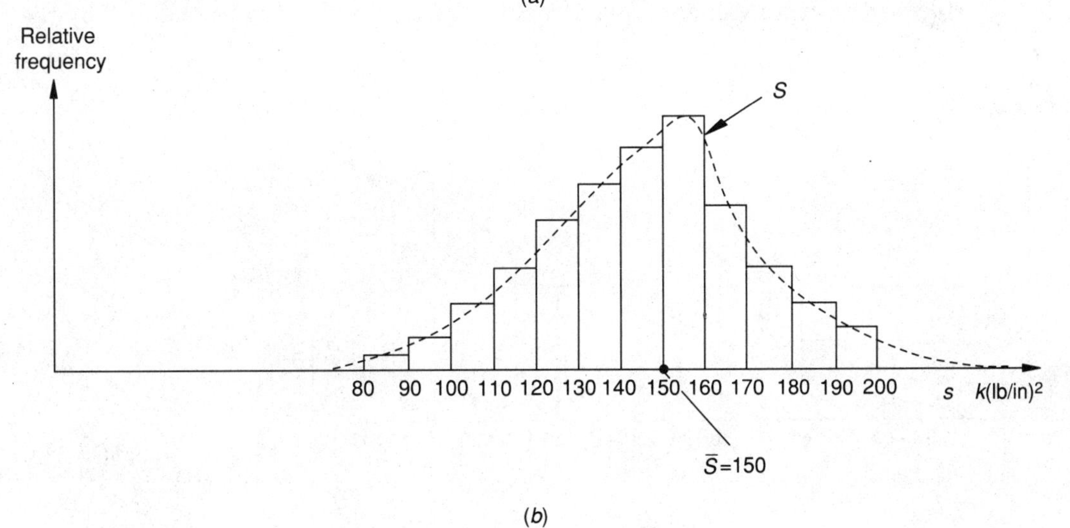

Figure 1.3

According to traditional design procedures, the computed factor of safety is greater than one ($n = \bar{S}/\bar{L} = 150/75 = 2$) and hence the bolt will not fail (in theory). However, when the distributions of bolt strength and load are shown on the same graph, the result will be indicated as in Fig. 1.4. Since there is an overlap of the distributions, there is a finite probability of the load exceeding any specified value of the strength, that is, there is a finite probability of failure of the bolt.

For example, let a different material, having the same distribution but with the mean strength equal to a constant multiplied by that of the previous material ($c\,\bar{S}$), be used for the bolt. If the mean value of the applied load is taken as $c\,\bar{L}$ (changed from the previous value by the factor c) without any change in the distribution, the new factor of safety is given by

$$n = \frac{c\,\bar{S}}{c\,\bar{L}} = \frac{\bar{S}}{\bar{L}}$$

The distributions of strength and load, for the case of $c > 1$, are shown in Fig. 1.5 [1.119]. Since the probability of failure increases with the overlapping (or shaded) area between the strength and load distribution curves, the probability of failure of the component can be seen to be different in the two cases for the same factor of safety. Next, consider the case in which the strength of the bolt and the applied load follow different distributions with the mean values remaining the same as shown in Fig. 1.6 [1.119]. Here also the factor of safety remains the same, but the probability of failure changes because the overlapping area between the strength and load distributions reduces when both the standard deviations are reduced. Finally, it is possible for the mean values of bolt strength and applied load to be changed by the

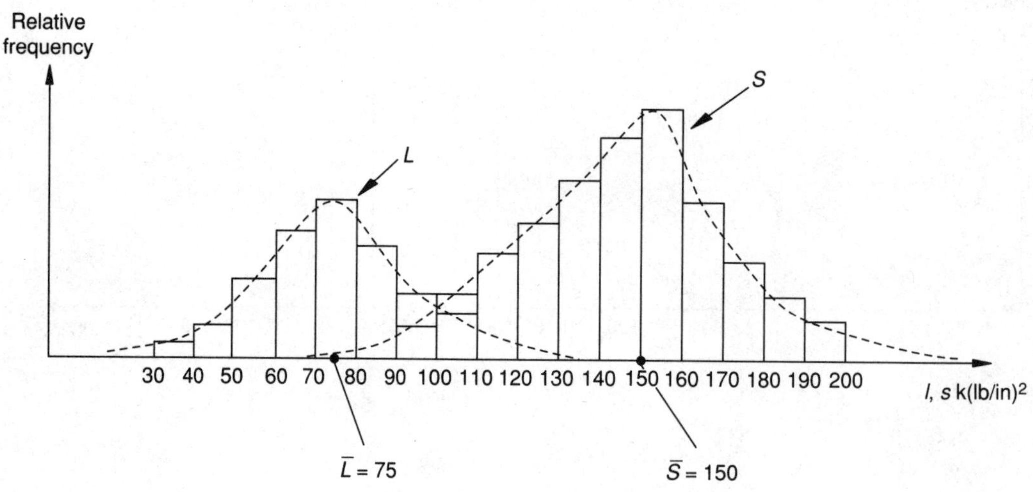

Figure 1.4

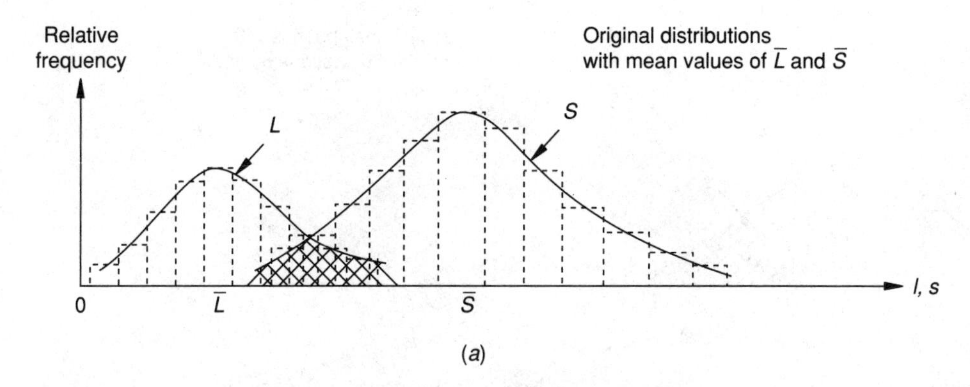

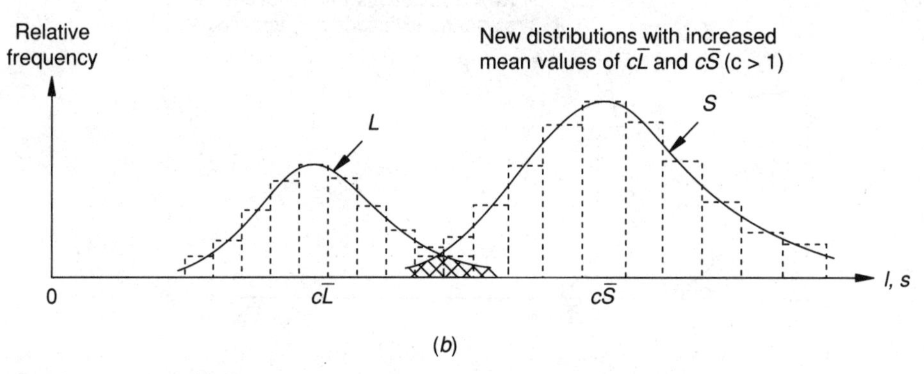

Figure 1.5 Change in probability of failure due to changes in the mean values of load (L) and strength (S).

same factor with different values of standard deviations so that the factor of safety may remain unchanged but the probability of failure will be different [1.13, 1.119].

1.5 Reliability Management

Reliability of a system or component is not achieved accidentally; it is to be built into the system or component. Reliability is an inherent characteristic of the system, similar to the system's capacity or power rating. It needs to be addressed at every stage of the product or system development including design, manufacturing, testing and maintenance phases. In the design phase, proper design methods related to the components, materials, processes, tolerances, and so on, have to be carefully selected. The objectives at this stage are to ensure that well-established design procedures are applied, known materials and processes are used, and the areas of uncertainty are highlighted for further action. When the initial hardware manufacturing is completed, tests are to be carefully planned, executed, and data

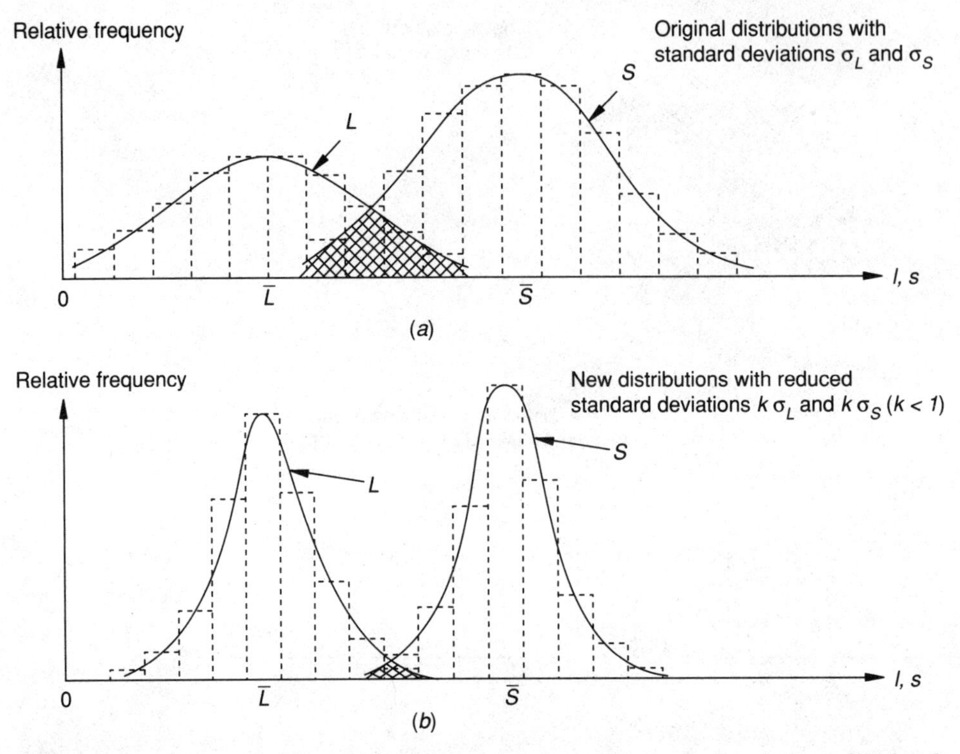

Figure 1.6 Change in probability of failure due to changes in the standard deviations of load and strength.

collected to generate confidence in the design. During the regular production stage, suitable quality control techniques are to be used to ensure that the proven design is repeated. Further testing may be done to eliminate substandard items and to improve confidence. The data collection, analysis, and both preventive and corrective maintenance actions are continued through the product-life phase. The reliabilities observed in practical use are compared with the initial predictions and, if necessary, corrective actions are taken. The data is also used to provide guidelines for the improvement and development of future products.

In many cases, a system is controlled and maintained by human operators. In such cases, human reliability [1.70] must be considered in the evaluation of the overall system reliability. In general, human reliability can be improved by designing such that the assembly, operation, and maintenance of the system are as straightforward as possible. Other factors that contribute to the improvement of human reliability include proper training of operators, low stress levels for operators, and proper positioning of gauge displays and controls. The development, design, and manufacture of a product should also take into account the product liability laws that hold the manufacturer responsible for the injuries and damages caused by a product that is judged to be defective and dangerous.[1]

Reliability can be estimated only by testing the product until it fails. Obviously, the more tests are conducted, the more confidence one can have in the estimated reliability value. Since testing involves money and time, one must consider a tradeoff between the confidence level in the estimated reliability and the cost of additional testing. Thus the implementation and management of a total reliability program for the development of a new system requires a reliability group that possesses not only an understanding of the basic mathematics of reliability, but also a knowledge of design methods, economics, system interface problems, quality control and testing techniques, and human factors.

1.6 History of Reliability Engineering

Reliability engineering emerged as a separate discipline in the USA during the early 1950s. The complexity of problems associated with the electronic systems designed during the early 1940s for use in the war effort led to the growth of the theory of reliability. It was reported that during World War II nearly 60 percent of the airborne equipment shipped to the Far East was damaged on arrival, and 50 percent of the spare parts and equipment in storage became unserviceable before they were ever used. In 1949, about 70 percent of the electronic equipment possessed by the Navy was not operating properly.

In view of these difficulties with electronic components, the Air Force formed the ad hoc Group on Reliability of Electronic Equipment in December 1950 to study the situation and recommend measures that would increase the reliability of equipment and reduce maintenance. The Navy began a lengthy study on radio tubes in April 1951. The Army started a similar investigation in June 1951. In late 1952, the Department of Defense (DOD), with a view of coordinating the efforts of the Air Force, Navy, and Army, established the Advisory Group on Reliability of Electronic Equipment (AGREE). In June 1957 AGREE published its first report on reliability [1.118]. The report concluded that reliability testing must be made an integral part in the development of new systems. The new equipment was required to be tested for several hours in a high-stress environment that involves on and off switching, low and high temperatures, and steady and vibratory conditions. The idea was to detect most of the design defects at a sufficiently early stage so as

[1]An introduction to product liability, including the methods of preventing product liability, is given in Appendix D.

to be able to correct them before the start of production. The report also recommended that the procuring agencies should accept the equipment only after getting the reliability demonstrated by the manufacturer. The reliability demonstration was to prove that a desired statistical confidence level has been achieved for the specified MTBF (mean time before failure) of the equipment. The report also provided detailed test plans for demonstrating various levels of statistical confidence. The AGREE report was accepted by the DOD and soon the AGREE recommendations became an accepted procedure [1.69]. Military equipment contractors and the related manufacturing companies were forced to invest in expensive environmental test equipment for demonstrating the reliability of their products. They very soon realized that levels of reliability could be achieved that far exceed the levels attained earlier by traditional methods. The DOD reissued the AGREE report on testing as the military standard, MIL-STD-781: Reliability Demonstration—Exponential Distribution, which was revised as MIL-STD-781B in 1967 [1.106].

The AGREE report and its recommendations on reliability were later adopted by NASA and many other major contractors, suppliers, and purchasers of high technology equipment [1.106–1.109]. In 1965 the DOD issued MIL-STD-785B: Reliability Programs for Systems and Equipment, which was revised in 1980 [1.108]. This standard made compulsory the integration of reliability activity with the traditional engineering activities of design, development, and production. This was intended to detect and eliminate potential reliability problems at an early stage in the development of a new system.

Meanwhile, mechanical structures were becoming much more complex and started posing more difficult problems, especially in aerospace and military applications. Unlike electronic components and equipment for which large amount of failure data was available, reliability data was relatively scarce on mechanical structures. The safety aspects of structures were studied as early as 1929. Similarly, the study of fatigue life in materials and the related subject of extreme value theory applicable to material strengths and loads was started in the mid-1930s. Contributions to the reliability approach to static structural design were made by Freudenthal and Pugsley. Largely motivated by vibration problems encounted in the commercial jet aircraft, Birnbaum and Saunders presented an ingenious statistical model for life lengths of structures under dynamic loading [1.44].

While dealing with reliability-based design of machines and structures, we can study the relative importance of mechanical and structural failures from the point of view of loss of human lives (see Table 1.2). It can be seen that the risk to life from mechanical and structural failures is not substantial. In spite of this good record, designers endeavor to achieve even higher reliabilities for machines and structures.

TABLE 1.2 Representative Risks of Death [1.105, 1.115]

Activity/Cause	Number of deaths per year per 10^6 persons
Air travel	9
Rail travel	4
Water transport	9
Motor vehicles	300
Poison	20
Lightning	0.5
Fires	40
Machinery	10
Structural failures	0.2

1.7 Reliability Literature

The literature on reliability engineering is large and diverse. Several text books are available [1.1–1.71] and numerous conference proceedings were published on the subject [1.72–1.105]. Also, several technical periodicals regularly or occasionally publish papers relating to reliability and reliability-based design. This is primarily because reliability affects so many disciplines, from electronic devices to machinery to airplane structures. The journals that regularly publish papers relating to reliability are:

IEEE Transactions on Reliability (published by IEEE Reliability Society)

Microelectronics and Reliability (published by Pergamon Press)

Risk Analysis (published by Plenum Press)

Probabilistic Engineering Mechanics (published by Computational Mechanics Publications)

Structural Safety (published by Elsevier Science Publishers)

Journal of Mechanical Design (published by ASME)

Reliability Engineering & System Safety (published by Elsevier Science Publishers)

Journal of Structural Engineering (published by ASCE)

The journals such as *Nuclear Engineering and Design, Nuclear Safety,* and *Engineering Optimization* occasionally publish papers dealing with reliability.

References and Bibliography

System reliability and design

1.1 I. Bazovsky, *Reliability Theory and Practice,* Prentice-Hall, Englewood Cliffs, New Jersey, 1961.
1.2 R. Billinton and R. N. Allan, *Reliability Evaluation of Engineering Systems: Concepts and Techniques,* Plenum Press, New York, 1983.
1.3 H. P. Bloch and F. K. Geitner, *An Introduction to Machinery Reliability Assessment,* Van Nostrand Reinhold, New York, 1990.
1.4 J. H. Bompas-Smith, *Mechanical Survival: The Use of Reliability Data,* McGraw-Hill, London, 1973.
1.5 A. D. S. Carter, *Mechanical Reliability,* Wiley, New York, 1972.
1.6 B. S. Dhillon, *Reliability Engineering in Systems Design and Operation,* Van Nostrand Reinhold, New York, 1983.
1.7 L. A. Doty, *Reliability for the Technologies,* Industrial Press, New York, 1985.
1.8 G. W. Dummer and R. C. Winton, *An Elementary Guide to Reliability,* 2nd ed., Pergamon Press, Oxford, 1974.
1.9 E. G. Frankel, *Systems Reliability and Risk Analysis*, Martinus Nijhoff Publishers, The Hague, The Netherlands, 1984.
1.10 H. Goldberg, *Extending the Limits of Reliability Theory,* John Wiley, New York, 1981.
1.11 D. L. Grosh, *A Primer of Reliability Theory,* John Wiley, New York, 1989.
1.12 R. P. Haviland, *Engineering Reliability and Long Life Design,* D. Van Nostrand Co., Princeton, New Jersey, 1964.
1.13 E. B. Haugen, *Probabilistic Approaches to Design,* John Wiley, New York, 1968.
1.14 E. B. Haugen, *Probabilistic Mechanical Design,* John Wiley, New York, 1980.
1.15 E. J. Henley and H. Kumamoto, *Designing for Reliability and Safety Control,* Prentice-Hall, Englewood Cliffs, New Jersey, 1985.
1.16 K. C. Kapur and L. R. Lamberson, *Reliability in Engineering Design,* John Wiley, New York, 1977.
1.17 G. Kivenson, *Durability and Reliability in Engineering Design,* Hayden Book Co., New York, 1971.
1.18 R. R. Landers, *Reliability and Product Assurance,* Prentice-Hall, Englewood Cliffs, New Jersey, 1963.
1.19 E. E. Lewis, *Introduction to Reliability Engineering,* John Wiley, New York, 1987.
1.20 C. Lipson, J. Kerawalla, and L. Mitchell, *Engineering Applications of Reliability,* The University of Michigan, Ann Arbor, 1963.
1.21 W. Nelson, *How to Analyze Reliability Data,* American Society for Quality Control, Milwaukee, Wisc., 1983.
1.22 P. D. T. O'Connor, *Practical Reliability Engineering,* Heyden & Son Ltd., London, 1981.
1.23 S. Osaki, *Stochastic System Reliability Modeling,* World Scientific Publishing Co., Singapore, 1985.
1.24 A. Pages and M. Gondran, *System Reliability: Evaluation and Prediction in Engineering,* Springer-Verlag, Berlin, 1986.
1.25 E. Pieruschka, *Principles of Reliability,* Prentice-Hall, Englewood Cliffs, New Jersey, 1963.
1.26 A. M. Polovko, *Fundamentals of Reliability Theory* (English translation), Academic Press, New York, 1968.
1.27 A. S. Pronikov, *Dependability and Durability of Engineering Products* (English translation), John Wiley, New York, 1973.
1.28 G. H. Sandler, *System Reliability Engineering,* Prentice-Hall, Englewood Cliffs, New Jersey, 1963.
1.29 M. L. Shooman, *Probabilistic Reliability: An Engineering Approach,* McGraw-Hill, New York, 1968.
1.30 J. N. Siddall, *Probabilistic Engineering Design: Principles and Applications,* Marcel Dekker, Inc., New York, 1983.
1.31 C. O. Smith, *Introduction to Reliability,* McGraw-Hill, New York, 1976.
1.32 E. J. Tangerman (ed.), "A Manual of Reliability," *Product Engineering,* May 16, 1960 pp. 65–96.

1.33 F. A. Tillman, C. L. Hwang and W. Kuo, *Optimization of Systems Reliability,* Marcel Dekker Inc., New York, 1980.
1.34 P. A. Tobias and D. C. Trindade, *Applied Reliability,* Van Nostrand Reinhold, New York, 1986.
1.35 W. H. Von Alven (ed.), *Reliability Engineering,* Prentice-Hall, Englewood Cliffs, New Jersey, 1964.

Structural reliability

1.36 G. Augusti, A. Baratta, and F. Casciati, *Probabilistic Methods in Structural Engineering,* Chapman and Hall, London, 1984.
1.37 V. V. Bolotin, *Statistical Methods in Structural Mechanics,* English translation, Holden Day, Inc., Oakland, Calif., 1969.
1.38 P. T. Christensen and M. J. Baker, *Structural Reliability Theory and Its Applications,* Springer-Verlag, Berlin, 1982.
1.39 P. T. Christensen and Y. Murotsu, *Application of Structural Systems Reliability Theory,* Springer-Verlag, Berlin, 1986.
1.40 I. Elishakoff, *Probabilistic Methods in the Theory of Structures,* Wiley, New York, 1983.
1.41 G. C. Hart, *Uncertainty Analysis, Loads, and Safety in Structural Engineering,* Prentice-Hall, Englewood Cliffs, New Jersey, 1982.
1.42 E. Leporati, *The Assessment of Structural Safety,* English translation, Research Studies Press, Forest Grove, Ore., 1979.
1.43 Y. K. Lin, *Probabilistic Theory of Structural Dynamics,* McGraw-Hill, New York, 1967.
1.44 H. O. Madsen, S. Krenk, and N. C. Lind, *Methods of Structural Safety,* Prentice-Hall, Englewood Cliffs, New Jersey, 1986.
1.45 R. E. Melchers, *Structural Reliability, Analysis and Prediction,* Ellis Horwood, Chichester, England, 1987.
1.46 T. Nakagawa, H. Ishikawa, and A. Tsurui (eds.), *Recent Studies on Structural Safety and Reliability,* Elsevier Applied Science, London, 1989.
1.47 J. W. Provan, *Probabilistic Fracture Mechanics and Reliability,* Martinus Nijhoff Publishers, Dordrecht, The Netherlands, 1987.

Reliability handbooks

1.48 "Automotive Electronics Reliability Handbook," AE-9, Society of Automotive Engineers, Inc., Warrendale, Penn., 1987.
1.49 W. G. Ireson, *Reliability Handbook,* McGraw-Hill, New York, 1966.
1.50 B. A. Kozlov and I. A. Ushakov, *Reliability Handbook,* Holt, Rinehart & Winston Inc., New York, 1970.

Reliability and maintainability

1.51 H. Ascher and H. Feingold, *Repairable Systems Reliability,* Marcel Dekker Inc., New York, 1984.
1.52 H. J. Bajaria, *Integration of Reliability, Maintainability and Quality Parameters in Design,* SP-533, Soc. of Automotive Eng., Warrendale, Penn., 1983.
1.53 B. S. Dhillon, *Systems Reliability, Maintainability and Management,* Petrocelli Books Inc., New York, 1983.
1.54 B. S. Dhillon and H. Reiche, *Reliability and Maintainability Management,* Van Nostrand Reinhold, New York, 1985.
1.55 J. W. Foster, D. T. Phillips, and T. R. Rogers, *Reliability, Availability and Maintainability,* M/A Press, Beaverton, Ore., 1981.
1.56 M. O. Locks, *Reliability, Maintainability, and Availability Assessment,* Spartan Books, dist. by Haydon Books Co., Rochelle Park, New Jersey, 1973.

Mathematical methods of reliability

1.57 L. J. Bain, *Statistical Analysis of Reliability and Life-Testing Models: Theory and Methods,* Marcel Dekker Inc., New York, 1978.
1.58 R. E. Barlow, F. Proschan, and L. C. Hunter, *Mathematical Theory of Reliability,* Wiley, New York, 1965.
1.59 B. V. Gnedenko, Y. K. Belyayev, and A. D. Solovyev, *Mathematical Methods of Reliability Theory,* English translation, Academic Press, New York, 1969.
1.60 D. K. Lloyd and M. Lipow, *Reliability: Management, Methods, and Mathematics,* Prentice-Hall, Englewood Cliffs, New Jersey, 1962.
1.61 N. H. Roberts, *Mathematical Methods in Reliability Engineering,* McGraw-Hill, New York, 1964.

Reliability and mechanisms

1.62 B. Z. Sandler, *Probabilistic Approach to Mechanisms,* Elsevier Science Publishing Co. Inc., Amsterdam, The Netherlands, 1984.

Reliability and power systems

1.63 G. J. Anders, *Probability Concepts in Electric Power Systems,* John Wiley, New York, 1990.
1.64 P. W. Becker and F. Jensen, *Design of Systems and Circuits for Maximum Reliability or Maximum Production Yield,* McGraw-Hill, New York, 1977.
1.65 B. S. Dhillon, *Power System Reliability, Safety and Management,* Ann Arbor Science, Ann Arbor, Mich., 1983.

Reliability and quality control

1.66 R. H. W. Brook, *Reliability Concepts in Engineering Manufacture,* John Wiley, New York, 1972.
1.67 N. L. Enrick, *Quality Control and Reliability,* 7th ed., Industrial Press, New York, 1977.
1.68 B. S. Dhillon, *Quality Control, Reliability, and Engineering Design,* Marcel Dekker Inc., New York, 1985.
1.69 S. Halpern, *The Assurance Sciences: An Introduction to Quality Control and Reliability,* Prentice-Hall, Englewood Cliffs, New Jersey, 1978.

Reliability, human factors and safety systems

1.70 B. S. Dhillon, *Human Reliability: With Human Factors,* Pergamon Press, Oxford, 1986.
1.71 A. E. Green, *Safety Systems Reliability,* Wiley, New York, 1983.

Conference proceedings

1.72 *Advanced Techniques in Failure Analysis Symposium—1976,* Institute of Electrical and Electronic Engineers, New York, 1976.
1.73 A. H. S. Ang and M. Shinozuka (eds.), *Probabilistic Mechanics and Structural Reliability,* American Society of Civil Engineers, New York, 1979.
1.74 *Annals of Assurance Sciences,* Proceedings of the Annual Symposium on Reliability, published by Inst. of Electrical and Electronic Engineers, Inc., New York, 1969–1974.
1.75 G. Apostolakis, S. Garribba, and G. Volta (eds.), *Synthesis and Analysis Methods for Safety and Reliability Studies,* Plenum Press, New York, 1980.
1.76 *Avionics, Reliability, Its Techniques and Related Disciplines,* Adv. Group for Aerospace Research & Dev., AGARD-CP-261, 1979.
1.77 A. P. Basu (ed.), *Reliability and Quality Control,* North-Holland, Amsterdam, 1986.

1.78 S. B. Bennett, A. L. Ross, and P. Z. Zemanick (eds.), *Failure Prevention and Reliability,* Papers presented at the Design Eng. Tech. Conf., Chicago, IL, Sept. 26–28, 1977, American Soc. of Mech. Eng., New York, 1977.

1.79 K. L. Carper (ed.), *Forensic Engineering: Learning from Failures,* Amer. Soc. of Civil Eng., New York, 1986.

1.80 P. T. Christensen (ed.), *Reliability Theory and Its Application in Structural and Soil Mechanics,* Martinus Nijhoff Publishers, The Hague, The Netherlands, 1983.

1.81 *Composite Reliability,* ASTM STP-580, Amer. Soc. for Testing and Materials, Philadelphia, Penn., 1975.

1.82 P. E. Doepker (ed.), *Failure Prevention and Reliability-1985,* Amer. Soc. of Mech. Eng., New York, 1985.

1.83 A. C. Gangadharan and S. J. Brown, Jr. (eds.), *Failure Data and Failure Analysis In Power and Processing Industries,* Amer. Soc. of Mech. Eng., New York, 1977.

1.84 K. Gibble (ed.), *Management Lessons from Engineering Failures,* Amer. Soc. of Civil Eng., New York, 1986.

1.85 K. Hennig (ed.), *Random Vibrations and Reliability,* Academie-Verlag, Berlin, 1983.

1.86 M. Grigoriu (ed.), *Risk, Structural Engineering and Human Error,* University of Waterloo Press, Waterloo, Canada, 1984.

1.87 D. Grouchko (ed.), *Operations Research and Reliability,* Gordon and Breach, New York, 1971.

1.88 E. J. Henley and J. W. Lynn (eds.), *Generic Techniques in Systems Reliability Assessment,* Noordhoff, Neyden, The Netherlands, 1976.

1.89 *International Seminar on Probabilistic and Extreme Load Design of Nuclear Plant Facilities,* Amer. Soc. of Civil Eng., New York, 1979.

1.90 G. M. Kurajian (ed.), *Failure Prevention and Reliability—1983,* Amer. Soc. of Mech. Eng., New York, 1983.

1.91 F. T. C. Loo (ed.), *Failure Prevention and Reliability—1981,* Amer. Soc. of Mech. Eng., New York, 1981.

1.92 S. M. Ma (ed.), *Effects of Deterioration on Safety and Reliability of Structures,* Amer. Soc. of Civil Eng., New York, 1986.

1.93 *Mechanical Reliability Concepts,* Amer. Soc. of Mech. Eng., New York, 1965.

1.94 *Methods for Assessing the Structural Reliability of Brittle Materials,* ASTM STP-844, Amer. Soc. for Testing and Materials, Philadelphia, Penn., 1984.

1.95 W. D. Milestone (ed.), *Reliability, Stress Analysis and Failure Prevention Methods in Mechanical Design,* Amer. Soc. of Mech. Eng., New York, 1980.

1.96 T. Moan and M. Shinozuka (eds.), *Structural Safety and Reliability,* Elsevier Science Publishing Co. Inc., Amsterdam, The Netherlands, 1981.

1.97 A. S. Nowak (ed.), *Modeling Human Errors in Structural Design and Construction,* Amer. Soc. of Civil Eng., New York, 1986.

1.98 "Power Plant Reliability," Adv. Group for Aerospace Research and Dev., AGARD-CP-215, 1977.

1.99 *Prevention of Structural Failures,* American Society for Metals, Metals Park, Ohio, 1978.

1.100 *Reliability Problems of Reactor Pressure Components,* vols. I and II, International Atomic Energy Agency, Vienna, 1978.

1.101 M. Shinozuka and J. T. P. Yao (eds.), *Proceedings of the Symposium on Probabilistic Methods in Structural Engineering,* Amer. Soc. of Civil Eng., New York, 1981.

1.102 C. Sundararajan (ed.), *Reliability and Safety of Pressure Components,* Amer. Soc. of Mech. Eng., New York, PVP-Vol. 62, 1982.

1.103 J. L. Swedlow, T. A. Cruse, and J. C. Halpin (eds.), *Proceedings of the Colloquium on Structural Reliability: The Impact of Advanced Materials on Engineering Design,* Carnegie-Mellon University, Pittsburgh, Penn., 1972.

1.104 C. P. Tsokos and I. N. Shimi (eds.), *The Theory and Applications of Reliability: With Emphasis on Bayesian and Nonparametric Methods,* vols. I and II, Academic Press, New York, 1977.

1.105 H. J. Wingerder (ed.), *Reliability Data Collection and Use in Risk and Availability Assessment,* Proceedings of the 5th EuRe Data Conference, Heidelberg, Springer-Verlag, Berlin 1986.

Military standards

1.106 MIL-STD-781B, Military Standard, Reliability Tests Exponential Distribution, Dept. of Defense, Washington, D.C., November 15, 1967.
1.107 MIL-STD-781C, Military Standard, Reliability Design Qualification and Production Acceptance Tests: Exponential Distribution, Dept. of Defense, Washington, D.C., October 21, 1977.
1.108 MIL-STD-785B, Military Standard, Reliability Program for Systems and Equipment, Development and Production, Dept. of Defense, Washington, D.C., September 1980.
1.109 MIL-STD-1629A, Military Standard, Procedures for Performing a Failure Mode Effects and Criticality Analysis, Dept. of Defense, Washington, D.C., November 1980.

Additional references

1.110 "Rationalization of Safety and Serviceability Factors in Structural Codes," CIRIA (Construction Industry Research and Information Association), Report No. 63, London, 1977.
1.111 "Digital Evaluation and Generic Failure Analysis Data," MDR-10, Reliability Analysis Center, Rome Air Development Center, Griffiss Air Force Base, New York, 1978.
1.112 Y. Fukumoto and Y. Itoh, "Statistical Study of Experiments on Welded Beams," ASCE J. of the Structural Division, vol. 107, no. ST1, pp. 89–103, Jan. 1981.
1.113 W. H. Middendorf, *Design of Devices and Systems,* 2nd ed., Marcel Dekker, New York, 1990.
1.114 "Non-Electronic Parts Reliability Data," NPRD-1, Reliability Analysis Center, Rome Development Center, Griffiss Air Force Base, New York, 1978.
1.115 N. Rasmussen, et al., "Reactor Safety Study," WASH-1400-D, U.S. Atomic Energy Commission, Washington, D.C., 1974.
1.116 N. A. Tiner, *Failure Analysis with the Electron Microscope,* Fox-Mathis Publications, Los Angeles, CA, 1973.
1.117 *Selected Papers by Alfred M. Freudenthal. Civil Engineering Classics,* American Society of Civil Engineers, New York, 1981.
1.118 "Reliability of Military Electronic Equipment," *Report of the Advisory Group on Reliability of Electronic Equipment (AGREE)*, Office of the Assistant Secretary of Defense, Washington, D.C., June 4,1957.
1.119 D. Kececioglu and D. Cormier, "Designing a Specified Reliability Directly into a Component," *Proc. of the Third Annual Aerospace Reliability and Maintainability Conf.*, Soc. of Automotive Engineers, Inc., New York, June 1964, pp. 546–565.

Review Questions

1.1 Define reliability.

1.2 What is the difference between a component and a system?

1.3 What is a bathtub curve?

1.4 Why does the failure rate decrease in the early stages of a component's life?

1.5 Why are the failure rates constant during the useful lives of most components and systems?

1.6 Give reasons for the increasing failure rate of a component or a system after its useful life.

1.7 Give the failure rates of three mechanical components.

1.8 Indicate the failure rates of three electrical components.

1.9 What are the various modes of failure of a mechanical or a structural component?

1.10 Define the term "factor of safety." Why is it considered to be inadequate to indicate the safety of a system?

1.11 What is the AGREE report?

1.12 State two common activities of a human being and the associated risks of death.

1.13 Name two journals that publish papers on reliability.

1.14 Identify a specific engineering system and state possible causes of its failure.

Problems

1.1 The following sample data represent the Young's modulus of steel in 10^6 lb/in^2:

25.1	26.5	27.4	28.4	29.1
29.9	31.2	32.3	34.7	27.2
28.1	29.2	29.8	30.1	31.3
32.5	26.9	30.3	25.9	27.6
28.5	29.3	30.4	31.4	32.7
29.4	30.5	31.6	32.8	28.3
25.4	28.6	29.5	30.6	31.8
33.4	28.3	28.7	29.6	30.8
31.9	33.8	30.9	29.6	27.7
26.6	26.8	27.8	28.9	29.7

a. Construct the histogram using intervals of 10^6 lb/in^2 with 25×10^6 lb/in^2 as the lower limit of the first interval.
b. Plot the relative frequency diagram.
c. Find the mean value and standard deviations of the Young's modulus.

1.2 The eccentricity of the applied load (in inches) in a sample of nominally identical columns is shown as follows:

0.410	0.050	0.090	0.195	0.345
0.155	0.320	0.120	0.290	0.065
0.275	0.230	0.140	0.265	0.215
0.070	0.115	0.305	0.435	0.130
0.535	0.110	0.205	0.085	0.135
0.125	0.185	0.480	0.175	0.145
0.380	0.165	0.255	0.180	0.240
0.220	0.105			

Plot the histogram and the corresponding frequency-distribution curve.

1.3 The maximum load carried by 34 nominally identical welded beams (in kilonewtons) before failure [1.112] are given in the following table:

176.0	157.0	131.0	143.1	154.1
164.8	124.6	150.0	161.9	129.3
168.0	140.9	123.1	165.1	139.1
170.9	151.3	141.3	152.8	128.2
186.9	129.3	153.6	166.9	139.2
147.4	156.9	136.0	147.3	

Construct the histogram and the corresponding relative frequency diagram.

1.4 The compressive strengths of a sample of concrete test cylinders are given in the following table [in k(lb/in^2)]:

5.9	6.2	5.8	7.8	6.5
6.3	8.9	5.3	3.7	1.4
2.1	6.8	9.1	4.3	3.2
7.2	6.1	5.7	4.9	2.6
3.4	6.8	8.3	5.1	7.3
8.2	7.7	5.4	3.7	4.5
4.1	5.6	6.4	6.7	7.9
6.9	7.5	5.2	4.3	6.6
5.4	6.4			

Plot the histogram and the corresponding probability density function. Also determine the mean and standard deviations of the compressive strength.

1.5 Bars made of two different grades of steel are used for reinforcing concrete in a construction project. Experiments conducted on a sample of the reinforcing bars gave the following results for the yield strength in k(lb/in^2):

35.7	40.9	43.4	30.2	35.5
31.1	43.3	40.8	38.1	33.7
33.2	38.8	39.6	41.5	32.5
42.5	40.4	33.8	31.2	40.3
41.2	42.9	36.6	31.7	41.8
42.8	38.4	39.9	34.6	32.2
37.5	41.7	32.3	41.1	40.6
40.7	42.7	32.6	37.2	33.4
42.3	40.1	32.9	39.5	41.6
42.2	41.4	34.5	39.3	41.3
34.1	39.2			

- **a.** Plot the histogram.
- **b.** Plot the relative-frequency diagram.
- **c.** How do you interpret the shape of the relative frequency diagram?
- **d.** What are the mean value and standard deviations of the yield strength?

Chapter

2

Basic Probability Theory

Biographical Note

Jerome Cardan

Jerome Cardan (or Gerolamo Cardano), born in Pavia, Italy on September 24, 1501, is considered to be the pioneer in the field of probability. Encouraged by his father, Fazio Cardano, who was a lawyer and professor of medicine in Milan as well as a friend of Leonardo da Vinci, Cardan began his university studies in 1520 at Pavia and completed them in 1526 at Padua with a doctorate in medicine. Cardan has an interesting personality. In 1534, while practicing medicine, he joined as a teacher of mathematics. He wrote on many subjects such as mathematics, astronomy, physics, music, and games of chance. He was put in prison in 1570 for publishing the horoscope of Christ. The book he wrote on gambling Liber de Ludo Aleae (Book on Games of Chance), *is considered to be the first book on probability. The book describes numerous problems relating to the games of dice and cards, and includes tips on how to cheat and avoid being cheated. The book also introduces the concepts of probability, mathematical expectation, and the law of large numbers. Cardan discovered the method of solving a cubic equation which has become known as the "Cardan's method." His multi-faceted personality ended dramatically with his prediction of the day of his own death. When the day arrived he was still alive, so he committed suicide to preserve his reputation [2.6–2.9].*

2.1 Introduction

Every phenomenon in real life has a certain element of uncertainty. For example, the strength of a beam, the wind velocity at a particular locality, the number of vehicles crossing a bridge, and the life of a machine cannot be predicted exactly. These phenomena are chance dependent and one has to resort to *probability theory* to describe the characteristics of such phenomena [2.1–2.3]. This chapter provides the basic concepts of the probability theory that are necessary for the continuity of presentation in this book.

Before introducing the concept of probability, it is necessary to define the terms *experiment* and *event*. An experiment denotes the act of performing something the outcome of which is subject to uncertainty and not known exactly. For example, tossing a coin, rolling a die, and measuring the yield strength of steel can be called experiments. The number of possible outcomes in an experiment may be finite or infinite depending on the nature of the experiment. The outcome is a head or a tail in the case of tossing a coin, and any one of the numbers 1, 2, 3, 4, 5, and 6 in the case of rolling a die. On the other hand, the outcome may be any real number, say, between 28,000 and 32,000 lb/in^2 in the case of measuring the yield strength of steel.

An event represents the outcome of a single experiment. For example, realizing a head on tossing a coin, getting an even number (2 or 4 or 6) on rolling a die and observing the yield strength of steel to be greater than 30,000 lb/in^2 in a measurement can be called events. Several possible outcomes (events) may be realized for a given experiment. For example, in the case of rolling a die, some of the events that may be realized are as follows:

1. getting number 1,
2. getting number 4,
3. getting an odd number (1 or 3 or 5),
4. getting a number greater than 3,
5. getting numbers 2 and 3 simultaneously (impossible event), and
6. getting any of the numbers 1, 2, 3, 4, 5, and 6 (certain event).

2.2 Mutually Exclusive Events

If the occurrence of an event precludes the occurrence of other events in a given experiment, the events are called *mutually exclusive*. This means that two events are mutually exclusive if both events could not occur simultaneously. The event of tail and the event of head occurring simultaneously in a toss of a coin is an example of mutually exclusive events.

2.3 Set Theory

The outcome of an experiment can be represented conveniently using the relationships of set theory. A set is a well defined collection of objects, so that given any object, we can say whether it belongs to the set or not. Each object of the set is called an element of the set. The set of all elements under consideration is called a universal set and the set containing no element is called an empty or null set.

Union of two sets. The union of two sets A and B is defined as the set of all elements that belong to A or B or both, and is denoted by the symbol $A \cup B$, which reads as "A-union-B." For example, if A denotes the set of numbers (1, 2, 3) and B denotes the numbers (1, 3, 4), the union of A and B represents the set of numbers (1, 2, 3, 4).

Intersection of two sets. The intersection of two sets A and B is defined as the set of elements that belong to both A and B, and is denoted by the symbol $A \cap B$, which reads as "A-intersection-B." For example, if A denotes the set of numbers (1, 2, 3) and B denotes the numbers (1, 3, 4), the intersection of A and B represents the set of numbers (1, 3).

Complement of a set. The complement of a set A, denoted as $\overline{A}$, is the set of all those elements of the universal set which do not belong to A. For example, if the universal set contains the hundred numbers $1, 2, \ldots, 100$, and the set A consists of the numbers $1, 2, \ldots, 10$, the complement of A consists of the remaining ninety numbers so that $\overline{A} = (11, 12, \ldots, 100)$.

A set A is said to be a subset of a set B if whenever an element x belongs to A, it also belongs to B, and we denote it as $A \subset B$ which reads as "A belongs to B or A is contained in B."

Commutative rule. The union and intersection of sets are commutative, that is, if A and B are two sets, then $A \cup B = B \cup A$ and $A \cap B = B \cap A$

Associative rule. The union and intersection of sets are associative, that is, if A, B and C are three sets, then $(A \cup B) \cup C = A \cup (B \cup C)$ and $(A \cap B) \cap C = A \cap (B \cap C)$

Distributive rule. The union and intersection of sets are distributive, that is, if A, B and C are three sets, then $(A \cup B) \cap C = (A \cap C) \cup (B \cap C)$ and $(A \cap B) \cup C = (A \cup C) \cap (B \cup C)$

de Morgan's rule. This rule states that the complement of unions (intersections) is equal to the intersections (unions) of the respective complements. Thus, if A and B are two events and $\overline{A}$ and $\overline{B}$ are their complements, then de Morgan's rule states that $\overline{A \cup B} = \overline{A} \cap \overline{B}$ and $\overline{A \cap B} = \overline{A} \cup \overline{B}$

2.4 Sample Points and Sample Space

The outcome of an experiment or trial can be represented conveniently with the help of set theory. A sample space defines the set which includes all the possible outcomes of an experiment. For example, in the die-rolling experiment, the sample space consists of the numbers {1, 2, 3, 4, 5, 6}. Similarly, if S represents the yield strength of steel, then its sample space is given by $\{0 < s < \infty\}$. Thus the sample space may be finite or infinite depending on whether the number of outcomes is finite or infinite, respectively. Each element of the sample space is called a sample point. Thus a sample point represents a possible outcome of a trial. An event can be defined as a subset of the sample space. It is often convenient to associate a geometric figure called a Venn diagram with the ideas of sample space (set) and event (subset).

Example 2.1 A thermal power plant buys four boilers. If there is a 50 percent chance that each boiler works for a year without any breakdown, identify the possible events at the end of the first year.

solution Since each boiler can be either in good or operating condition (G) or in failed condition (B) after one year, the possible states at the end of one year can be represented as follows:

Boiler Condition after One Year of Operation

Condition of first boiler	Condition of second boiler	Condition of third boiler	Condition of fourth boiler	Total number of boilers in good condition
G	G	G	G	4
G	G	G	B	3
G	G	B	G	3
G	B	G	G	3
B	G	G	G	3
G	G	B	B	2
G	B	B	G	2
B	B	G	G	2
B	G	G	B	2
G	B	G	B	2
B	G	B	G	2
G	B	B	B	1
B	G	B	B	1
B	B	G	B	1
B	B	B	G	1
B	B	B	B	0

Each of these outcomes, namely, GGGG, GGGB, . . . , BBBB can be called a sample point. Thus, the sample space of the experiment (of observing the condition of four boilers at the end of one year) consists of sixteen sample points, and it can be represented in a Venn diagram as

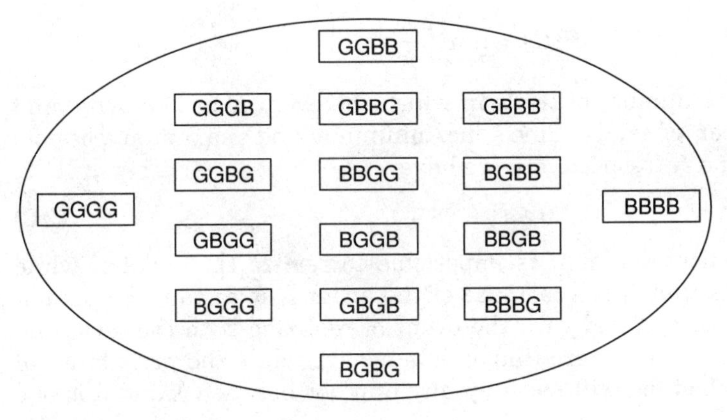

Figure 2.1

shown in Fig. 2.1. Some of the possible events and the sample points corresponding to them can be identified as follows:

Event	Physical meaning of the event	Sample points corresponding to the event				
E1	All four boilers in operating condition	GGGG	——	——	——	——
E2	Two boilers in operating condition	GGBB BGBG	GBBG	BBGG	BGGB	GBGB
E3	One boiler in operating condition	GBBB	BGBB	BBGB	BBBG	——
E4	None in operating condition	BBBB	——	——	——	——
E5	At least three in operating condition	GGGG	GGGB	GGBG	GBGG	BGGG
E6	At most two in operating condition	GGBB BGBG BBBB	GBBG GBBB ——	BBGG BGBB ——	BGGB BBGB ——	GBGB BBBG ——

2.5 Definition of Probability

Probability can be defined in several ways. We shall consider two definitions, namely, the relative frequency and the axiomatic definitions in this section.

2.5.1 Relative frequency (statistical) definition

The probability of occurrence of an event E is defined as the ratio of the number of occurrences of the event E to the total number of trials. Thus $P(E)$ is defined as

$$P(E) = \lim_{N \to \infty} \left(\frac{n}{N} \right) \tag{2.1}$$

where n denotes the number of trials in which the event E has occurred and N the total number of trials. Since the minimum and maximum possible values of n are 0 and N, respectively, we have

$$0 \le P(E) \le 1 \tag{2.2}$$

where $P(E) = 0$ indicates that it is impossible to realize the event E while $P(E) = 1$ represents that the occurrence of the event E is certain. For example, the probability associated with the event of realizing both the head and the tail on tossing a coin is zero (impossible event), while the probability of the event that a rolled die will show up any number between 1 and 6 is one (certain event).

2.5.2 Axiomatic definition

According to axiomatic approach, the probability of occurrence of an event E is defined as a number $P(E)$ such that $P(E)$ obeys the following three axioms or postulates

1. $P(E) \ge 0$
2. $P(E) = 1$ if E is a certain event
3. If E_1 and E_2 are mutually exclusive events, then

$$P(E_1 \cup E_2) = P(E_1) + P(E_2) \tag{2.3}$$

Since the foregoing three axioms are essentially assumptions, they need no proof. However, these axioms and the resulting theory must be useful and consistent with the practical problems.

Although any of the two definitions of probability can be used for the development of probability theory, the axiomatic approach has been used widely in the literature.

2.6 Laws of Probability

2.6.1 Union and intersection of two events

If E_1 and E_2 denote two events, the union of E_1 and E_2 (denoted as $E_1 \cup E_2$ or less commonly as $E_1 + E_2$) means the occurrence of E_1 or E_2 or both. The intersection of E_1 and E_2 (denoted as E_1E_2 or $E_1 \cap E_2$) means the joint occurrence of E_1 and E_2. The union and the intersection operations are most easily explained in terms of the Venn diagram as shown in Fig. 2.2. Here the set E_1 is composed of the disjoint subsets A_1 and A_2 and set E_2 is composed of the disjoint subsets A_2 and A_3. The subset A_2 denotes the points common to E_1 and E_2, while the subset A_1 represents points in E_1 but not in E_2, and A_3 represents points that belong to E_2 but not E_1. The union of the sets E_1 and E_2 denotes the points belonging to the subsets A_1, A_2, and A_3 and the intersection of the sets E_1 and E_2 denotes the points belonging to the subset A_2. For example, if E_1 denotes the shortage of materials and E_2 denotes the shortage of skilled workers in an automobile plant, $E_1 \cup E_2$ is the shortage of materials or skilled workers, or both, and $E_1 \cap E_2$ is the shortage of materials and skilled workers. Mathematically, we can write that

$$P(E_1 \cup E_2) = P(E_1) + P(E_2) - P(E_1 E_2) = \text{total crossed area in Fig. 2.2} \qquad (2.4)$$

and

$$P(E_1 \cap E_2) = \text{double crossed area in Fig. 2.2} \qquad (2.5)$$

2.6.2 Mutually exclusive events

When the two sets E_1 and E_2 have no common elements, the areas do not overlap in the Venn diagram as shown in Fig. 2.3, and they are called disjoint or mutually exclusive sets. In this case, the two sets contain no common elements, and hence their intersection is a *null set*.[1] In other words, the two

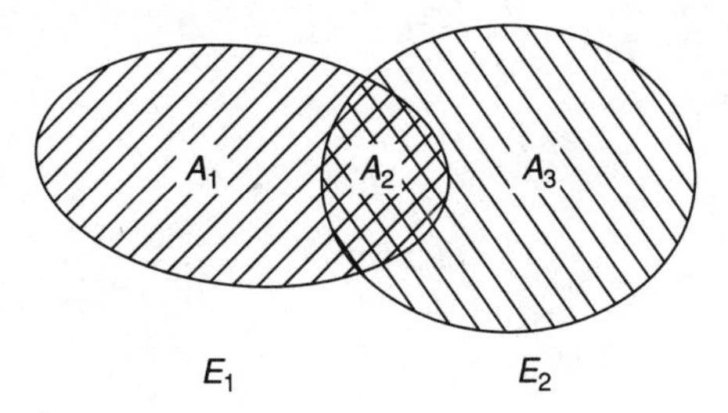

Figure 2.2

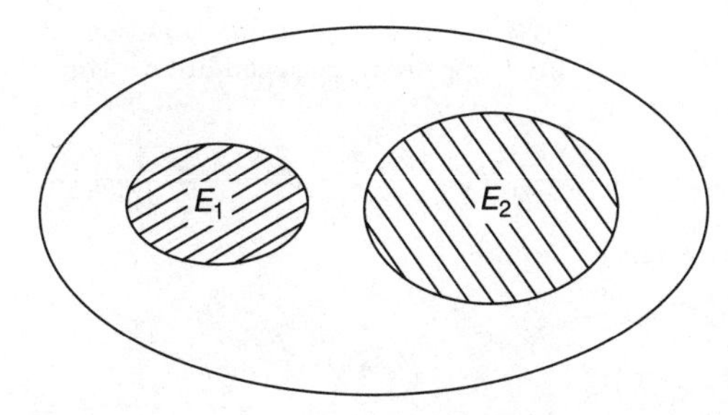

Figure 2.3

[1]A set with no elements is called a *null set*.

events E_1 and E_2 cannot occur simultaneously if they are mutually exclusive. For example, if E_1 denotes rain in a locality and E_2 denotes drought in the same locality, then the Venn diagram has no overlapping region. In this case, E_1 and E_2 are disjoint or mutually exclusive. Hence the intersection $E_1 \cap E_2$ will be a null set, that is, the probability of simultaneous occurrence of E_1 and E_2 will be zero, that is

$$P(E_1 \cap E_2) = 0. \tag{2.6}$$

and, from Eq. (2.4), we get

$$P(E_1 \cup E_2) = P(E_1) + P(E_2) \tag{2.7}$$

Example 2.2 Find (1) the probability of having one boiler left operative at the end of the first year and (2) the probability of having at least three boilers operative at the end of the first year in Example 2.1.

solution Among the sixteen possible outcomes listed in Example 2.1, only one can be realized at the end of the first year. If one possibility (event) is realized, none of the other fifteen possibilities (events) can be realized. Hence, the sixteen events are mutually exclusive. Because $P(G) = P(B) = 1/2$, all the sixteen events are equally likely to occur and hence the probability of realizing any one of the sixteen possibilities is $1/16$.

1. Now, if we define the event E, as "exactly one boiler operating at the end of the first year," it consists of the four sample points GBBB, BGBB, BBGB, and BBBG.

$$\therefore P(E_1) = \frac{4}{16} = \frac{1}{4}$$

Similarly, we can obtain that

$$P(E_2) = P\,(\text{Exactly two boilers operating}) = \frac{6}{16} = \frac{3}{8}$$

$$P(E_3) = P\,(\text{Exactly three boilers operating}) = \frac{4}{16} = \frac{1}{4}$$

$$P(E_4) = P\,(\text{Exactly four boilers operating}) = \frac{1}{16}$$

$$P(E_0) = P\,(\text{No boiler operating}) = \frac{1}{16}$$

2. If E denotes the event that "at least three boilers in operating condition," it implies that either all four are in good condition or three are in good condition. Thus the event E is equivalent to the union of E_3 and E_4.

$$\therefore P(E) = P(E_3 \cup E_4) = P(E_3) + P(E_4) - P(E_3 \cap E_4) = \frac{1}{4} + \frac{1}{16} = \frac{5}{16}$$

since E_3 and E_4 are mutually exclusive events.

2.6.3 Complementary events

Associated with any event E there is another event known as the complementary event (designated as $\overline{E}$). If the sample space corresponding to the event E is represented as S in the Venn diagram of Fig. 2.4, the remainder of the space $\overline{S}$ denotes the complement or negation of E, $\overline{E}$. It indicates that E did

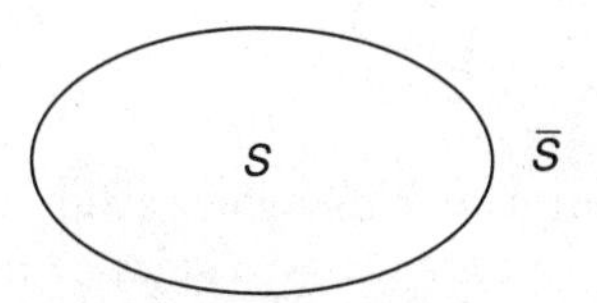

S = sample space of E
$\overline{S}$ = sample space of $\overline{E}$

Figure 2.4

not occur. Hence the whole space represents the sample spaces of the two events E and $\overline{E}$. If $P(E)$ and $P(\overline{E})$ are their respective probabilities, then

$$P(E) + P(\overline{E}) = 1 \tag{2.8}$$

For example, if E denotes the event "appearance of number 6 upon rolling of a die," then $\overline{E}$ represents the event "appearance of any other number." Hence, for a fair die,

$$P(E) = \frac{1}{6} \qquad \text{and} \qquad P(\overline{E}) = 1 - P(E) = \frac{5}{6}$$

Example 2.3 A small town has two banks A and B. It is estimated that 45 percent of the potential customers do business only with bank A, 30 percent only with bank B, and 15 percent with both banks A and B. The remaining 10 percent of the customers do business with none of the banks. If E_1 (E_2) denotes the event of a randomly selected customer doing business with bank A (B), find the following probabilities: (i) $P(\overline{E}_1)$, (ii) $P(E_2)$, (iii) $P(E_1 \cap E_2)$, (iv) $P(E_1 \cup E_2)$, (v) $P(\overline{E}_1\,\overline{E}_2)$, (vi) $P(\overline{E}_1 \cup E_2)$

solution

(i) The customers doing business with bank A include those who do business exclusively with bank A and those who do business with both banks A and B. Thus, $P(E_1) = 0.45 + 0.15 = 0.6$ and $P(\overline{E}_1) = 1 - P(E_1) = 1 - 0.6 = 0.4$.

(ii) The customers doing business with bank B include those who do business exclusively with bank B and those who do business with both banks A and B. Hence $P(E_2) = 0.30 + 0.15 = 0.45$.

(iii) $P(E_1 \cap E_2) = P(E_1E_2) = 0.15$

(iv) $P(E_1 \cup E_2) = P(E_1) + P(E_2) - P(E_1E_2) = 0.60 + 0.45 - 0.15 = 0.90$

(v) $P(\overline{E}_1\overline{E}_2) = P(\overline{E_1 \cup E_2}) = 1 - P(E_1 \cup E_2) = 1 - 0.9 = 0.1$

(vi) $P(\overline{E}_1 \cup E_2) = P(\overline{E}_1) + P(E_2) - P(\overline{E}_1E_2) = [1 - P(E_1)] + P(E_2) - P(\overline{E}_1 \cap E_2)$ where the event $\overline{E}_1 \cap E_2$ denotes the customers who do not do business with bank A, but do business with bank B (i.e., 30 percent).
Hence, $P(\overline{E}_1 \cup E_2) = (1 - 0.6) + 0.45 - 0.30 = 0.55$.

2.6.4 Conditional probability

The occurrence of the event E_2 when it is known that E_1 has occurred, is known as the conditional event and is denoted as $E_2|E_1$. The probability of this conditional event is defined as

$$P\left(E_2 \middle| E_1\right) = \frac{P(E_1 E_2)}{P(E_1)}; \qquad P(E_1) > 0 \tag{2.9}$$

For example, consider an applied mathematics class consisting of 35 male and 15 female students. Among the male students, 15 are science majors and 20 are engineering majors. Among the female students, 5 major in science and 10 major in engineering. Let us consider the following events: E_1 = student is a female and E_2 = student is a science major so that $P(E_1) = 15/(15+35) = 3/10$ and $P(E_2) = (5+15)/(15+35) = 2/5$. The situation is shown in Fig. 2.5*a*. If a randomly selected student is known to be a female (i.e., given the event E_1), the probability of that student majoring in science is given by $5/15 = 1/3$. This is shown graphically in Fig. 2.5*b*. As soon as the student is known to be a female, the area corresponding to male students will be zero and the proportion of science majors among female students will be $5/15 = 1/3$. The probability of finding a randomly selected student to be a female majoring in science can be obtained as $P(E_1 E_2) = P(E_2|E_1)\,P(E_1) = (1/3)(3/10) = 1/10$. This is a general rule and the probability of occurrence of both E_1 and E_2 is given by the product of

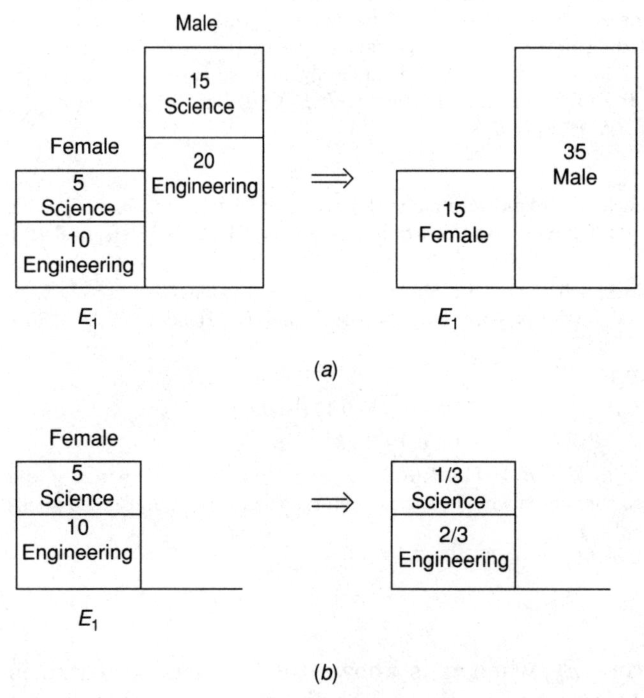

Figure 2.5

$P(E_1)$ and $P(E_2)$:

$$\therefore P(E_1E_2) = P(E_1) \cdot P\left(E_2 \middle| E_1\right) = P(E_2) \cdot P\left(E_1 \middle| E_2\right) \tag{2.10}$$

From Eq. (2.10) we find that if $P(E_1) = 0$ or $P(E_2) = 0$, then $P(E_1E_2) = 0$. Further, if E_1 and E_2 are mutually exclusive, $P(E_1E_2) = 0$ and hence $P(E_1 | E_2) = 0$ from Eq. (2.9).

Example 2.4 Experiments conducted to find the yield and ultimate strengths of bolted timber joints yielded the following results:

Specimen	Yield strength (lb)	Ultimate strength (lb)
1	3410	4120
2	3520	4240
3	2800	4010
4	2950	4050
5	3100	4230
6	3080	3970
7	3250	4290
8	2900	3980
9	2880	3770
10	3140	4060
11	2580	3730
12	2990	4020
13	3210	4310
14	3160	4190
15	2770	3790
16	2890	4040
17	3380	4420
18	3470	4410
19	3090	4080
20	2630	3680
21	2860	3990
22	2740	3810
23	3130	4270
24	2880	3780
25	3210	4310

Find the conditional probabilities $P(U_a | Y_a)$ and $P(Y_a | U_a)$ where Y_a and U_a denote the events of realizing acceptable values of yield strength (≥ 3000 lb) and ultimate strength (≥ 4000 lb), respectively.

solution From the data, we find that

$P(Y_aU_a)$ = Probability of realizing yield strength greater than or equal to 3000 pounds and ultimate strength greater than or equal to 4000 pounds = 12 / 25.

Similarly, we find that

$$P(Y_a \bar{U}_a) = \frac{1}{25}$$

$$P(\bar{Y}_a U_a) = \frac{4}{25}$$

$$P(\bar{Y}_a \bar{U}_a) = \frac{8}{25}$$

$P(Y_a)$ = Probability of realizing the yield strength greater than or equal to 3000 lb = 13 / 25.

$P(U_a)$ = Probability of realizing the ultimate strength greater than or equal to 4000 lb = 16 / 25.

Thus, we can compute the required probabilities as

$$P\left(U_a \middle| Y_a\right) = \frac{P(U_a Y_a)}{P(Y_a)} = \frac{\left(\frac{12}{25}\right)}{\left(\frac{13}{25}\right)} = \frac{12}{13}$$

and

$$P\left(Y_a \middle| U_a\right) = \frac{P(Y_a U_a)}{P(U_a)} = \frac{\left(\frac{12}{25}\right)}{\left(\frac{16}{25}\right)} = \frac{3}{4}$$

2.6.5 Statistically independent events

If the occurrence of the event E_1 in no way affects the probability of occurrence of the event E_2, then the events E_1 and E_2 are said to be statistically independent. In this case, we have,

$$P\left(E_1 \middle| E_2\right) = \frac{P(E_1 E_2)}{P(E_2)} = P(E_1) \tag{2.11}$$

and

$$P(E_1 E_2) = P(E_1) \cdot P(E_2) \tag{2.12}$$

For example, if $P(E_1) = P$ (shortage of steel in a project) = 0.4, and $P(E_2) =$ (realizing the head on tossing a coin) = 0.7; then obviously E_1 and E_2 are statistically independent and $P(E_1 E_2) = P(E_1) \cdot P(E_2) = 0.28$.

2.6.6 General laws

The union, intersection and conditional events defined in terms of two events can be generalized for more than two events. For example, the union of three events E_1, E_2, and E_3 can be obtained as follows:

$$P(E_1 \cup E_2 \cup E_{3)} = P(E_1 \cup W), \qquad \text{where} \qquad W = E_2 \cup E_3$$

$$= P(E_1) + P(W) - P(E_1 W)$$

$$= P(E_1) + P(E_2) + P(E_3) - P(E_2 E_3) - P(E_1) \cdot P(W)$$

$$= P(E_1) + P(E_2) + P(E_3) - P(E_1 E_2) - P(E_2 E_3)$$

$$- P(E_3 E_1) + P(E_1 E_2 E_3) \tag{2.13}$$

In terms of the Venn diagram, the union of E_1, E_2, and E_3 is given by the area bounded by the thick line and their intersection is given by the shaded region shown in Fig. 2.6. If the sample space of E_1 is denoted by $A_1 + A_4 + A_6 + A_7$, that of E_2 by $A_2 + A_4 + A_5 + A_7$, and that of E_3 by $A_3 + A_5 + A_6 + A_7$, the area bounded by the thick line can be obtained as follows:

Area bounded by the thick line

$$= A_1 + A_2 + A_3 + A_4 + A_5 + A_6$$

$$= (A_1 + A_4 + A_6 + A_7) + (A_2 + A_4 + A_5 + A_7)$$

$$+ (A_3 + A_5 + A_6 + A_7) - (A_4 + A_5 + A_6 + 2A_7)$$

$$= P(E_1) + P(E_2) + P(E_3)$$

$$- [\underbrace{(A_4 + A_7)}_{P(E_1E_2)} + \underbrace{(A_5 + A_7)}_{P(E_2E_3)} + \underbrace{(A_6 + A_7)}_{P(E_1E_3)} - \underbrace{A_7}_{(E_1E_2E_3)}] \tag{2.14}$$

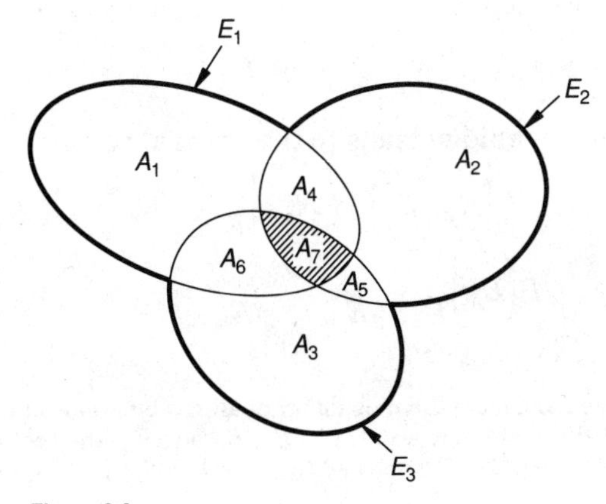

Figure 2.6

Thus, Eq. (2.13) can also be derived geometrically. The probability of a union of n events $E_1, E_2, \ldots, E_n$ can be derived in a similar manner to be[2]

$$
\begin{aligned}
P(E_1 \cup E_2 \cup E_3 \cup \cdots \cup E_n) = & \underbrace{\left[P(E_1) + P(E_2) + P(E_3) + \cdots + P(E_n)\right]}_{\binom{n}{1} = n \text{ terms}} \\
& - \underbrace{\left[P(E_1 E_2) + P(E_1 E_3) + \cdots + P(E_i E_j)\right]}_{\binom{n}{2} \text{ terms, } i \neq j} \\
& + \underbrace{\left[P(E_1 E_2 E_3) + P(E_1 E_2 E_4) + \cdots + P(E_i E_j E_k)\right]}_{\binom{n}{3} \text{ terms, } i \neq j \neq k} \\
& + \cdots + (-1)^{n-1} \underbrace{\left[P(E_1 E_2 E_3 \cdots E_n)\right]}_{\binom{n}{n} = 1 \text{ term}}
\end{aligned}
\tag{2.15}
$$

Thus there are

$$\binom{n}{1} + \binom{n}{2} + \binom{n}{3} + \cdots + \binom{n}{n-1} + \binom{n}{n} = 2^n - 1 \tag{2.16}$$

terms in the expansion of $P(E_1 \cup E_2 \cup \cdots \cup E_n)$.

In the case of mutually exclusive or disjoint events, all possible intersections are zero, and Eq. (2.15) simplifies to

$$P(E_1 \cup E_2 \cup E_3 \cup \cdots \cup E_n) = P(E_1) + P(E_2) + P(E_3) + \cdots + P(E_n) \tag{2.17}$$

We can define the intersection of three events E_1, E_2 and E_3 as

$$P(E_1 E_2 E_3) = P(W) \cdot P\left(E_3 \middle| W\right) \tag{2.18}$$

where W represents the simultaneous occurrence of E_1 and E_2. Thus

$$P(E_1 E_2 E_3) = P(E_1) \cdot P\left(E_2 \middle| E_1\right) \cdot P\left(E_3 \middle| E_1 E_2\right) \tag{2.19}$$

Successive application of this technique leads to the general result:

$$
\begin{aligned}
P(E_1 E_2 E_3 \ldots E_n) = P(E_1)\, & P\left(E_2 \middle| E_1\right) \cdot P\left(E_3 \middle| E_1 E_2\right) \\
& \cdots P\left(E_n \middle| E_1 E_2 E_3 \cdots E_{n-1}\right)
\end{aligned}
\tag{2.20}
$$

---

[2] The notation $P(E_i E_j)$ $i \neq j$, is used to denote the probability of all combinations of events taken two at a time. For example, if there are four events E_1, E_2, E_3 and E_4, the second term of Eq. (2.15) is $-[P(E_1E_2) + P(E_1E_3) + P(E_1E_4) + P(E_2E_3) + P(E_2E_4) + P(E_3E_4)]$.

Thus, the probability of intersection of n events is expressed as the product of one independent probability and $(n-1)$ conditional probabilities.

2.7 Total Probability Theorem

If $B_1, B_2, \ldots, B_n$ are a set of mutually exclusive, collectively exhaustive events, the probability of another event A can always be expanded as

$$P(A) = P(A \cap B_1) + P(A \cap B_2) + \cdots + P(A \cap B_n)$$

$$= \sum_{i=1}^{n} P(A \cap B_i)$$

$$= \sum_{i=1}^{n} P\left(A \middle| B_i\right) \cdot P(B_i) \tag{2.21}$$

This is called the theorem of total probability and $P(A)$ is given by the crossed area in the Venn diagram shown in Fig. 2.7.

Example 2.5 In example 2.1, find the probability of realizing at least three boilers in operating condition after one year.

solution Let $E_1 = \text{GGGG}$, $E_2 = \text{GGGB}$, $E_3 = \text{GGBG}$, $E_4 = \text{GBGG}$, $E_5 = \text{BGGG}$, $E_6 = \text{GGBB}$, $E_7 = \text{GBBG}$, $E_8 = \text{BBGG}$, $E_9 = \text{BGBG}$, $E_{10} = \text{BGGB}$, $E_{11} = \text{GBGB}$, $E_{12} = \text{GBBB}$, $E_{13} = \text{BGBB}$, $E_{14} = \text{BBGB}$, $E_{15} = \text{BBBG}$, $E_{16} = \text{BBBB}$, and $P(E_i) = 1/16$ for $i = 1$ to 16.

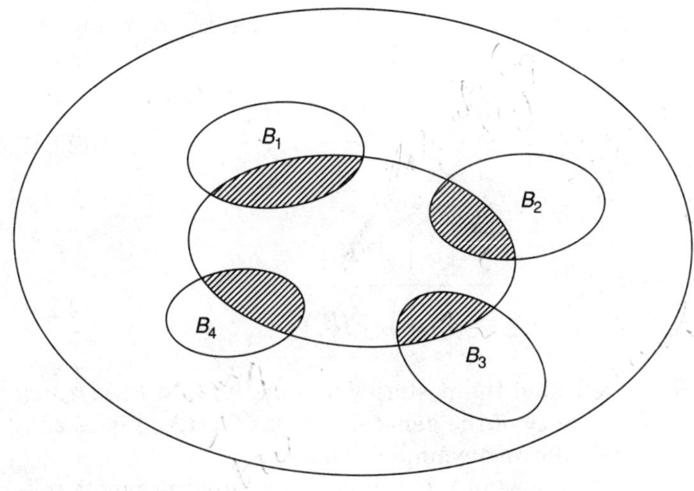

Figure 2.7

Now we define the event A as realizing at least three boilers in operating condition after one year. Applying the total probability theorem, we obtain

$$P(A) = \sum_{i=1}^{16} P\left(A \middle| E_i\right) P(E_i)$$

$$= \left(1 \cdot \frac{1}{16}\right) + \left(1 \cdot \frac{1}{16}\right) + \left(1 \cdot \frac{1}{16}\right) + \left(1 \cdot \frac{1}{16}\right) + \left(1 \cdot \frac{1}{16}\right)$$

$$+ \left(0 \cdot \frac{1}{16}\right) + \left(0 \cdot \frac{1}{16}\right) + \left(0 \cdot \frac{1}{16}\right) + \left(0 \cdot \frac{1}{16}\right) + \left(0 \cdot \frac{1}{16}\right)$$

$$+ \left(0 \cdot \frac{1}{16}\right) + \left(0 \cdot \frac{1}{16}\right) + \left(0 \cdot \frac{1}{16}\right) + \left(0 \cdot \frac{1}{16}\right) + \left(0 \cdot \frac{1}{16}\right)$$

$$+ \left(0 \cdot \frac{1}{16}\right) = \frac{5}{16}$$

2.8 Bayes' Rule

One of the most useful formulas in the theory of applied probability can be derived by using the concept of conditional probability as follows [2·4, 2·5]. If A and B are two events, since $P(A \cap B) = P(B \cap A)$ we obtain

$$P\left(A \middle| B\right) P(B) = P\left(B \middle| A\right) P(A) \tag{2.22}$$

that is,

$$P\left(B \middle| A\right) = \frac{P\left(A \middle| B\right) \cdot P(B)}{P(A)} \tag{2.23}$$

Equation (2.23) is called the Bayes' rule. This can be generalized by using the total probability theorem given by Eq. (2.21). Thus,

$$P\left(B_j \middle| A\right) = \frac{P\left(A \middle| B_j\right) \cdot P(B_j)}{P(A)} \tag{2.24}$$

and, this can be rewritten, using Eq. (2.21), as

$$P\left(B_j \middle| A\right) = \frac{P\left(A \middle| B_j\right) \cdot P(B_j)}{\sum_{i=1}^{n} P\left(A \middle| B_i\right) \cdot P(B_i)} \tag{2.25}$$

Here the quantity $P(B_j|A)$ is called the posterior probability and $P(A|B_i)$ is called prior probability. The utility of the generalized Bayes' rule, Eq. (2.25), can be explained through the following example.

Let A denote the event of "collapse of a building." The building might collapse due to an earthquake (event B_1), wind loading (event B_2) or any other cause (event B_3). Let the probabilities of occurrence of these events at a locality, in a particular year, be known as $P(B_1)$, $P(B_2)$ and $P(B_3)$. From the details of design, we will be knowing the probabilities $P(A|B_i)$, $i = 1,2,3$, i.e., the probability of the collapse of the building given the occurrence of an earthquake, wind load or any other cause. Now, let us assume that the event A has already occurred and we would like to know the probability of earthquake being the cause of collapse of the building. In such a case, we can find $P(B_1|A)$ by applying the generalized Bayes' rule.

Example 2.6 In an engine manufacturing plant, 70 percent of the crankshafts are ground by machinist R and 30 percent by machinist S. It is known from past experience that the crankshafts ground by machinists R and S contain 2 percent and 3 percent defective units, respectively. If a randomly selected crankshaft is found to be defective, find the probability that it was ground by the machinist R.

solution We define the following events: A = finding the crankshaft to be defective, B_1 = crankshaft is ground by the machinist R, and B_2 = crankshaft is ground by the machinist S. From the given data, we have $P(B_1) = 0.7$, $P(B_2) = 0.3$, $P(A|B_1) = 0.02$, and $P(A|B_2) = 0.03$. The application of the generalized Bayes rule gives

$$P\left(B_1 \middle| A\right) = \frac{P\left(A \middle| B_1\right) P(B_1)}{\sum_{i=1}^{2} P\left(A \middle| B_i\right) P(B_i)}$$

$$= \frac{(0.02)(0.7)}{(0.02)(0.7) + (0.03)(0.3)}$$

$$= \frac{14}{23}$$

References and Bibliography

2.1. T. T. Soong, *Probabilistic Modeling and Analysis in Science and Engineering,* John Wiley, New York, 1981.
2.2. R. L. Scheaffer and J. T. McClave, *Probability and Statistics for Engineers,* 2d ed., Duxbury Press, Boston, 1986.
2.3. A. H. S. Ang and W. H. Tang, *Probability Concepts in Engineering Planning and Design*: vol. 1: *Basic Principles,* John Wiley, New York, 1975.
2.4. L. L. Lapin, *Probability and Statistics for Modern Engineering,* Brooks/Cole Publishing Co.–Engineering Div., Monterey, CA, 1983.
2.5. W. W. Hines and D. C. Montgomery, *Probability and Statistics in Engineering and Management Science,* 2d ed., John Wiley, New York, 1980.
2.6. I. Todhunter, *A History of the Mathematical Theory of Probability,* G. E. Stechert & Co., New York, 1931.
2.7. I. Asimov, *Asimov's Biographical Encyclopedia of Science and Technology,* Doubleday & Co., Garden City, New York, 1972.
2.8. A. C. King and C. B. Read, *Pathways to Probability,* Holt, Rinehart & Winston, New York, 1963.
2.9. C. C. Gillispie (ed.-in-chief), *Dictionary of Scientific Biography,* vol. 3, Charles Scribner's Sons, New York, 1980.

Review Questions

2.1 Give five engineering examples involving uncertainty.

2.2 What is the difference between an experiment and an event?

2.3 What are mutually exclusive events? Give two examples of mutually exclusive events.

2.4 What are universal and null sets?

2.5 Define the union and intersection of two sets.

2.6 What is the complement of a set?

2.7 Explain the commutative, associative, and distributive rules of set theory?

2.8 What is de Morgan's rule?

2.9 What is a sample point?

2.10 Define sample space.

2.11 What is a Venn diagram?

2.12 What is the statistical definition of probability?

2.13 State the axiomatic definition of probability.

2.14 What is conditional probability?

2.15 What is meant by statistically independent events?

2.16 What is the total probability theorem?

2.17 What is Bayes' rule?

2.18 Give two engineering applications of Bayes' theorem.

Problems

2.1 Explain the validity of the cumulative, associative, distributive, and de Morgan's rules stated in Section 2.3 using Venn diagrams.

2.2 The following events are defined after observing the scores of five students in a mathematics examination: A = at most four students failed to secure a score of 90 percent, B = exactly three students secured a score of 90 percent, and C = all five students failed to secure a score of 90 percent. Interpret the following events: (i) $A \cup B$, (ii) $A \cup C$, (iii) $B \cup C$, (iv) $A \cap B$, (v) $A \cap C$, (vi) $B \cap C$, (vii) $\overline{A} \cap \overline{C}$, (viii) $\overline{A} \cup \overline{B}$, (ix) $\overline{A} \cup \overline{C}$, (x) $\overline{B} \cap \overline{C}$.

2.3 A machinist produced 100 shafts according to the specification 1" ± 0.001". During inspection, the diameters of 85 shafts were found to be within the tolerance limits and 15 were found to be outside the tolerance band. If 6 shafts are randomly selected, find the probability of finding the diameter of at least one shaft falling outside the tolerance limits.

2.4 One hundred castings were inspected for defects and 85 were found to be free of defects. Among the 15 defective castings, 9 were found to have blow holes, and 12 were found to have sand inclusions. (a) How many castings have exactly one type of defect? (b) How many castings have at least one type of defect? (c) How many castings have both types of defects? 15

2.5 The computer laboratory in a university has 40 microcomputers. Of these, 25 are connected to a dot matrix printer, 10 are connected to a laser printer and 8 are not connected to any printer. (a) How many microcomputers are connected to at least one type of printer? (b) How many are connected to both types of printers?

2.6 Three girls—Kathy, Cindy, and Nancy—compete in a swimming race. If each girl is equally likely to win the race, find (a) the probability of Kathy or Cindy winning the race, (b) the probability of Nancy winning the race, and (c) the probability of Kathy not winning the race.

2.7 Six gear boxes are assembled in a machine shop. It is known that two have defects. If two gear boxes are selected at random, what is the probability that (a) both will be defective, (b) both will not be defective, (c) one will be defective and the other will not be defective, (d) the second one will be defective given the first one is not defective, and (e) the second one will not be defective given that the first one is not defective?

2.8 In a high school senior class, 15 percent of the students secured an SAT score of 1200 or higher. It is known from past experience that 60 percent of the students whose SAT score is at least 1200 select engineering as major while 10 percent of those students whose SAT score is less than 1200 select engineering as major in college. If a randomly selected student is known to choose engineering, what is the probability that his/her SAT score is at least 1200?

2.9 The castings for the engine block in a diesel engine manufacturing facility are supplied by two different companies. Company A supplies 200 castings each day out of which 2 percent do not meet the specifications. Company B supplies 300 castings each day out of which 3 percent do not satisfy the specifications. (a) What is the probability that a randomly selected casting was supplied by company *B*? (b) What is the probability that a randomly selected casting will not meet the specifications? (c) If a casting does not meet the specifications, what is the probability that it was supplied by company *A*? (d) If a casting meets the specifications, what is the probability that it was supplied by company *B*?

2.10 Two boilers are used to supply steam for the operation of a steam turbine. Each boiler can supply upto a maximum of 80 percent of the steam required for the turbine. The probability of failure of each boiler is 0.05 and that of both boilers failing simultaneously is 0.03. Find the probability of the steam turbine getting the complete supply of steam in case one of the boilers fails.

2.11 Suppose that 15 percent of all people have a high-cholesterol level. It is found that 5 percent of people with a high-cholesterol level get a heart attack while only 1 percent of those without a high-cholesterol level get a heart attack. If a randomly selected person gets a heart attack, find the probability of that person having a high-cholesterol level.

2.12 Among the 30 students enrolled in an Engineering Design course, 23 belong to mechanical engineering branch and 7 belong to other branches of engineering. If 2 students are randomly chosen from this class, what is the probability of both the students belonging to a branch other than mechanical engineering?

2.13 Suppose that 15 percent of all people have a high-cholesterol level. A particular type of cholesterol test, when used on a person having a high-cholesterol level, detects it correctly 95 percent of the time and fails to detect it 5 percent of the time. The same test, when used on a person without a high-cholesterol level, wrongly indicates a high-cholesterol level 4 percent of the time. (a) If the test shows a high-cholesterol level for a randomly selected person, what is the probability of that person actually not having a high-cholesterol level? (b) What is the probability of a randomly selected person having a high-cholesterol level and being indicated as such by the test? (c) What is the probability of a randomly selected person not having a high-cholesterol level but the test indicating the person as having a high-cholesterol level? (d) Do you consider the test to be reliable?

2.14 A boiler has three independent safety valves which open and let steam off to the atmosphere in case of a high-pressure buildup. The boiler is considered to be safe if at least two valves open in the case of a high-pressure buildup. If the probability of each valve opening in case of high-pressure buildup is 0.9, find the probability of the boiler to be in the safe condition in case of a high-pressure buildup.

2.15 A gear can fail either due to excessive bending stress or excessive surface wear. Let *B* and *W* denote the failures in these modes. If it is known that in a particular application, $P(B) = 0.0005$, $P(W) = 0.001$, and $P(B \mid W) = 0.1$, find (a) the probability of failure of the gear, and (b) the probability of a gear having only excessive bending stress but not excessive surface wear.

2.16 A small river is found to be polluted due to the chemical waste from two major industries, A and B. Due to the strict standards enforced by the EPA, the probabilities of controlling the pollutions from A and B are 0.7 and 0.85, respectively, in the next two years. It is estimated that, if only one of the industries complies with the strict standards of EPA, there is a 75 percent probability of controlling the pollution in the river to an acceptable level. (a) Find the probability of controlling the pollution in the river in the next two years. (b) If the pollution in the river is not controlled in the next two years, what is the probability that it is caused entirely by the industry A?

2.17 An automatic spot-welding machine is used in an assembly plant. It is known that 5 percent of the welds made by the machine are defective. If 10 welds made by the machine are inspected at random, what is the probability of finding (a) no defective welds, (b) exactly one defective weld, (c) at least one defective weld, and (d) no more than one defective weld?

2.18 Among the graduate students of a university, 26 percent receive loans, 55 percent receive research assistantships, and 8 percent receive both. What percent of students receive either a loan or an assistantship? What percent receive neither a loan nor an assistantship?

2.19 The percentage of mechanical-, civil-, aeronautical-, industrial-, and chemical-engineering students in the course, Engineering Mathematics, is, respectively, 25, 20, 15, 22, and 18. The probability of a mechanical-, civil-, aeronautical-, industrial-, and chemical-engineering student getting the grade A is 0.7, 0.6, 0.8, 0.5, and 0.65, respectively. If a student chosen at random at the end of the semester earned an A in the course, find the probability of the student belonging to mechanical engineering.

2.20 The machine used to detect blow holes in castings detects not only 84 percent of all blow holes, but also incorrectly indicates a blow hole in 2 percent of good castings. It is known, from past experience, that 5 percent of all castings will have blow holes. Find the probability that a casting which the machine indicates to have a blow hole is in fact a good one.

2.21 A crane has two cables C_1 and C_2 to lift a load as shown in Fig. 2.8. The cable C_2 is designed to be slightly longer than cable C_1 so that C_1 carrys the load under normal circumstances while cable C_2 is intended to carry the load only when C_1 fails. From past experience, it is known that the probability of failure of cable C_1 is 0.01 and the probability of failure of cable C_2 when it is required to carry the load all by itself is 0.25. (a) What is the probability of the crane

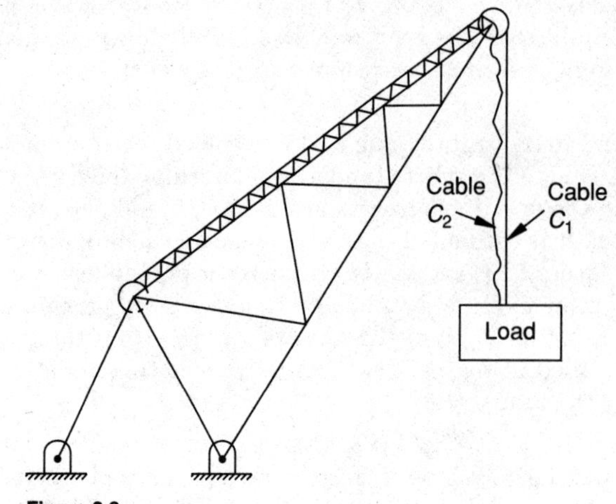

Figure 2.8

lifting the load? (b) What is the probability of the crane failing to lift the load? (c) If the load is being lifted by the crane, what is the probability that none of the cables have failed?

Chapter

3

Random Variables and Probability Distributions

Biographical Note

Pafnutii Lvovitch Chebyshev

Pafnutii Lvovitch Chebyshev (or Tchebysheff or Tshebysheff) was born on May 16, 1821, and died on December 8, 1894 in Russia. He graduated from Moscow University with a degree in mathematics in 1841. He was one of the most famous Russian mathematicians and he made numerous important contributions to the theory of numbers, algebra, theory of probability, analysis, and applied mathematics. Chebyshev won a silver medal in 1841 for deriving an error estimate in the Newton-Raphson iterative method. He was very curious about mechanical inventions during his childhood and it was stated that during his very first lesson in geometry he saw its application to mechanics. His technological inventions include a calculating machine built in the late 1870s. When his father became very poor during the famine of 1840, Chebyshev helped support his family. He devoted his life to scientific teaching and never married. He became interested in the theory of numbers and stated the Chebyshev problem relating probability to the theory of numbers. The Chebyshev inequality, which gives the probability of an event occurring not more than a specified distance from the average, is one of the popular theorems in the theory of probability. [3.6, 3.9]

3.1 Introduction

An event has been defined as the possible outcome of an experiment. Let us assume that a random event is the measurement of a quantity X, which takes on various values x within the range $-\infty < x < \infty$. Such a quantity (like X) is called a *random variable.* We denote a random variable by a capital letter and the particular value taken by it by a lower case letter. Random variables are of two types: (i) discrete and (ii) continuous. If the random variable is allowed to take only discrete values $x_1, x_2, \ldots, x_n$, it is called a *discrete random variable.* On the other hand, if the random variable is permitted to take any real value in a specified range, it is called a *continuous random variable.* For example, the number of people who report for work on a particular day at a factory is a discrete random variable, whereas, the weight of a machined metal part can be treated as a continuous random variable.

3.2 Probability Mass Function for Discrete Random Variables

Corresponding to each value x_i that a discrete random variable X can take, we can associate a probability of occurrence $P(x_i)$. We can describe the probabilities associated with the random variable X by a table of values, but it will be easier to write a general formula that permits one to calculate $P(x_i)$ by substituting the appropriate value of x_i. Such a formula is called the *probability mass function* of the random variable X, and is usually denoted as $p_X(x_i)$ or simply as $p(x_i)$. Thus the function which gives the probability of realizing the random variable $X = x_i$ is called the probability mass function $p_X(x_i)$

$$p(x_i) = p_X(x_i) = P(X = x_i) \tag{3.1}$$

3.3 Cumulative Distribution Function for Discrete Random Variables

Although a random variable X is completely described by the probability mass function, it is often convenient to deal with another related function known as the probability distribution function. The probability that the value of the random variable X is less than or equal to some number x is defined as the cumulative distribution function $F_X(x)$. Thus,

$$F_X(x) = P(X \leq x) = \sum_i p_X(x_i) \tag{3.2}$$

where the summation extends over those values of i for which $x_i \leq x$. Since the distribution function is a *cumulative probability*, it is also termed as the *cumulative distribution function.*

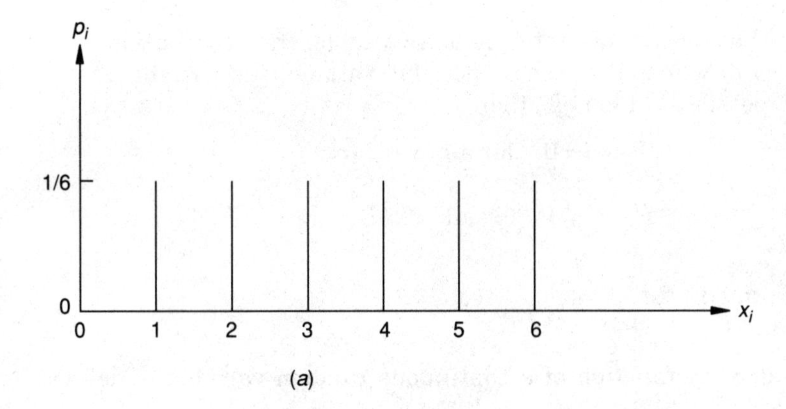

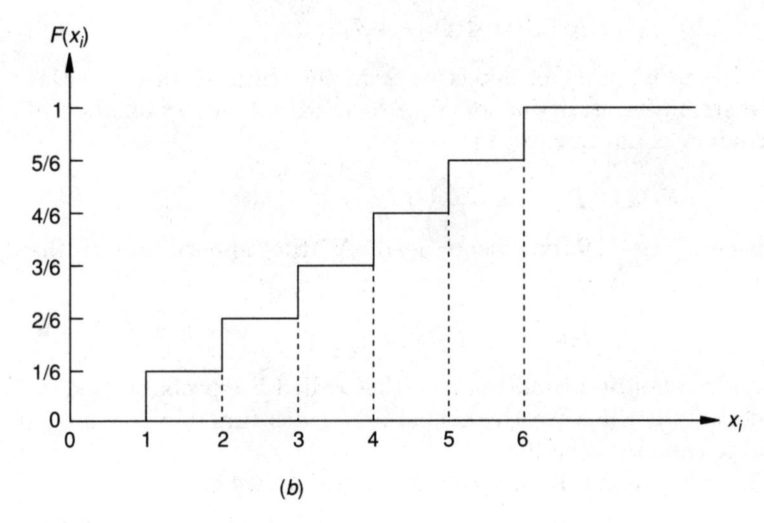

Figure 3.1 (*a*) Probability mass function, (*b*) Cumulative distribution function.

Example 3.1 Find the probability mass and distribution functions for the number realized when a fair die is thrown.

solution Since each face is equally likely to show up, the probability of realizing any number between 1 and 6 is 1/6.

$$P(X = 1) = P(X=2) = \ldots = P(X = 6) = \frac{1}{6}$$

and hence,

$$p_X(1) = p_X(2) = \ldots = p_X(6) = \frac{1}{6}$$

The analytical form of $F_X(x)$ is

$$F_X(x) = x / 6 \quad \text{for} \quad 1 \le x \le 6$$

Correspondingly, the probability mass and distribution functions are shown in Fig. 3.1 where X is the random variable denoting the number realized after throwing the die.

It can be seen that for any discrete random variable, the distribution function will be a *step function*. If the least possible value of the variable X is L and the greatest possible value is U, then

$$F_X(x) = 0 \quad \text{for all} \quad x < L$$

and

$$F_X(x) = 1 \quad \text{for all} \quad x > U$$

3.4 Probability Density Function for Continuous Random Variables

The probability density function of a continuous random variable is defined by[1]

$$f_X(x)\,dx = P(x \le X \le x + dx) \tag{3.3}$$

which is equal to the probability of detecting X in the infinitesimal interval $(x, x + dx)$. The distribution function of X is defined as the probability of detecting X less than or equal to x, that is,

$$F_X(x) = P(X \le x) = \int_{-\infty}^{x} f(x')\,dx' \tag{3.4}$$

where the condition $F(-\infty) = 0$ has been used. As the upper-limit of the integral goes to infinity, we have

$$\int_{-\infty}^{\infty} f_X(x)\,dx = F_X(\infty) = 1 \tag{3.5}$$

This is called the normalization condition. $f_X(x)$ is called the probability density function to distinguish it from the probability mass function defined in the case of a discrete random variable.

From the definition of probability distribution function, we get

$$F_X(x + \Delta x) - F_X(x) = P(x \le X \le x + \Delta x) \tag{3.6}$$

and the probability of detecting the variable X in the finite interval $[x_1, x_2]$ may be written as

$$P(x_1 \le X \le x_2) = \int_{x_1}^{x_2} f_X(x)\,dx = P(X \le x_2) - P(X \le x_1)$$

$$= F_X(x_2) - F_X(x_1) \tag{3.7}$$

It can be seen that if the random variable X is continuous and if the first derivative of the distribution function exists, the probability density function $f_X(x)$ is given by the first derivative of $F_X(x)$.

---

[1]The probability density function can also be defined as $f_X(x)\,dx = P(x - dx/2 \le X \le x + dx/2)$ = Probability of realizing X in the infinitesimal interval $(x - dx/2,\ x + dx/2)$.

$$f_X(x) = \frac{dF_X(x)}{dx} = \lim_{\Delta x \to 0} \frac{F_X(x + \Delta x) - F_X(x)}{\Delta x} \tag{3.8}$$

A typical probability density function and the corresponding distribution function are shown in Fig. 3.2.

Example 3.2 The probability density function of the wind load acting on a water tank (X, in kips) is given by

$$f_X(x) = \begin{cases} \dfrac{3x}{50 \times 10^4}(100 - x); & 0 \le x \le 100 \text{ kips} \\ 0; & \text{otherwise} \end{cases}$$

Find the probability of realizing the wind load to be less than 20 kips or greater than 80 kips.

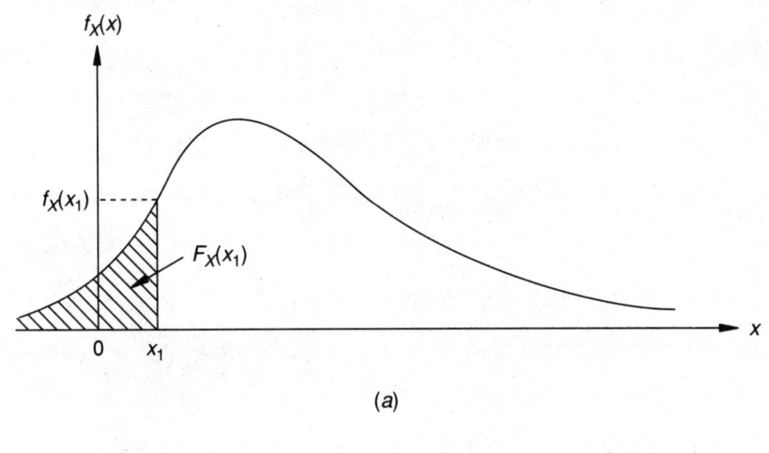

(a)

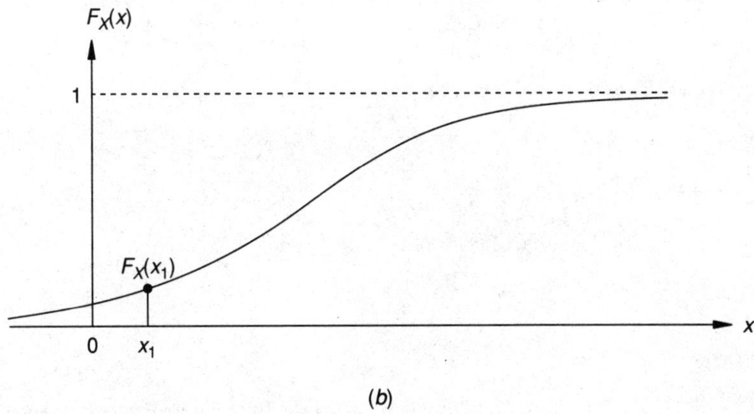

(b)

Figure 3.2 (a) Probability density function, (b) Probability distribution function.

solution If $E_1(E_2)$ denotes the event of realizing the wind load to be less than 20 kips (greater than 80 kips), the required probability is given by

$$P(E_1 \cup E_2) = P(E_1) + P(E_2) - P(E_1 E_2) = P(E_1) + P(E_2)$$

since the probability of simultaneous occurrence of E_1 and E_2 is zero. Here

$$P(E_1) = P(X \le 20) = \int_{x=0}^{20} f_X(x)\,dx$$

$$= \frac{3}{50 \times 10^4} \int_0^{20} x(100 - x)\,dx = \frac{3}{50 \times 10^4} \left(50\,x^2 - \frac{1}{3}\,x^3\right) \Big|_0^{20}$$

$$= \frac{3}{50 \times 10^4} \left[50(400) - \frac{1}{3}\,(8000)\right]$$

$$= 0.104$$

$$P(E_2) = P(X \ge 80) = \int_{80}^{100} f_X(x)\,dx$$

$$= \frac{3}{50 \times 10^4} \int_{80}^{100} x(100 - x)\,dx = \frac{3}{50 \times 10^4} \left(50\,x^2 - \frac{1}{3}\,x^3\right) \Big|_{80}^{100}$$

$$= \frac{3}{50 \times 10^4} \left[50(10^4) - \frac{1}{3}(10^6) - 50(80)^2 + \frac{1}{3}(80)^3\right]$$

$$= 0.104$$

Hence the probability of realizing X less than 20 kips or greater than 80 kips is given by $0.104 + 0.104 = 0.208$.

Example 3.3 Consider the probability density function of the wind load given in Example 3.2. If the wind load is known to be greater than 20 kips, find the probability of its value being larger than 80 kips.

solution The desired conditional probability can be expressed as

$$P\left(X \ge 80 \,\Big|_{X \ge 20}\right) = \frac{P(X \ge 80)}{P(X \ge 20)}$$

Expressing $P(X \ge 20)$ as $1 - P(X < 20)$, we find that

$$P\left(X \ge 80 \,\Big|_{X \ge 20}\right) = \frac{P(X \ge 80)}{1 - P(X < 20)} = \frac{0.104}{1 - 0.104} = \frac{0.104}{0.896} = 0.1161$$

3.5 Mean, Mode, and Median

The probability density or distribution function of a random variable contains all the information about the variable. However, in many cases we require only the gross properties but not the entire information about the random variable. In such cases we can specify the range of the variable, a measure of its central value, or an indicator of its spread.

3.5.1 Mean

The *mean value*, also termed as the *expected value*, *mathematical expectation*, or *average*, is used to describe the central tendency of a random variable. The mode and median are also used for the same purpose.

Discrete case. Let us assume that there are n trials of an experiment in which the random variable X is observed to take on the value x_1 (n_1 times), x_2 (n_2 times), . . . , x_m (n_m times) so that $n_1 + n_2 + \ldots + n_m = n$. Then the mean value or the arithmetic mean of X, denoted as $\overline{X}$, is given by

$$\overline{X} = \frac{\sum_{k=1}^{m} x_k n_k}{n} = \sum_{k=1}^{m} x_k \left(\frac{n_k}{n}\right) = \sum_{k=1}^{m} x_k \cdot p_X(x_k) \tag{3.9}$$

where (n_k / n) is the relative frequency of occurrence of x_k and is same as the probability mass function $p_X(x_k)$. Hence, in general, the mean value, which is also denoted as $E(X)$, the expected value, of the discrete random variable can be expressed as

$$\overline{X} = E(X) = \sum_{\text{all } i} x_i \, p_X(x_i) \tag{3.10}$$

Continuous case. If $f_X(x)$ is the density function of a continuous random variable X, the mean is given by[2]

$$\mu = E(X) = \int_{-\infty}^{\infty} x \, f_X(x) \, dx \tag{3.11}$$

According to the definition of a random variable, any function of a random variable is itself a random variable. Thus if $h(x)$ is an arbitrary function of x, the expected value of $h(x)$ is defined as

$$E[h(x)] = \sum_i h(x_i) \, p_X(x_i), \quad \text{if } X \text{ is discrete}$$

$$= \int_{-\infty}^{\infty} h(x) \, f_X(x) \, dx, \quad \text{if } X \text{ is continuous} \tag{3.12}$$

Example 3.4 What is the expected value of the upturned face of a single die?

---

[2]The symbol μ is generally used to denote the mean value of a continuous random variable and $\overline{X}$ that of a discrete random variable.

solution

$$\bar{X} = E(X) = \sum_{\text{all } i} x_i \, p_X(x_i)$$

$$= 1(1/6) + 2(1/6) + 3(1/6) + 4(1/6) + 5(1/6) + 6(1/6) = 3.5$$

Example 3.5 Find the mean value of the Rockwell hardness number X of a particular type of steel for which the density function is given by

$$f_X(x) = \begin{cases} 0, & \text{for } x < 50 \\ 1/20, & \text{for } 50 \le x \le 70 \\ 0, & \text{for } x > 70 \end{cases}$$

solution

$$\mu = E[X] = \int_{-\infty}^{\infty} x \, f_X(x) \, dx$$

$$= \int_{50}^{70} x(1/20) \, dx = (1/20) \left(\frac{x^2}{2}\right)_{50}^{70} = 60$$

3.5.2 Mode

Discrete case. Mode is the value of X at which the probability mass function $p_X(x_i)$ has the highest value. In essence, mode is the most probable value of the random variable.

Continuous case. Mode is the value of X at which $df_X(x)/dx = 0$, that is, the value of x corresponding to the peak value of the probability density function.

3.5.3 Median

Discrete case. Median is the number in the middle.

Continuous case. Median (x_{50}) is the value of X at which the cumulative distribution function has a value of 0.5:

$$0.5 = \int_{-\infty}^{x_{50}} f_X(x) \, dx$$

The significance of mean, mode, and median of a random variable is illustrated in Fig. 3.3.

Example 3.6 Find the mean, mode and median of the random variable whose distribution function is given by[3]

$$F_X(x) = 1 - \exp\left[-\frac{1}{2} kx^2\right], \qquad 0 \le x < \infty$$

where k is the parameter (constant) of the distribution.

---

[3]This is a one-parameter distribution called the Rayleigh distribution.

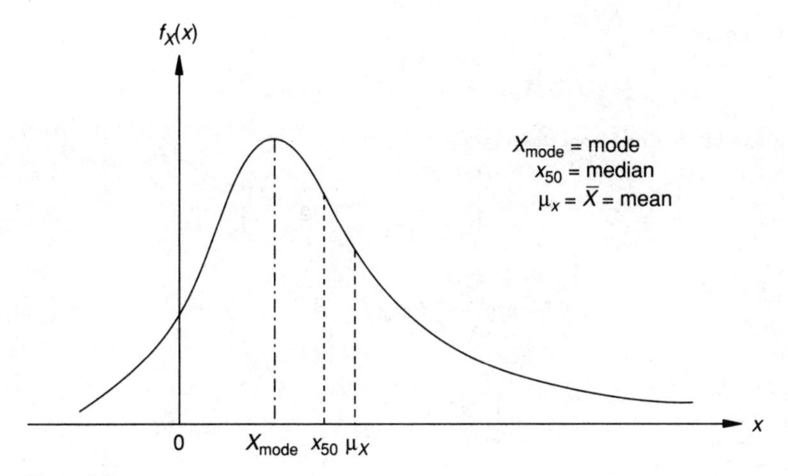

Figure 3.3

solution The probability density function can be obtained by differentiating the distribution function as

$$f_X(x) = \frac{d\,F_X(x)}{dx} = kx \exp\left[-\frac{1}{2}kx^2\right]$$

We take the integral of this density function from 0 to ∞ as

$$\int_0^\infty f_X(x)\,dx = F_X(x)\Big|_0^\infty = 1$$

and find that it satisfies the normality condition. The mean of X is given by

$$\mu_X = \int_0^\infty x\,f_X(x)\,dx = 2\int_0^\infty \frac{1}{2}kx^2 \exp\left[-\frac{1}{2}kx^2\right]dx$$

By making a change of variable as

$$\frac{1}{2}kx^2 = y$$

the expression for μ_X can be written as[4]

$$\mu_X = \sqrt{\frac{2}{k}}\int_0^\infty \sqrt{y}\,e^{-y}dy = \sqrt{\frac{\pi}{2k}}$$

To obtain the mode of X, we set

$$\frac{d\,f_X(x)}{dx} = \frac{d}{dx}\left[k \cdot x \cdot \exp\left\{-\frac{1}{2}kx^2\right\}\right] = k \cdot \exp\left\{-\frac{1}{2}kx^2\right\} - k^2x^2 \cdot \exp\left\{-\frac{1}{2}kx^2\right\} = 0$$

---

[4]From calculus, the value of the definite integral can be obtained as $\int_0^\infty \sqrt{x}\,e^{-nx}\,dx = \frac{1}{2n}\sqrt{\pi/n}$.

and solve for x to obtain

$$X_{\text{mode}} = \sqrt{\frac{1}{k}}$$

The median of X can be obtained from the relation

$$0.5 = \int_0^{x_{50}} f_X(x)\,dx = F_X(x)\Big|_0^{x_{50}} = \left\{1 - \exp\left[-\frac{1}{2}kx^2\right]\right\}_0^{x_{50}}$$

that is,

$$0.5 = \exp\left(-\frac{1}{2}kx_{50}^2\right)$$

Thus

$$x_{50} = \left\{\frac{\log_e 4}{k}\right\}^{1/2}$$

3.6 Standard Deviation and Skewness Coefficient

The expected value or mean is a measure of the central tendency, indicating the location of the distribution on the coordinate axis representing the random variable. A measure of the variability of the random variable is usually given by a quantity known as the standard deviation. Another quantity, which not only gives a measure of the variability but also a measure of the symmetry of the density function is called the *skewness coefficient.*

3.6.1 Standard deviation

The mean square deviation or variance of a random variable X is defined as

$$\begin{aligned}
\sigma_X^2 = \text{var}(X) &= E\,[(X - \mu_X)^2] \\
&= E\,[X^2 - 2X\,\mu_X + \mu_X^2] \\
&= E\,[X^2] - 2\,\mu_X\,E\,[X] + E\,[\mu_X^2] \\
&= E\,[X^2] - \mu_X^2
\end{aligned} \tag{3.13}$$

and the standard deviation as

$$\sigma_X = +\sqrt{\text{var}\,(X)} = \sqrt{E\,(X^2) - \mu_X^2} \tag{3.14}$$

The coefficient of variation is a measure of dispersion in nondimensional form and is defined as

$$\text{coefficient of variation of } X = \gamma_X = \frac{\text{standard deviation}}{\text{mean}} = \frac{\sigma_X}{\mu_X} \tag{3.15}$$

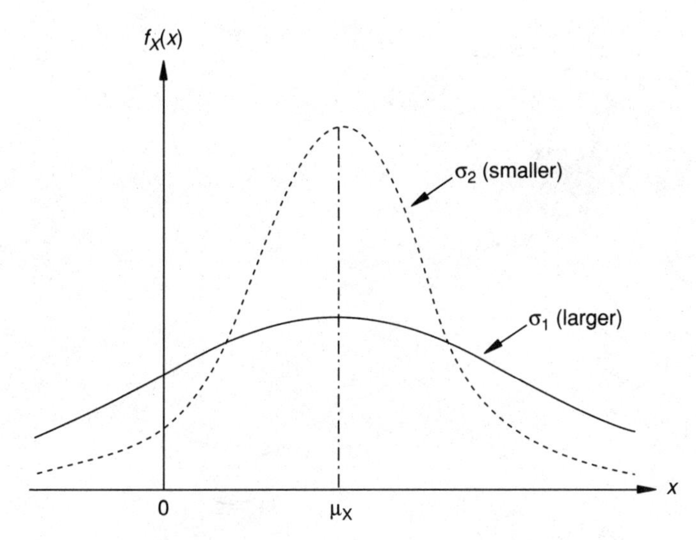

Figure 3.4

The variance of a random variable is analogous to the moment of inertia of a weight about its centroid. Figure 3.4 shows two density functions with the same mean μ_X, but with different variances. As can be seen, the variance or standard deviation is a measure of the variability of a random variable (or breadth of the density function).

3.6.2 Skewness coefficient

The expected value of the cube of the deviation of the random variable from its expected value (also known as the third moment of the distribution about the mean) is taken as a measure of the skewness of the distribution

$$E\,[(X-\mu_X)^3] = \int_{-\infty}^{\infty} (X-\mu_X)^3\, f_X(x)\, dx, \quad \text{if } X \text{ is continuous}$$

$$= \sum_{\text{all } i} (x_i - \overline{X})^3\, p_X(x_i), \quad \text{if } X \text{ is discrete} \tag{3.16}$$

The value of $E\,[(X-\mu_X)^3]$ can be positive or negative as illustrated in Fig. 3.5. The skewness coefficient is defined as

$$\text{skewness coefficient} = \frac{E\,[(X-\mu_X)^3]}{\sigma_X^3} \tag{3.17}$$

Example 3.7 Find the coefficient of variation and the skewness coefficient of the wind-induced load (X) whose density function is given by

$$f_X(x) = \frac{3}{50 \times 10^4}\, x\,(100-x); \qquad 0 \le x \le 100 \text{ kips}$$

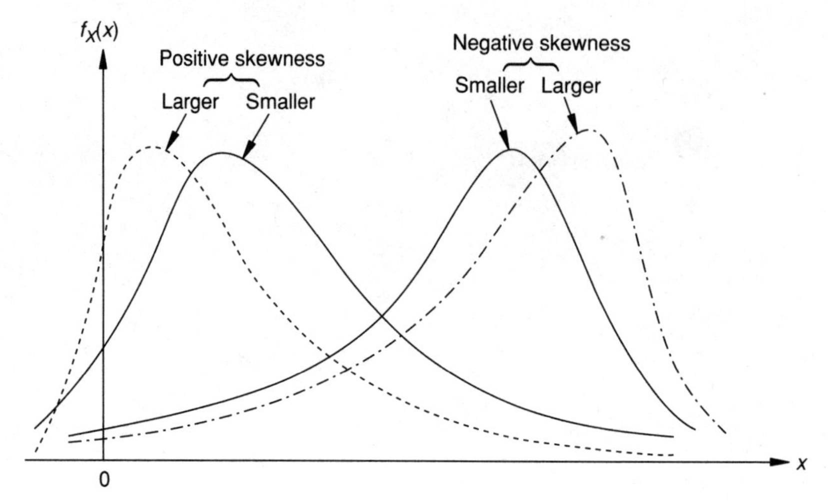

Figure 3.5

solution The mean value of X is given by

$$\mu_X = \int_0^{100} x\, f_X(x)\, dx = \frac{3}{50 \times 10^4} \int_0^{100} (100x^2 - x^3)\, dx$$

$$= \frac{3}{50 \times 10^4} \left(\frac{100}{3} x^3 - \frac{1}{4} x^4 \right) \Bigg|_0^{100}$$

$$= \frac{3}{50 \times 10^4} \left[\frac{100}{3}(10^6) - \frac{1}{4}(10^8) \right] = 50$$

The expected value of X^2 is

$$E[X^2] = \int_0^{100} x^2\, f_X(x)\, dx = \frac{3}{50 \times 10^4} \int_0^{100} (100\, x^3 - x^4)\, dx$$

$$= \frac{3}{50 \times 10^4} \left(\frac{100}{4} x^4 - \frac{1}{5} x^5 \right) \Bigg|_0^{100}$$

$$= \frac{3}{50 \times 10^4} \left[25(10^8) - \frac{1}{5}(10^{10}) \right] = 3000$$

$$\therefore \sigma_X = \sqrt{E[X^2] - \mu_X^2} = \sqrt{3000 - 2500} = 22.3607$$

Thus the coefficient of variation of X will be

$$\gamma_X = \frac{\sigma_X}{\mu_X} = \frac{22.3607}{50} = 0.4472$$

The third moment of the distribution about the mean is given by

$$E[(X-\mu_X)^3] = \int_0^{100} (x-50)^3 f_X(x)\,dx$$

$$= \frac{3}{50\times 10^4}\int_0^{100} (x-50)^3 (100x - x^2)\,dx$$

$$= \frac{3}{50\times 10^4}\int_0^{100} (-x^5 + 250\,x^4 - 22{,}500\,x^3 + 875{,}000\,x^2 - 12{,}500{,}000x)\,dx$$

$$= \frac{3}{50\times 10^4}\left[-\frac{x^6}{6} + \frac{250}{5}x^5 - \frac{22{,}500}{4}x^4 + \frac{875{,}000}{3}x^3 - \frac{12{,}500{,}000}{2}x^2\right]\Bigg|_0^{100}$$

$$= 0$$

The skewness coefficient of X is given by Eq. (3.17)

$$\text{skewness coefficient} = \frac{0}{\sigma_X^3} = 0$$

3.7 Moments of Random Variables

The moments of a distribution are the expected values of the power of the random variable that have the given distribution. The rth moment about the origin, denoted as μ_r', is given by

$$\mu_r' = \int_{-\infty}^{\infty} x^r \cdot f_X(x)\cdot dx \tag{3.18}$$

which can be termed as the expected value of x^r, $E(X^r)$. The zeroth order moment μ_o' is the area under the density function which is, of course, unity. The first order moment can be seen to be same as the mean. It can be seen that if the density function $f_X(x)$ is an even function about the origin, the moments of order, 1, 3, 5, . . . , are zero.

The moments about the origin for a discrete random variable that takes on the values $x_1, x_2, x_3, \ldots, x_n$ are given by

$$\mu_r' = \sum_{i=1}^{n} x_i^r\, p(x_i) \tag{3.19}$$

The moments about an arbitrary point k are defined as

$$E\,[(x-k)^r] = \int_{-\infty}^{\infty} (x-k)^r \cdot f_X(x)\cdot dx, \qquad \text{for a continuous random variable}$$

$$= \sum_{i=1}^{n} (x_i - k)^r \cdot p\,(x_i), \qquad \text{for a discrete random variable} \tag{3.20}$$

If $k = \mu_1' =$ mean, the moments are called central moments and are given by

$$\mu_r = E\left[(x-\mu_1')^r\right] = \int_{-\infty}^{\infty} (x-\mu_1')^r f_X(x)\,dx \tag{3.21}$$

Accordingly,

$$\mu_1 = \text{first moment about the mean} = \int_{-\infty}^{\infty} (x - \mu_1')\, f_X(x)\, dx$$

$$= \mu_1' - \mu_1' = 0 \tag{3.22}$$

and

$$\mu_2 = \text{second moment about the mean} = \int_{-\infty}^{\infty} (x - \mu_1')^2\, f_X(x)\, dx$$

$$= \int_{-\infty}^{\infty} (x^2 - 2x\, \mu_1' + \mu_1'^2)\, f_X(x)\, dx$$

$$= \mu_2' - (\mu_1')^2 = \text{var}\,(X) \tag{3.23}$$

It can be seen that for a density function which is symmetrical about the mean, the moments μ_1, μ_3, μ_5, . . . , are zero. Although the higher-order moments give additional information about the random variable, they are generally not required since μ_X and σ_X reveal enough information about the variable for most engineering purposes.

3.8 Importance of Moment Functions—Chebyshev Inequality

The moment function defined by Eqs. (3.18) to (3.21) represent the gross properties of the probabilistic nature of a random variable. In most applications, only the first few moments are generally used. To illustrate the application of the moment functions, we consider an inequality known as the *Chebyshev inequality*. According to the Chebyshev inequality, the following relation is valid for any random variable X regardless of its distribution

$$P\left[|x - \mu_X| \geq t\, \sigma_X\right] \leq \frac{1}{t^2} \tag{3.24}$$

where t is any positive constant. In words, the Chebyshev inequality states that the probability of the absolute value of the difference between the random variable and its mean exceeding a constant t times the standard deviation is not greater than t^{-2}. For $t = 2$, Eq. (3.24) gives

$$P\left[|x - \mu_X| \geq 2\, \sigma_X\right] \leq \frac{1}{2^2} = 0.2500$$

and for $t = 4$, we get

$$P\left[|x - \mu_X| \geq 4\, \sigma_X\right] \leq \frac{1}{4^2} = 0.0625$$

Thus, the probability of exceeding a stated magnitude of deviation from the mean tends to decrease as σ_X decreases. The inequality (3.24) can be applied to simple design problems as indicated in the following example.

Example 3.8 Find the area of cross-section of the uniform bar shown in Fig. 3.6 under a probabilistic load P with $\mu_P = 0$ and $E(P^2) = \sigma_P^2$. Assume the cross-sectional area of the bar, A, and the permissible stress, x^*, to be deterministic.

solution If X denotes the uniform stress at any section in the bar, we have

$$X = \frac{P}{A} \tag{E1}$$

Since A is deterministic ($\mu_A = A$ and $\sigma_A = 0$), we obtain

$$\mu_X = 0, \quad \sigma_X = \frac{\sigma_P}{A} \tag{E2}$$

The bar will fail whenever $|X| \geq x^*$. Thus the probability of failure of the bar can be estimated using the inequality (3.24) by setting

$$x^* = t\,\sigma_X \tag{E3}$$

Equations (E3) and (E2) give

$$t = \frac{x^*}{\sigma_X} = \frac{x^* A}{\sigma_P} \tag{E4}$$

The application of inequality (3.24) yields

$$P\left[|X| \geq x^*\right] \leq \frac{\sigma_P^2}{x^{*2} A^2}$$

This shows that in order to design the bar for a specified upper bound on the probability of failure, say $1/t^2$, the area of cross section of the bar has to be equal to

$$A = \frac{t \cdot \sigma_P}{x^*}$$

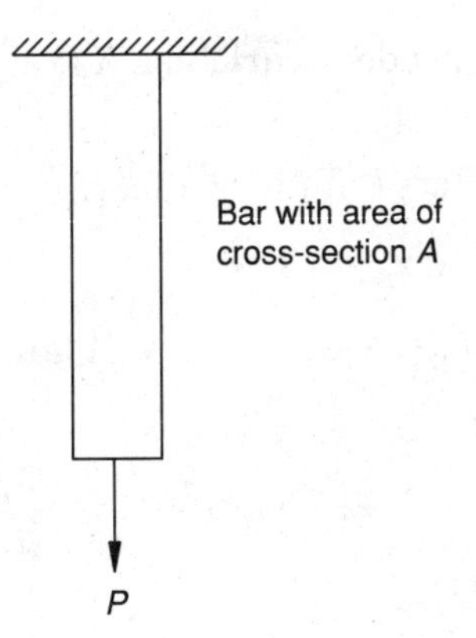

Figure 3.6

For example, a probability of failure of 1 in 100 implies that $t = 10$ and the corresponding area is given by

$$A = \frac{10 \cdot \sigma_P}{x^*}$$

The Chebyshev inequality (3.24) applies to any distribution. It can be shown that if the probability density function is unimodal,[5] the following inequality, known as *Gauss inequality*, holds [3.1]

$$P\left[|x - \mu_X| \geq t\ \sigma_X\right] \leq \frac{4(1 + \alpha^2)}{9(t - |\alpha|^2)} \tag{3.25}$$

for $t > |\alpha|$ where $\alpha = (\mu_X - X_{\text{mode}}) / \sigma_X$. If the probability density function is symmetric and unimodal, the mean and the mode coincide ($\alpha = 0$) and the Gauss inequality reduces to the *Camp-Meidell inequality* [3.1]:

$$P\left[|x - \mu_X| \geq t\ \sigma_X\right] \leq \frac{4}{9}\frac{1}{t^2} \text{ for } t > 0 \tag{3.26}$$

It can be seen that this inequality is sharper than Chebyshev inequality by a factor of 4 / 9.

3.9 Jointly Distributed Random Variables

When two or more random variables are being considered simultaneously, their joint behavior is determined by a joint probability distribution function. The probability distributions of single random variables are called *univariate* distributions and the distributions that involve two random variables are called *bivariate* distributions. In general, if a distribution involves more than one random variable, it is called a *multivariate* distribution.

3.9.1 Joint density and distribution functions

The joint density function of n continuous random variables $X_1, X_2, \ldots, X_n$ is defined [3.8] as

$$\begin{aligned} f_{X_1,\ldots,X_n}(x_1, x_2, \ldots, x_n) \cdot dx_1 \ldots dx_n &= P(x_1 \leq X_1 \leq x_1 + dx_1, \\ &\quad x_2 \leq X_2 \leq x_2 + dx_2, \ldots, x_n \leq X_n \leq x_n + dx_n) \\ &= \text{Probability that a random point } x_1', x_2', \ldots, x_n' \text{ will fall in the} \\ &\quad \text{infinitesimal region defined by} \\ &\quad x_1 \leq x_1' \leq x_1 + dx_1, \ldots, x_n \leq x_n' \leq x_n + dx_n \end{aligned} \tag{3.27}$$

[5] A *unimodal distribution* is one that has a single peak in its range of definition.

For independent random variables, the joint density function is given by the product of individual or marginal density functions as

$$f_{X_1,\ldots,X_n}(x_1,\ldots,x_n) = f_{X_1}(x_1)\ldots f_{X_n}(x_n) \tag{3.28}$$

The joint distribution function

$$F_{X_1,X_2,\ldots,X_n}(x_1, x_2, \ldots, x_n)$$

associated with the density function of Eq. (3.27) is given by

$$F_{X_1,\ldots,X_n}(x_1,\ldots,x_n) = P[X_1 \le x_1,\ldots, X_n \le x_n]$$

$$= \int_{-\infty}^{x_1}\cdots\int_{-\infty}^{x_n} f_{X_1,\ldots,X_n}(x_1', x_2', \ldots, x_n')\cdot dx_1'\cdot dx_2'\ldots dx_n' \tag{3.29}$$

If $X_1, X_2, \ldots, X_n$ are independent random variables, we have

$$F_{X_1,\ldots,X_n}(x_1,\ldots,x_n) = F_{X_1}(x_1)\cdot F_{X_2}(x_2)\ldots F_{X_n}(x_n) \tag{3.30}$$

It can be seen that the joint density function can be obtained by differentiating the joint distribution function as

$$f_{X_1,\ldots,X_n}(x_1,\ldots,x_n) = \frac{\partial^n}{\partial x_1\,\partial x_2\ldots\partial x_n}\left[F_{X_1,\ldots,X_n}(x_1,\ldots,x_n)\right] \tag{3.31}$$

3.9.2 Obtaining the marginal or individual density function from the joint density function

Let the joint density function of two random variables X and Y be denoted by $f(x, y)$ and the marginal or individual density functions of X and Y by $f_X(x)$ and $f_Y(y)$, respectively. Take the infinitesimal rectangle with corners located at the points (x, y), $(x + dx, y)$, $(x, y + dy)$, and $(x + dx, y + dy)$. The probability of a random point (x', y') falling in this rectangle is $f_{X,Y}(x, y)\cdot dx\cdot dy$. The integral of such probability elements with respect to y (for a fixed value of x) is the sum of the probabilities of all the mutually exclusive ways of obtaining the points lying between x and $x + dx$. Let the lower- and upper-limits of y be $a_1(x)$ and $b_1(x)$. Then

$$P[x \le x' \le x + dx] = \left[\int_{a_1(x)}^{b_1(x)} f_{X,Y}(x, y)\cdot dy\right]\cdot dx = f_X(x)\cdot dx$$

$$\therefore\ f_X(x) = \int_{y_1=a_1(x)}^{y_2=b_1(x)} f_{X,Y}(x, y)\cdot dy \tag{3.32}$$

Similarly, we can show that

$$f_Y(y) = \int_{a_2(y)}^{b_2(y)} f_{X,Y}(x, y)\cdot dx \tag{3.33}$$

where $a_2(y)$ and $b_2(y)$ denote the lower- and upper-limits of x for the specified value of y.

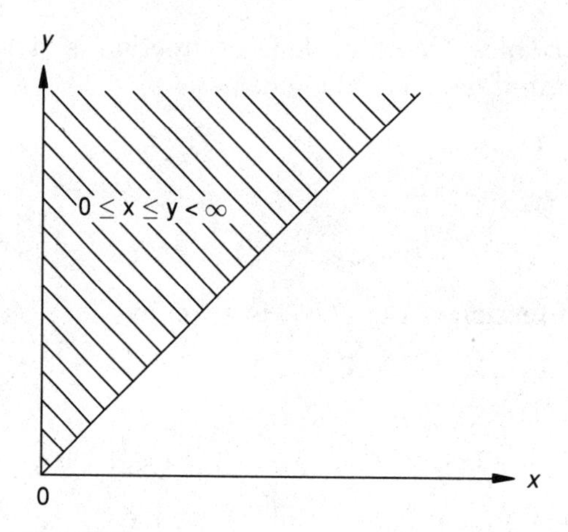

Figure 3.7

Example 3.9 The joint density function of two random variables X and Y is given by

$$f(x,y) = \begin{cases} 12\exp(-4x - 2y); & 0 \le x \le y,\ 0 \le y < \infty \\ 0; & \text{otherwise} \end{cases}$$

Determine whether X and Y are independent.

solution The marginal density functions of X and Y can be computed as (see Fig. 3.7 for the ranges of x and y):

$$f(x) = 12\int_{y=x}^{\infty} e^{-4x-2y}\,dy = 12\,e^{-4x}\int_{x}^{\infty} e^{-2y}\,dy$$

$$= 12\,e^{-4x}\left(\frac{e^{-2y}}{-2}\right)\Bigg|_{x}^{\infty} = 6\,e^{-6x}$$

and

$$f(y) = 12\int_{x=0}^{y} e^{-4x-2y}\,dx = 12e^{-2y}\int_{0}^{y} e^{-4x}\,dx$$

$$= 12\,e^{-2y}\left(\frac{e^{-4x}}{-4}\right)\Bigg|_{0}^{y} = 3e^{-2y}\,(1 - e^{-4y})$$

Since $f(x, y) \neq f(x)f(y)$, the variables X and Y are not independent.

3.10 Moments of Jointly Distributed Random Variables

If $Z = g(X, Y)$ is a function of two jointly distributed random variables, its expected value is given by

$$E[Z] = E[g(X, Y)] = \sum_{\text{all } x_i}\sum_{\text{all } y_j} g(x_i, y_j)\,p_{X,Y}(x_i, y_j)$$

if X and Y are discrete, and

$$E[Z] = \int_{-\infty}^{\infty}\int_{-\infty}^{\infty} g(x, y)\cdot f_{X,Y}(x, y)\cdot dx\cdot dy \tag{3.34}$$

if X and Y are continuous.

In particular, if we take

$$g(x, y) = x^l y^m \tag{3.35}$$

the moment about the origin of order $l + m$ of the random variables X and Y is given by

$$E[X^l Y^m] = \int_{-\infty}^{\infty} \int_{-\infty}^{\infty} x^l y^m f_{X,Y}(x, y) \cdot dx \cdot dy \tag{3.36}$$

Similarly, by taking

$$g(x, y) = (x - \mu_X)^l (y - \mu_Y)^m \tag{3.37}$$

the central moment of order $l + m$ of the variables X and Y can be obtained as

$$\begin{aligned} E[g(x, y)] &= E[(X - \mu_X)^l (Y - \mu_Y)^m] \\ &= \int_{-\infty}^{\infty} \int_{-\infty}^{\infty} (x - \mu_X)^l (y - \mu_Y)^m f_{X,Y}(x, y)\, dx\, dy \end{aligned} \tag{3.38}$$

The most valuable central moments are the second-order moments, namely, the moments obtained by taking ($l = 2$, $m = 0$), ($l = 0$, $m = 2$) and ($l = 1$, $m = 1$). For $l = 2$, $m = 0$, we obtain the variance of X as

$$E[(X - \mu_X)^2] = \text{var}[X] = \int_{-\infty}^{\infty} (x - \mu_X)^2 \cdot f_X(x) \cdot dx \tag{3.39}$$

For $l = 0$, $m = 2$, we obtain the variance of Y as

$$E[(Y - \mu_Y)^2] = \text{var}[Y] = \int_{-\infty}^{\infty} (y - \mu_Y)^2 \cdot f_Y(y) \cdot dy \tag{3.40}$$

For $l = 1$, $m = 1$, we obtain the covariance of X and Y as

$$\begin{aligned} E[(X - \mu_X)(Y - \mu_Y)] &= \text{cov}(X, Y) \\ &= \int_{-\infty}^{\infty} \int_{-\infty}^{\infty} (x - \mu_X)(y - \mu_Y) \cdot f_{X,Y}(x, y) \cdot dx \cdot dy = \sigma_{X,Y} \end{aligned} \tag{3.41}$$

The correlation coefficient $\rho_{X,Y}$ for the random variables is defined as

$$\rho_{X,Y} = \frac{\text{cov}(X, Y)}{\sigma_X \cdot \sigma_Y} \tag{3.42}$$

and it can be proven that $-1 \le \rho_{X,Y} \le 1$. The physical meaning of the correlation coefficient is that its value will be nearly unity provided the two random variables are linearly related, and will be nearly zero if the two random variables are not linearly related. The significance of the correlation coefficient is illustrated in Fig. 3.8.[6]

---

[6]If experiments are conducted to find the values of two related random variables X and Y (such as the yield strength and ultimate strength of steel), we realize the values x_i and y_i in the ith experiment. If the experimental values realized in all the experiments, (x_i, y_i), are plotted in the (X, Y) space, the resulting diagram is known as the *scattergram*.

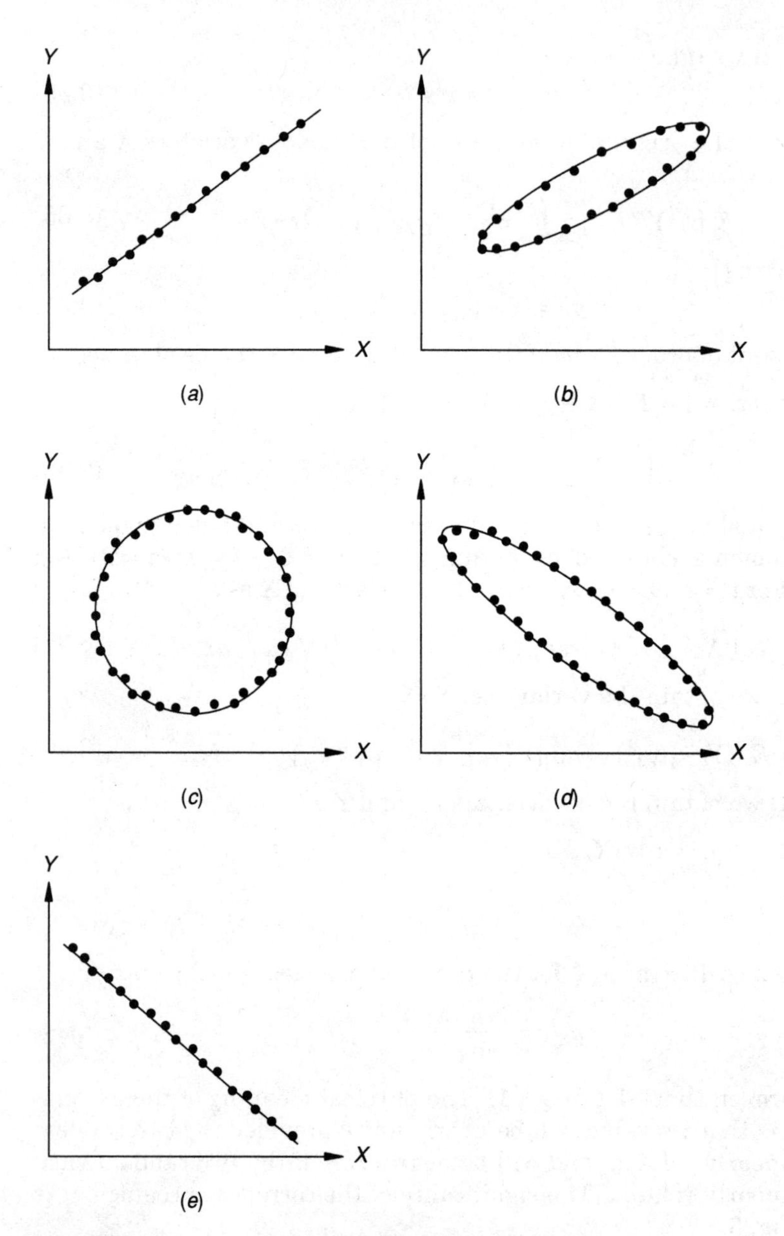

Figure 3.8 (a) $\rho_{X,Y} = 1$ (X and Y are linearly related with positive slope), (b) $\rho_{X,Y}$ = positive (X and Y are slightly linearly related with positive slope), (c) $\rho_{X,Y} = 0$ (X and Y are not linearly related), (d) $\rho_{X,Y}$ = negative (X and Y are slightly linearly related with negative slope), (e) $\rho_{X,Y} = -1$ (X and Y are linearly related with negative slope).

3.11 Probability Distributions

There are several types of probability distributions for describing various types of discrete and continuous random variables. Some of the common distributions and their characteristics are summarized in Table 3.1. In any physical problem, one chooses a particular type of probability distribution depending on:

1. the nature of the problem;
2. the underlying assumptions associated with the distribution;
3. the shape of the graph between $f(x)$ or $F(x)$ and x obtained after plotting the available data, and
4. the convenience and simplicity afforded by the distribution in subsequent computations.

The properties of some of the more commonly used distributions are presented in the following sections and the characteristics of some additional distributions are summarized in Table 3.2.

3.11.1 Binomial distribution

Repeated independent trials of a random experiment are called Bernoulli trials if there are only two possible outcomes for each trial and their probabilities remain the same throughout the trials. Let S (success) and F (failure) denote the two outcomes and p and q the corresponding probabilities. Clearly

$$p + q = 1 \tag{3.43}$$

The sample space of each individual trial is formed by the two sample points S and F. The sample space of n-Bernoulli trials contains 2^n sample points. Since the trials are independent, the probabilities multiply. For example,

$$P\,(SSFS) = ppqp = p^3 q$$

Let S_n denote the number of successes in n trials. Then it can be shown that

$$P(S_n = k) = \binom{n}{k} p^k q^{n-k} \tag{3.44}$$

Since Eq. (3.44) represents the kth term in the Binomial expansion of $(p+q)^n$, it is called the binomial distribution. It can also be shown that

$$E\,[S_n] = \overline{S}_n = np \tag{3.45}$$

$$\text{var}\,[S_n] = E\,[(S_n - \overline{S}_n)^2] = npq \tag{3.46}$$

Example 3.10 A thermal power plant buys four boilers. If the probability of a boiler functioning without failure for a year is 0.7, find the probability mass and distribution functions of the status of the boilers at the end of the year.

TABLE 3.1 Characteristics of Common Probability Distributions

Distribution	Probability mass or density function, $p(x)$ or $f(x)$	Range of x	Expected value, $E(x)$	Variance, σ^2	Comments
Binomial	$\frac{n!}{x!(n-x)!} p^x (1-p)^{n-x}$	$x = 1, 2, \ldots, n$; $0 \le p \le 1$	np	$np(1-p)$	n = number of trials; p = probability of success in each trial
Poisson	$\frac{e^{-\lambda} \lambda^x}{x!}$	$x = 0, 1, 2, \ldots$; $\lambda > 0$	λ	λ	λ = average number of successes
Uniform (rectangular)	$\frac{1}{b-a}$	$a \le x \le b$	$\frac{a+b}{2}$	$\frac{(b-a)^2}{12}$	$-\infty < a, b < \infty$
Triangular	$\frac{2}{b-a}\left(\frac{x-a}{c-a}\right)$ for $a \le x \le c$ and $\frac{2}{b-a}\left(\frac{b-x}{b-c}\right)$ for $c \le x \le b$	$a \le x \le b$	$\frac{1}{3}(a+b+c)$	$\frac{1}{18}(a^2+b^2+c^2-ab-bc-ca)$	$a \le c \le b$
Exponential	$\lambda e^{-\lambda x}$	$o \le x < \infty$	$\frac{1}{\lambda}$	$\frac{1}{\lambda^2}$	$\lambda > 0$
Gaussian (normal)	$\frac{1}{\sqrt{2\pi}\,\sigma} \exp\left\{-\frac{(x-\mu)^2}{2\sigma^2}\right\}$	$-\infty < x < \infty$	μ	σ^2	$-\infty < \mu < \infty$; $\sigma > 0$

TABLE 3.2 Characteristics of Additional Probability Distributions

Distribution	Probability density function, $f(x)$	Range of x	Expected value, $E(x)$	Variance, σ^2	Comments
Rayleigh	$kx \exp\left\{-\frac{k}{2}x^2\right\}$	$0 \le x < \infty$	$\sqrt{\frac{\pi}{2k}}$	$\frac{2}{k}\left(1-\frac{\pi}{4}\right)$	$k > 0$
Lognormal	$\frac{1}{\sqrt{2\pi}\sigma x}\exp\left\{-\frac{(\ln x-\mu)^2}{2\sigma^2}\right\}$	$0 < x < \infty$	$\exp\left(\mu+\frac{\sigma^2}{2}\right)$	$\exp(2\mu+2\sigma^2)-\exp(2\mu+\sigma^2)$	$-\infty < \mu < \infty$
Weibull	$\frac{b}{\theta-x_o}\left(\frac{x-x_o}{\theta-x_o}\right)^{b-1}\exp\left\{-\left(\frac{x-x_o}{\theta-x_o}\right)^b\right\}$	$x_o \le x < \infty$	$x_o+(\theta-x_o)\,\Gamma\left(1+\frac{1}{b}\right)$	$(\theta-x_o)^2\left\{\Gamma\left(1+\frac{2}{b}\right)-\Gamma^2\left(1+\frac{1}{b}\right)\right\}$	$\Gamma(\,)$ = gamma function; b = shape parameter; θ = characteristic value
Gamma	$\frac{\lambda^k}{\Gamma(k)}x^{k-1}e^{-\lambda x}$	$0 \le x < \infty$	$\frac{k}{\lambda}$	$\frac{k}{\lambda^2}$	$k > 0$; $\lambda > 0$; $\Gamma(\,)$ = gamma function
Beta	$\frac{1}{\beta(p,q)}\frac{(x-a)^{p-1}(b-x)^{q-1}}{(b-a)^{p+q-1}}$	$a \le x \le b$	$a+\frac{p}{p+q}(b-a)$	$\frac{pq\,(b-a)^2}{(p+q)^2\,(p+q-1)}$	p, q = positive. $\beta(p,q)$ = beta function $=\frac{\Gamma(p)\cdot\Gamma(q)}{\Gamma(p+q)}$

solution Here $p = 0.7$, $q = 0.3$ and $n = 4$.

$$P(S_n = 0) = 1(0.7)^0(0.3)^4 = 0.0081$$

$$P(S_n = 1) = 4(0.7)^1(0.3)^3 = 0.0756$$

$$P(S_n = 2) = 6(0.7)^2(0.3)^2 = 0.2646$$

$$P(S_n = 3) = 4(0.7)^3(0.3)^1 = 0.4116$$

$$P(S_n = 4) = 1(0.7)^4(0.3)^0 = 0.2401$$

The probability mass and distribution functions are shown in Figs. 3.9*a* and 3.9*b*, respectively.

3.11.2 Normal Distribution

The density function of a normally distributed random variable X (also known as Gaussian distribution) is given by

$$f_X(x) = \frac{1}{\sqrt{2\pi}\,\sigma_X} \exp\left[-\frac{1}{2}\left(\frac{x - \mu_X}{\sigma_X}\right)^2\right] \tag{3.47}$$

and X is identified as $N(\mu_X, \sigma_X)$. The parameters of the distribution μ_X and σ_X denote, respectively, the mean value and standard deviation of the variable X. The density function and the corresponding distribution function are shown in Fig. 3.10. The normal distribution has the following properties:

1. Any linear function of a normally distributed random variable is also normally distributed.
2. If

$$E\,[X] = 0, \qquad \text{then}$$

$$E\,[X^{2n+1}] = 0, \qquad \text{and}$$

$$E\,[X^{2n}] = 1 \cdot 3 \cdot 5 \ldots \cdot (2n-1) \cdot \sigma_X^{2n} \tag{3.48}$$

Hence all the moments can be deduced from the second moment of a normal variate.[7]

Joint normal law for n independent random variables.

$$f_{X_1, X_2, \ldots, X_n}(x_1, x_2, \ldots, x_n)$$

$$= \frac{1}{\sqrt{(2\pi)^n} \cdot \sigma_1 \cdot \sigma_2 \cdot \sigma_n} \exp\left[-\frac{1}{2}\sum_{k=1}^{n}\left(\frac{x_k - \mu_k}{\sigma_k}\right)^2\right]$$

---

[7]A normally distributed random variable is also known as a *normal variate*.

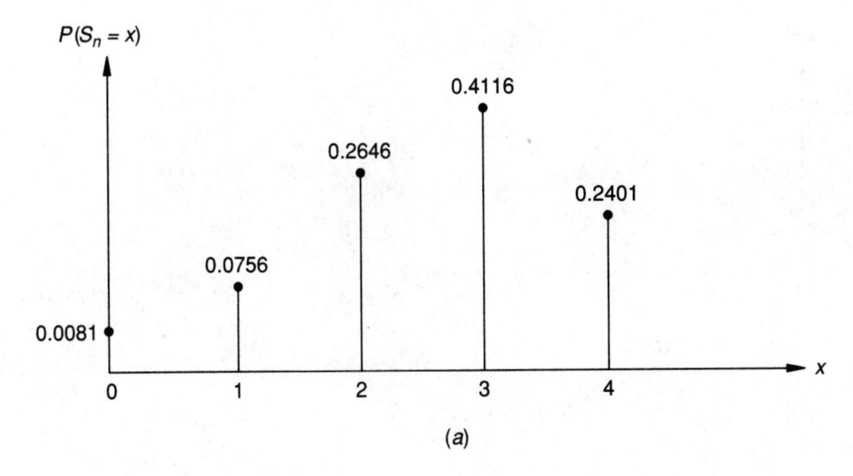

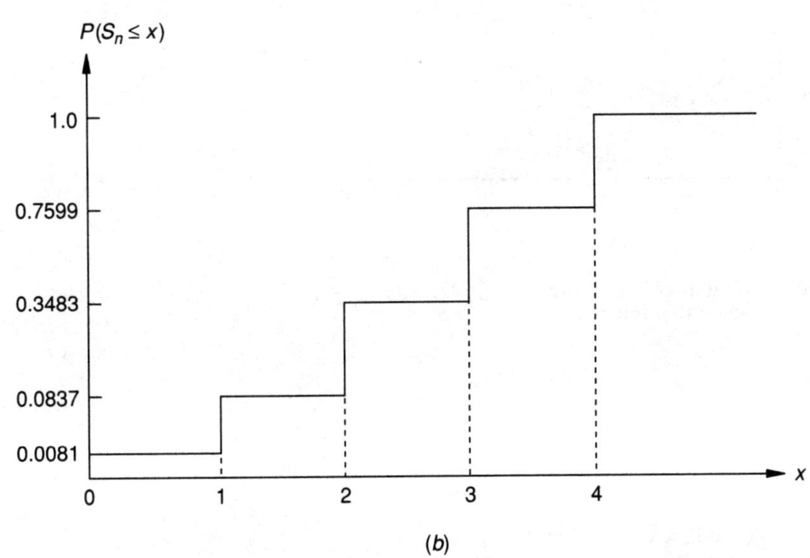

Figure 3.9

$$= f_{X_1}(x\,1) \cdot f_{X_2}(x\,2) \cdots f_{X_n}(x_n) \tag{3.49}$$

where $\sigma_i = \sigma_{X_i}$ and $\mu_i = \mu_{X_i}$.

If correlation between X_k and X_j is not zero. In this case, the joint density function is given by

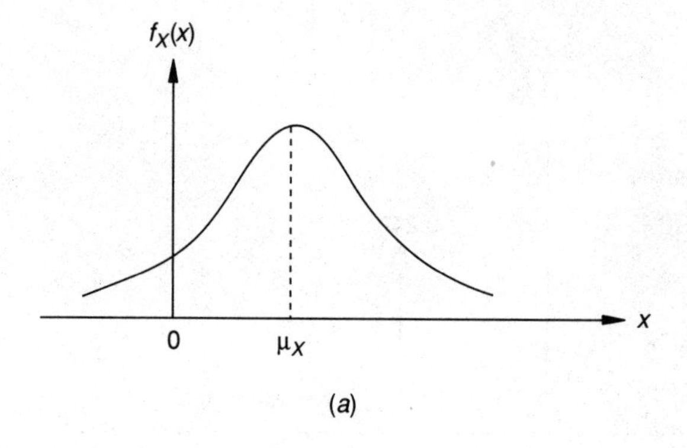

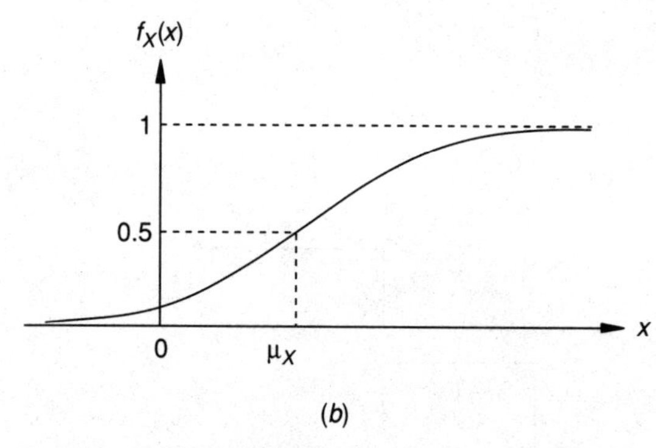

Figure 3.10 Normal distribution function. (*a*) Density function, (*b*) Distribution function.

$$f_{X_1, X_2, \ldots, X_n}(x1, x2, \ldots, xn)$$

$$= \frac{1}{\sqrt{(2\pi)^n} \mid [K] \mid} \exp\left[-\frac{1}{2} \sum_{j=1}^{n} \sum_{k=1}^{n} \left\{[K]^{-1}\right\}_{jk} (x_j - \mu_{X_j}) \cdot (x_k - \mu_{X_k})\right] \quad (3.50)$$

where

$$K_{X_j X_k} = E\,[(x_j - \mu_{X_j}) \cdot (x_k - \mu_{X_k})]$$

$$= \int_{-\infty}^{\infty} \int_{-\infty}^{\infty} (x_j - \mu_{X_j}) \cdot (x_k - \mu_{X_k}) \cdot f_{X_j, X_k}(x_j, x_k) \cdot dx_j \cdot dx_k$$

$$= \text{covariance between } X_j \text{ and } X_k,$$

$$[K] = \text{correlation matrix} = \begin{bmatrix} K_{11} & K_{12} & \cdots & K_{1n} \\ K_{21} & K_{22} & \cdots & K_{2n} \\ \cdot & \cdot & & \cdot \\ \cdot & \cdot & & \cdot \\ \cdot & \cdot & & \cdot \\ K_{n1} & K_{n2} & \cdots & K_{nn} \end{bmatrix}$$

and

$$\left\{[K]^{-1}\right\}_{jk} = jk\text{th element of } [K]^{-1}.$$

It is to be noted that

$$K_{X_j,X_k} = \begin{cases} 0, & \text{for } j \neq k \\ \sigma^2_{X_j}, & \text{for } j = k \end{cases}$$

in case there is no correlation between X_j and X_k.

Standard normal distribution. A Gaussian distribution with parameters $\mu = 0$ and $\sigma = 1$ is called the *standard normal distribution* and is identified as $N(0, 1)$. The density function of a standard normal variate Z is given by

$$f_Z(z) = \frac{1}{\sqrt{2\pi}} e^{-1/2z^2}; \qquad -\infty < z < \infty \tag{3.51}$$

and is symmetric about its mean $\mu = 0$, as shown in Fig. 3.11. The distribution function of the standard normal variate Z is commonly denoted as $\Phi(z)$ and is given by

$$\Phi(z) = F_Z(z) = \int_{-\infty}^{z} \frac{1}{\sqrt{2\pi}} e^{-1/2z^2}\, dz \tag{3.52}$$

This is shown in Fig. 3.12. If $\Phi(z_p) = p$ is given, the standard normal variate z_p corresponding to the cumulative probability (p) is denoted as

$$z_p = \Phi^{-1}(p) \tag{3.53}$$

The values of the distribution function $\Phi(z)$ of a standard normal variate are given as tables of normal distribution (see Appendix A). Usually the probabilities are given in tables only for positive values of z and

$$\Phi(-z) = 1 - \Phi(z) \tag{3.54}$$

due to the symmetry of the density function about zero. Similarly, we can find that

$$z_p = \Phi^{-1}(p) = -\Phi^{-1}(1 - p) \tag{3.55}$$

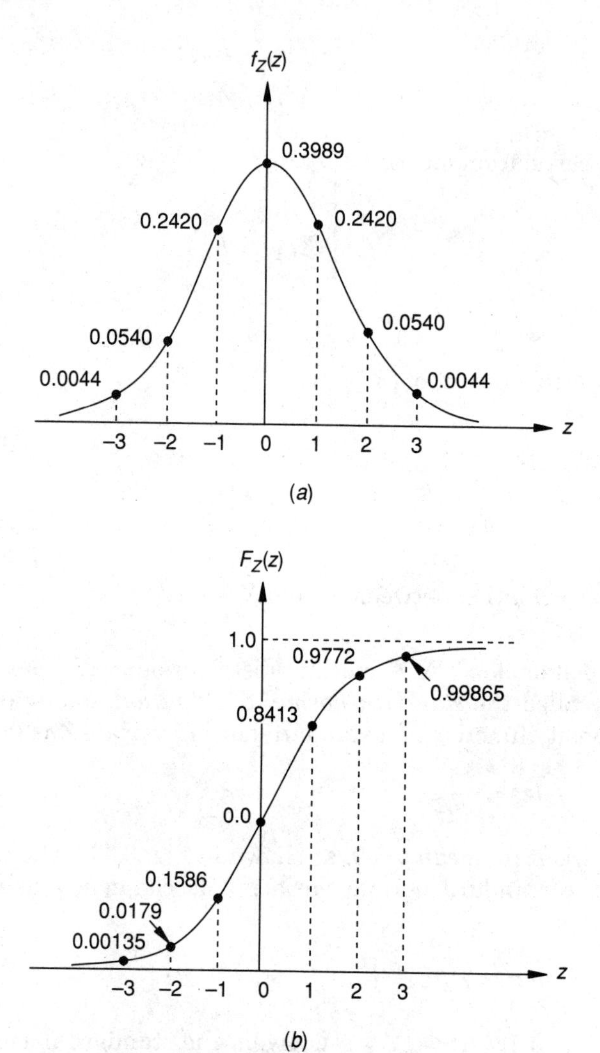

Figure 3.11 (*a*) Standard normal density function, (*b*) Standard normal distribution function.

In many practical applications, it will be of interest to find probabilities either up to a particular value of standard deviation from $-\infty$ or within a specified number of standard deviations from the mean. Such values can be found from tables of normal distribution (or from Fig. 3.11*b* and are shown in Fig. 3.13. Once the standard normal table of $\Phi(z)$ is available, the probabilities of any other normal distribution can be determined readily as follows. For a nonstandard X with $N(\mu, \sigma)$,

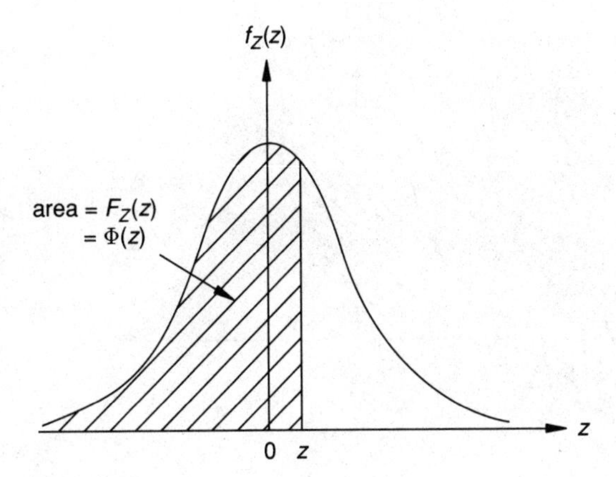

Figure 3.12

$$P(l \le X \le u) = \frac{1}{\sqrt{2\pi} \cdot \sigma} \int_l^u e^{-1/2[(x-\mu)/\sigma]^2} \cdot dx \tag{3.56}$$

This represents the area under the density function between l and u as shown in Fig. 3.14. By defining a new variable (standard normal variate) z as

$$z = \frac{x - \mu}{\sigma}, \quad \text{we find} \quad dz = \frac{dx}{\sigma},$$

and Eq. (3.56) can be rewritten as

$$P(l \le X \le u) = \frac{1}{\sigma\sqrt{2\pi}} \int_{(l-\mu)/\sigma}^{(u-\mu)/\sigma} e^{-1/2z^2} \cdot \sigma\, dz = \frac{1}{\sqrt{2\pi}} \int_{(l-\mu)/\sigma}^{(u-\mu)/\sigma} e^{-1/2z^2} \cdot dz \tag{3.57}$$

which can be recognized to be the area under the standard normal density function between $(u-\mu)/\sigma$ and $(l-\mu)/\sigma$. Thus the required probability can be found as

$$P(l \le X \le u) = \Phi\left(\frac{u-\mu}{\sigma}\right) - \Phi\left(\frac{l-\mu}{\sigma}\right) \tag{3.58}$$

Example 3.11. Tests conducted on a sample of 100 automobile brakes have yielded a mean value of 56,669.5 miles and a standard deviation of 12,393.5 miles for the life of brakes (see Example 1.1). Assuming normal distribution, find the probability of realizing the life of brakes less than 50,000 miles.

solution Denoting the life of brakes as X, we have μ_X = 56,669.5 miles and σ_X = 12,393.64 miles. The corresponding standard normal variate (z) is defined as

$$z = \frac{x - \mu_X}{\sigma_X} = \frac{x - 56{,}669.5}{12{,}393.64}$$

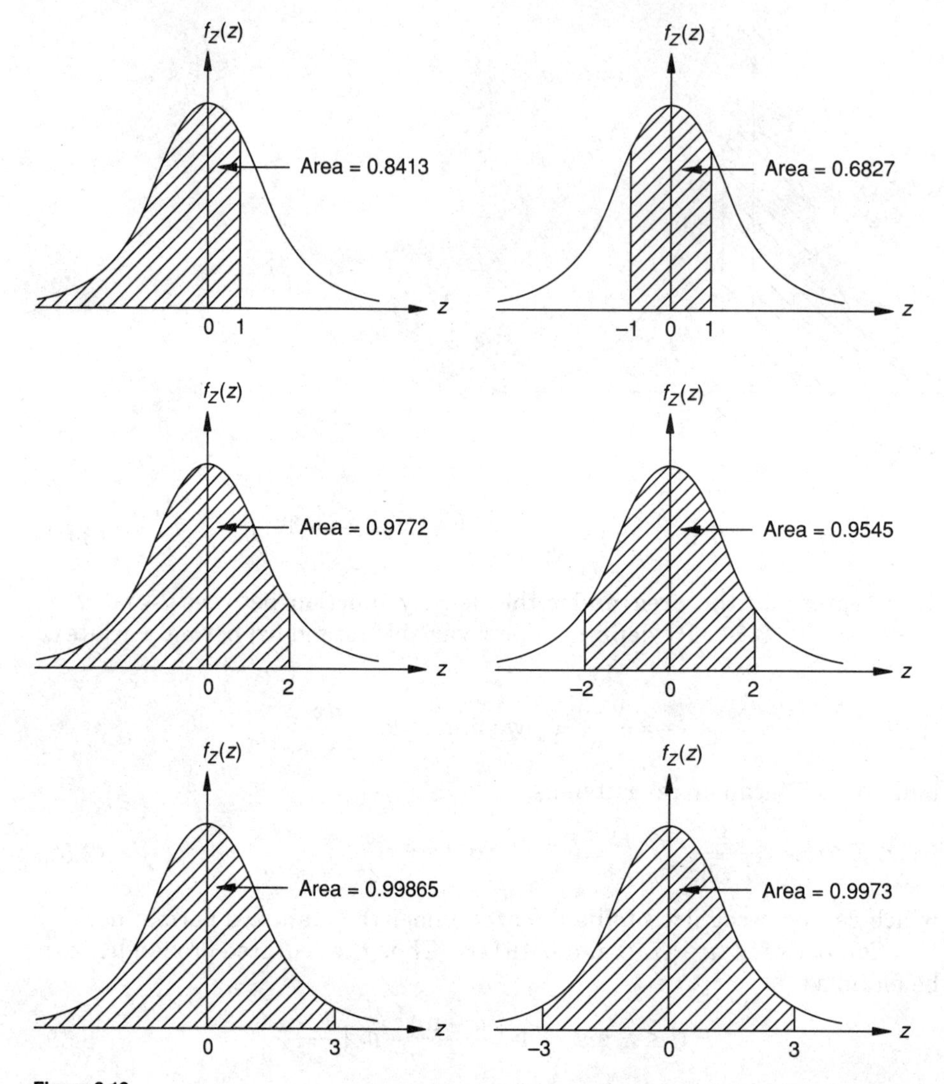

Figure 3.13

and the required probability is given by

$$P(X < 50{,}000) = P\left(z < \frac{50{,}000 - 56{,}669.5}{12{,}393.64}\right) = P(z < -0.5381)$$

Since the probability density function is symmetric about $z = 0$, the probability of realizing z less than -0.5381 is same as the probability of realizing z greater than $+0.5381$. This gives

$$P(z < -0.5381) = P(z > 0.5381) = 1 - P(z < 0.5381) = 1 - 0.7054 = 0.2946$$

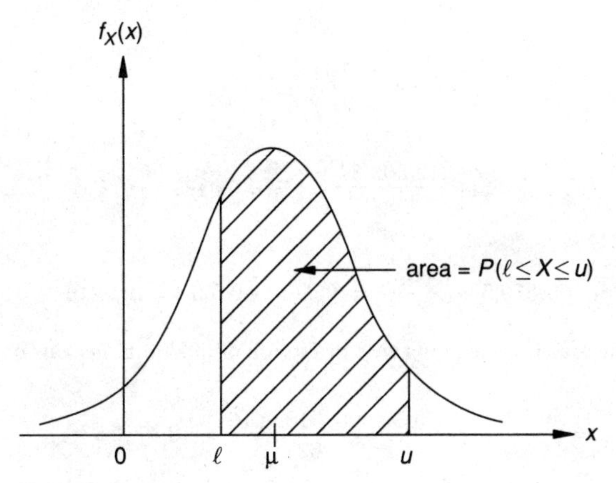

Figure 3.14

3.11.3 Lognormal distribution

A random variable X is said to follow lognormal distribution if $Y = \ln X$ follows normal distribution. Thus

$$f(y) = \frac{1}{\sqrt{2\pi}\,\sigma_Y} \exp\left[-\frac{1}{2}\left(\frac{y-\mu_Y}{\sigma_Y}\right)^2\right]; \qquad -\infty < y < \infty \tag{3.59}$$

Since $Y = \ln X$, Eq. (3.59) can be rewritten, in terms of X, as [3.2]

$$f(x) = \frac{1}{\sqrt{2\pi}x\;\sigma_Y} \exp\left[-\frac{1}{2}\left(\frac{\ln x-\mu_Y}{\sigma_Y}\right)^2\right]; \qquad x \ge 0 \tag{3.60}$$

where

$$\sigma_Y^2 = \ln\left[\left(\frac{\sigma_X}{\mu_X}\right)^2 + 1\right] \tag{3.61}$$

and

$$\mu_Y = \ln \mu_X - \frac{1}{2}\sigma_Y^2 \tag{3.62}$$

The following example illustrates the application of lognormal distribution.

Example 3.12 If the life of brakes described in Example 3.11 is assumed to follow lognormal distribution with mean and standard deviations of 56,669.5 and 12,393.64 miles, respectively, find the probability of realizing the life of brakes less than 50,000 miles.

solution Let X denote the life of brakes with mean and standard deviations of $\mu_X = 56{,}669.5$ miles and $\sigma_X = 12{,}393.64$ miles. Since $Y = \ln X$ follows normal distribution, the mean and standard deviations of the variable Y are given by

$$\mu_Y = \ln \mu_X - \frac{\sigma_Y^2}{2}$$

where

$$\sigma_Y^2 = \ln\left[\left(\frac{\sigma_X}{\mu_X}\right)^2 + 1\right]$$

In the present case,

$$\sigma_Y^2 = \ln\left[\left(\frac{12{,}393.64}{56{,}669.5}\right)^2 + 1\right] = 0.0467$$

and hence $\sigma_Y = 0.2162$ and

$$\mu_Y = \ln 56{,}669.5 - \frac{0.0467}{2} = 10.9450 - 0.0234 = 10.9216$$

The standard normal variate corresponding to a brake life of 50,000 miles can be determined as

$$z = \frac{\ln X - \mu_Y}{\sigma_Y} = \frac{\ln 50{,}000 - 10.9216}{0.2162} = -0.4710$$

Thus the required probability is given by

$$P(X < 50{,}000) = P(z < -0.4710) = P(z > 0.4710)$$

$$= 1 - P(z \le 0.4710) = 1 - 0.6808 = 0.3192$$

3.12 Central Limit Theorem

If $X_1, X_2, \ldots, X_n$ are n mutually independent random variables with finite mean and variance, then the sum

$$S_n = \sum_{i=1}^{n} X_i \tag{3.63}$$

tends to a normal variable if no single variable contributes significantly to the sum as n tends to infinity. This is known as the *central limit theorem* and is useful in several engineering applications [3.2, 3.3]. The central limit theorem explains why many physical phenomena can be described, approximately, by normal distribution. For example, the tensile strength of an alloy steel component can be considered to be influenced by the percentages of the alloying elements such as manganese, chromium, nickel and silicon, the heat treatment it received, and the machining process used during its production. If each of these effects tends to add to the others in determining the value of the tensile strength, then the tensile strength can be approximated by normal distribution according to the central limit theorem. Another important use of the central limit theorem is in the area of statistical inference. Many estimators such as sample mean and sample variance are used to make inferences about the characteristics of the population. Since these estimators involve sums or averages of the sample measurements, the estimators follow normal distribution according to the central limit theorem when the sample size is sufficiently large. This result is useful when repeated samplings are used to make inferences.

3.13 Normal Approximation to Binomial Distribution

In binomial experiments, each trial yields either 0 or 1 success with a probability of $q = 1 - p$ and p, respectively. If there are a total of n trials, we can consider each trial as an independent observation and hence the total number of successes (r) in n trials can be considered as a sum of n independent observations. If n is large, according to the central limit theorem, we can approximate the binomial variate r as a normal variate with mean np and variance $np(1 - p)$. Thus standard normal distribution tables can be used to approximate the binomial probabilities[8] (see Problem 3.41).

References

3.1. G. P. Wadsworth and J. G. Bryan, *Introduction to Probability and Random Variables,* McGraw-Hill, New York, 1960.
3.2. G. C. Hart, *Uncertainty Analysis, Loads, and Safety in Structural Engineering,* Prentice-Hall, Englewood Cliffs, New Jersey, 1982.
3.3. W. Mendenhall, *Introduction to Probability and Statistics,* 3d ed., Duxbury Press, Belmont, Calif., 1971.
3.4. I. Olkin and J. W. Pratt, *A Multivariate Chebyshev Inequality,* Annals Math. Stat., vol. 29, 1958, pp. 201–211.
3.5. E. Parzen, *Modern Probability Theory and Its Applications,* John Wiley, New York, 1960.
3.6. A. C. King and C. B. Read, *Pathways to Probability,* Holt, Rinehart & Winston, New York, 1963.
3.7. S. S. Rao and C. P. Reddy, "Mechanism Design by Chance Constrained Programming Techniques," *Mechanism and Machine Theory,* vol. 14, 1979, pp. 413–424.
3.8. A. H. S. Ang and W. H. Tang, *Probability Concepts in Engineering Planning and Design,*vol. 1: *Basic Principles,* John Wiley, New York, 1975.
3.9. C. C. Gillispie (ed.-in-chief), *Dictionary of Scientific Biography,* vol. 3, Charles Scribner's Sons, New York, 1980.

Review Questions

3.1 Define the various measures of central tendency of a random variable.

3.2 Define the terms standard deviation and variance.

3.3 What is the measure used to indicate the degree of nonsymmetry of a distribution?

3.4 How is the skewness coefficient defined for (a) a discrete distribution, and (b) a continuous distribution?

3.5 What is a random variable?

---

[8]This result was presented on November 12, 1733 by Abraham De Moivre (1667–1754, English) in a paper titled, Approximatio ad Summam Terminorum Binomii $(a + b)^n$ in Seriem Expansi. As such, November 12, 1733 is usually taken as the date of the origin of the "normal probability curve" [3.6].

3.6 Explain the difference between a discrete and a continuous random variable. Give two examples of each.

3.7 What is the difference between a histogram and a density function?

3.8 Define the terms: mean, mode, and median.

3.9 What is the difference between unimodal and multimodal distributions?

3.10 What is a probability mass function?

3.11 How are the probability density and probability distribution functions related?

3.12 What is the significance of probability distribution function?

3.13 Define the rth central moment of a distribution function.

3.14 Why is the area under a density function equal to one?

3.15 Can the probability density function have a value greater than one for any distribution?

3.16 Under what conditions will the mean, mode, and median of a random variable be identical?

3.17 Explain the notation: $N(\mu, \sigma)$.

3.18 What is a standard normal variate?

3.19 How is the root mean square value of a random variable related to its standard deviation?

3.20 What are the characteristics of Gaussian distribution?

3.21 What is central limit theorem and what is its use?

3.22 What is a Bernoulli trial?

3.23 Define binomial distribution.

3.24 What is the difference between variance and covariance?

3.25 Define correlation coefficient.

3.26 What is Chebyshev inequality? Give two practical uses of this inequality.

3.27 What are univariate, bivariate and multivariate distributions?

3.28 Define a marginal density function.

3.29 How do you find the marginal density function from a joint density function?

Problems

3.1 An experiment consists of tossing a fair coin 5 times. (a) Find the probability mass and distribution functions for the number of heads realized. (b) Find the probability of realizing heads at least 3 times out of the 5 trials.

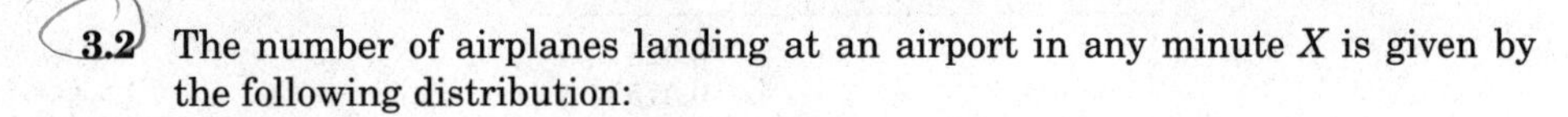

3.2 The number of airplanes landing at an airport in any minute X is given by the following distribution:

x	$p(x)$
0	0.06
1	0.14
2	0.20
3	0.25
4	0.18
5	0.12
6	0.05

(a) Find the mean, standard deviation, and skewness coefficient of X. (b) What is the probability of having more than 3 landings in any minute? (c) What is the probability of having the number of landings between 2 and 5 in any minute?

3.3 Depending on the surface finish requirements, a machinist uses different cutting speeds on a lathe. From past experience, the various cutting speeds used during machining operations and the probabilities of using them are given as follows

Cutting speed, X(m/min)	Probability
20	0.04
30	0.10
40	0.16
50	0.15
60	0.14
70	0.12
80	0.09
90	0.08
100	0.06
110	0.03
120	0.02
130	0.01

(a) Plot the probability mass and distribution functions. (b) Find the mean, mode and median of the random variable.

3.4 For the probability mass function given in Problem 3.3, determine $E(X^2)$, $E(X^3)$, standard deviation and skewness coefficient.

3.5 The number of cars arriving at a toll booth in a minute and their probabilities are given below:

Number of cars arriving in a minute (X)	Probability
1	0.025
2	0.075
3	0.125
4	0.150
5	0.200
6	0.275
7	0.100
8	0.050

(a) Find the probability distribution function of X. (b) Find the mean, mode, and median of the number of cars arriving in a minute.

3.6 For the probability mass function given in Problem 3.5, determine $E(X^2)$, $E(X^3)$, standard deviation and skewness coefficient.

3.7 The diameter of a forged crankshaft is specified as 1.2 ± 0.02 in. Past experience with the forging process gives a mean value of 1.2 in and a standard deviation of 0.005 in. Find the percent of discarded crankshafts. Assume normal distribution for the diameter of the crankshafts.

3.8 The life of a heat exchanger (in thousands of hours) is represented by the probability density function

$$f(x) = ae^{-ax}; \qquad x \geq 0$$

Determine (a) mean, (b) mode, (c) median, (d) standard deviation, and (e) skewness coefficient of the life of the heat exchanger.

3.9 An O-ring used to seal the leakage of fluid in a pressure vessel is subjected to a random pressure X [in k(lb/in^2)] which can be approximated by the distribution

$$f(x) = \begin{cases} \dfrac{x}{10}; & 0 \leq x \leq 2 \\ \dfrac{10-x}{40}; & 2 \leq x \leq 10 \end{cases}$$

Find (a) mean, (b) mode, (c) median, (d) standard deviation, and (e) skewness coefficient of the pressure induced on the O-ring.

3.10 The force applied on an engine brake, X (in kips), can be modeled as a random variable following the density function

$$f(x) = \begin{cases} \dfrac{x}{40}; & 0 \leq x \leq 8 \\ \dfrac{10-x}{10}; & 8 \leq x \leq 10 \end{cases}$$

Find: (a) mean, (b) mode, (c) median, (d) standard deviation, and (e) skewness coefficient of the force applied on the brake.

3.11 The coefficient of kurtosis (flatness) of a random variable X is defined as

$$\delta = \frac{E[(X-\mu)^4]}{\sigma_X^4}$$

Find the coefficient of kurtosis for the life of the heat exchanger stated in Problem 3.8.

3.12 Find the coefficient of kurtosis for the cutting speed described in Problem 3.3.

3.13 Find the probability mass and distribution functions for the sum of numbers realized when two fair dice are thrown. Also find the mean and standard deviations of the sum of numbers realized.

3.14 Experiments conducted on the fatigue life of helical springs indicated a mean value of 6×10^4 cycles and a standard deviation of 2×10^3 cycles. Assuming that the probability density function for fatigue life is not known, find the ranges within which the fatigue life will lie with probabilities of at least 0.5, 0.6, 0.7, 0.8, 0.9, and 0.9999. Use the Chebyshev inequality.

3.15 Plot a graph between

$$P\left[\mu - k\sigma \le X \le \mu + k\sigma\right]$$

and k for the following cases:

1. Normal distribution for X with $0 \le k \le 10$;
2. Exponential distribution for X with $0 \le k \le 10$;
3. Uniform distribution for X with $0 \le k \le 10$; and
4. Chebyshev inequality with $1 \le k \le 10$. What do you observe from these graphs?

3.16 The probabilistic characteristics of the bending strength (X_1) and the surface wear strength (X_2) of a gear tooth are given by $\mu_1 = 20{,}000$ psi, $\sigma_1 = 2{,}000$ psi, $\mu_2 = 60{,}000$ psi and $\sigma_2 = 3{,}000$ psi. If the bending and surface wear strengths are required to lie in the ranges 16,000 to 24,000 psi and 54,000 to 66,000 psi, respectively, with a minimum probability of 0.6, determine the necessary correlation coefficient between the bending and surface wear strengths of the gear tooth. Use the generalized Chebyshev inequality which can be stated in the case of two random variables X_1 and X_2 as [3.4]:

$$P\left[|X_1 - \mu_1| \ge t\,\sigma_1 \quad \text{or} \quad |X_2 - \mu_2| \ge t\,\sigma_2\right] \le \frac{1}{t^2}\left(+\sqrt{1-\rho_{1,2}^2}\right)$$

where μ_i and σ_i are, respectively, the mean value and the standard deviation of X_i ($i = 1, 2$) and $\rho_{1,2}$ is the correlation coefficient of X_1 and X_2.

3.17 The bathroom scales manufactured by a company exhibit a standard deviation of 0.5 lb when the mean weight is 100 lb. Determine the probability of a 100 lb weight being in error less than or equal to ±2.0 lb using the following:

(a) Chebyshev inequality, (b) Camp-Meidell inequality, and (c) Gauss inequality assuming that the mean and mode are separated by one standard deviation.

3.18 The times to failure, in hours of operation, of the differential and the rear axle of a truck (T_1 and T_2) can be approximated by the joint density function

$$f(t_1, t_2) = 1.05 \times 10^{-4} \exp\left\{-(10^{-4} t_1 + 5 \times 10^{-6} t_2)\right\}; \qquad t_1 \geq 0; \qquad t_2 \geq 0$$

a. Find the marginal density functions of T_1 and T_2.
b. Find the mean values of T_1 and T_2.

3.19 The joint density function of two random variables is given by

$$f(x, y) = \frac{1}{102}(x + 3x^2 y)$$

for $1 \leq x \leq 4$ and $1 \leq y \leq 2$. (a) Find the marginal density functions of X and Y. (b) Find the mean values of X and Y. (c) Find the correlation coefficient of X and Y.

3.20 Experiments are conducted on fiber-reinforced plastic specimens to find the load at which the first crack appeared and the load at which the specimen broke and the following values are observed:

Specimen number	Load at which first crack appeared, X_1 (newtons)	Load at which specimen broke, X_2 (newtons)
1	5025	6075
2	5050	6225
3	5075	6125
4	5100	6175
5	5125	6150
6	5150	6225
7	5175	6175
8	5200	6200
9	5225	6175
10	5250	6100
11	5275	6125
12	5300	6175
13	5325	6200
14	5350	6175
15	5375	6125
16	5400	6150
17	5425	6225
18	5450	6300
19	5475	6275
20	5500	6300

(a) Plot the scattergram. (b) Find μ_{X_1}, μ_{X_2}, σ_{X_1}, σ_{X_2} and ρ_{X_1,X_2}.

3.21 The following numbers indicate the grades received by the students in two mid-semester examinations of the course, "Reliability-Based Design":

Student	Examination 1 (X_1)	Examination 2 (X_2)
1	30	50
2	35	40
3	40	45
4	45	55
5	50	45
6	50	70
7	55	60
8	60	50
9	65	70
10	70	45
11	70	60
12	75	60
13	75	70
14	80	60
15	80	70
16	85	60
17	85	75
18	90	85
19	90	40
20	95	60

(a) Plot the data in a scattergram. (b) Find μ_{X_1}, μ_{X_2}, σ_{X_1} and σ_{X_2}. (c) Find the correlation coefficient between X_1 and X_2.

3.22 The surface finish achieved in a machining operation at different cutting speeds of the machine tool is indicated in the following table:

Cutting speed (rpm)	Surface finish achieved (the higher the number, the better the finish)
200	4
220	11
240	5
260	3
280	6
300	2
320	12
340	12
360	15
380	12
400	17
420	20
440	18
460	21
480	18
500	24

(a) Plot the data in a scattergram. (b) Find the correlation between the surface finish and the cutting speed.

3.23 The presence of clearances in joints of a linkage makes it necessary to consider the link lengths and the clearances as random variables [3.7]. Figure 3.15*a* shows two adjacent links in their nominal (ideal) positions while Fig. 3.15*b* shows their positions with clearances. If r denotes the radius of the clearance zone at the joint, the pin of link 2 (P_2) can lie anywhere in the circular zone. Then the equivalent length of link 1 can be expressed as

$$l_{1e} = \sqrt{(l_1 + x)^2 + y^2} \approx l_1 + x$$

since $y \ll l_1 + x$ where x and y denote the coordinates of the pin P_2 with respect to the origin O_2 shown in Figs. 3.15*b* and 3.15*c*. Assuming that the pin P_2 can lie anywhere in the clearance zone with equal probability, the joint density function of x and y can be expressed as

$$f(x,y) = \begin{cases} \dfrac{1}{\pi r^2}; & x^2 + y^2 \le r^2 \\ 0; & \text{otherwise} \end{cases} \tag{E1}$$

Since the computation of the mean and standard deviation of l_{1e} requires the mean and standard deviation of x, determine these quantities (μ_x and σ_x) using the probability density function of Eq. (E1).

3.24 Find the probability distribution function corresponding to the following density functions:

$$(a)\quad f(x) = \begin{cases} \dfrac{1}{b-a}; & a \le x \le b \\ 0; & \text{otherwise} \end{cases}$$

$$(b)\quad f(x) = \begin{cases} \lambda e^{-\lambda x}; & x \ge 0 \\ 0; & \text{otherwise} \end{cases}$$

Also sketch the density and distribution functions in each case.

3.25 The electric motors manufactured by a company are found to have an average life of 10^4 hours. Assuming an exponential distribution, find the fraction of motors that would fail before (a) the average life and (b) half the average life.

3.26 The welds produced by an automatic welding machine are found to be defective 1 percent of the time. If six welds are randomly selected, find the probability of finding the number of defective welds to be (a) equal to four, (b) at least one, (c) no more than two, and (d) equal to three.

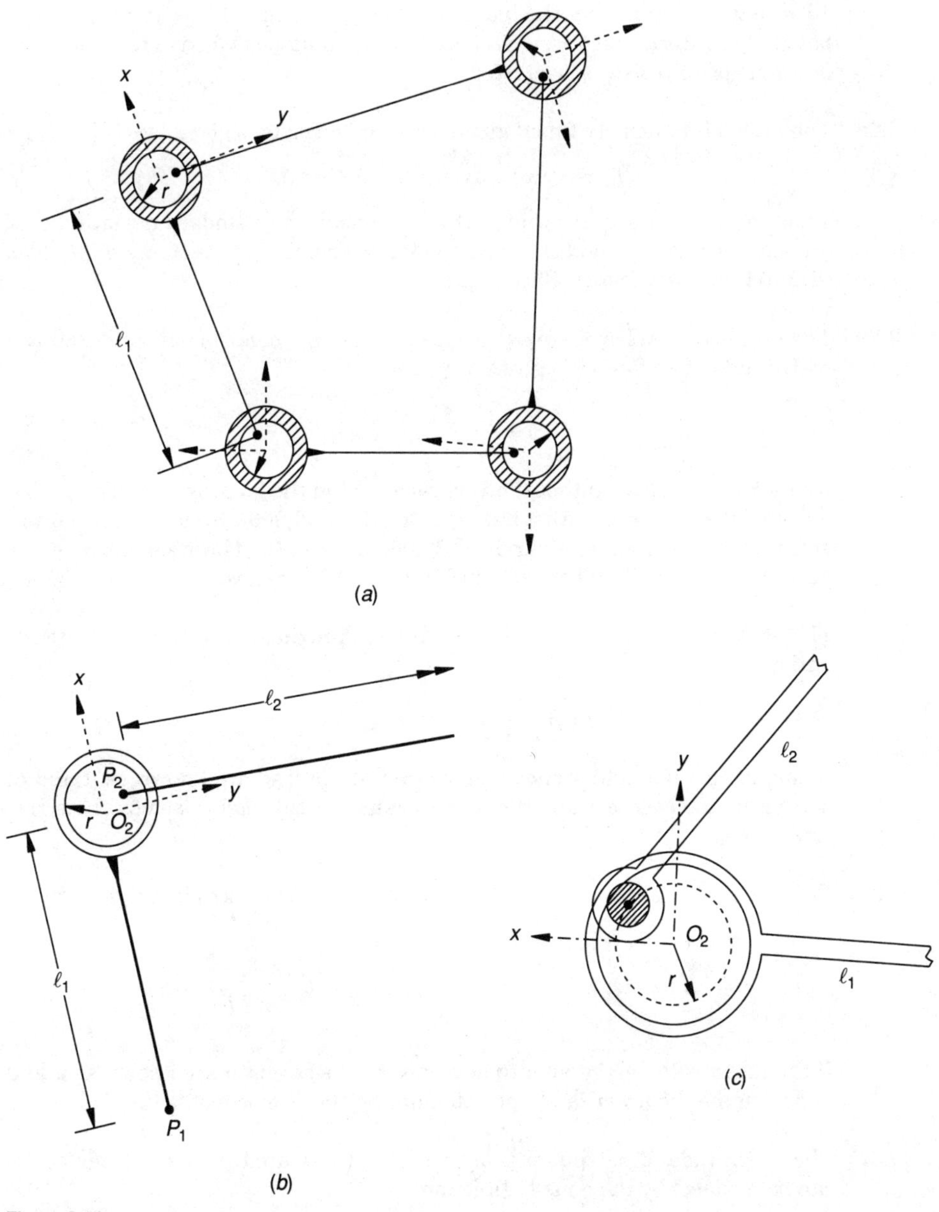

Figure 3.15

3.27 The quality control department rejects the whole batch of resistors whenever it finds two or more defective units in a sample of 15 resistors tested. If a batch of resistors contain 1 percent defective units, determine the following: (a) the mean and standard deviation of the defective resistors in a sample of

15 resistors chosen from the batch, (b) the probability of rejecting the entire batch of resistors, and (c) the probability of finding no defective resistor when a sample of 15 resistors are tested.

3.28 The probability density function of a random variable is given by

$$f_X(x) = ax^2 + bx + c; \qquad 0 \le x \le 1$$

where a, b and c are constants. (i) If the mean and standard deviation of X are specified as 0.6 and 0.1, respectively, determine the constants a, b and c. (ii) Find the skewness coefficient of X.

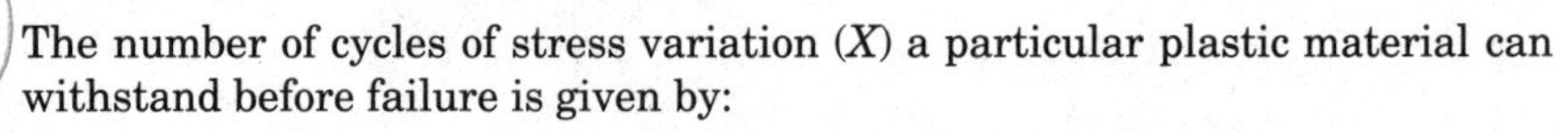

3.29 The number of cycles of stress variation (X) a particular plastic material can withstand before failure is given by:

$$f(x) = \frac{1}{4} e^{-\frac{x}{4}}; \qquad x \ge 0$$

where x is measured in thousands of cycles. Find the following: (a) The probability of the material withstanding more than 40,000 cycles without failure given that it has already survived 30,000 cycles. (b) The probability of the plastic material failing between 30,000 and 40,000 cycles.

3.30 The load raised by a screw jack (X) can be approximated by a uniform distribution:

$$f(x) = \frac{1}{b - a}; \qquad a \le x \le b$$

If the mean value and standard deviation of the load are known as μ and σ, determine the smallest and the largest values of the load raised by the screw jack.

3.31 The strength of a leather belt (X) follows a triangular distribution given by

$$f(x) = \begin{cases} \dfrac{4}{(b-a)^2}(x - a); & a \le x \le \dfrac{a+b}{2} \\ \dfrac{4}{(b-a)^2}(b - x); & \dfrac{a+b}{2} \le x \le b \end{cases}$$

If the mean value and standard deviation of the strength are known as μ and σ, determine the lower- and upper-bounds on the strength of belt.

3.32 The magnitude of earthquakes in a region (measured in the Richter scale) can be modeled by the density function

$$f(x) = \frac{4}{\pi(4 + x^2)}$$

Find (a) the average magnitude of the earthquakes, (b) the probability of the magnitude of the next earthquake exceeding a value of 4.0, and (c) the mode and the median of X.

3.33 The stress induced in a machine during its operation has a mean value of 25,000 lb/in^2 with a standard deviation of 1,000 lb/in^2. The material of the machine has a mean strength of 35,000 lb/in^2 with a standard deviation of 2,000 lb/in^2. Find the probability of failure of the machine assuming that both the induced stress and the material strength follow normal distribution.

3.34 An elevator is supported by a wire rope that can withstand a mean stress of 25,000 lb/in^2 with a standard deviation of 500 lb/in^2. The load on the elevator is a random variable that causes a stress with a mean value of 20,000 lb/in^2 and a standard deviation of 2,000 lb/in^2. If the load on the elevator and the strength of the rope are normally distributed, find the probability of failure of the rope.

3.35 The life of an automobile battery is known to follow normal distribution with a mean of 2000 days and a standard deviation of 200 days. (a) Find the percentage of batteries which fail before 1500 days. (b) What should be the guarantee period to be specified by the manufacturer so that the manufacturer will encounter only 10 percent failures of the battery?

3.36 The tolerances on the diameter of a shaft are specified as 2 ± 0.05 in. The shafts produced by a machinist are found to have a mean value of 2.01 in and a standard deviation of 0.02 in. Find the fraction of the machined shafts that need to be discarded.

3.37 The point reached by the end effector of a robot manipulator along a line is given by 10 ± 0.5 in. What is the probability of the end effector reaching a point beyond 10.3 in? What is the probability that the point reached is beyond 10.3 in given that it is beyond 10.1 in?

3.38 The lives of tires produced by a tire manufacturer *A* are approximately normally distributed with $\mu = 60{,}000$ miles and $\sigma = 5000$ miles. An automobile manufacturer *B* requires that at least 95 percent of the tires should have a life greater than 50,000 miles. Can the automobile manufacturer *B* use the tires produced by the tire manufacturer *A*?

3.39 The error in the altitude of a helicopter predicted by an altimeter can be approximated as a normal variate with mean μ and standard deviation 100 m. Find the following probabilities for the error: (a) less than or equal to −200 m, (b) greater than or equal to 300 m, (c) between −200 m and 150 m, (d) less than −100 m or greater than 250 m.

3.40 The diameters of a batch of ball bearings are known to follow normal distribution with a mean 4.0 in and a standard deviation 0.1 in. If a ball bearing is chosen randomly, find the probability of realizing the following event: (a) diameter between 3.8 in and 4.3 in, (b) diameter smaller than 3.9 in, (c) diameter larger than 4.2 in.

3.41 The performance of a certain batch of electric resistance strain gauges is being tested using a random sample of 1000 strain gauges. If past manufacturing experience indicates the reliability of the strain gauges as 0.97, find the probability of observing 25 or more defective strain gauges in the sample of 1000 tested using a normal approximation to the binomial distribution [3.5].

3.42 Write a computer program, in the form of a subroutine, to find the value of a standard normal variate z given the corresponding cumulative distribution function value, $\Phi(z)$. Using this program, generate the values of z when $\Phi(z)$ varies between 0.1 and 0.9 in increments of 0.1.

3.43 A missile is fired to hit a target point O. Due to uncertainties associated with the firing, the missile can hit any point within radius a of the target point with equal probability, (see Figure 3.16), so that

$$f(x, y) = \begin{cases} \dfrac{1}{\pi a^2}; & \text{for } x^2 + y^2 \le a^2 \\ 0; & \text{otherwise} \end{cases}$$

(a) Find the marginal density functions $f(x)$ and $f(y)$. (b) What is the probability of hitting any point within a radius of $a/4$ around the target point O? (c) What is the probability of hitting any point within a distance of $a/4$ from the y-axis?

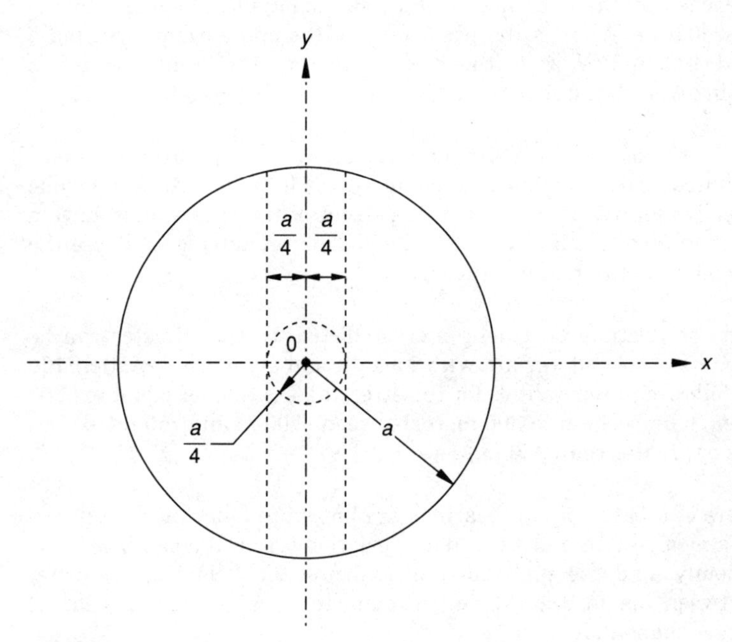

Figure 3.16

3.44 From the sample data of the Young's modulus given in Problem 1.1, determine the probability of realizing the Young's modulus greater than 30×10^6 lb/in^2 assuming (a) normal distribution, and (b) lognormal distribution.

3.45 From the compressive strength data of concrete cylinders given in Problem 1.4, find the probability of realizing the compressive strength between 3 and 6 k(lb/in^2) assuming (a) normal distribution, and (b) lognormal distribution.

3.46 Find the probability of realizing the first crack below a load of 5200 N and the breakage of the specimen above a load of 6200 N for the fiber-reinforced plastic specimens in Problem 3.20.

3.47 The beta density function is given by

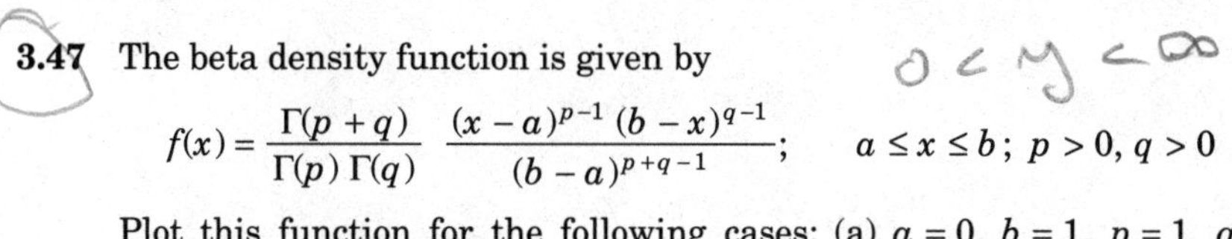

$$f(x) = \frac{\Gamma(p+q)}{\Gamma(p)\,\Gamma(q)} \; \frac{(x-a)^{p-1}\,(b-x)^{q-1}}{(b-a)^{p+q-1}}; \qquad a \le x \le b;\; p > 0,\, q > 0$$

Plot this function for the following cases: (a) $a = 0$, $b = 1$, $p = 1$, $q = 2$; (b) $a = 0$, $b = 1$, $p = 2$, $q = 2$; (c) $a = 0$, $b = 1$, $p = 0.5$, $q = 2$; (d) $a = 0$, $b = 1$, $p = 2$, $q = 4$; (e) $a = 0$, $b = 1$, $p = 3$, $q = 3$; (f) $a = 0$, $b = 1$, $p = 4$, $q = 2$.

3.48 The gamma density function is given by

$$f(x) = \frac{\lambda^k}{\Gamma(k)}\, x^{k-1}\, e^{-\lambda x}; \qquad 0 \le x < \infty; \qquad \lambda > 0$$

Plot this function for the following cases: (a) $k = 3$, $\lambda = 1$; (b) $k = 3$, $\lambda = 2$; (c) $k = 3$, $\lambda = 4$; (d) $k = 2$, $\lambda = 1$; (e) $k = 1$, $\lambda = 1$; (f) $k = 0.5$, $\lambda = 1$.

3.49 Find the parameters of beta distribution for the automobile brake life data given in Example 1.1.

3.50 Find the parameters of gamma distribution for the automobile brake life data given in Example 1.1.

Chapter

4

Extremal Distributions

Biographical Note

Abraham De Moivre

Abraham De Moivre was born on May 26, 1667, in France and died on November 27, 1754, in London. He was the son of a surgeon and the family left France in 1685. He came across Newton's Principia Mathematica *(1687) while he was tutoring for a living and became interested in mathematics. He used to tear a few pages at a time from* Principia, *stuff them in his pocket and study them between his tutoring hours. Later, he became an intimate friend of Newton and it was said that when Newton was asked questions about* Principia, *he would direct them to De Moivre for the answers. De Moivre published his first book on probability, entitled* Doctrine of Chances, *in 1718 and dedicated it to Newton. On November 12, 1733, De Moivre presented a paper on binomial theorem that led him to the idea of what is now called the "normal probability curve." As a result, this date is usually taken as the date of origin of normal distribution theory. De Moivre was chosen to decide the controversy over the invention of differential calculus between Newton and Leibniz. His contributions to trigonometry reflect in two well-known theorems, which bear his name, on the expansion of trigonometric functions. According to an interesting story, which may not be true, shortly before his death De Moivre wanted to sleep 15 minutes more every day. He was getting 6 hours of sleep every day when he announced this and naturally he would reach the 24 hour limit in 72 days. Surprisingly he died on the 73rd day, in his sleep. [4.13, 4.14]*

4.1 Introduction

In many engineering applications, we will be concerned with the largest- or smallest-value of a random variable. Success or failure of a system may rest solely on its ability to function under the *maximum demand* (load) or *minimum capacity* (strength), and not simply on the typical values. Floods, wind, temperature, solar radiation, and floor loadings are variables for which the largest value in a sequence may be critical to several engineering systems. Similarly, the smallest value of strength of materials will also be critical in some situations. The following examples indicate situations where a knowledge of the minimum- or maximum-value of a random variable is required.

1. In structural- and mechanical-design, we need to know the minimum strength of the material and the maximum load acting on the structure for a safe design.
2. In flood control programs, such as the design of dams, we need to know the intensity of the maximum flood of a river, whereas in irrigation and water supply programs a knowledge of the minimum flow will be a matter of concern.
3. The largest amount of traffic experienced in a day is important in the design of highways.
4. The corrosion of a metal is associated with small pits caused by chemical agents. Thus the corrosion resistance of a component is governed by the size of the largest pit present.
5. The safe design of offshore platforms involves a knowledge of the maximum wave heights experienced in the ocean.
6. The development of a new suburb is encouraged when the maximum pollution level caused by the industrial and automobile exhausts in the region is limited to small values.
7. The design of high-rise buildings is influenced by the maximum wind or earthquake load expected in the region.
8. The *size* plays a major role in the strength of any material. According to a widely accepted theory, the body of material (such as a machine component) can be considered to be composed of several small pieces or elemental volumes. The material (component) is considered to have failed whenever any one of the small pieces (elemental volumes) fails. This is called the *weakest link model* since the strength of the total volume is governed by that of the elemental volume that has the smallest strength.
9. The maximum vertical gust velocity encountered by an aircraft is important for a safe design.

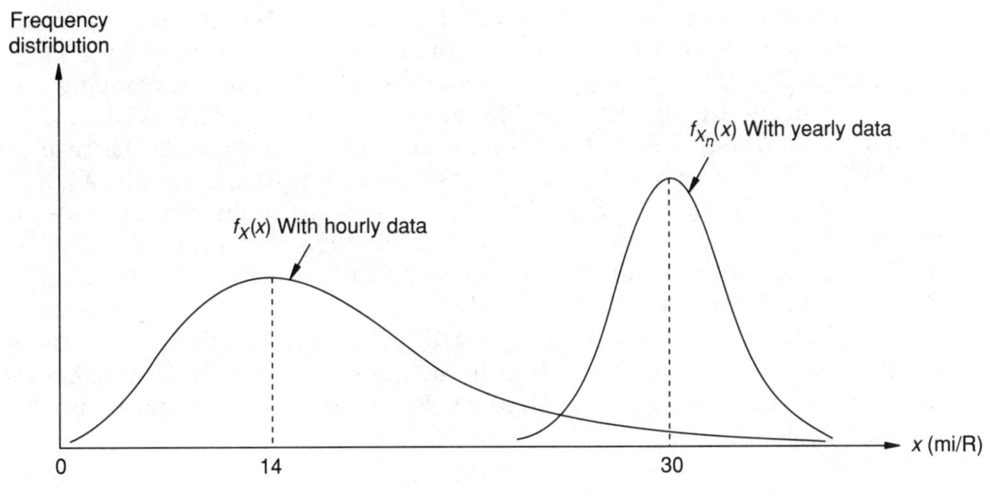

Figure 4.1

A systematic study of the maximum- or minimum-values of the random variables influencing the design may lead to a better design. Further, the concept of finite-life design can also be incorporated and thus a better use of materials can be made and technological obsolescence can be cared for. Gumbel [4.1] has found that the use of extreme values is much more reliable than the median whose value increases with the sample size. Wirsching and Jones [4.7] have used extreme values in reliability analysis and design.

To illustrate the concept of extreme value distribution, assume that we have the wind velocity record for the past 40 years on an hourly basis. This means that we have a total of $40 \times 365 \times 24 = 350{,}400$ data points (assuming 365 days per year). Let the mean value corresponding to this data be 14 mph, and the distribution of the wind velocity be $f_X(x)$ as shown in Fig. 4.1. Suppose we want to know about the distribution followed by the largest wind velocity measured on a yearly basis. This time, we have one maximum wind velocity for each year and a total of 40 data points only. Naturally, the distribution of this extreme wind velocity will be shifted towards the right with a higher mean (30 mph) as indicated in Fig. 4.1. Our interest is to find the distribution of the largest value, X_n (and sometimes, the smallest value, X_1) in terms of the original (parent) distribution, $f_X(x)$.

4.2 Extreme Value Distributions in Terms of Parent Distribution

To express the density functions of the smallest- and largest-values of a random variable in terms of the parent density function, consider a random variable X denoting the yield strength of steel. Let n tests be conducted to find the yield strength of steel as X_i', $i = 1, 2, \ldots, n$. Arrange the results in an increasing order so that the smallest value is designated as X_1' and the largest value as X_n'. If the n tests are repeated, we obtain, after rearranging the values in an increasing order, a different set of values as $X_1'', X_2'', \ldots, X_n''$. Similarly, the repetition of the n tests a number of times yields the results $(X_1''', X_2''', \ldots, X_n''')$, and so on. Let the results of all the tests, that is, $X_1', X_1'', X_1''', \ldots, X_2', X_2'', X_2''', \ldots, X_n', X_n'', X_n''', \ldots$, fit into a distribution, $F_X(x)$, which is known as the (parent) distribution function of the random variable X. If we fit separate distributions for the smallest values only (i.e., for $X_1', X_1'', X_1''', \ldots$) and for the largest values only (i.e., for $X_n', X_n'', X_n''', \ldots$), the resulting distributions, $F_{X_1}(x)$ and $F_{X_n}(x)$ are called the extreme value distributions for the smallest and the largest value of X, respectively. It is possible to express the extreme value distributions $F_{X_1}(x)$ and $F_{X_n}(x)$ in terms of the parent distribution $F_X(x)$.

To find the distribution function of the smallest value X_1, we use the relation

$$1 - F_{X_1}(x) = P(X_1 > x) = P(X_1 > x)P(X_2 > x) \cdots P(X_n > x) \tag{4.1}$$

if all X_i are independent. This equation indicates that the probability of realizing X_1 to be greater than x is same as the probability of realizing all $X_i (i = 1, 2, \ldots, n)$ to be greater than x. If all the X_i are identically distributed with a common distribution function $F_X(x)$, then

$$1 - F_{X_1}(x) = (P(X > x))^n = (1 - F_X(x))^n$$

or

$$F_{X_1}(x) = 1 - (1 - F_X(x))^n \tag{4.2}$$

Similarly the distribution function of the largest value X_n is given by

$$F_{X_n}(x) = P(X_n \le x) = P(\text{all } X_i \le x)$$

$$= P(X_1 \le x)P(X_2 \le x) \cdots P(X_n \le x)$$

if all X_i are independent. This means that the probability of realizing X_n less than or equal to x is same as the probability of realizing all $X_i (i = 1,2, \ldots, n)$ to be less than or equal to x. If all the X_i are identically distributed with the distribution function $F_X(x)$,

$$F_{X_n}(x) = (F_X(x))^n \tag{4.3}$$

Order statistics. When the sample values $(X_1, X_2, \ldots, X_n)$ of a random variable are arranged in an increasing order as $X_1 \le X_2 \le \cdots \le X_i \le \cdots \le X_n$, then the ith element X_i is called the ith order statistic of the sample. Thus the distributions of the smallest and the largest values can be called the distributions of the first and the nth order statistics, respectively.

Example 4.1 The time elapsed between the arrival of customers at an automobile repair shop can be assumed to follow an exponential distribution with a mean of 10 minutes. A knowledge of the maximum time elapsed between the arrival of customers will be of interest in deciding the number of mechanics to be made available in the shop. Assuming that the time elapsed between the arrival of customers is independent from one customer to the next, compute the probability that the maximum time elapsed between customers exceeds 12 minutes over a period covering the arrival of 15 customers.

solution The parent distribution for the time elapsed between the arrival of customers X is given by

$$F_X(x) = 1 - e^{-\lambda x}$$

where $\lambda = 1/\mu_X = 1/10 = 0.1$. Since 15 customers imply 14 inter-customer periods ($n = 14$), the probability of realizing the maximum intercustomer time ($Y \equiv X_{14}$) to be less than 12 minutes is given by Eq. (4.3):

$$F_Y(12) = (F_X(12))^{14} = (1 - e^{-0.1(12)})^{14} = 0.006622$$

Hence the probability that the maximum time elapsed between the arrival of customers exceeding 12 minutes is $1.0 - 0.006622 = 0.993378$.

Example 4.2 The strength of a welded joint, in k(lb/in^2), is distributed normally as

$$f_X(x) = \frac{1}{\sqrt{2\pi}\, 7.5} \exp\left[-\frac{1}{2}\left(\frac{x-75}{7.5}\right)^2\right]$$

Find the probability distribution and density functions for the largest strength realized in a sample of n welded joints. Plot these functions for $n = 1$, 10, and 20.

solution The probability distribution function corresponding to the largest strength in a sample of n welded joints is given by Eq. (4.3)

$$F_{X_n}(x_n) = [F_X(x_n)]^n \tag{E1}$$

where

$$F_X(x_n) = \int_{-\infty}^{x_n} f(x)\, dx = \int_{-\infty}^{x_n} \frac{1}{7.5\sqrt{2\pi}} \exp\left[-\frac{1}{2}\left(\frac{x-75}{7.5}\right)^2\right] dx \tag{E2}$$

This gives

$$F_{X_n}(x_n) = \left[\int_{-\infty}^{x_n} \frac{1}{7.5\sqrt{2\pi}} \exp\left[-\frac{1}{2}\left(\frac{x-75}{7.5}\right)^2\right] dx\right]^n \tag{E3}$$

Defining a standard normal variate z_n as

$$z_n = \frac{x_n - 75}{7.5}$$

Eq. (E3) can be expressed as

$$F_{X_n}(x_n) = F_{Z_n}(z_n) = [\Phi(z_n)]^n \tag{E4}$$

The probability density function, $f_{X_n}(x_n)$, can be obtained from Eq. (E4) as

$$f_{X_n}(x_n) = n[\Phi(z_n)]^{n-1} f_{Z_n}(z_n) \tag{E5}$$

With the help of standard normal tables, Eqs. (E4) and (E5) can be plotted for any specific value of n. The result is shown in Fig. 4.2.

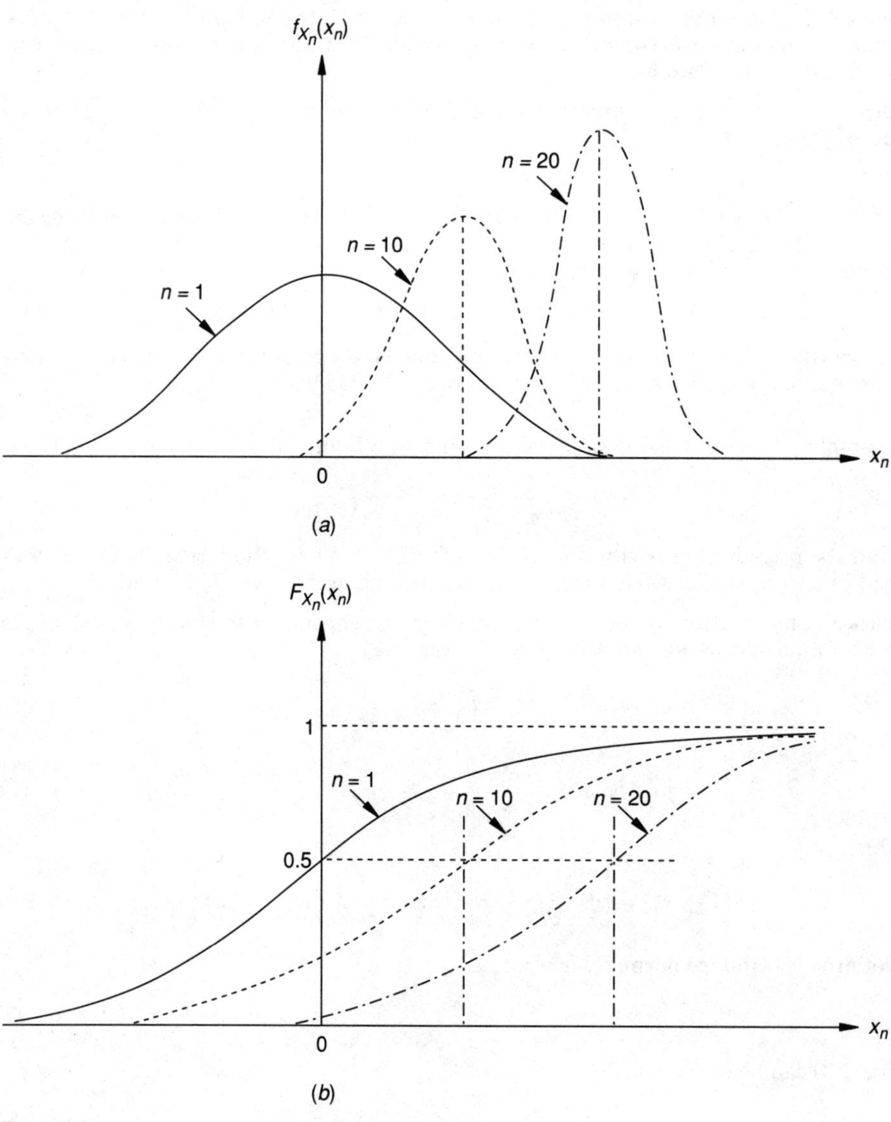

Figure 4.2

4.3 Asymptotic Distributions

If the random variables $X_i, i = 1, 2, \ldots, n$ are independent and follow a common distribution $F_X(x)$, the shape of the distribution function of the extreme value X_1 or X_n becomes increasingly insensitive to the exact shape of the common distribution function $F_X(x)$, as n tends to infinity. The limiting forms of the distribution functions of the extreme values X_1 or X_n as $n \to \infty$, are known as asymptotic distributions. The asymptotic distributions often describe the behavior of the random variable X_1 or X_n reasonably well even when the exact shape of the parent distribution function $F_X(x)$ is not known precisely. The asymptotic forms of $F_{X_1}(x)$ and $F_{X_n}(x)$ are classified into three types based on the general features of the tail part of the distribution of the random variable X [4.1]. Different types of asymptotic distributions have been used for different engineering applications.

4.4 Type-I Asymptotic Distributions

4.4.1 Maximum value

The type-I asymptotic distribution for the maximum values is useful whenever the right tail of the parent distribution $F_X(x)$ is unbounded ($x \to \infty$) and is of an exponential type. In such a case, we can express $F_X(x)$ as

$$F_X(x) = 1 - e^{-h(x)} \tag{4.4}$$

where $h(x)$ increases with x monotonically. The distributions such as normal, lognormal and gamma distributions belong to this category. If $F_X(x)$ is of the form given by Eq. (4.4), the asymptotic distribution function for the largest value $Y \equiv X_n$ is given by

$$F_Y(y) = \exp[-e^{-a(y-w)}]; \qquad -\infty < y < \infty \tag{4.5}$$

where the parameters of the distribution, a and w, can be determined from the observed data. The probability density function of Y can be obtained as

$$f_Y(y) = a \exp[-a(y-w) - e^{-a(y-w)}]; \qquad -\infty < y < \infty \tag{4.6}$$

It can be shown that the mode (Y_{mode}), mean (μ_Y), and the variance (σ_Y^2) of Y are given by

$$Y_{\text{mode}} = w \tag{4.7}$$

$$\mu_Y = w + \frac{\gamma}{a} \tag{4.8}$$

$$\sigma_Y^2 = \frac{\pi^2}{6a^2} \tag{4.9}$$

where $\gamma \approx 0.577$ is the Euler's constant. This distribution has a positive skewness coefficient (γ_1) with $\gamma_1 \approx 1.1396$ as shown in Fig. 4.3.

4.4.2 Smallest Value

The type-I asymptotic distribution for the minimum values is useful whenever the left tail of the parent distribution is unbounded ($-\infty < x$) and decreases to zero towards the left in an exponential form. For example, the normal distribution belongs to this category. In this case, the asymptotic distribution function for the smallest value, $Z \equiv X_1$, is given by

$$F_Z(z) = 1 - \exp[-e^{a(z-w)}]; \qquad -\infty < z < \infty \tag{4.10}$$

where a and w are the parameters of the distribution. The density function can be obtained by differentiating Eq. (4.10) as

$$f_Z(z) = a \exp[a(z-w) - e^{a(z-w)}]; \qquad -\infty < z < \infty \tag{4.11}$$

The mode (Z_{mode}), mean (μ_Z), and the variance ($\sigma_Z{}^2$) of Z can be found as

$$Z_{\text{mode}} = w \tag{4.12}$$

$$\mu_Z = w - \frac{\gamma}{a} \tag{4.13}$$

$$\sigma_Z{}^2 = \frac{\pi^2}{6a^2} \tag{4.14}$$

This distribution has a negative skewness coefficient (γ_1) with $\gamma_1 \approx -1.1396$, as shown in Fig. 4.3.

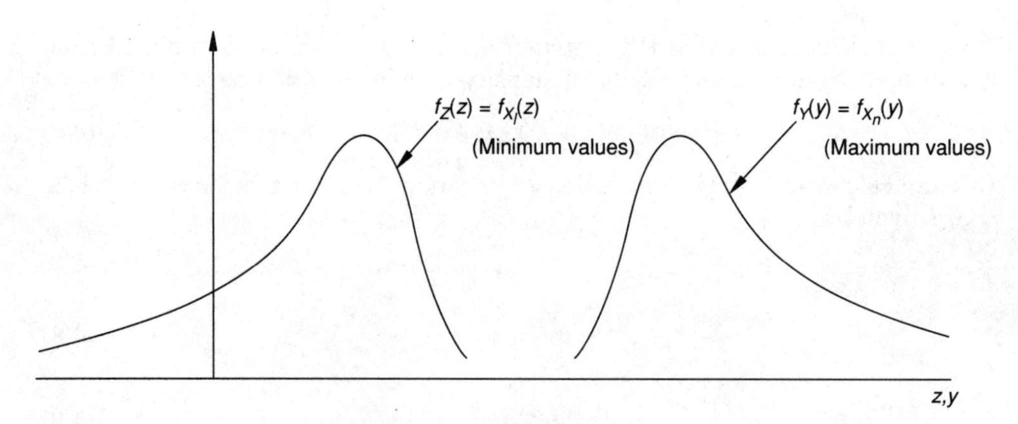

Figure 4.3 Typical shapes of type-I asymptotic distributions.

Notes:

1. It can be seen that the type-I asymptotic distributions for largest and smallest values are mirror images of each other. Typical shapes of $f_Y(y)$ and $f_Z(z)$ are shown in Fig. 4.3.
2. The type-I distributions are also known as Gumbel distributions.
3. The type-I maximum value distribution is useful, for example, to describe the maximum flood in a river.
4. The type-I minimum value distribution is useful, for example, to describe the strength of brittle materials where the specimen is assumed to consist of several microscopic volumes of material. In this case, the specimen fails whenever the weakest microscopic volume fails.

4.5 Type-II Asymptotic Distributions

4.5.1 Maximum value

The type-II asymptotic distribution for the maximum values is useful whenever the parent distribution $F_X(x)$ is defined over the range $0 < x < \infty$ and approaches one as $x \to \infty$ according to the relation

$$F_X(x) = 1 - \alpha\left(\frac{1}{x}\right)^m; \qquad x > 0 \tag{4.15}$$

where $\alpha > 0$ and $m > 0$ are the parameters of the distribution. In this case, $F_Y(y)$ can be expressed as

$$F_Y(y) = \exp\left[-\left(\frac{w}{y}\right)^m\right]; \qquad m > 0,\ w > 0,\ y > 0 \tag{4.16}$$

so that

$$f_Y(y) = \frac{m}{w}\left(\frac{w}{y}\right)^{m+1} e^{-(w/y)^m}; \qquad m > 0,\ w > 0,\ y > 0 \tag{4.17}$$

The mode, mean, and variance of Y are given by

$$Y_{\text{mode}} = \frac{w\,m}{m+1} \tag{4.18}$$

$$\mu_Y = w\,\Gamma\left(1 - \frac{1}{m}\right); \qquad m > 1 \tag{4.19}$$

and

$$\sigma_Y^2 = w^2\left[\Gamma\left(1 - \frac{2}{m}\right) - \Gamma^2\left(1 - \frac{1}{m}\right)\right]; \qquad m > 2 \tag{4.20}$$

where $\Gamma(x)$ is the gamma function defined as

$$\Gamma(x) = \int_0^\infty y^{x-1} e^{-y}\, dy \tag{4.21}$$

If x is a positive integer, the gamma function is given by

$$\Gamma(x) = (x - 1)! \tag{4.22}$$

4.5.2 Smallest value

The type-II asymptotic distribution for the minimum values is useful whenever the parent distribution $F_X(x)$ is defined over the range $-\infty < x \le 0$ and has a form similar to that of Eq. (4.15). In this case, $F_Z(z)$ can be expressed as

$$F_Z(z) = 1 - \exp\left[-\left|\frac{w}{z}\right|^m\right]; \qquad z \le 0 \tag{4.23}$$

where $w > 0$ and $m > 0$ are the parameters of the distribution.

Notes:

1. The type-II maximum value distribution is useful for representing annual maximum winds and other meteorological and hydrological phenomena.
2. The type-II minimum value distribution is not commonly used since the required parent distribution shape is not commonly observed in practical applications.

4.6 Type-III Asymptotic Distributions

4.6.1 Maximum value

The type-III asymptotic distribution for the maximum values is useful whenever the parent distribution $F_X(x)$ is defined over the range $-\infty < x \le \omega$ and has the form

$$F_X(x) = 1 - a(\omega - x)^m; \qquad x \le \omega \tag{4.24}$$

where $a > 0$ and $m > 0$ are the parameters of the distribution. Notice that Eq. (4.24) denotes a uniform distribution when $m = 1$ and a triangular distribution when $m = 2$. In this case the type-III maximum value distribution can be derived to be of the form [4.1, 4.10]

$$F_Y(y) = \exp\left[-\left(\frac{\omega - y}{\omega - v}\right)^m\right]; \qquad y \le \omega \tag{4.25}$$

with

$$f_Y(y) = \frac{m}{\omega - v}\left(\frac{\omega - y}{\omega - v}\right)^{m-1} \exp\left[-\left(\frac{\omega - y}{\omega - v}\right)^m\right]; \qquad y \le \omega \tag{4.26}$$

where $v < \omega$.

4.6.2 Smallest value

The type-III minimum value distribution can be used whenever the parent distribution is defined over the range $\varepsilon \le x < \infty$ and has the form

$$F_X(x) = a(x - \varepsilon)^m; \qquad x \ge \varepsilon \tag{4.27}$$

where $a > 0$ and $m > 0$ are the parameters of the distribution. In this case, the type-III asymptotic distribution for the minimum value can be obtained as [4.1, 4.10]

$$F_Z(z) = 1 - \exp - \left\{ \left(\frac{z - \varepsilon}{v - \varepsilon} \right)^m \right\}; \qquad z \ge \varepsilon,\ m > 0,\ v > \varepsilon \tag{4.28}$$

By differentiating Eq. (4.28), we obtain the density function as

$$f_Z(z) = \frac{m}{v - \varepsilon} \left(\frac{z - \varepsilon}{v - \varepsilon} \right)^{m-1} \exp \left[- \left(\frac{z - \varepsilon}{v - \varepsilon} \right)^m \right]; \qquad z \ge \varepsilon \tag{4.29}$$

In this case the median, mode, mean, and variance of Z are given by

$$Z_{50} = \varepsilon + (v - \varepsilon)(\ln 2)^{1/m} \tag{4.30}$$

$$Z_{\text{mode}} = \varepsilon + (v - \varepsilon) \left(\frac{m - 1}{m} \right)^{1/m} \tag{4.31}$$

$$\mu_Z = \varepsilon + (v - \varepsilon)\, \Gamma\left(1 + \frac{1}{m}\right) \tag{4.32}$$

and

$$\sigma_Z^2 = (v - \varepsilon)^2 \left\{ \Gamma\left(1 + \frac{2}{m}\right) - \Gamma^2\left(1 + \frac{1}{m}\right) \right\} \tag{4.33}$$

where $\Gamma(\,)$ is the gamma function defined in Eqs. (4.21) and (4.22).

Note: Equation (4.27) denotes gamma distribution when $\varepsilon = 0$. For this case ($\varepsilon = 0$), the type-III minimum value distribution denotes Weibull distribution, which is useful in the study of material strength in tension and fatigue.

4.7 Return Period

The *return period* is defined as the average elapsed time between occurrences of an event with a specified magnitude or greater. For example, a 20-year earthquake intensity is an intensity that is equaled or exceeded, on the average, once every 20 years over a long period of time. It does not mean that an exceedance occurs every 20 years, but that the average time between exceedances is 20 years. An *exceedance* is an event with a magnitude equal to or greater than a certain value. The concept of return period can also be applied to minimum values such as droughts, low flows, shortages, and so on.

In this case, the return period would be the average time between events with a certain magnitude or less. Such an event can still be called an exceedance in the sense that the severity of a drought (or an earthquake) exceeds some specified level. Regardless of whether the return period refers to an event greater than a specified value or to an event less than a specified value, it can be related to the probability of exceedance. If an exceedance occurs every 20 years, on the average, then the probability that the event occurs in any given year is $1/20 = 0.05$ or 5 percent. Thus the return period T and the probability P are related as

$$T = \frac{1}{P} \tag{4.34}$$

Example 4.3 The probability distribution function of the yearly maximum wind velocity, in mi/h, at a particular locality is represented as

$$F(x) = \exp\left[-\exp\left(-\frac{x-60}{20}\right)\right]$$

Find the return period of a yearly maximum wind velocity of 100 mi/h.

solution The probability of realizing the yearly maximum wind velocity greater than 100 mi/h is given by

$$P(x > 100) = 1 - P(x \le 100) = 1 - F(100)$$

$$= 1 - \left[\exp\left(-\exp\left(-\frac{100-60}{20}\right)\right)\right] = 1 - \exp[-\exp(-2)] = 0.1266$$

Thus the return period is given by

$$T = \frac{1}{P(x > 100)} = \frac{1}{0.1266} = 7.8989 \text{ years}$$

4.8 Characteristic Value

If the random variable T denotes the time between consecutive occurrences of an event E with a specified magnitude x or greater, then the mean value of T, $\overline{T}(x)$, gives the return period:

$$\overline{T}(x) = \frac{1}{1 - F(x)} \tag{4.35}$$

where $F(x)$ is the cumulative distribution function indicating the probability of occurrence of the event E with a magnitude smaller than x. Thus the expected number of occurrences of event E with magnitude x or greater in one time period will be $1/\overline{T}(x) = 1 - F(x)$, and in k time periods it will be

$$k\left(\frac{1}{\overline{T}(x)}\right) = k[1 - F(x)] \tag{4.36}$$

The particular value of the random variable X, denoted as x_{cl}, is called the *characteristic largest value* over a period of k time periods if the mean value of the number of exceedances of x_{cl} in k time periods is unity. The characteristic largest value gives an idea of the central location of the possible largest values. By setting Eq. (4.36) to unity, we obtain

$$k[1 - F(x_{cl})] = 1$$

or

$$F(x_{cl}) = 1 - \frac{1}{k} \tag{4.37}$$

If we consider k time periods, the distribution of the largest value of x_{cl} is given by Eq. (4.3):

$$F_{X_k}(x_{cl}) = F^k(x_{cl}) = \left(1 - \frac{1}{k}\right)^k \tag{4.38}$$

Thus the probability of exceeding the characteristic largest value in k time periods will be

$$1 - F_{X_k}(x_{cl}) = 1 - \left(1 - \frac{1}{k}\right)^k \tag{4.39}$$

As k tends to infinity, Eq. (4.39) converges to the value $1 - e^{-1} \approx 0.6321$. In a similar manner, the *characteristic smallest value* x_{cs} can be obtained from

$$kF(x_{cs}) = 1$$

or

$$F(x_{cs}) = \frac{1}{k} \tag{4.40}$$

If we consider k time periods, the probability of realizing a value smaller than the characteristic smallest value in k time periods is given by Eq. (4.2):

$$F_{X_1}(x_{cs}) = 1 - [1 - F(x_{cs})]^k = 1 - \left(1 - \frac{1}{k}\right)^k \tag{4.41}$$

Again, it can be seen that for large values of k (k tending to infinity), Eq. (4.41) converges to the value, $1 - e^{-1} \approx 0.6321$.

4.9 Fitting Extremal Distributions to Experimental Data

In practice, the observed data is plotted on a graph paper, called the probability paper. This paper provides scaled ordinates such that the cumulative distribution function plots as a straight line. Thus the probability paper reduces the comparison between the data and the probability model to a comparison between the data points and a straight line. The probability paper can also be used to find the parameters of the distribution from the observed data.

Chapter 15 describes the construction and use of probability papers in a greater detail.

To derive the necessary relationships, we first consider the extremal distributions corresponding to the maximum values. Taking natural logarithms of Eq. (4.5) for type-I largest values, we obtain

$$\ln F_Y(y) = -\exp[-a(y-w)] \tag{4.42}$$

By taking natural logarithms once again, Eq. (4.42) gives

$$y = -\frac{1}{a}\ln\ln\frac{1}{F_Y(y)} + w \tag{4.43}$$

This equation represents a straight line of the form

$$v = ay + b \tag{4.44}$$

where $v = -\ln\ln[1/F_Y(y)]$ and $b = -aw$. Hence a plot of v versus y will be a straight line for type-I distribution for the largest values. Similarly, the type-I distribution for minimum values, Eq. (4.10), gives

$$\ln[1 - F_Z(z)] = -e^{a(z-w)}$$

or

$$\ln\frac{1}{1 - F_Z(z)} = e^{a(z-w)} \tag{4.45}$$

By taking natural logarithms, Eq. (4.45) leads to

$$z = \frac{1}{a}\ln\ln\frac{1}{1 - F_Z(z)} + w$$

or

$$u = az + b \tag{4.46}$$

where $u = \ln\ln\{1/[1 - F_Z(z)]\}$ and $b = -aw$. Thus a plot of u versus z will be a straight line for type-I distribution for the smallest values.

To plot the experimental data, let there be N (observed) values for the random variable y. Arrange the data in an increasing order such that $y_1 \le y_2 \le \cdots \le y_i \le \cdots \le y_N$ where y_1 and y_N denote, respectively, the smallest and the largest observed values. Then the fractions $\{i/(N+1)\}$, $i = 1, 2, \ldots, N$ are computed and the points $\{y_i, i/(N+1)\}$, $i = 1, 2, \ldots, N$ are plotted on the extremal probability paper after suitable transformation. The fraction, $\{i/(N+1)\}$, is considered the cumulative probability, $F(y_i)$ corresponding to the observed value y_i for the purpose of plotting. Once the different points are plotted, then a straight line can be fitted through these points using the least squares method. The error between the actual, experimental, point and the corresponding point on the straight line can also be computed. The magnitude of this error can be used as a measure to determine which of the distributions fits the observed data best. The procedure is illustrated in Example 4.4.

4.9.1 Least squares fit

Let the equation approximating the data points (x_i, y_i), $i = 1, 2, \ldots, N$ be represented as

$$y = ax + b \tag{4.47}$$

The error between the value predicted by Eq. (4.47) and the experimental value observed at x_i, namely y_i, is given by

$$y_i - ax_i - b$$

and the function $f(a, b)$ to be minimized in the least squares method can be constructed as

$$f(a, b) = \sum_{i=1}^{N} (y_i - ax_i - b)^2 \tag{4.48}$$

The necessary conditions for minimizing f yield

$$\frac{\partial f}{\partial a} = \sum_{i=1}^{N} 2(y_i - ax_i - b)(-x_i) = 0$$

or,

$$a\left(\sum_{i=1}^{N} x_i^2\right) + b\left(\sum_{i=1}^{N} x_i\right) = \left(\sum_{i=1}^{N} x_i\, y_i\right) \tag{4.49}$$

and

$$\frac{\partial f}{\partial b} = \sum_{i=1}^{N} 2(y_i - ax_i - b)(-1) = 0$$

or,

$$a\left(\sum_{i=1}^{N} x_i\right) + b(N) = \left(\sum_{i=1}^{N} y_i\right) \tag{4.50}$$

The solution of Eqs. (4.49) and (4.50) gives

$$a = \frac{\left(\sum_{i=1}^{N} x_i\right)\left(\sum_{i=1}^{N} y_i\right) - N\left(\sum_{i=1}^{N} x_i\, y_i\right)}{\left(\sum_{i=1}^{N} x_i\right)^2 - N\left(\sum_{i=1}^{N} x_i^2\right)} \tag{4.51}$$

$$b = \frac{\left(\sum_{i=1}^{N} x_i\right)\left(\sum_{i=1}^{N} x_i y_i\right) - \left(\sum_{i=1}^{N} x_i^2\right)\left(\sum_{i=1}^{N} y_i\right)}{\left(\sum_{i=1}^{N} x_i\right)^2 - N\left(\sum_{i=1}^{N} x_i^2\right)} \tag{4.52}$$

Example 4.4 The following load data (in newtons) has been obtained experimentally: 13,800; 18,875; 10,450; 12,550; 15,175; 14,025; 11,875; 13,150; 16,850; 12,300; 13,425; 14,700. Find the extremal distribution corresponding to the type-I largest value for this data.

solution

1. Rearrange the experimental data in an increasing order, $y_1 < y_2 < \cdots < y_N$, as shown in the second column of Table 4.1.

TABLE 4.1

Serial number i	Ordered value of load, Newtons (y_i)	Cumulative probability $F(y_i) = \frac{i}{N+1}$	$\frac{1}{F(y_i)}$	$-\ln\ln\frac{1}{F(y_i)}$
1	10,450	1/13	13.00	–0.9419
2	11,875	2/13	6.50	–0.6269
3	12,300	3/13	4.33	–0.3828
4	12,550	4/13	3.25	–0.1644
5	13,150	5/13	2.60	0.0455
6	13,425	6/13	2.17	0.2572
7	13,800	7/13	1.86	0.4796
8	14,025	8/13	1.63	0.7226
9	14,700	9/13	1.44	1.0004
10	15,175	10/13	1.30	1.3380
11	16,850	11/13	1.18	1.7894
12	18,875	12/13	1.08	2.5252

2. Calculate the probability distribution function $F_Y(y)$:

$$F_Y(y) = \frac{i}{N+1}$$

as shown in the third column of Table 4.1.

3. Compute the ordinates v_i corresponding to the data (abscissa) y_i as

$$v_i = -\ln\ln\frac{1}{F_Y(y_i)}$$

and are shown in Table 4.1.

4. The least squares fit is given by

$$v = ay + b$$

where the constants a and b are given by equations similar to Eqs. (4.51) and (4.52). This results in

$$v = 0.000446322y - 5.714323$$

5. The parameters of the distribution are given by

$$a = 0.000446322$$

and

$$w = -\frac{b}{a} = \frac{5.714323}{0.000446322} = 12803.139$$

Thus the type-I extremal distribution corresponding to the largest values for the given load data can be expressed as

$$F_Y(y) = \exp(-\exp[-0.000446322(y - 12803.139)])$$

References and Bibliography

4.1. E. J. Gumbel, *Statistics of Extremes,* Columbia University Press, New York, 1958.
4.2. W. Weibull, "A Statistical Distribution Function of Wide Applicability," *ASME J. of Appl. Mech.*, vol. 18, 1951, pp. 293–297.
4.3. T. T. Soong, *Probabilistic Modeling and Analysis in Science and Engineering*, John Wiley, New York, 1981.
4.4. J. R. Benjamin and C. A. Cornell, *Probability, Statistics and Decision for Civil Engineers*, McGraw-Hill, New York, 1970.
4.5. H. C. S. Thom, "Distributions of Extreme Winds in the United States," *J. of Structural Div., Proc. ASCE*, vol. 86, no. ST4, Ap. 1960, pp. 11–24.
4.6. S. S. Rao and B. D. Gupta, "Application of Extremal Distributions in the Design of Thermal Systems," *J. of Mech. Design, Trans. of ASME*, vol. 102, 1980, pp. 481–489.
4.7. P. H. Wirsching and L. H. Jones, "On the Use of the Extreme Value Distribution in Reliability Analysis and Design," *J. of Eng. for Industry, Trans. of ASME*, vol. 98, Aug. 1976, pp. 1080–1085.
4.8. C. T. Haan, *Statistical Methods in Hydrology*, The Iowa State University Press, Ames, Iowa, 1977.
4.9. E. J. Gumbel and P. G. Carlson, "Extreme Values in Aeronautics," *J. of the Aeronautical Sciences*, vol. 21, no. 6, June 1954, pp. 389–398.
4.10. E. Castillo, *Extreme Value Theory in Engineering*, Academic Press, San Diego, 1988.
4.11. A. H. S. Ang and W. H. Tang, *Probability Concepts in Engineering Planning and Design*, Vol. II: *Decision, Risk, and Reliability*, John Wiley, New York, 1984.
4.12. E. E. Lewis, *Introduction to Reliability Engineering*, John Wiley, New York, 1987.
4.13. A. C. King and C. B. Read, *Pathways to Probability*, Holt, Rinehart & Winston, New York, 1963.
4.14. C. C. Gillispie (ed.-in-chief), *Dictionary of Scientific Biography*, vol. 9, Charles Scribner's Sons, New York, 1980.

Review Questions

4.1 What is an extremal distribution? Give five applications of extremal distributions.

4.2 Express the extreme value distributions in terms of the parent distribution.

4.3 What is the significance of an asymptotic distribution?

4.4 What is Gumbel distribution?

4.5 Where are extremal distributions of type-I used?

4.6 Give a practical application of type-II maximum value distribution.

4.7 What are the practical applications of type-III distribution for the smallest value?

4.8 What is the significance of the return period?

4.9 Express the return period in terms of the probability distribution function.

4.10 How do you construct an extremal probability paper?

4.11 What is a characteristic largest value?

4.12 Under what conditions Weibull distribution can be used to represent the smallest value of a random variable?

Problems

4.1 The load required to be lifted by a screw jack is uniformly distributed between 2000 lb and 3000 lb. Determine the probability distribution and density functions, $F(x_n)$ and $f(x_n)$, for the largest load lifted in a sample of n screw jacks. Plot the functions $F(x_n)$ and $f(x_n)$ for $n = 1, 5, 10, 15,$ and 20.

4.2 For the screw jack stated in Problem 4.1, determine the probability distribution and density functions, $F(x_1)$ and $f(x_1)$, for the smallest load lifted in a sample of n screw jacks. Plot the functions $F(x_1)$ and $f(x_1)$ for $n = 1, 5, 10, 15$ and 20.

4.3 The fatigue life of a bolted joint, in thousands of cycles, is given by

$$f(x) = \begin{cases} \dfrac{x - 600}{10000} & \text{for } 600 \le x \le 700 \\ \dfrac{800 - x}{10000} & \text{for } 700 \le x \le 800 \end{cases}$$

Find the probability distribution- and density-functions, $F(x_n)$ and $f(x_n)$, corresponding to the largest fatigue in a sample of n bolted joints. Plot the functions $F(x_n)$ and $f(x_n)$ for $n = 1, 10$ and 20.

4.4 For the bolted joints described in Problem 4.3, find the probability distribution and density functions, $F(x_1)$ and $f(x_1)$, corresponding to the smallest fatigue life in a sample of n bolted joints and plot them for $n = 1, 10$ and 20.

4.5 The probability density function of the time to failure, in hours, of a disk brake is given by

$$f(x) = 2 \times 10^{-5}\, e^{-2\times10^{-5}x}$$

Find the probability distribution and density functions, $F(x_n)$ and $f(x_n)$ for the largest failure time realized in a sample of n disk brakes. Plot the functions $F(x_n)$ and $f(x_n)$ for $n = 1$, 10, and 20.

4.6 For the disk brakes considered in Problem 4.5, find the probability distribution and density functions, $F(x_1)$ and $f(x_1)$, for the smallest failure time realized in a sample of n disk brakes. Plot the functions $F(x_1)$ and $f(x_1)$ for $n = 1$, 10, and 20.

4.7 The strength of structural steel, z in k(lb/in^2), is described by Weibull distribution with parameters $\varepsilon = 100$, $m = 5$, and $v = 150$ so that

$$f(z) = \frac{1}{10}\left(\frac{z-100}{50}\right)^4 \exp\left[-\left(\frac{z-100}{50}\right)^5\right]$$

Determine the mean and standard deviations of the strength of steel.

4.8 A mainframe computer has a constant breakdown rate of one breakdown every 96 hours of continuous operation. Find the probability of solving a problem that requires 120 hours of continuous operation of the computer over a period of 30 days.

4.9 The probability distribution function of the yearly maximum flood, in m^3/s, in a river is given by

$$F(x) = \exp\left[-\exp\left(-\frac{x-25}{5}\right)\right]$$

Find the intensity of the flood that corresponds to a return period of 50 years.

4.10 The maximum daily demand of electric power from a thermal power station is known to follow extremal distribution for the largest value with a mean of 5 MW and a standard deviation of 1 MW. What is the probability of the daily demand of electric power exceeding a value of 8 MW if the extremal distribution is of (a) type-I, (b) type-II, and (c) type-III?

4.11 For the thermal power station described in Problem 4.10, determine the particular value of the daily demand of electric power which will be met 99 percent of the time.

4.12 The minimum monthly flow rate of water in a river follows extremal distribution for the smallest value with a mean of 3-million gallons per hour and a standard deviation of 1-million gallons per hour. Find the probability of having a flow rate of less than 1-million gallons per hour in any month assuming the extremal distribution to be of (a) type-I, (b) type-II, and (c) type-III.

4.13 For the river considered in Problem 4.12, determine the value of the monthly minimum flow rate which will be guaranteed to occur 95 percent of the time.

4.14 It is required to design a landing gear of an aircraft to withstand an impact load which occurs no more than 1 percent of the time during its 20-year design life. Find the probability of occurrence of one or more impact loads of this magnitude or larger in any one year. Also find the probability of occurrence of more than one impact load of this or larger magnitude during the design life of the landing gear.

4.15 Derive linear relationships for plotting type-I, -II, and -III asymptotic distributions corresponding to the largest value.

4.16 Derive linear relationships for plotting type-I, -II, and -III asymptotic distributions corresponding to the smallest value.

4.17 The AISI 1060 annealed steel has a mean tensile strength of 620 MPa with a standard deviation of 40 MPa. Assuming a minimum tensile strength of 500 MPa, determine the Weibull parameters.

4.18 The following data represent the yearly maximum value of wind-induced load on a building frame in kilonewtons:

43.30	54.49	58.71	48.18
53.02	47.38	42.96	47.92
63.52	40.78	55.77	55.61
45.93	45.05	41.31	53.95
48.26	50.37	58.83	46.08
50.51	54.91	48.21	49.76
49.57	51.28	44.67	57.31
43.93	39.91	67.72	37.59
46.77	53.29	43.11	48.84
59.12	67.59	34.93	

Fit this data into the following distributions and compare the results:

1. Normal distribution
2. Type-I, -II, and -III distributions for largest values.

4.19 The following data indicate the minimum values of strength, in k(lb/in^2), found for various batches of steel:

48.71	79.87	56.66	51.17	60.17
52.92	59.37	61.78	75.47	64.32
65.32	53.71	48.31	57.48	44.99
57.37	52.26	56.34	71.44	68.89
69.31	58.62	66.91	56.78	82.14
40.27	54.23	54.82	56.11	60.37
59.98	44.31	47.32	67.12	67.80
55.18	50.89	63.77	53.34	52.18
43.12	73.18	70.22	58.71	59.41
62.33	57.93	55.83	49.24	

Fit this data into the following distributions and compare the results:

1. Normal distribution
2. Type-I, -II, and -III distributions for smallest values.

4.20 The hourly data of outside dry bulb temperature in degrees C in New Delhi, India are given as follows:

45.1	33.8	20.1
43.2	32.7	18.9
42.4	31.9	17.9
41.8	30.4	16.8
41.3	29.3	41.0
28.2	40.1	27.0
39.2	26.1	38.0
25.2	36.9	23.9
35.8	23.1	34.7
22.2	34.1	21.6

Find the extremal distributions corresponding to type-I, -II, and -III largest values for this data.

4.21 The hourly data of outside solar radiation in New Delhi, India are shown as follows:

86.2	66.8	29.9
85.0	63.2	26.1
84.5	59.9	22.6
83.1	56.3	19.5
82.4	52.0	15.4
81.5	48.2	11.3
79.0	44.8	9.6
76.6	40.7	7.2
72.7	37.4	70.3
32.5		

Determine the extremal distributions corresponding to type-I, -II, and -III largest values for the data.

4.22 The yearly maximum values of dry bulb temperature in degrees C in Poona, India over 18 years are given as follows:

40.25	40.48	40.61
40.79	41.02	41.20
41.43	41.64	41.80
42.00	42.24	42.41
42.63	42.82	43.05
43.24	43.39	43.67

Fit the data into the following distributions and compare the results:

1. Normal distribution
2. Type-I, -II, and -III distributions for the largest values.

4.23 The maximum gust accelerations, measured in g units, experienced by airplanes during flight periods of 250 hours each in a particular region are given as follows [4.9]:

0.73	0.61	0.50	0.65	0.66	0.80
1.80	1.35	1.10	1.00	1.05	1.00
1.30	1.00	1.65	1.30	0.98	0.91
0.80	1.20	0.91	0.82	1.05	1.40
0.70	1.25				

If a precision instrument which can withstand a maximum acceleration of 2.00 g's is transported by an airplane in the region, find the probability that the instrument will be damaged using the following distributions:

1. Normal distribution
2. Type-I distribution for the largest value
3. Type-II distribution for the largest value
4. Type-III distribution for the largest value.

4.24 Write a computer program to find the parameters of type-I, -II, and -III distributions corresponding to the largest value using experimental data. Use this program to determine the extremal distributions for the data of Problem 4.18.

4.25 Write a computer program to find the parameters of type-I, -II, and -III distributions corresponding to the smallest value using experimental data. Use this program to determine the extremal distributions for the data of Problem 4.19.

Chapter

5

Functions of Random Variables

Biographical Note

Reverend Thomas Bayes

Reverend Thomas Bayes was born in London in 1702 and died in Tunbridge Wells, England in 1761.He was the eldest son of Ann Bayes and Joshua Bayes and was educated privately, as was usual with Nonconformists at that time. Bayes' father was a respected theologian of dissent and Bayes became first his father's assistant and later spent most of his life as minister at the chapel in Tunbridge Wells. His mathematical work, although small in quantity, is of very high quality. He was elected as a Fellow of the Royal Society of London in 1742. He devoted much time to probability while preaching in the church, and is well known for introducing the concept of inverse probability, more commonly known as the "Bayes' rule." This was presented in his brief Essay Towards Solving a Problem in the Doctrine of Chances *published in 1763. This concept helps in estimating the probabilities of the causes by which an observed event may have been produced. Bayes' rule can be used in several applications including inspection processes [5.10, 5.11].*

5.1 Introduction

In engineering, most phenomena involve functional relationships in which a dependent variable is expressed in terms of one or more independent variables. For example, the stress in a bar is a function of the load and the cross-sectional area, the work done is a function of the force and the distance through which the force is moved, and so on. If any of the independent variables are random, the dependent variable will also be random. The probability distribution of the dependent variable as well as its moments can be derived from those of the independent random variables. The techniques presented in this chapter will permit us to determine the probability characteristics of random variables that are functionally dependent on some others with known probability characteristics.

5.2 Functions of a Single Random Variable

If X is a random variable, any other variable Y defined as a function of X will also be a random variable. If $f_X(x)$ and $F_X(x)$ denote, respectively, the probability density and distribution functions of X, the problem is to find the density function $f_Y(y)$ and the distribution function $F_Y(y)$ of the random variable Y. Let the functional relation be expressed as

$$Y = g(X) \tag{5.1}$$

By definition, the distribution function of Y is the probability of realizing Y less than or equal to y:

$$F_Y(y) = P(Y \leq y) = P(g(x) \leq y) = \int_{g(x) \leq y} f_X(x)\, dx \tag{5.2}$$

where the integration is to be carried over all values of x for which $g(x) \leq y$. For example, if the functional relation between x and y is as shown in Fig. 5.1, the range of integration will be $\Delta x_1 + \Delta x_2 + \Delta x_3 + \cdots$. The probability density function of Y is given by

$$f_Y(y) = \frac{d}{dy}\left[F_Y(y)\right] \tag{5.3}$$

Equations (5.2) and (5.3) represent general relationships and are applicable to any function $g(x)$. Specifically, if $g(x)$ is a monotonically increasing or decreasing function as shown in Fig. 5.2, the density function $f_Y(y)$ can be expressed in a more convenient way. In this case, the inverse relation of Eq. (5.1) can be denoted as

$$x = g^{-1}(y) = h(y) \tag{5.4}$$

where x can be seen to be a single valued function of y. Differentiation of Eq. (5.4) gives

$$dx = \left|\frac{dh}{dy}\right| dy = |h'(y)|\, dy \tag{5.5}$$

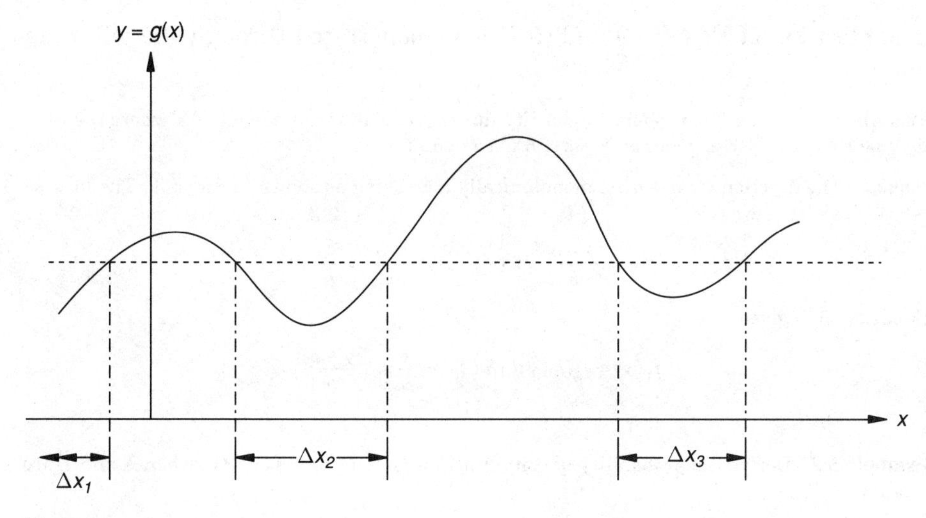

Figure 5.1

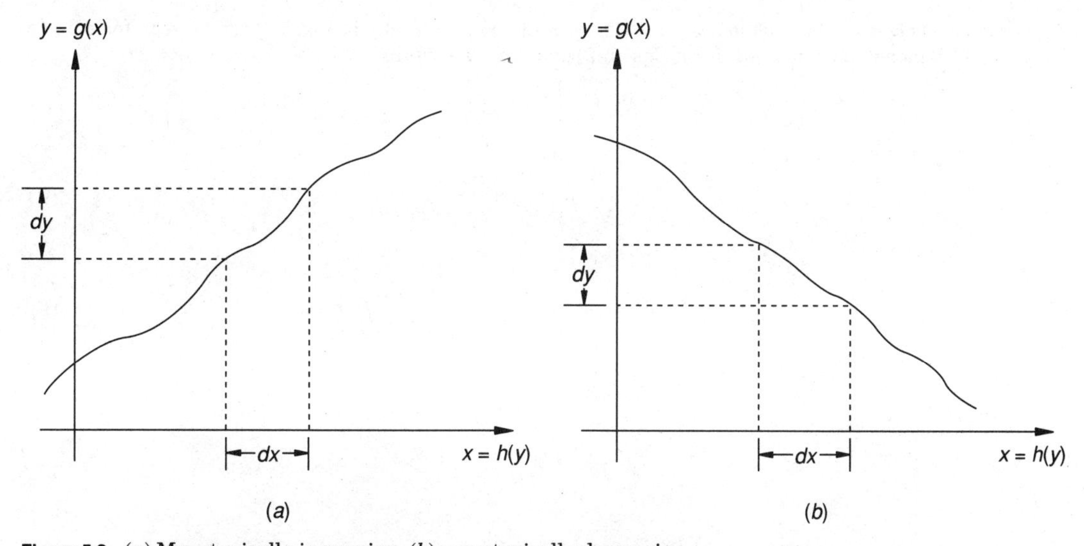

Figure 5.2 (a) Monotonically increasing, (b) monotonically decreasing.

where the absolute sign is used to take care of both the increasing and the decreasing type of functions (to avoid negative values for the density and distribution functions). Equations (5.2) and (5.5) lead to

$$F_Y(y) = \int_{-\infty}^{y} f_X[h(y)] \, |h'(y)| \, dy \tag{5.6}$$

and the differentiation of $F_Y(y)$ gives

$$f_Y(y) = f_X[h(y)] \, |h'(y)| \tag{5.7}$$

The application of Eqs. (5.2) and (5.7) is demonstrated through the following examples.

Example 5.1 Derive the probability density function of Y in terms of that of X where the relation between X and Y is given by $Y = a + bX$, $a > 0$, $b > 0$.

solution The function $y = a + bx$ is monotonically increasing as shown in Fig. 5.3. The inverse relation can be expressed as

$$x = \frac{y-a}{b} = h(y) \quad \text{and} \quad \frac{dh}{dy} = \frac{1}{b}$$

Equation (5.7) gives

$$f_Y(y) = f_X[h(y)]\,|h'(y)| = \frac{1}{b} f_X\left(\frac{y-a}{b}\right)$$

Example 5.2 Derive the probability density function $f_Y(y)$ in terms of $f_X(x)$ when X and Y are related as

$$Y = (X+1)^2$$

solution The function $y = (x+1)^2$, represented graphically in Fig. 5.4, can be seen to be a nonmonotonic function. Using Eqs. (5.2) and (5.6), we obtain

$$F_Y(y) = P[(X+1)^2 \le y] = \int_{\text{all } x \text{ where } (x+1)^2 \le y} f_X(x)\,dx$$

that is,

$$F_Y(y) = \int_{-\sqrt{y}-1}^{+\sqrt{y}-1} f_X(x)\,dx \tag{E1}$$

and

$$f_Y(y) = \frac{dF_Y(y)}{dy} = \frac{d}{dy}\left\{\int_{-\sqrt{y}-1}^{+\sqrt{y}-1} f_X(x)\,dx\right\} \tag{E2}$$

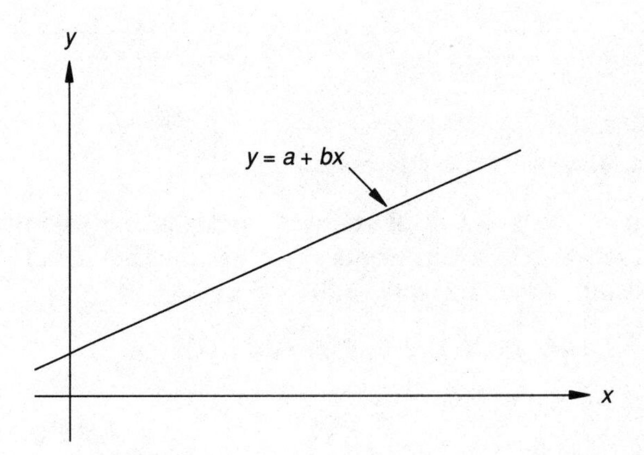

Figure 5.3

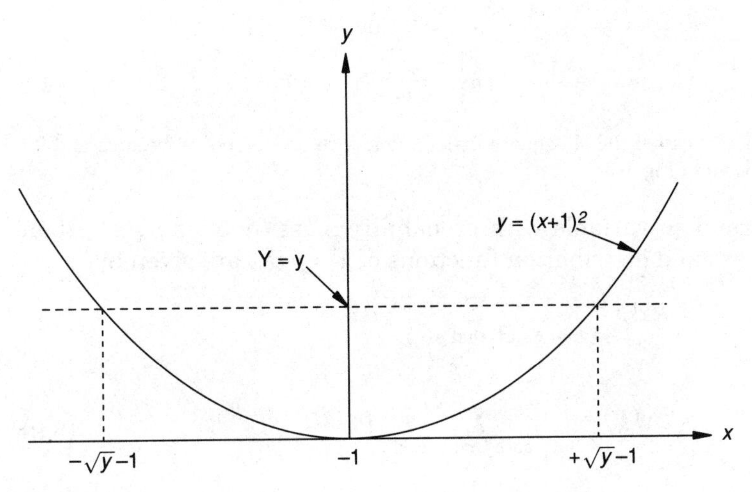

Figure 5.4

Using the formula of differentiation under integral sign

$$\frac{d}{da}\int_p^q f_X(x)\,dx = \int_q^p \frac{\partial}{\partial a}[f_X(x)]\,dx + f_X(q)\frac{dq}{da} - f_X(p)\frac{dp}{da}$$

Eq. (E2) gives

$$f_Y(y) = \begin{cases} \dfrac{f_X(\sqrt{y}-1) + f_X(-\sqrt{y}-1)}{2\sqrt{y}}; & y > 0 \\ 0; & y \le 0 \end{cases} \tag{E3}$$

Example 5.3 The energy (E) stored in a flywheel of mass moment of inertia I rotating at an angular velocity Ω is given by

$$E = \frac{1}{2} I\, \Omega^2$$

If I is deterministic and Ω is a random variable following normal distribution $N(0,1)$ determine the probability density function of the energy E.

solution The energy stored in the flywheel can be expressed as $E = a\,\Omega^2$ where $a = I/2$. This gives

$$\omega = \pm\sqrt{\frac{e}{a}} \qquad \text{and} \qquad \left|\frac{d\omega}{de}\right| = \frac{1}{2\sqrt{ae}} \tag{E1}$$

The probability density function of the energy E can be obtained from Eq. (5.7):

$$f_E(e) = \left\{ f_\Omega\left(\sqrt{\frac{e}{a}}\right) + f_\Omega\left(-\sqrt{\frac{e}{a}}\right) \right\} \frac{1}{2\sqrt{ae}} \tag{E2}$$

Since

$$f_\Omega(\omega) = \frac{1}{\sqrt{2\pi}} \exp\left\{ -\frac{1}{2}\,\omega^2 \right\} \tag{E3}$$

Eq. (E2) gives

$$f_E(e) = \frac{1}{\sqrt{2\,\pi\,a\,e}}\,\exp\left\{-\frac{e}{2\,a}\right\}; \qquad e \geq 0 \tag{E4}$$

This density function is called the chi-square distribution with one degree of freedom and its general shape is shown in Fig. 5.5.

If x is a discrete random variable with probability mass function $p_X(x)$, then the probability mass and distribution functions of $Y = g(X)$ are given by

$$p_Y(y) = \sum_{\text{all } x \text{ such that } g(x_i) = y} p_X(x_i) \tag{5.8}$$

and

$$F_Y(y) = \sum_{\text{all } x \text{ such that } g(x_i) \leq y} p_X(x_i) \tag{5.9}$$

Example 5.4 Find the probability mass function of y when $y = x^2$ and

$$p_Y(y) = \begin{cases} \dfrac{27}{1093}\left(\dfrac{1}{3}\right)^x; & x = -3, -2, -1, 0, 1, 2, 3 \\ 0; & \text{otherwise} \end{cases}$$

solution Since $x = -3, -2, -1, 0, 1, 2, 3$, and $y = x^2$, the possible values of y are 0, 1, 4, and 9. Using Eq. (5.8), we obtain

$$p_Y(0) = \sum_{x^2 = 0} p_X(x) = p_X(0) = \frac{27}{1093}$$

$$p_Y(1) = \sum_{x^2 = 1} p_X(x) = p_X(-1) + p_X(1) = \frac{27}{1093}(3) + \frac{27}{1093}\left(\frac{1}{3}\right) = \frac{90}{1093}$$

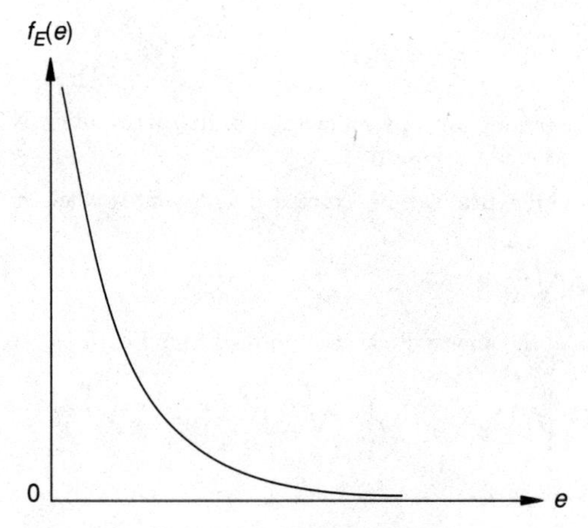

Figure 5.5 Chi-square distribution with one degree of freedom.

$$p_Y(4) = \sum_{x^2=4} p_X(x) = p_X(-2) + p_X(2) = \frac{27}{1093}(9) + \frac{27}{1093}\left(\frac{1}{9}\right) = \frac{246}{1093}$$

$$P_Y(9) = \sum_{x^2=9} p_X(x) = p_X(-3) + p_X(3) = \frac{27}{1093}(27) + \frac{27}{1093}\left(\frac{1}{27}\right) = \frac{730}{1093}$$

and $p_Y(y) = 0$ for all other values of y.

5.3 Functions of Two Random Variables

Consider a function of two random variables X_1 and X_2 as

$$Y = g(X_1, X_2) \tag{5.10}$$

~~If X_1 and X_2 are discrete random variables with~~ If X_1 and X_2 are discrete random variables with joint probability mass function $p_{X_1, X_2}(x_{1i}, x_{2j})$, the probability mass function of Y can be found as

$$p_Y(y) = P(Y = y) = P[g(X_1, X_2) = y] = \sum_{\substack{i \quad j \\ g(x_{1i},\, x_{2j}) = y}}\sum = p_{X_1, X_2}(x_{1i}, x_{2j}) \tag{5.11}$$

and the cumulative distribution function as

$$F_Y(y) = \sum_{\substack{i \quad j \\ g(x_{1i},\, x_{2j}) \le y}}\sum = p_{X_1, X_2}(x_{1i}, x_{2j}) \tag{5.12}$$

If X_1 and X_2 are continuous random variables with the joint density function $f_{X_1, X_2}(x_1, x_2)$, the probability distribution function of Y is given by

$$F_Y(y) = \iint_{g(x_1, x_2) \le y} f_{X_1, X_2}(x_1, x_2)\, dx_1\, dx_2 \tag{5.13}$$

If Eq. (5.10) can be solved for X_1 as

$$X_1 = g^{-1}(Y, X_2) \tag{5.14}$$

Eq. (5.13) can be expressed as

$$F_Y(y) = \int_{x_2=-\infty}^{\infty} \int_{y=-\infty}^{y} f_{X_1, X_2}(g^{-1}, x_2) \left| \frac{\partial g^{-1}}{\partial y} \right| dy\, dx_2 \tag{5.15}$$

The probability density function of Y can be obtained from Eq. (5.15) as

$$f_Y(y) = \int_{-\infty}^{\infty} f_{X_1, X_2}(g^{-1}, x_2) \left| \frac{\partial g^{-1}}{\partial y} \right| dx_2 \tag{5.16}$$

5.3.1 Sum of two random variables

If

$$Y = X_1 + X_2 \tag{5.17}$$

the probability distribution function of Y is given by

$$F_Y(y) = P[Y \le y] = \iint_{x_1 + x_2 \le y} f_{X_1, X_2}(x_1, x_2)\, dx_1\, dx_2$$

$$= \int_{-\infty}^{\infty} \int_{-\infty}^{(y - x_2)} f_{X_1, X_2}(x_1, x_2)\, dx_1\, dx_2 \tag{5.18}$$

where $f_{X_1, X_2}(x_1, x_2)$ is the joint density function of X_1 and X_2 and the integration is over the crossed region shown in Fig. 5.6. Differentiation of Eq. (5.18) gives

$$f_Y(y) = \frac{dF_Y(y)}{dy} = \int_{-\infty}^{\infty} f_{X_1, X_2}(y - x_2, x_2)\, dx_2 \tag{5.19}$$

If X_1 and X_2 are independent, $f_{X_1, X_2}(x_1, x_2) = f_{X_1}(x_1)\, f_{X_2}(x_2)$ and Eq. (5.19) becomes

$$f_Y(y) = \int_{-\infty}^{\infty} f_{X_1}(y - x_2)\, f_{X_2}(x_2)\, dx_2 \tag{5.20}$$

Example 5.5 The number of automobiles produced per day in each of the two plants by an automobile manufacturer is a random variable following Poisson distribution. Find the probability distribution of the total number of automobiles produced per day.

solution Let X_1 and X_2 denote the number of automobiles produced per day in the two plants with

$$p_{X_1}(x_1) = \frac{(\alpha)^{x_1}}{x_1!} e^{-\alpha} \tag{E1}$$

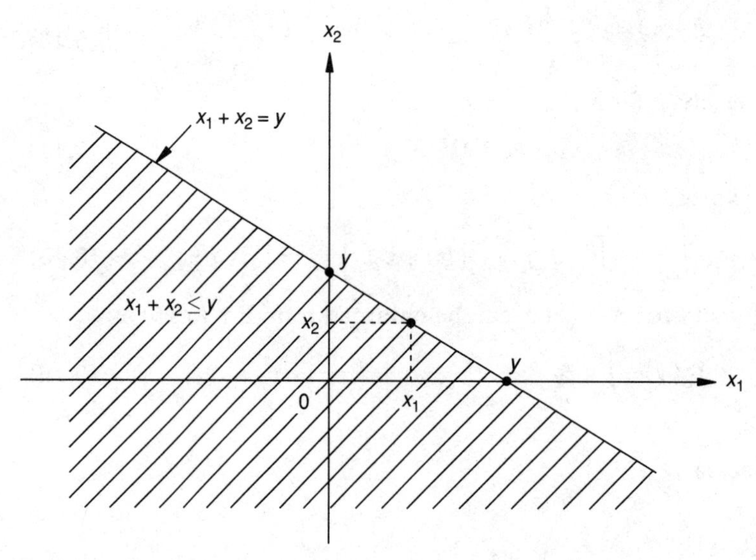

Figure 5.6 Region defined by $x_1 + x_2 \le y$.

and

$$p_{X_2}(x_2) = \frac{(\beta)^{x_2}}{x_2!} e^{-\beta} \tag{E2}$$

where α and β are the parameters of the distribution indicating the mean values (and also the variances) of X_1, X_2, respectively. If $Y = X_1 + X_2$, the probability mass function of Y can be determined as

$$p_Y(y) = \underset{x_{1i} + x_{2j} = y}{\sum_i \sum_j} p_{X_1, X_2}(x_{1i}, x_{2j}) = \sum_{\text{all } x_{1i}} p_{X_1, X_2}(x_{1i}, y - x_{1i}) \tag{E3}$$

If X_1 and X_2 are assumed to be independent random variables,

$$p_Y(y) = \sum_{\text{all } x_1} p_{X_1}(x_1)\, p_{X_2}(y - x_1)$$

$$= \sum_{\text{all } x_1} \left[\frac{(\alpha)^{x_1}}{x_1!} e^{-\alpha}\right] \left[\frac{(\beta)^{y - x_1}}{(y - x_1)!} e^{-\beta}\right]$$

$$= e^{-(\alpha+\beta)} \left\{ \sum_{\text{all } x_1} \frac{(\alpha)^{x_1} (\beta)^{y - x_1}}{x_1!\,(y - x_1)!} \right\} \tag{E4}$$

Since the quantity in brackets in Eq. (E4) is the binomial expansion of $(\alpha + \beta)^y/y!$, we obtain

$$p_Y(y) = \frac{(\alpha + \beta)^y}{y!} e^{-(\alpha+\beta)} \tag{E5}$$

This shows that the total number of automobiles produced per day also follows Poisson distribution with the parameter $(\alpha + \beta)$.

Example 5.6 Find the probability density function of sum of two independent normally distributed random variables.

solution Let X_1 and X_2 be the normal variates with

$$f_{X_1}(x_1) = \frac{1}{\sqrt{2\pi}\,\sigma_{X_1}} e^{-\frac{1}{2}\{(x_1 - \mu_{X_1})/\sigma_{X_1}\}^2} \tag{E1}$$

and

$$f_{X_2}(x_2) = \frac{1}{\sqrt{2\pi}\,\sigma_{X_2}} e^{-\frac{1}{2}\{(x_2 - \mu_{X_2})/\sigma_{X_2}\}^2} \tag{E2}$$

The substitution of Eqs. (E1) and (E2) into Eq. (5.20) gives

$$f_Y(y) = \frac{1}{2\pi\sigma_{X_1}\sigma_{X_2}} \int_{-\infty}^{\infty} \exp\left[-\frac{1}{2}\left\{\left(\frac{y - x_2 - \mu_{X_1}}{\sigma_{X_1}}\right)^2 + \left(\frac{x_2 - \mu_{X_2}}{\sigma_{X_2}}\right)^2\right\}\right] dx_2 \tag{E3}$$

which can be rewritten as

$$f_Y(y) = \frac{1}{2\pi\sigma_{X_1}\sigma_{X_2}} \exp\left[-\frac{1}{2}\left\{\left(\frac{y - \mu_{X_1}}{\sigma_{X_1}}\right)^2 + \left(\frac{\mu_{X_2}}{\sigma_{X_2}}\right)^2\right\}\right]$$

$$\times \int_{-\infty}^{\infty} \exp\left[-\frac{1}{2}\left\{\left(\frac{x_2}{\sigma_{X_1}}\right)^2 - 2x_2\left(\frac{y - \mu_{X_1}}{\sigma_{X_1}^2}\right) + \left(\frac{x_2}{\sigma_{X_2}}\right)^2 - 2x_2\left(\frac{\mu_{X_2}}{\sigma_{X_2}^2}\right)\right\}\right] dx_2 \tag{E4}$$

The integral in Eq. (E4) can be expressed as

$$\int_{-\infty}^{\infty} \exp\left[-\frac{1}{2}\left\{c_1 x_2^2 - 2\, x_2\, c_2\right\}\right] dx_2$$

$$= \int_{-\infty}^{\infty} \exp\left[-\frac{1}{2}\, c_1 \left\{x_2^2 - 2\,\frac{c_2}{c_1}\, x_2 + \left(\frac{c_2}{c_1}\right)^2 - \left(\frac{c_2}{c_1}\right)^2\right\}\right] dx_2 \tag{E5}$$

where

$$c_1 = \left(\frac{1}{\sigma_{X_1}^2} + \frac{1}{\sigma_{X_2}^2}\right) \tag{E6}$$

and

$$c_2 = \left(\frac{\mu_{X_2}}{\sigma_{X_2}^2} + \frac{y - \mu_{X_1}}{\sigma_{X_1}^2}\right) \tag{E7}$$

By defining a new random variable w as

$$w = x_2 - \frac{c_2}{c_1} \tag{E8}$$

the integral of Eq. (E5) can be written as

$$e^{(c_2^2/2\,c_1)} \int_{-\infty}^{\infty} e^{-1/2\, c_1\, w^2}\, dw = \sqrt{2\frac{\pi}{c_1}}\, e^{(c_2^2/2\,c_1)} \tag{E9}$$

Thus Eqs. (E4) and (E9) lead to

$$f_Y(y) = \frac{1}{2\,\pi\,\sigma_{X_1}\,\sigma_{X_2}} \exp\left[-\frac{1}{2}\left\{\left(\frac{y - \mu_{X_1}}{\sigma_{X_1}}\right)^2 + \left(\frac{\mu_{X_2}}{\sigma_{X_2}}\right)^2\right\}\right]$$

$$\times \sqrt{\frac{2\,\pi\,\sigma_{X_1}^2\,\sigma_{X_2}^2}{(\sigma_{X_1}^2 + \sigma_{X_2}^2)}} \exp\left\{\frac{1}{2}\left[\frac{\mu_{X_2}\,\sigma_{X_1}^2 + (y - \mu_{X_1})\,\sigma_{X_2}^2}{\sigma_{X_1}^2\,\sigma_{X_2}^2}\right]^2 \left(\frac{\sigma_{X_1}^2\,\sigma_{X_2}^2}{\sigma_{X_1}^2 + \sigma_{X_2}^2}\right)\right\} \tag{E10}$$

which upon simplification yields

$$f_Y(y) = \frac{1}{\sqrt{2\,\pi\,(\sigma_{X_1}^2 + \sigma_{X_2}^2)}} \exp\left[-\frac{1}{2}\left\{\frac{y - (\mu_{X_1} + \mu_{X_2})}{\sqrt{(\sigma_{X_1}^2 + \sigma_{X_2}^2)}}\right\}^2\right] \tag{E11}$$

This equation shows that Y is also a normal variate with

$$\mu_Y = \mu_{X_1} + \mu_{X_2} \tag{E12}$$

and

$$\sigma_Y^2 = \sigma_{X_1}^2 + \sigma_{X_2}^2 \tag{E13}$$

Note: This result can be generalized to the case of a linear sum of n random variables. If

$$Y = \sum_{i=1}^{n} a_i\, X_i \tag{5.21}$$

where a_i are constants and X_i are independent normal variates $N(\mu_{X_i}, \sigma_{X_i})$, then Y also is a normal variate $N(\mu_Y, \sigma_Y)$ with

$$\mu_Y = \sum_{i=1}^{n} a_i \, \mu_{X_i} \tag{5.22}$$

and

$$\sigma_Y^2 = \sum_{i=1}^{n} a_i^2 \, \sigma_{X_i}^2 \tag{5.23}$$

5.3.2 Product of two random variables

Let

$$Y = X_1 X_2 \tag{5.24}$$

From Eq. (5.13), the probability distribution function of Y can be expressed as

$$F_Y(y) = \iint_{x_1 x_2 \le y} f_{X_1, X_2}(x_1, x_2)\, dx_1\, dx_2 \tag{5.25}$$

The region of integration, in which $x_1 x_2 \le y$, is shown in Fig. 5.7. Thus Eq. (5.25) can be rewritten as

$$F_Y(y) = \int_0^{\infty} \int_{-\infty}^{(y/x_2)} f_{X_1, X_2}(x_1, x_2)\, dx_1\, dx_2$$

$$+ \int_{-\infty}^{0} \int_{(y/x_2)}^{\infty} f_{(X_1, X_2)}(x_1, x_2)\, dx_1\, dx_2 \tag{5.26}$$

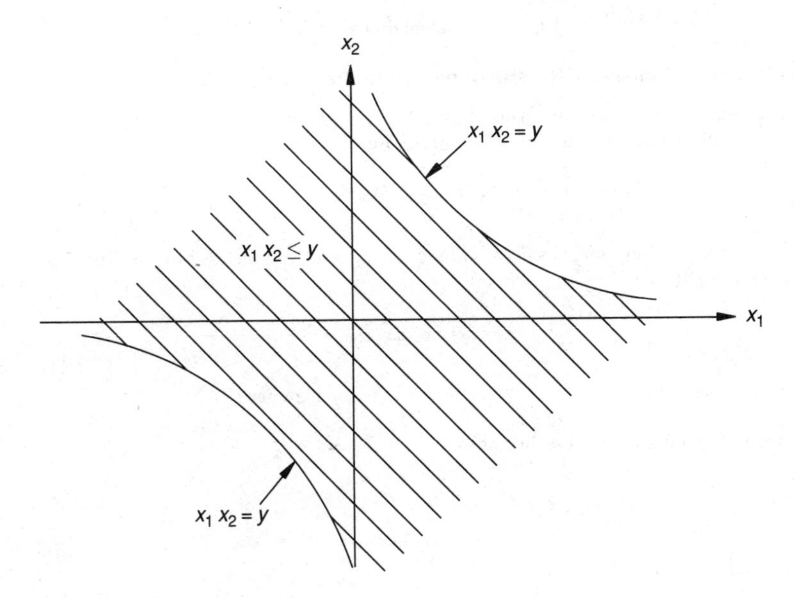

Figure 5.7 Region in which $x_1 x_2 \le y$.

If X_1 and X_2 are independent, Eq. (5.26) reduces to

$$F_Y(y) = \int_0^{\infty} F_{X_1}\left(\frac{y}{x_2}\right) f_{X_2}(x_2)\, dx_2 + \int_{-\infty}^{0} \left[1 - F_{X_1}\left(\frac{y}{x_2}\right)\right] f_{X_2}(x_2)\, dx_2 \qquad (5.27)$$

Differentiation of Eq. (5.27) gives the probability density function of Y as

$$f_Y(y) = \frac{dF_Y(y)}{dy} = \int_{-\infty}^{\infty} f_{X_1}\left(\frac{y}{x_2}\right) f_{X_2}(x_2) \left|\frac{1}{x_2}\right| dx_2 \qquad (5.28)$$

5.3.3 Quotient of two random variables

Let

$$Y = \frac{X_1}{X_2} \qquad (5.29)$$

Then Eq. (5.16) gives

$$f_Y(y) = \int_{-\infty}^{\infty} |x_2|\, f_{X_1, X_2}(y\, x_2, x_2)\, dx_2 \qquad (5.30)$$

The use of Eq. (5.30) is illustrated through the following example.

Example 5.7 The tensile load acting on a uniform bar (L) and the cross sectional area of the bar (A) are assumed to be independent random variables with the distributions

$$f_L(l) = \begin{cases} \alpha\, e^{-\alpha l}; & l > 0 \\ 0; & \text{elsewhere} \end{cases} \qquad \text{(E1)}$$

and

$$f_A(a) = \begin{cases} \beta\, e^{-\beta a}; & a > 0 \\ 0; & \text{elsewhere} \end{cases} \qquad \text{(E2)}$$

Find the probability density function of the stress induced in the bar.

solution Since the stress induced in the bar (Y) is given by $Y = X_1 / X_2$ where $X_1 = L$ and $X_2 = A$, the distribution function of Y is given by

$$F_Y(y) = \iint_{x_1/x_2 \le y} f_{X_1, X_2}(x_1, x_2)\, dx_1\, dx_2 \qquad \text{(E3)}$$

where the region of integration, for possible values of x_1 and x_2, is shown in Fig. 5.8. Equation (E3) can be rewritten as

$$F_Y(y) = \begin{cases} \int_0^{\infty} \int_0^{(x_2 y)} f_{X_1, X_2}(x_1, x_2)\, dx_1\, dx_2; & y > 0 \\ 0; & \text{elsewhere} \end{cases} \qquad \text{(E4)}$$

Since X_1 and X_2 are independent, Eq. (E4) becomes

$$F_Y(y) = \int_0^{\infty} \int_0^{(x_2 y)} f_{X_1}(x_1)\, f_{X_2}(x_2)\, dx_1\, dx_2$$

$$= \begin{cases} \int_0^{\infty} F_{X_1}(x_2\, y)\, f_{X_2}(x_2)\, dx_2; & y > 0 \\ 0; & \text{elsewhere} \end{cases} \qquad \text{(E5)}$$

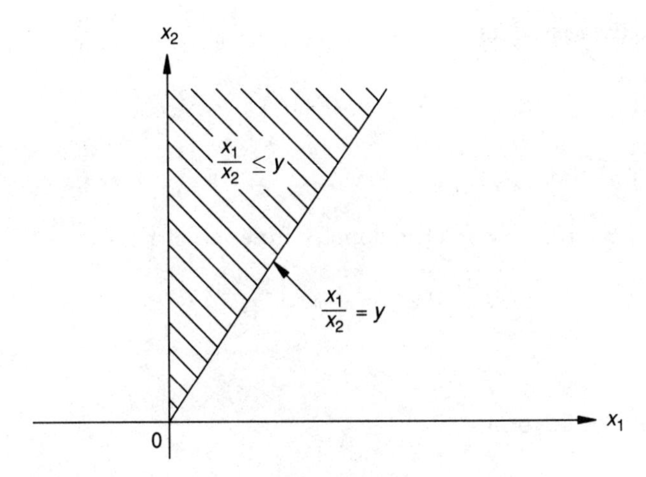

Figure 5.8 Region of integration in Eq. (E3).

The probability density function can be found by differentiating Eq. (E5) as

$$f_Y(y) = \begin{cases} \int_0^\infty x_2\, f_{X_1}(x_2\, y)\, f_{X_2}(x_2)\, dx_2; & y > 0 \\ 0; & \text{elsewhere} \end{cases} \tag{E6}$$

Substitution of Eqs. (E1) and (E2) into Eq. (E6) gives, for $y > 0$,

$$f_Y(y) = \int_0^\infty a\, \alpha\, e^{-\alpha a y}\, \beta e^{-\beta a}\, da = \alpha \beta \int_0^\infty a\, e^{-(\alpha y + \beta) a} da$$

$$= \alpha\, \beta\, \frac{1}{(\alpha\, y + \beta)^2} \tag{E7}$$

5.4 Function of Several Random Variables

Consider the function

$$Y = g(X_1, X_2, \ldots, X_n) \tag{5.31}$$

The distribution function of Y can be obtained by generalizing Eq. (5.13) as

$$F_Y(y) = \underset{g(x_1, x_2, \ldots, x_n) \le y}{\int \cdots \int} f_{X_1, X_2, \ldots, X_n}(x_1, x_2, \ldots, x_n)\, dx_1\, dx_2 \ldots dx_n \tag{5.32}$$

By solving Eq. (5.31) as

$$x_1 = g^{-1} = g^{-1}(y, x_2, x_3, \ldots, x_n) \tag{5.33}$$

Eq. (5.32) can be expressed as

$$F_Y(y) = \int_{-\infty}^{\infty} \cdots \int_{-\infty}^{\infty} \int_{-\infty}^{g^{-1}} f_{X_1,\ldots,X_n}(x_1,\ldots,x_n)\,dx_1\ldots dx_n \tag{5.34}$$

$$= \int_{-\infty}^{\infty} \cdots \int_{-\infty}^{\infty} \int_{-\infty}^{y} f_{X_1,X_2,\ldots,X_n}(g^{-1},x_2,\ldots,x_n)\left|\frac{\partial g^{-1}}{\partial y}\right| dy\,dx_2\ldots dx_n$$

Differentiation of Eq. (5.34) gives the density function of Y as

$$f_Y(y) = \int_{-\infty}^{\infty} \cdots \int_{-\infty}^{\infty} f_{X_1,X_2,\ldots,X_n}(g^{-1},x_2,\ldots,x_n)\left|\frac{\partial g^{-1}}{\partial y}\right| dx_2\ldots dx_n \tag{5.35}$$

5.5 Moments of a Function of Several Random Variables

As seen in Sections 5.3 and 5.4, the probability distribution of a function of several random variables can be derived from the known probability distributions of the original random variables. However, the derivations are quite involved, particularly when the functions are nonlinear. In such cases, the moments of the function of the random variables can be determined in terms of the moments of the basic random variables rather easily. In most cases, the task will be simple because only the first two moments, namely the mean and the variance, are of primary importance.

5.5.1 Mean and variance of a linear function

Let

$$Y = a_1 X_1 + a_2 X_2 + \cdots + a_n X_n \tag{5.36}$$

where a_i are constants and X_i are random variables with means $\mu_i = \mu_{X_i}$ and standard deviations $\sigma_i = \sigma_{X_i}$. The mean value of Y is, by definition,

$$\mu_Y = E[Y] = \int_{-\infty}^{\infty} y\, f_Y(y)\,dy$$

$$= \int_{-\infty}^{\infty} \cdots \int_{-\infty}^{\infty} (a_1x_1 + \cdots + a_nx_n) f_{X_1,\ldots,X_n}(x_1,\ldots,x_n)\,dx_1\ldots dx_n$$

$$= a_1 \int_{-\infty}^{\infty} x_1\, f_{X_1}(x_1)\,dx_1 + \cdots + a_n \int_{-\infty}^{\infty} x_n\, f_{X_n}(x_n)\,dx_n$$

$$= a_1\,\mu_1 + \cdots + a_n\,\mu_n \tag{5.37}$$

where the marginal density functions, $f_{X_i}(x_i)$, were obtained from the joint density function $f_{X_1,X_2,\ldots,X_i,\ldots,X_n}(x_1,x_2,\ldots,x_i,\ldots,x_n)$ as

$$f_{X_i}(x_i) = \int_{x_1=-\infty}^{\infty} \cdots \int_{x_{i-1}=-\infty}^{\infty} \int_{x_{i+1}=-\infty}^{\infty}$$

$$\cdots \int_{x_n=-\infty}^{\infty} f_{X_1,\ldots,X_{i-1},X_i,X_{i+1},\ldots,X_n}(x_1,\ldots,x_{i-1},x_i,x_{i+1},\ldots,x_n)$$

$$dx_1 \ldots dx_{i-1}\, dx_{i+1} \ldots dx_n \tag{5.38}$$

Equation (5.37) shows that the mean value of a linear sum of several random variables is the linear sum of the mean values. The variance of Y is given by

$$\begin{aligned}
\text{var}[Y] = \sigma_Y^2 &= E[(Y-\mu_Y)^2] \\
&= E[\{(a_1X_1 + \cdots + a_nX_n) - (a_1\mu_1 + \cdots + a_n\mu_n)\}^2] \\
&= E[\{a_1(X_1-\mu_1) + \cdots + a_n(X_n-\mu_n)\}^2] \\
&= E[a_1^2(X_1-\mu_1)^2 + \cdots + a_n^2(X_n-\mu_n)^2 + 2a_1a_2(X_1-\mu_1)(X_2-\mu_2) \\
&\quad + 2a_1a_3(X_1-\mu_1)(X_3-\mu_3) + \cdots + 2a_1a_n(X_1-\mu_1)(X_n-\mu_n) \\
&\quad + 2a_2a_3(X_2-\mu_2)(X_3-\mu_3) + \cdots]
\end{aligned} \tag{5.39}$$

It can be seen that the first set of n terms on the right hand side of Eq. (5.39) denotes the variances and the remaining terms represent the covariances. This gives

$$\text{var}[Y] = \sum_{i=1}^{n} a_i^2 \,\text{var}[X_i] + \sum_{\substack{i=1 \\ i\neq j}}^{n} \sum_{j=1}^{n} a_i a_j \,\text{cov}[x_i, x_j]$$

or,

$$\sigma_Y^2 = \sum_{i=1}^{n} a_i^2 \sigma_i^2 + \sum_{\substack{i=1 \\ i\neq j}}^{n} \sum_{j=1}^{n} a_i a_j \rho_{ij} \sigma_i \sigma_j \tag{5.40}$$

where σ_i^2 is the variance of X_i and ρ_{ij} is the correlation coefficient between X_i and X_j .

5.5.2 Mean and variance of sum of two random variables

If $Y = X_1 \pm X_2$, Eqs. (5.37) and (5.40) can be specialized to obtain

$$\mu_Y = \mu_{X_1} \pm \mu_{X_2} \tag{5.41}$$

and

$$\sigma_Y = \begin{cases} (\sigma_{X_1}^2 + \sigma_{X_2}^2)^{1/2}; & X_1 \text{ and } X_2 \text{ uncorrelated} \\ (\sigma_{X_1}^2 + \sigma_{X_2}^2 \pm 2\rho\sigma_{X_1}\sigma_{X_2})^{1/2}; & X_1 \text{ and } X_2 \text{ correlated} \end{cases} \tag{5.42}$$

where ρ is the correlation coefficient between X_1 and X_2.

5.5.3 Mean and variance of product of two random variables

If $Y = X_1 X_2$, then the mean value of Y is given by

$$\mu_Y = E[X_1 X_2] = \int_{-\infty}^{\infty}\int_{-\infty}^{\infty} x_1 x_2 f_{X_1,X_2}(x_1, x_2)\, dx_1\, dx_2 \tag{5.43}$$

If X_1 and X_2 are independent,

$$f_{X_1,X_2} = f_{X_1}(x_1)\, f_{X_2}(x_2) \tag{5.44}$$

and Eq. (5.43) can be rewritten as

$$\mu_Y = \int_{-\infty}^{\infty} x_1 f_{X_1}(x_1)\, dx_1 \int_{-\infty}^{\infty} x_2 f_{X_2}(x_2)\, dx_2 = \mu_{X_1}\mu_{X_2} \tag{5.45}$$

The variance of Y can be determined as

$$\sigma_Y^2 = E[(Y - \mu_Y)^2] = E\left[X_1^2 X_2^2 + \mu_{X_1}^2\mu_{X_2}^2 - 2\mu_{X_1}\mu_{X_2}X_1X_2\right]$$

$$= E\left[X_1^2 X_2^2\right] + \mu_{X_1}^2\mu_{X_2}^2 - 2\,\mu_{X_1}\mu_{X_2}E[X_1X_2] \tag{5.46}$$

where, for independent variables,

$$E\left[X_1^2 X_2^2\right] = \int_{-\infty}^{\infty}\int_{-\infty}^{\infty} x_1^2 x_2^2 f_{X_1,X_2}(x_1, x_2)\, dx_1 dx_2 = \mu_{X_1^2}\mu_{X_2^2} \tag{5.47}$$

For normally distributed variables, we have (see [5.2] and also Problem 5.27)

$$E\left[X^2\right] = \mu_X^2 + \sigma_X^2 \tag{5.48}$$

and hence Eq. (5.46) yields

$$\sigma_Y = \left[\mu_{X_1}^2\sigma_{X_2}^2 + \mu_{X_2}^2\sigma_{X_1}^2 + \sigma_{X_1}^2\sigma_{X_2}^2\right]^{1/2} \tag{5.49}$$

If X_1 and X_2 are correlated with a correlation coefficient of ρ, the standard deviation of Y [5.2] is given by

$$\sigma_Y = \mu_{X_1}\mu_{X_2}\left\{\left(\frac{\sigma_{X_1}^2}{\mu_{X_1}^2} + \frac{\sigma_{X_2}^2}{\mu_{X_2}^2} + \frac{\sigma_{X_1}^2\sigma_{X_2}^2}{\mu_{X_1}^2\mu_{X_2}^2}\right)\left(1 + \rho^2\right)\right\}^{1/2} \tag{5.50}$$

5.5.4 Mean and variance of quotient of two random variables

If $Y = X_1 / X_2$, the mean and variance of Y can be obtained, assuming normal distribution for X_1 and X_2 as [5.2]

$$\mu_Y = \frac{\mu_{X_1}}{\mu_{X_2}} \tag{5.51}$$

and

$$\sigma_Y = \begin{cases} \dfrac{1}{\mu_{X_2}}\left[\dfrac{\mu_{X_1}^2\,\sigma_{X_2}^2 + \mu_{X_2}^2\,\sigma_{X_1}^2}{\mu_{X_2}^2 + \sigma_{X_2}^2}\right]^{1/2}; & X_1 \text{ and } X_2 \text{ uncorrelated} \\ \dfrac{\mu_{X_1}^2}{\mu_{X_2}^2}\left[\dfrac{\sigma_{X_1}^2}{\mu_{X_1}^2} + \dfrac{\sigma_{X_2}^2}{\mu_{X_2}^2} - 2\rho\dfrac{\sigma_{X_1}\sigma_{X_2}}{\mu_{X_1}\mu_{X_2}}\right]^{1/2}; & X_1 \text{ and } X_2 \text{ correlated} \end{cases} \tag{5.52}$$

5.5.5 Mean and variance of a general nonlinear function of several random variables

Consider a nonlinear function of several random variables

$$Y = g(X_1, X_2, \ldots, X_n) \tag{5.53}$$

The mean and variance of Y are, by definition,

$$\mu_Y = E[Y] = \int_{x_1=-\infty}^{\infty} \cdots \int_{x_n=-\infty}^{\infty} g(x_1, x_2, \ldots, x_n) \times f_{X_1,X_2,\ldots,X_n}(x_1, x_2, \ldots, x_n)\, dx_1\, dx_2 \ldots dx_n \tag{5.54}$$

and

$$\sigma_Y^2 = E\left[(Y - \mu_Y)^2\right] = \int_{x_1=-\infty}^{\infty} \cdots \int_{x_n=-\infty}^{\infty} [g(x_1, x_2, \ldots, x_n) - \mu_Y]^2 \times f_{X_1,X_2,\ldots,X_n}(x_1, x_2, \ldots, x_n)\, dx_1\, dx_2 \cdots dx_n \tag{5.55}$$

In many engineering applications, the joint density function of $X_1, X_2, \ldots, X_n$ may not be known; but the mean and variance of the random variables X_i may be known. Furthermore, even when the joint density function is known, the multiple integrations involved in Eqs. (5.54) and (5.55) may be difficult to perform. In these cases, we would like to find the approximate mean and variance of the function g. For this, we expand the function $g(X_1, X_2, \ldots, X_n)$ in a Taylor's series about the mean values $\mu_1 = \mu_{X_1}$, $\mu_2 = \mu_{X_2}, \ldots, \mu_n = \mu_{X_n}$ as

$$Y = g(\mu_1, \mu_2, \ldots, \mu_n) + \sum_{i=1}^{n} \frac{\partial g}{\partial X_i}\bigg|_{\mu_1,\mu_2,\ldots,\mu_n} (X_i - \mu_i) + \frac{1}{2}\sum_{i=1}^{n}\sum_{j=1}^{n} \frac{\partial^2 g}{\partial X_i\,\partial X_j}\bigg|_{\mu_1,\mu_2,\ldots,\mu_n} (X_i - \mu_i)(X_j - \mu_j) + \cdots \tag{5.56}$$

where the partial derivatives are evaluated at the mean values of X_i as

$$\left.\frac{\partial g}{\partial X_i}\right|_{\mu_1,\mu_2,\ldots,\mu_n} = \frac{\partial g}{\partial X_i}(X_1=\mu_1,\ldots,X_n=\mu_n) \tag{5.57}$$

By neglecting terms involving second and higher order derivatives of g in Eq. (5.56), we obtain

$$Y \approx g(\mu_1,\mu_2,\ldots,\mu_n) + \sum_{i=1}^{n} \left.\frac{\partial g}{\partial X_i}\right|_{\mu_1,\ldots,\mu_n} (X_i-\mu_i) \tag{5.58}$$

Since $g(\mu_1,\ldots,\mu_n)$ and $\left.\frac{\partial g}{\partial X_i}\right|_{\mu_1,\ldots,\mu_n}$ are constants, Eq. (5.58) gives the mean value of Y as

$$\mu_Y = E[Y] \approx g(\mu_1,\ldots,\mu_n) \tag{5.59}$$

Similarly, the variance of Y can be obtained as

$$\sigma_Y^2 \approx \sum_{i=1}^{n} a_i^2 \,\text{var}[X_i] + \sum_{\substack{i=1\\ i\neq j}}^{n}\sum_{j=1}^{n} a_i\, a_j \,\text{cov}[X_i,X_j] \tag{5.60}$$

where

$$a_i = \left.\frac{\partial g}{\partial X_i}\right|_{\mu_1,\ldots,\mu_n} \tag{5.61}$$

and

$$a_j = \left.\frac{\partial g}{\partial X_j}\right|_{\mu_1,\ldots,\mu_n} \tag{5.62}$$

If the variables X_i are statistically independent, then

$$\sigma_Y^2 \approx \sum_{i=1}^{n} a_i^2\, \sigma_{X_i}^2 \tag{5.63}$$

The procedure adopted in deriving Eqs. (5.59), (5.60), and (5.63) is known as the *partial derivative rule.*

Note: The accuracy of expressions for the mean and variance of Y can be improved by including higher order derivative terms in Eq. (5.56). For example, if the second derivative terms are also included, the mean value of Y can be determined as

$$\mu_Y = E[Y] \approx g(\mu_1,\ldots,\mu_n) + \frac{1}{2}\sum_{i=1}^{n}\sum_{j=1}^{n} \left.\frac{\partial^2 g}{\partial X_i\,\partial X_j}\right|_{\mu_1,\ldots,\mu_n} \text{cov}[X_i,X_j] \tag{5.64}$$

If X_i are independent, then Eq. (5.64) becomes

$$\mu_Y = E[Y] \approx g(\mu_1,\ldots,\mu_n) + \frac{1}{2}\sum_{i=1}^{n} \left.\frac{\partial^2 g}{\partial X_i^2}\right|_{\mu_1,\ldots,\mu_n} \text{var}[X_i] \tag{5.65}$$

Example 5.8 The inertia force developed in a reciprocating engine (F) is given by (see Fig. 5.9):

$$F = \underbrace{mr\omega^2 \cos \omega t}_{\text{primary force}} + \underbrace{m \frac{r^2\omega^2}{l} \cos 2\omega t}_{\text{secondary force}} \tag{E1}$$

where m = mass of the piston, r = length of the crank, l = length of the connecting rod, ω = angular velocity of the crank and t = time. If the mean values and coefficients of variation of the parameters are given by $\overline{m}$ = 1 kg, γ_m = 0.10, $\overline{r}$ = 7.5 cm, γ_r = 0.05, $\overline{l}$ = 30 cm, γ_l = 0.05, $\overline{\omega}$ = 1500 rpm, and γ_ω = 0.10, find the first order mean and standard deviation of F when $t = 0.003333$ s. Assume that all the random variables are statistically independent. cov = 0

solution The first order approximate mean value of the inertia force at $t = 0.003333$ s can be obtained as

$$\overline{F} = \overline{m}\,\overline{r}\,\overline{\omega}^2 \cos \overline{\omega}\, t + \overline{m} \frac{\overline{r}^2\, \overline{\omega}^2}{\overline{l}} \cos 2\, \overline{\omega}\, t$$

$$= (1)(0.075)(157.08)^2 \cos (30^o) + \left(\frac{1}{0.3}\right)(0.075)^2\, (157.08)^2 \cos (60^o)$$

$$= 1833.9515\ N$$

where the mean value of the angular velocity of the crank is

$$\overline{\omega} = \frac{1500(2\,\pi)}{60} = 50\,\pi = 157.08\ \frac{\text{rad}}{\text{s}}$$

The partial derivatives of F with respect to the random variables, along with their values at the mean values of random variables, are given by

$$\frac{\partial F}{\partial m} = r\omega^2 \cos \omega t + \frac{r^2\omega^2}{l} \cos 2\omega t = 1833.9514$$

$$\frac{\partial F}{\partial r} = m\omega^2 \cos \omega t + \frac{2mr\omega^2}{l} \cos 2\omega t = 27536.9522$$

$$\frac{\partial F}{\partial \omega} = 2mr\omega \cos \omega t - mr\omega^2 t \sin \omega t + \frac{2mr^2\omega}{l} \cos 2\omega t$$

$$- \frac{2mr^2\omega^2 t}{l} \sin 2\omega t = 17.5951$$

$$\frac{\partial F}{\partial l} = - \frac{mr^2\omega^2}{l^2} \cos 2\omega t = -771.0664$$

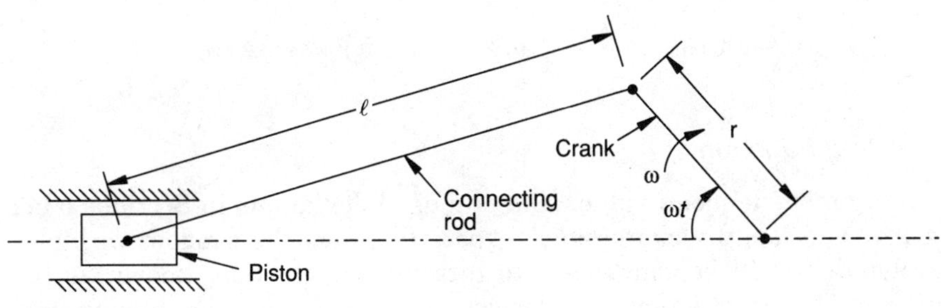

Figure 5.9

Noting that the standard deviations of the random variables are $\sigma_m = \overline{m}\,\gamma_m = (1)(0.1) = 0.1$ kg, $\sigma_r = \overline{r}\,\gamma_r = (0.075)(0.05) = 0.00375$ m, $\sigma_\omega = \overline{\omega}\,\gamma_\omega = (157.08)(0.1) = 15.708$ rad/s, and $\sigma_l = \overline{l}\,\gamma_l = (0.3)(0.05) = 0.015$ m, the first order approximate variance of F can be obtained as

$$\sigma_F^2 = \left(\frac{\partial F}{\partial m}\right)^2 \sigma_m^2 + \left(\frac{\partial F}{\partial r}\right)^2 \sigma_r^2 + \left(\frac{\partial F}{\partial \omega}\right)^2 \sigma_\omega^2 + \left(\frac{\partial F}{\partial l}\right)^2 \sigma_l^2$$

$$= (1833.9514)^2\,(0.1)^2 + (27536.9522)^2\,(0.00375)^2 + (17.5951)^2\,(15.708)^2$$

$$+ (-771.0664)^2\,(0.015)^2 = 120818.9366$$

Thus

$$\sigma_F = 347.5902$$

Note: The second order approximate mean value of the inertia force can be determined as follows:

$$\frac{\partial^2 F}{\partial m^2} = 0, \quad \frac{\partial^2 F}{\partial r^2} = \frac{2m\omega^2}{l} \cos 2\omega t = 82247.088,$$

$$\frac{\partial^2 F}{\partial \omega^2} = 2mr \cos \omega t - 4mr\omega t \sin \omega t - mr\omega^2 t^2 \cos \omega t$$

$$+ \frac{2mr^2}{l} \cos 2\omega t - \frac{8mr^2\omega t}{l} \sin 2\omega t - \frac{4mr^2\omega^2 t^2}{l} \cos 2\omega t$$

$$= 0.4656$$

$$\frac{\partial^2 F}{\partial l^2} = \frac{2mr^2\omega^2}{l^3} \cos 2\omega t = 5140.443$$

$$\overline{F} = \overline{m}\,\overline{r}\,\overline{\omega}^2 \cos \overline{\omega}\, t + \frac{\overline{m}\,\overline{r}^2\,\overline{\omega}^2}{\overline{l}} \cos 2\,\overline{\omega}\, t$$

$$+ \frac{1}{2}\left(\frac{\partial^2 F}{\partial m^2}\right)\sigma_m^2 + \frac{1}{2}\left(\frac{\partial^2 F}{\partial r^2}\right)\sigma_r^2 + \frac{1}{2}\left(\frac{\partial^2 F}{\partial \omega^2}\right)\sigma_\omega^2 + \frac{1}{2}\left(\frac{\partial^2 F}{\partial l^2}\right)\sigma_l^2$$

$$= 1602.6315 + 231.3200 + 0 + \frac{1}{2}(82247.088)(0.00375)^2$$

$$+ \frac{1}{2}(0.4656)(15.708)^2 + \frac{1}{2}(5140.443)(0.015)^2 = 1949.9904N$$

5.6 Moment Generating Function

The evaluation of the mean values and standard deviations in Sections 5.5.1 to 5.5.4 involved lengthy expressions, especially when the random variables are correlated. It will be convenient to introduce a function, known as the moment generating function, to derive expressions for the means and standard deviations of functions of random variables.[1] If X is a random variable, then the quantity $e^{X\theta}$ is a function of X for each fixed value of θ. The moment generating function of X, denoted as $M(\theta)$, is defined as the expected value of the function $e^{X\theta}$:

$$M(\theta) = E[e^{X\theta}] = \int_{-\infty}^{\infty} e^{x\theta} f_X(x)\, dx \tag{5.66}$$

Differentiating Eq. (5.66) r times with respect to θ gives

$$\frac{d^r M(\theta)}{d\theta^r} = \int_{-\infty}^{\infty} x^r e^{x\theta} f_X(x)\, dx \tag{5.67}$$

and setting $\theta = 0$,

$$\frac{d^r M(0)}{d\theta^r} = \int_{-\infty}^{\infty} x^r f_X(x)\, dx = E[X^r] = \mu_r'$$

$$= r\text{th moment of } X \text{ about the origin} \tag{5.68}$$

Equation (5.68) can be used to obtain moments of any order of the distribution. The moments of a function of a random variable $g(X)$ can also be evaluated using the moment generating function of X. In this case, the moment generating function of $g(X)$ is defined as

$$M(\theta) = E\left[e^{\theta g(x)}\right] = \int_{-\infty}^{\infty} e^{\theta g(x)} f_X(x)\, dx \tag{5.69}$$

The rth moment of $g(X)$ about the origin can be obtained as

$$E[\{g(X)\}^r] = \frac{d^r M(0)}{d\theta^r} = \int_{-\infty}^{\infty} \{g(x)\}^r f_X(x)\, dx \tag{5.70}$$

The moment generating function can also be defined for multiple random variables. Let X_1 and X_2 be two random variables with joint density function $f_{X_1,X_2}(x_1, x_2)$. Then

$$M(\theta_1, \theta_2) = E\left[e^{\theta_1 X_1} e^{\theta_2 X_2}\right]$$

$$= \int_{-\infty}^{\infty}\int_{-\infty}^{\infty} e^{\theta_1 x_1 + \theta_2 x_2} f_{X_1,X_2}(x_1, x_2)\, dx_1\, dx_2 \tag{5.71}$$

The rth moment of X_1, for example, can be found by differentiating the moment generating function r times with respect to θ_1 and then setting $\theta_1 = \theta_2 = 0$. Similarly, the joint moment $E[X_1^p X_2^q]$ can be obtained by differentiating $M(\theta_1, \theta_2)$ p times with respect to θ_1 and q times with respect to θ_2 and then setting $\theta_1 = \theta_2 = 0$.

---

[1]In fact, the moment generating function of a random variable can also be used to generate all the moments of the random variable. Thus the moment generating function can be considered as an alternate way of describing the probability law of a random variable.

5.6.1 Moments of normally distributed variables

Since normal distribution is commonly used in many engineering applications, we shall now consider the moments of normally distributed variables using moment generating functions. For a normal variate X with mean μ and standard deviation σ, the moment generating function is given by

$$M(\theta) = E\left[e^{\theta X}\right] = \frac{1}{\sqrt{2\pi}\,\sigma}\int_{-\infty}^{\infty} e^{\theta x - 1/2\{(x-\mu)/\sigma\}^2} \tag{5.72}$$

The term in the exponent can be rewritten as

$$\theta x - \frac{1}{2}\left(\frac{x-\mu}{\sigma}\right)^2 = -\frac{1}{2\sigma^2}[x - (\mu + \theta\,\sigma^2)]^2 + \mu\,\theta + \frac{1}{2}\,\sigma^2\,\theta^2 \tag{5.73}$$

Thus Eq. (5.72) becomes

$$M(\theta) = \exp\left(\mu\,\theta + \frac{1}{2}\,\sigma^2\,\theta^2\right)$$

$$\times\int_{-\infty}^{\infty}\frac{1}{\sqrt{2\,\pi}\,\sigma}\,\exp\left\{-\frac{1}{2\,\sigma^2}\,[x - (\mu + \sigma^2\,\theta)]^2\right\}dx \tag{5.74}$$

The expression under the integral sign can be identified as the probability density function of a normal variate with mean $(\mu + \sigma^2\theta)$ and standard deviation σ, and thus integrates to 1. Thus the moment generating function of the normal variate X is given by

$$M(\theta) = e^{(\mu\theta + 1/2\sigma^2\theta^2)} \tag{5.75}$$

Example 5.9 Derive an expression for the mean value of $Y = X^2$ when X is normally distributed with mean μ and standard deviation σ.

solution The mean value of the function Y is given by

$$E[X^2] = \mu_{X^2} = \frac{\partial^2 M(0)}{\partial\theta^2}$$

The moment generating function of X is given by Eq. (5.75):

$$M(\theta) = \exp(\theta\,\mu + 1/2\,\sigma^2\,\theta^2) = e^w$$

where

$$w = \theta\,\mu + 1/2\,\sigma^2\,\theta^2$$

Differentiation of $M(\theta)$ gives

$$\frac{\partial M(\theta)}{\partial\theta} = (\mu + \sigma^2\,\theta)\,e^w$$

$$\frac{\partial^2 M(\theta)}{\partial\theta^2} = \sigma^2\,e^w + (\mu + \sigma^2\,\theta)^2\,e^w$$

Thus

$$E[X^2] = \mu_{X^2} = \frac{\partial^2 M(0)}{\partial \theta^2} = \sigma^2 + \mu^2$$

5.7 Functions of Several Random Variables

In Sections 5.2 and 5.3, we considered the derivation of the probability density function of a random variable defined as a function of other random variable(s). In many cases, we encounter several functions involving multiple random variables. The joint probability density and distribution functions of variables defined as functions of several other random variables can be derived as follows. Let

$$Y_k = g_k(X_1, X_2, \ldots, X_n), \quad k = 1, 2, \ldots, m, \quad m \leq n \tag{5.76}$$

Then the joint distribution function $F_{Y_1, Y_2, \ldots, Y_m}(y_1, y_2, \ldots, y_m)$, by definition, is given by

$$F_{Y_1, Y_2, \ldots, Y_m}(y_1, y_2, \ldots, y_m) = P(Y_1 \leq y_1, Y_2 \leq y_2, \ldots, Y_m \leq y_m)$$

$$= \int_{x_1} \int_{x_2} \cdots \int_{x_n} f_{X_1, X_2, \ldots, X_n}(x_1, x_2, \ldots, x_n)\, dx_1\, dx_2 \ldots dx_n \tag{5.77}$$

where the integrals have to be evaluated over the domain of the n-dimensional $(X_1, X_2, \ldots, X_n)$–space in which the inequalities

$$g_k(x_1, x_2, \ldots, x_n) \leq y_k, \qquad k = 1, 2, \ldots, m$$

are satisfied. The joint density function is given by

$$f_{Y_1, Y_2, \ldots, Y_m}(y_1, y_2, \ldots, y_m) = \frac{\partial^m F_{Y_1, Y_2, \ldots, Y_m}(y_1, y_2, \ldots, y_m)}{\partial y_1\, \partial y_2 \ldots \partial y_m} \tag{5.78}$$

Equations (5.77) and (5.78) are valid for any kind of functions g_k, $k = 1, 2, \ldots, m$. If the set of Eqs. (5.76) has a unique solution for some m-variables as

$$X_k = h_k(Y_1, Y_2, \ldots, Y_m; X_{m+1}, X_{m+2}, \ldots, X_n), \quad k = 1, 2, \ldots, m \tag{5.79}$$

then by changing the variables on the right hand side of Eq. (5.77), we obtain

$$F_{Y_1, Y_2, \ldots, Y_m}(y_1, y_2, \ldots, y_m) = \int_{-\infty}^{y_1} dy_1 \int_{-\infty}^{y_2} dy_2 \ldots \int_{-\infty}^{y_m} dy_m \int_{x_{m+1}=-\infty}^{\infty} \cdots \int_{x_n=-\infty}^{\infty}$$

$$f_{X_1, X_2, \ldots, X_n}(h_1, h_2, \ldots, h_m; x_{m+1}, \ldots, x_n) |J|\, dx_{m+1} \ldots dx_n \tag{5.80}$$

where J denotes the Jacobian matrix given by

$$J = \text{determinant of} \begin{bmatrix} \frac{\partial h_1}{\partial y_1} & \frac{\partial h_2}{\partial y_1} & \cdots & \frac{\partial h_m}{\partial y_1} \\ \frac{\partial h_1}{\partial y_2} & \frac{\partial h_2}{\partial y_2} & \cdots & \frac{\partial h_m}{\partial y_2} \\ \cdot & \cdot & \cdots & \cdot \\ \cdot & \cdot & \cdots & \cdot \\ \cdot & \cdot & \cdots & \cdot \\ \frac{\partial h_1}{\partial y_m} & \frac{\partial h_2}{\partial y_m} & \cdots & \frac{\partial h_m}{\partial y_m} \end{bmatrix} \tag{5.81}$$

By differentiating Eq. (5.80), we can find the joint probability density function as

$$f_{Y_1, Y_2, \ldots, Y_m}(y_1, y_2, \ldots, y_m)$$
$$= f_{X_1, X_2, \ldots, X_n}(h_1, h_2, \ldots, h_m; x_{m+1}, x_{m+2}, \ldots, x_n)\,|J| \tag{5.82}$$

Example 5.10 Find the joint density function of Y_1 and Y_2 where $Y_1 = X_1 X_2$, $Y_2 = X_1 / X_2$ and the joint density function of X_1 and X_2 is given by

$$f_{X_1,X_2}(x_1, x_2) = \frac{1}{x_1^2\, x_2^2}; \qquad 1 \le x_1 < \infty; \qquad 1 \le x_2 < \infty$$

solution Since $Y_1 = g_1(X_1, X_2) = X_1 X_2$ and $Y_2 = g_2(X_1, X_2) = X_1 / X_2$, we obtain

$$X_1 = h_1(Y_1, Y_2) = \sqrt{Y_1 Y_2} \qquad \text{and} \qquad X_2 = h_2(Y_1, Y_2) = \sqrt{\frac{Y_1}{Y_2}}$$

The Jacobian matrix is given by

$$[J] = \begin{bmatrix} \frac{\partial h_1}{\partial y_1} & \frac{\partial h_2}{\partial y_1} \\ \frac{\partial h_1}{\partial y_2} & \frac{\partial h_2}{\partial y_2} \end{bmatrix} = \begin{bmatrix} \frac{1}{2}\sqrt{\frac{y_2}{y_1}} & \frac{1}{2}\frac{1}{\sqrt{y_1 y_2}} \\ \frac{1}{2}\sqrt{\frac{y_1}{y_2}} & -\frac{1}{2}\frac{\sqrt{y_1}}{y_2^{3/2}} \end{bmatrix}$$

so that

$$|[J]| = \left| -\frac{1}{4}\frac{1}{y_2} - \frac{1}{4}\frac{1}{y_2} \right| = \frac{1}{2y_2}$$

The application of Eq. (5.82) gives

$$f_{Y_1,Y_2}(y_1, y_2) = f_{X_1,X_2}(h_1, h_2)\,|[J]| = \frac{1}{h_1^2 h_2^2}\,\frac{1}{2y_2} = \frac{1}{(y_1 y_2)\left(\frac{y_1}{y_2}\right)}\,\frac{1}{2\,y_2} = \frac{1}{2\,y_1^2 y_2}$$

References and Bibliography

5.1. W. W. Hines and D. C. Montgomery, *Probability and Statistics in Engineering and Management Science,* 2d ed., John Wiley, New York, 1980.
5.2. E. B. Haugen, *Probabilistic Approaches to Design,* John Wiley, New York, 1968.
5.3. G. P. Wadsworth and J. G. Bryan, *Applications of Probability and Random Variables,* 2d ed., McGraw-Hill, New York, 1974.
5.4. M. Kaufman and A. H. Seidman (eds.), *Handbook of Electronics Calculations for Engineers and Technicians,* 2d ed., McGraw-Hill, New York, 1988.
5.5. E. C. Guyer and D. L. Brownell (eds.), *Handbook of Applied Thermal Design,* McGraw-Hill, New York, 1989.
5.6. J. E. Shigley and C. R. Mischke, *Mechanical Engineering Design,* 5th ed., McGraw-Hill, New York, 1989.
5.7. C. E. Wilson, *Noise Control: Measurement, Analysis, and Control of Sound and Vibration,* Harper & Row, New York, 1989.
5.8. A. H. S. Ang and W. H. Tang, *Probability Concepts in Engineering Planning and Design,* vol. 1, *Basic Principles,* John Wiley, New York, 1975.
5.9. T. T. Soong, *Probabilistic Modeling and Analysis in Science and Engineering,* John Wiley, New York, 1981.
5.10. A. C. King and C. B. Read, *Pathways to Probability,* Holt, Rinehart & Winston, New York, 1963.
5.11. C. C. Gillispie (ed.-in-chief), *Dictionary of Scientific Biography,* vol. 1, Charles Scribner's Sons, New York, 1980.

Review Questions

5.1 How is the partial derivative rule derived?

5.2 Find the mean and variance of the function $Y = g(X_1, X_2, \ldots, X_n)$ using the partial derivative rule.

5.3 Express the probability distribution function of Y in terms of the probability density function of X where $Y = g(X)$.

5.4 Find the probability density function of $Y = 4X - 3$ when the density function of X is given by $f_X(x) = 6e^{-6x}$.

5.5 If the probability distributions of two independent normal variates X_1 and X_2 are given by $N(100,10)$ and $N(80,5)$, respectively, find the probability density function of $Y = X_1 + X_2$.

5.6 If the probability distributions of two independent normal variates X_1 and X_2 are given by $N(100,10)$ and $N(80,5)$, respectively, find the probability density function of $Y = X_1 - X_2$.

5.7 What is a Jacobian?

5.8 What is the use of a moment generating function?

5.9 Define the moment generating function of a random variable X.

5.10 What is a partial derivative rule?

5.11 Express the standard deviation of the function $g = X_1, X_2, \ldots, X_n)$ in terms of the joint density function of $X_1, X_2, \ldots, X_n$.

Problems

5.1 The ultimate strength of cast iron (S_u) is related to its Brinell hardness number (H_b) as

$$S_u = 0.23\, H_b - 12.5 \text{ k(lb/in}^2)$$

If H_b is uniformly distributed in the range 140 to 150, determine the probability density function of S_u.

5.2 The ratio of tensions in a belt drive is given by $r = e^{f\theta}$ where f is the coefficient of friction and θ is the angle of contact between the belt and the pulley. Find the probability density function of r when f follows uniform distribution in the range 0.2 to 0.3 and $\theta = 1$.

5.3 Find the probability density function of $x = \sin\theta$ when θ is uniformly distributed in the range $-\pi$ to π.

5.4 Find the probability density function of $y = \sqrt{x}$ when the probability density function of x is given by

$$f_X(x) = e^{-x}; \qquad x \geq 0$$

5.5 The probability distributions of two independent random variables are given by

$$f_{X_i}(x_i) = \begin{cases} e^{-x_i}; & x_i > 0 \\ 0; & \text{otherwise} \end{cases}$$

for $i = 1, 2$. Find the probability density function of $Y = 3\,X_1 + 5\,X_2$.

5.6 The joint probability mass function of three discrete random variables is given in the following table:

(x_1, x_2, x_3)	$p_{X_1, X_2, X_3}(x_1, x_2, x_3)$
(0, 0, 0)	1/8
(1, 0, 0)	3/32
(0, 1, 0)	3/16
(0, 0, 1)	1/32
(1, 1, 0)	5/32
(1, 0, 1)	1/16
(0, 1, 1)	7/32
(1, 1, 1)	1/8

Find the probability mass function of $Y = X_1 + X_2 + X_3$.

5.7 A function Y is defined as

$$Y = \sum_{i=1}^{n} c_i X_i$$

where c_i are constants and X_i are a set of independent random variables following exponential distribution:

$$f_{X_i}(x_i) = \begin{cases} \lambda_i e^{-\lambda_i x_i}; & x_i \geq 0 \\ 0; & \text{otherwise} \end{cases}$$

for $\lambda_i > 0$ and $i = 1, 2, \ldots, n$.

1. Find the probability density function of Y.
2. Find $f_Y(y)$ when $\lambda_i = \lambda$ for $i = 1, 2, \ldots, n$.
3. Find $f_Y(y)$ when $\lambda_i = \lambda$ and $c_i = 1$ for $i = 1, 2, \ldots, n$.
4. Show that the random variable Y follows gamma distribution in part 3.

5.8 If $X_1, X_2, \ldots, X_n$ denote a set of statistically independent random variables following the binomial distribution

$$f_{X_i}(x_i) = \binom{n_i}{x_i} p^{x_i} (1-p)^{n-x_i}; \quad x_i = 0, 1, 2, \ldots, n_i; \; i = 1, 2, \ldots, n$$

show that the function $Y = X_1 + X_2 + \cdots + X_n$ also follows binomial distribution. Also determine the mean value and standard deviation of Y.

5.9 A set of n mutually independent random variables X_i follows uniform distribution in the range 0 to 1 so that

$$f_{X_i}(x_i) = \begin{cases} 1; & 0 \leq x_i \leq 1 \\ 0; & \text{otherwise} \end{cases}$$

for $i = 1, 2, \ldots, n$. (*a*) Find the probability density function of $Y = X_1 + X_2$. (*b*) Find the probability density function of $Y = X_1 + X_2 + \cdots + X_n$.

5.10 The efficiency (η) of a V-threaded screw is given by

$$\eta = \frac{\cos\beta - \mu \tan\alpha}{\cos\beta + \mu \cot\alpha}$$

where α = helix angle of the screw, β = half of the thread angle, and μ = coefficient of friction. If α, β, and μ are random with mean values and coefficients of variation given by

$$\bar{\alpha} = 10^\circ, \; \bar{\beta} = 30^\circ, \; \bar{\mu} = 0.1, \; \gamma_\alpha = 0.05, \; \gamma_\beta = 0.05 \text{ and } \gamma_\mu = 0.1,$$

determine the first order mean and standard deviation of η.

5.11 The pressure, p, in kgf/cm^2, at which the oil film breaks down so that metal to metal contact occurs in a journal bearing is given by (see Fig. 5.10)

$$p = \frac{ZN}{475 \times 10^6} \left(\frac{d}{c}\right)^2 \frac{l}{d+l}$$

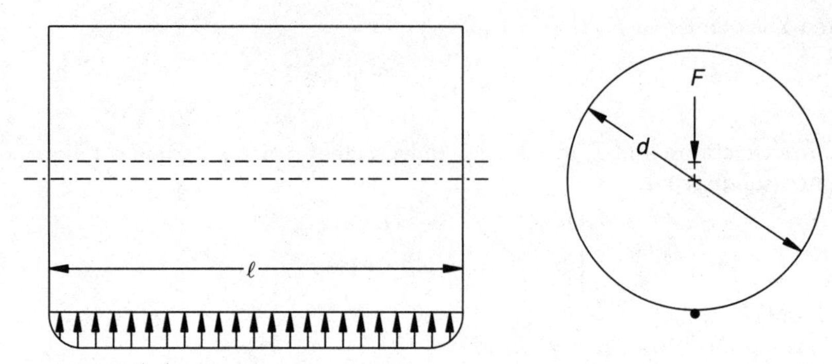

Figure 5.10 A journal bearing.

where Z = absolute viscosity of the lubricant in centipoise, N = speed of the journal in rev/min, d = diameter of the journal in cm, c = difference between the diameters of bushing and journal in cm and l = length of the bearing in cm. The bearing parameters are known to be random variables for a bearing used in a reciprocating compressor with

$$\bar{l} = 15 \text{ cm}, \bar{d} = 7.5 \text{ cm}, \bar{c} = 0.02 \text{ cm}, \bar{Z} = 60, \bar{N} = 300, \gamma_l = \gamma_d = \gamma_c = \gamma_Z = \gamma_N = 0.1.$$

Compute the first order mean and standard deviation of the pressure, p.

5.12 The length of the belt needed (l) in a belt drive (as shown in Fig. 5.11) is given by

$$l = \left[4c^2 - (d_2 - d_1)^2\right]^{1/2} + \frac{1}{2}(d_2\,\theta_2 + d_1\,\theta_1)$$

where

$$\theta_1 = \pi - 2\sin^{-1}\left(\frac{d_2 - d_1}{2c}\right)$$

and

$$\theta_2 = \pi + 2\sin^{-1}\left(\frac{d_2 - d_1}{2c}\right)$$

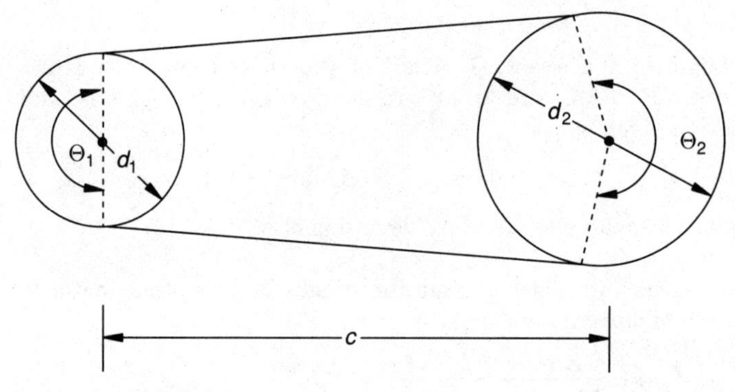

Figure 5.11 A belt drive.

where d_1 = diameter of the smaller pulley, d_2 = diameter of the larger pulley and c = center distance. If d_1, d_2 and c are random with

$\bar{d}_1 = 10$ cm, $\sigma_{d_1} = 0.1$ cm, $\bar{d}_2 = 20$ cm, $\sigma_{d_2} = 0.25$ cm, $\bar{c} = 100$ cm, and $\sigma_c = 0.5$ cm,

determine the mean and standard deviation of the length of belt.

5.13 The maximum contact pressure developed (p) between two cylinders under a load F can be expressed as

$$p = \sqrt{\frac{FE(d_1 + d_2)}{\pi l(1 - \nu^2)\, d_1 d_2}}$$

where d_1, d_2 = diameters of cylinders, ν = Poisson's ratio, E = Young's modulus and l = axial length of contact between the cylinders. If $(\bar{F}, \sigma_F) =$ (10,000, 2000) N, $(\bar{l}, \sigma_l) = (0.2,\ 0.02)$ m, $(\bar{E}, \sigma_E) = (2.06, 1.03) \times 10^6 N/\mathrm{m}^2$, $(\nu, \sigma_\nu) = (0.3,\ 0.015)$, $(\bar{d}_1, \sigma_{d_1}) = (0.1, 0.01)$ m and $(\bar{d}_2, \sigma_{d_2}) = (0.25, 0.025)$ m, determine the first order mean and standard deviations of the maximum contact pressure, p.

5.14 The actuating force (F) necessary to apply the brake shown in Fig. 5.12 is given by

$$F = \frac{M_n - M_f}{a}$$

where

$$M_n = p_a trb \left(\frac{\theta_2}{2} - \frac{1}{4} \sin 2\,\theta_2 \right)$$

$$M_f = p_a trf \left(r - r \cos \theta_2 - \frac{b}{2} \sin^2 \theta_2 \right)$$

with p_a = maximum pressure exerted on the brake lining, t = face width of the brake shoes, r = inner radius of the brake drum, b = radius of the brake shoe, f = coefficient of friction, a = linear distance between the brake shoe hinge and the line of application of the actuating force, and θ_2 = angle between the brake shoe hinge and the point of application of the actuating force (see Fig. 5.12). Find the mean value and standard deviation of F when $(\bar{p}_a, \sigma_{p_a}) = (10^5,\ 10^4)$ Pa, $(\bar{t}, \sigma_t) = (30, 1)$ mm, $(\bar{r}, \sigma_r) = (150, 10)$ mm, $(\bar{b}, \sigma_b) = (120, 4)$ mm, $(\bar{f}, \sigma_f) = (0.35,\ 0.035)$, $(\bar{a}, \sigma_a) = (200,\ 10)$ mm and $(\bar{\theta}_2, \sigma_{\theta_2}) = (120^\circ,\ 3^\circ)$.

5.15 An electrical network consists of three resistors in series as shown in Fig. 5.13. If the resistances are independent normal variates with $R_1 = N(50,\ 2)$ ohms, $R_2 = N(100,\ 5)$ ohms and $R_3 = N(150,\ 10)$ ohms, find the probability density function of the total resistance of the network.

5.16 A gear is to be mounted on a shaft using an interference fit as shown in Fig. 5.14. If the diameters of the shaft and the hole are given by $D = N(2,\ 0.05)$ inch and $d = N(1.95,\ 0.05)$ inch, determine the probability density function of the interference between the two parts.

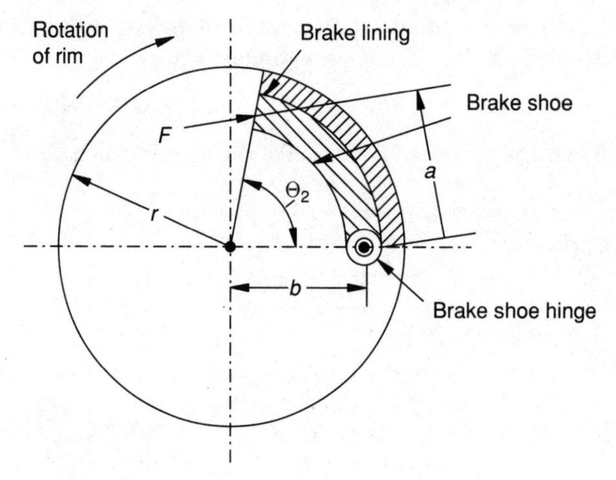

Figure 5.12

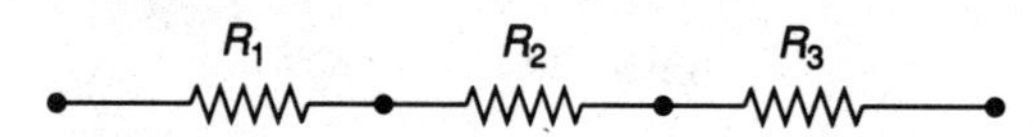

Figure 5.13

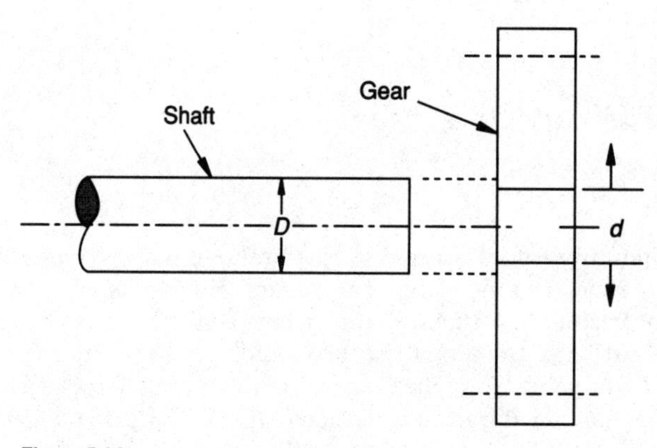

Figure 5.14

5.17 A method of determining the probability density function of the function $Y = g(X_1, X_2)$ in terms of the joint density function, $f_{X_1,X_2}(x_1, x_2)$, can be summarized [5.1] as follows:

1. Introduce a new random variable Z as $Z = h(X_1, X_2)$ where the function $h(X_1, X_2)$ is suitably selected so that X_1 and X_2 can be expressed in terms of Y and Z as $X_1 = p(Y, Z)$ and $X_2 = q(Y, Z)$.

2. Find the joint density function of Y and Z as

$$f_{Y,Z}(y, z) = f_{X_1,X_2}\left[p(y,z), q(y,z)\right] |J(y,z)|$$

where $|J(y, z)|$ is the Jacobian of the transformation defined as

$$J(y,z) = \begin{vmatrix} \dfrac{\partial x_1}{\partial y} & \dfrac{\partial x_1}{\partial z} \\ \dfrac{\partial x_2}{\partial y} & \dfrac{\partial x_2}{\partial z} \end{vmatrix}$$

3. Determine the probability density function of Y as

$$f_Y(y) = \int_{-\infty}^{\infty} f_{Y,Z}(y, z)\, dz$$

Using this method, find the probability density function of $Y = X_1 / X_2$ when the joint density function of X_1 and X_2 is given by

$$f_{X_1,X_2}(x_1, x_2) = \begin{cases} 25\, e^{-5(x_1+x_2)}; & x_1 \geq 0,\ x_2 \geq 0 \\ 0; & \text{otherwise} \end{cases}$$

Hint: Define Z as $Z = X_1 + X_2$.

5.18 An automobile repair shop performs different types of repair. The arrival rates of customers with different types of repair requests follow Poisson distribution with the following data:

Type of repair	Mean value of number of repairs per day
Brakes	5.1
Radiator	3.2
Air conditioning	2.3
Transmission	1.6
Fuel pump	4.5

Find the probability distribution function of the total number of repairs per day.

5.19 Two links, of lengths X_1 and X_2, are connected as shown in Fig. 5.15. If X_1 and X_2 follow uniform distributions over the ranges 0 to 2 and 0 to 3, respectively, determine the probability density function of the total length $X_1 + X_2$. Also find the probability of realizing $X_1 + X_2$ greater than or equal to 3.5.

5.20 The heat transfer rate q in BTU/h, through a uniform one-dimensional fin with an insulated end (Fig. 5.16) is given by

$$q = \sqrt{hPkA}\,(T_s - T_\infty) \tan hmL$$

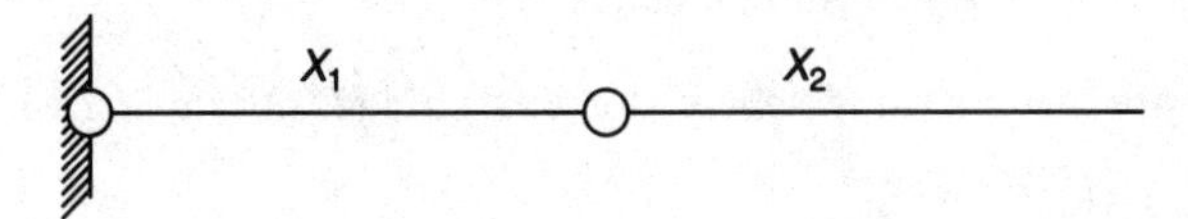

Figure 5.15

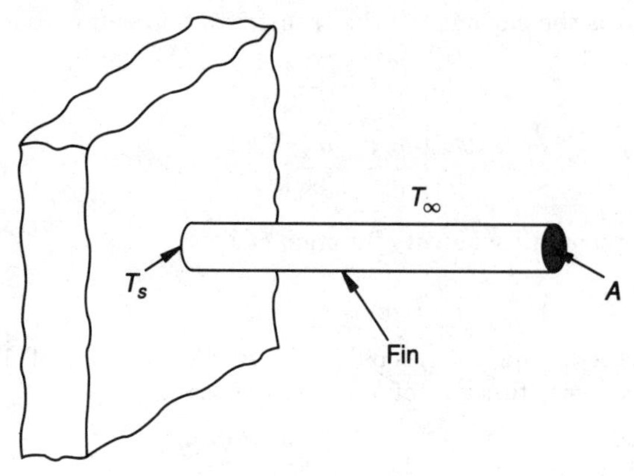

Figure 5.16

where h = heat transfer coefficient (BTU/h-ft^2-deg F), P = perimeter of the fin cross section (ft) , k = thermal conductivity of the fin (BTU/h-ft-deg F), A = area of cross section of the fin (ft^2), T_s = temperature at the root (deg F), T_∞ = atmospheric temperature (deg F), $m = \sqrt{hP / kA}$ ft^{-1}, and L = length of the fin (ft). For a steel circular fin with diameter, $(\overline{D}, \sigma_D) = (1, 0.1)$ in, $(\overline{h}, \sigma_h) = (20, 2)$ BTU/h-ft^2-deg F, $(\overline{k}, \sigma_k) = (40, 2)$ BTU/h-ft-deg F, $(\overline{T}_s, \sigma_{T_s}) = (200, 5)$ deg F, $(\overline{T}_\infty, \sigma_{T_\infty}) = (70, 3)$ deg F, $(\overline{L}, \sigma_L) = (40, 2)$ in, determine the mean value and standard deviation of q.

5.21 The pressure drop in the fluid flowing through a circular pipe (Δp) is given by

$$\Delta p = f \frac{L}{D} \frac{V^2}{2} \rho$$

where f = friction factor, L = length of pipe, D = diameter of pipe, V = velocity of fluid, and ρ = density of fluid. If $(\overline{f}, \sigma_f) = (0.02, 0.004)$, $(\overline{L}, \sigma_L) = (2000, 20)$ m, $(\overline{\rho}, \sigma_\rho) = (1000, 10)$ kg/m^3, $(\overline{D}, \sigma_D) = (0.1, 0.001)$ m, and $(\overline{V}, \sigma_V) = (10, 1)$ m/s, find the mean value and standard deviation of Δp.

5.22 The random motion of electrons in a resistor generates small noise terminal voltage, V_n, known as Johnson or white noise, and is given by [5.4]

$$V_n = \sqrt{4kTR\,\Delta f} \text{ volts}$$

where k = Boltzmann constant $= 1.38 \times 10^{-23} J/K$, T = absolute temperature (K), R = resistance (ohms), and Δf = bandwidth (Hz). If $(\overline{R}, \sigma_R) = (1, 0.01)$ MΩ, $(\overline{T}, \sigma_T) = (27, 1)$°C and $(\overline{\Delta f}, \sigma_{\Delta f}) = (10, 0.1)$ kHz, determine the mean value and standard deviation of the noise voltage generated.

5.23 The sound pressure level, L_p in decibels, due to an air jet nozzle is given by [5.7]

$$L_p = 168.6 + 20 \log \left(\frac{d}{r}\right) + 40 \log (1 - P^{-0.2857})$$

where d = nozzle diameter, r = distance from the nozzle where sound pressure is measured, P = pressure ratio $= p_i / p_a = 1 + p_i' / p_a$, p_i' = inlet gauge pressure, and p_a = atmospheric pressure. If $(\bar{d}, \sigma_d) = (0.25, 0.025)$ in, $(\bar{r}, \sigma_r) = (36, 4)$ in, $(\bar{p}_i', \sigma_{p_i'}) = (60, 1)$ lb/in^2, and $(\bar{p}_a, \sigma_{p_a}) = (14.696, 0.1)$ lb/in^2, find the mean and standard deviation of the sound pressure level.

5.24 Find the mean and variance of a product of n random variables.

5.25 Find the mean value and standard deviation of $Y = X_1 - X_2$.

5.26 If X_1 and X_2 denote independent discrete random variables following Poisson distribution, show that the difference, $Y = X_1 - X_2$, does not follow Poisson distribution.

5.27 Prove that, for a normally distributed variable X,

$$\mu_{X^2} = \mu_X{}^2 + \sigma_X{}^2$$

and

$$\sigma_{X^2} = 4\,\mu_X{}^2\,\sigma_X{}^2 + 2\sigma_X^4$$

using the moment generating function defined in Eq. (5.75).

5.28 The moment generating function corresponding to the normally distributed variables X_1 and X_2 is given by

$$M(\theta_1, \theta_2) = e^{(\theta_1\,\mu_1 + \theta_2\mu_2 + 1/2\,\theta_1^2\sigma_1^2 + 1/2\,\theta_2^2\,\sigma_2^2 + \rho\theta_1\theta_2\sigma_1\sigma_2)}$$

Using this, find the mean value and standard deviation of $Y = X_1 \pm X_2$.

5.29 Using the moment generating function of Problem 5.28, find the mean and standard deviation of $Y = X_1\,X_2$.

5.30 Derive the moment generating function of a random variable X which follows the distribution

$$f_X(x) = \begin{cases} \lambda\,e^{-\lambda x}; & x \geq 0 \\ 0; & \text{otherwise} \end{cases}$$

Also find the mean and variance of X using the moment generating function.

5.31 The moment generating function for a discrete random variable X is defined as

$$M(\theta) = \sum_{\text{summation on all } x \text{ for which } p_X(x) > 0} e^{\theta x}\,p_X(x)$$

where $p_X(x)$ is the probability mass function of X. Using this, find the moment generating function of X following binomial distribution. Also find the mean and standard deviation of X using the moment generating function.

5.32 Find the joint density function of Y_1 and Y_2 in terms of the joint density function of X_1 and X_2 when $Y_1 = X_1 + X_2$, $Y_2 = X_1 - X_2$, and

$$f_{X_1,X_2}(x_1,x_2) = \begin{cases} 25\,e^{-5(x_1+x_2)}; & x_1 \geq 0,\ x_2 \geq 0 \\ 0; & \text{otherwise} \end{cases}$$

Chapter

6

Time Dependent Reliability of Components and Systems

Biographical Note

Karl Friedrich Gauss

Karl Friedrich Gauss was born in Brunswick, Germany on 30 April 1777 and died in Göttingen, Germany on 23 February 1855. He is considered to be one of the three greatest mathematicians that history has produced; the other two being Archimedes and Newton. He was born to a poor family; his father was a bricklayer and a gardener. It is said that Gauss corrected an error in calculating his father's wages at the age of three. His doctoral thesis, in mathematics, dealt with the proof of the "Fundamental Theory of Algebra" which states that every polynomial equation of degree one or higher has a root. He made many contributions to mathematics, astronomy, and physics. The numbers of the form $x + i\,y$*, where* x *and* y *are integers and* $i = \sqrt{-1}$*, are known as Gaussian integers. The unit of magnetic flux density is named gauss in honor of his contributions to physics. His work on the theory of errors led him to discover the "method of least squares" and later the normal or Gaussian distribution curve. Although the normal curve was originally studied by De Moivre, Gauss was the first one to justify it by means of the theory of probability. Gauss was made the Director of Göttingen Observatory in 1807. He used to get involved with his work so deeply that even when a servant rushed to tell him that his wife was dying, he replied, "ask her to wait until I finish this work" [6.7, 6.8].*

6.1 Introduction

As stated earlier, reliability is the probability of successful performance of a system at any time. The system may be an electric fan which is expected to start and stop, and run perhaps for years with out failure, or a missile which is expected to be in standby condition for long periods of time and then respond just once without failure. The successful performance of any system depends on the extent to which reliability is designed and built in to it. In practice, it is observed that even seemingly identical systems, operating under similar conditions, fail at different times. A very simple example is a light bulb. Different light bulbs, even taken from the same batch of production, and put into operation under identical conditions, will have different lives. Thus the failure of such systems can only be described probabilistically. Since most engineering systems are composed of several components, the relationship between component reliability and system reliability must be understood. This chapter deals with the reliability analysis of components and systems, including the problem of reliability allocation in multicomponent systems.

6.2 Failure-Rate Time Curve

Experience with a large number of mechanical and electronic systems has shown that their failure characteristics follow definite patterns. A plot of failure rate versus time of a typical component such as an electric motor or a transistor is shown in Fig. 6.1. The curve shown in Fig. 6.1 is also called the

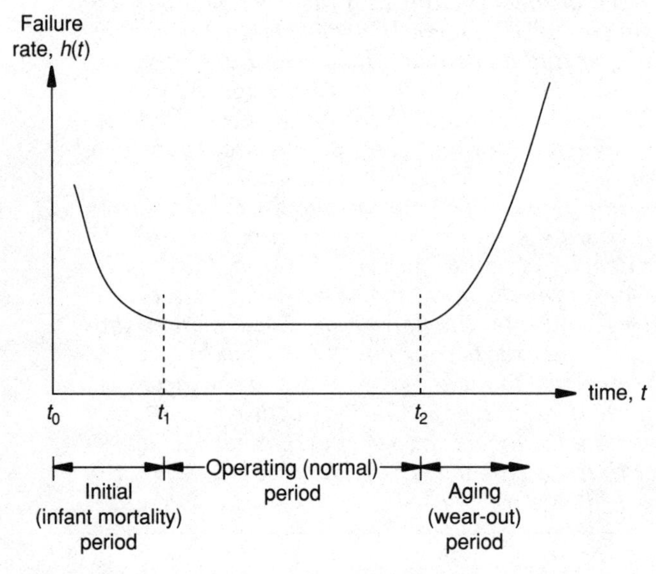

Figure 6.1

"bathtub curve" due to its shape. Surprisingly, the life of human beings also was found to follow a similar curve. In Fig. 6.1, the initial failures during the time 0 to t_1 are caused due to reasons such as defective materials, poor design, bad workmanship, poor quality control or poor transportation. This period is called the *infant mortality* or *burn-in* period and is characterized by a rapidly declining failure rate and is generally covered by the warranty or guarantee period given by the manufacturer. As the defective components are replaced by newer ones, the reliability improves. The next phase of failures occur in the *normal* (*operating*) period t_1 to t_2. These failures, also known as *random failures*, occur due to random overloads and other extreme operating conditions. The operating period is characterized by a relatively constant failure rate. After time t_2, aging or wearout failures occur due to excessive wear after the expected useful design life has been exceeded. The wearout period is characterized by a rapidly increasing rate of failure. From a design point of view, the operating period will be of interest.

6.3 Reliability and Hazard Functions

Consider a fixed number of identical components tested simultaneously under similar operating conditions. Let N = number of identical components tested, $N_s(t)$ = number of components surviving at time t, and $N_f(t)$ = number of components failed in time t. As the test proceeds, the number of surviving components gets smaller and the number of failed components gets larger. Then the reliability of the component at time t can be defined as

$$R(t) = \frac{N_s(t)}{N} = \frac{N - N_f(t)}{N} = 1 - \frac{N_f(t)}{N} \tag{6.1}$$

Since N is fixed, Eq. (6.1) gives

$$\frac{dR(t)}{dt} = -\frac{1}{N}\frac{dN_f(t)}{dt} \tag{6.2}$$

The rate at which components fail can be found as

$$\frac{dN_f(t)}{dt} = -N\frac{dR(t)}{dt} \tag{6.3}$$

The *failure rate*, also known as the *instantaneous rate of failure*, *hazard function*, or *hazard rate*, $h(t)$, is defined as

$$h(t) = \frac{1}{N_s(t)}\frac{dN_f(t)}{dt} \tag{6.4}$$

Dividing both sides of Eq. (6.3) by $N_s(t)$, we find the failure rate as

$$h(t) = \frac{1}{N_s(t)}\frac{dN_f(t)}{dt} = -\frac{N}{N_s}\frac{dR(t)}{dt} = -\frac{1}{R(t)}\frac{dR(t)}{dt} \tag{6.5}$$

Integration of Eq. (6.5) gives, using the relation $R(0) = 1$,

$$\int_0^t h(x)\,dx = -\int_0^t \frac{dR(x)}{R(x)} = -\ln R(t) \tag{6.6}$$

Thus the reliability of the component at time t can be expressed as

$$R(t) = \exp\left[-\int_0^t h(x)\,dx\right] \tag{6.7}$$

The failure rate can also be expressed in terms of the probability distribution function of life of the component as follows. The distribution function of life or time to failure T of the component is given by

$$F_T(t) = \int_0^t f_T(x)\,dx \tag{6.8}$$

and the reliability of the component by

$$R(t) = P(T > t) = 1 - P(T \le t) = 1 - F_T(t) = \exp\left[-\int_0^t h(x)\,dx\right] \tag{6.9}$$

This gives

$$\ln\left[1 - F_T(t)\right] = -\int_0^t h(x)\,dx \tag{6.10}$$

which, upon differentiation, yields

$$-\frac{f_T(t)}{1 - F_T(t)} = -h(t)$$

that is,

$$h(t) = \frac{f_T(t)}{R(t)} = \frac{f_T(t)}{1 - F_T(t)} \tag{6.11}$$

6.4 Modeling of Failure Rates

First we consider the portion of the failure-rate curve over the component life between times t_1 and t_2. During this time, chance or random failures occur and failure cannot be attributed to deterioration of the strength of the components over time. The failure rate during this period can be assumed as a constant (λ)

$$h(t) = \lambda \text{ failures per unit time} \tag{6.12}$$

From Eq. (6.11), we have[1]

$$f(t) = h(t)\,[1 - F(t)] = h(t)\exp\left[-\int_0^t h(\tau)d\tau\right] = \lambda e^{-\lambda t} \tag{6.13}$$

[1]The notations $f(t)$ and $F(t)$ are used in place of $f_T(t)$ and $F_T(t)$ for simplicity, to denote the probability density and distribution functions of T.

and

$$R(t) = \exp\left[-\int_0^t h(\tau)\, d\tau\right] = e^{-\lambda t} \tag{6.14}$$

Equation (6.13) denotes an exponential density function.

Next we consider the wearout portion of the failure-rate curve, (after time t_2). In general, this is a nonlinear curve. If a linear variation is assumed, for simplicity, we have

$$h(t) = c_1 t + c_2 \tag{6.15}$$

If λ and λ_{max} denote the failure rates at $t = t_2$ and $t = t_3$, respectively, we have

$$c_1 = \frac{\lambda_{max} - \lambda}{t_3 - t_2} \tag{6.16}$$

and

$$c_2 = \frac{\lambda t_3 - \lambda_{max} t_2}{t_3 - t_2} \tag{6.17}$$

The probability density function of the failure time can be determined, using Eqs. (6.11) and (6.9), as

$$f(t) = h(t) \exp\left[-\int_0^t h(\tau)\, d\tau\right] = (c_1 t + c_2) \exp\left[-\frac{1}{2} c_1 t^2 - c_2 t\right] \tag{6.18}$$

and

$$R(t) = \frac{f(t)}{h(t)} = \exp\left[-\frac{1}{2} c_1 t^2 - c_2 t\right] \tag{6.19}$$

The density function given by Eq. (6.18) degenerates to Rayleigh density function if $c_2 = 0$.

Similarly, the early failure portion of the failure-rate function, corresponding to failures due to manufacturing defects, can also be approximated, for simplicity, by a linear relation as

$$h(t) = c_1 t + c_2 \tag{6.20}$$

where c_1 and c_2 are constants. If λ_0 and λ denote the failure rates at times t_0 and t_1, respectively, the constants c_1 and c_2 can be evaluated as

$$c_1 = \frac{\lambda_0 - \lambda}{t_0 - t_1} \tag{6.21}$$

and

$$c_2 = \frac{\lambda t_0 - \lambda_0 t_1}{t_0 - t_1} \tag{6.22}$$

The probability density and reliability functions of the failure time can be expressed as in Eqs. (6.18) and (6.19) with c_1 and c_2 determined from Eqs. (6.21) and (6.22). The use of normal and Weibull distributions for representing the failure times is considered in Examples 6.3, 6.4, and 6.5.

6.5 Estimation of Failure Rate from Empirical Data

When the failure time data of a component or a system are available, we can find the reliability and failure rate functions numerically. For this, first the time interval or the duration of the test is divided into segments of length Δt each; then the statistical definition is used to estimate the reliability function at time t as

$$R(t) = \frac{N_s(t)}{N} \tag{6.23}$$

where the initial value of time t is taken as zero. From the known number of components that failed in the interval $(t, t + \Delta t)$, the failure rate can be computed as

$$h(t) = \frac{N_s(t) - N_s(t + \Delta t)}{N_s(t)\,\Delta t} \tag{6.24}$$

The following example is considered to illustrate the procedure.

Example 6.1 The experimental data obtained with a set of 10,000 identical components is shown in columns 1 and 2 of Table 6.1. Plot the corresponding failure-rate time curve.

solution The computational details are shown in columns 3, 4, and 5 of Table 6.1. The failure rate (entries in the last column of Table 6.1) versus time (entries in the first column of Table 6.1) curve is shown in Fig. 6.2.

6.6 Mean Time Before Failure (MTBF)

The mean time before failure (MTBF) or the *expected life* of the component is defined as the average value of T

$$\text{MTBF} = \int_0^\infty t\, f_T(t)\, dt = -\int_0^\infty t \frac{dR(t)}{dt} dt = -t\,R(t)\Big|_0^\infty + \int_0^\infty R(t)\, dt \tag{6.25}$$

Since all systems fail after a finite time, we have $R(t) \to 0$ as $t \to \infty$ and hence $tR(t) = 0$ at $t = \infty$. Also, $tR(t) = 0$ at $t = 0$. Thus Eq.(6.25) reduces to

$$\text{MTBF} = \int_0^\infty R(t)\, dt \tag{6.26}$$

Example 6.2 Find the failure rate of a component for which the time to failure is described by an exponential density function.

TABLE 6.1 Failure data of Example 6.1

Time of observation t (hours)	Number of components operating at time t	Reliability	Rate of change of reliability	Failure rate
		$R(t) = \dfrac{N_s(t)}{N}$	$-\dfrac{dR(t)}{dt} = -\dfrac{\Delta R}{\Delta t}$	$h(t) = -\dfrac{dR(t)}{dt}\dfrac{1}{R(t)}$
0	N = 10,000	1.0	——	——
50	8,880	0.8880	0.002240	0.002240
100	8,300	0.8300	0.001160	0.001306
150	7,918	0.7918	0.000764	0.000920
200	7,585	0.7585	0.000666	0.000842
250	7,274	0.7274	0.000622	0.000820
300	6,968	0.6968	0.000612	0.000842
350	6,668	0.6668	0.000600	0.000862
400	6,375	0.6375	0.000586	0.000878
450	6,088	0.6088	0.000574	0.000900
500	5,808	0.5808	0.000560	0.000920
550	5,535	0.5535	0.000546	0.000940
600	5,269	0.5269	0.000532	0.000962
650	5,011	0.5011	0.000516	0.000980
700	4,760	0.4760	0.000502	0.001002
750	4,517	0.4517	0.000486	0.001022
800	4,237	0.4237	0.000560	0.001240
850	3,864	0.3864	0.000746	0.001760
900	3,396	0.3396	0.000936	0.002422
950	2,819	0.2819	0.001154	0.003398
1,000	2,219	0.2219	0.001200	0.004256

solution The density function of failure time T is given by

$$f_T(t) = \lambda e^{-\lambda t}; \qquad t \geq 0,\ \lambda > 0$$

The reliability function can be found as

$$R(t) = 1 - F_T(t) = \int_t^{\infty} \lambda e^{-\lambda \tau}\, d\tau = e^{-\lambda t}; \qquad t \geq 0$$

and the failure rate as

$$h(t) = \frac{f_T(t)}{R(t)} = \frac{\lambda e^{-\lambda t}}{e^{-\lambda t}} = \lambda$$

Example 6.3 Plot the reliability and hazard functions for the cases (i) $\mu = 1$, $\sigma = 0.1$, (ii) $\mu = 5$, $\sigma = 0.1$, and (iii) $\mu = 1$, $\sigma = 0.5$ when the time to failure follows normal distribution.

solution The distribution function of the time to failure (T) is given by

$$F_T(t) = \int_{-\infty}^{t} \frac{1}{\sqrt{2\pi}\,\sigma} \exp\left[-\frac{1}{2}\left(\frac{\tau - \mu}{\sigma}\right)^2\right] d\tau \qquad \text{(E1)}$$

Since the integral in Eq. (E1) cannot be evaluated in a closed form, the value of $F_T(t)$ can only be found from standard normal tables. For this, we define the standard normal variate z as

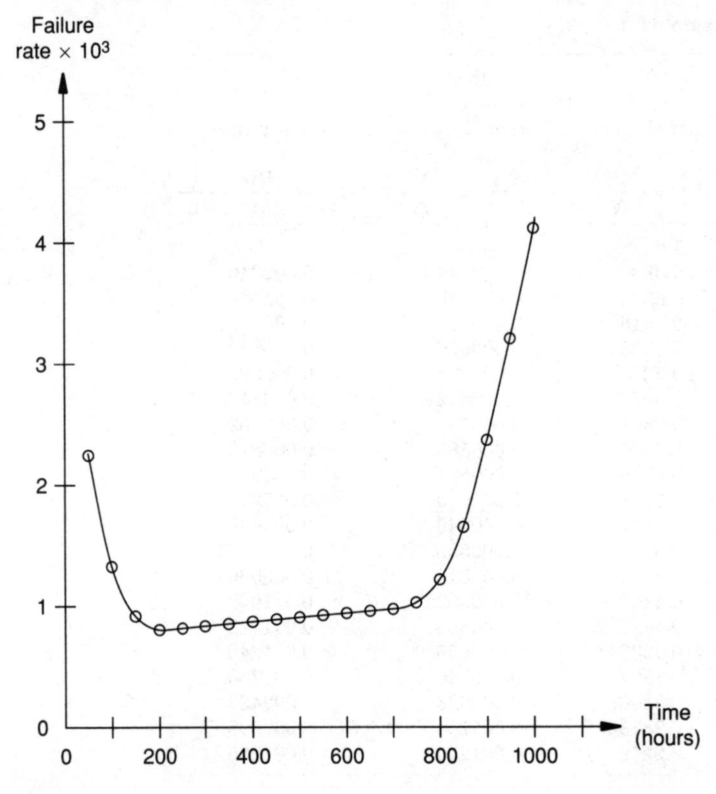

Figure 6.2

$$z = \frac{T - \mu}{\sigma} \tag{E2}$$

and find

$$F_T(t) = P(T \le t) = P\left(z \le \frac{t - \mu}{\sigma}\right) = \Phi\left(\frac{t - \mu}{\sigma}\right) \tag{E3}$$

The reliability function $R(t)$ is given by

$$R(t) = 1 - F_T(t) = 1 - \Phi\left(\frac{t - \mu}{\sigma}\right) \tag{E4}$$

The hazard function of T can be determined as

$$h(t) = \frac{f(t)}{1 - F(t)} = \frac{\frac{1}{\sigma} f\left(z = \frac{t - \mu}{\sigma}\right)}{1 - \Phi\left(\frac{t - \mu}{\sigma}\right)} \tag{E5}$$

The reliability and the hazard functions are shown plotted in Figs. 6.3a and b. It can be noted that different types of hazard functions (all increasing with t) can be achieved using normal distribution for the failure time T.

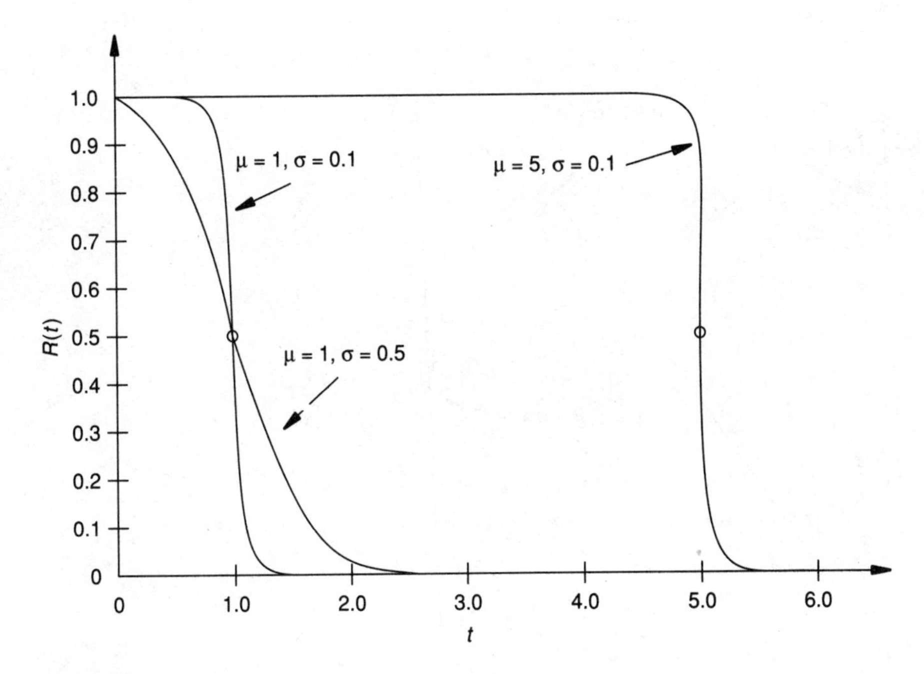

Figure 6.3*a*

Example 6.4 Find the failure rate of a mechanical component, the time to failure of which is given by Weibull distribution. Also plot the failure rate and reliability functions.

solution The time to failure (T), described by Weibull distribution, is given by

$$f_T(t) = \begin{cases} \dfrac{\beta}{\eta - l}\left(\dfrac{t-l}{\eta-l}\right)^{\beta-1} \exp\left[-\left(\dfrac{t-l}{\eta-l}\right)^{\beta}\right]; & t \geq l \\ 0; & t < l \end{cases} \tag{E1}$$

The distribution function of T is given by

$$F_T(t) = \int_l^t f_T(\tau)\,d\tau = \frac{\beta}{\eta - l}\int_{\tau=l}^{t}\left(\frac{\tau-l}{\eta-l}\right)^{\beta-1} \exp\left[-\left(\frac{\tau-l}{\eta-l}\right)^{\beta}\right]d\tau \tag{E2}$$

Let

$$x = \left(\frac{\tau - l}{\eta - l}\right)^{\beta}$$

so that

$$dx = \frac{\beta}{\eta - l}\left(\frac{\tau-l}{\eta-l}\right)^{\beta-1} d\tau$$

Hence,

$$F_T(t) = \int_{x'=0}^{x} e^{-x'}\,dx' = -e^{-x'}\Big|_0^x$$

$$= 1 - e^{-x} = 1 - \exp\left[-\left(\frac{t-l}{\eta-l}\right)^{\beta}\right] \tag{E3}$$

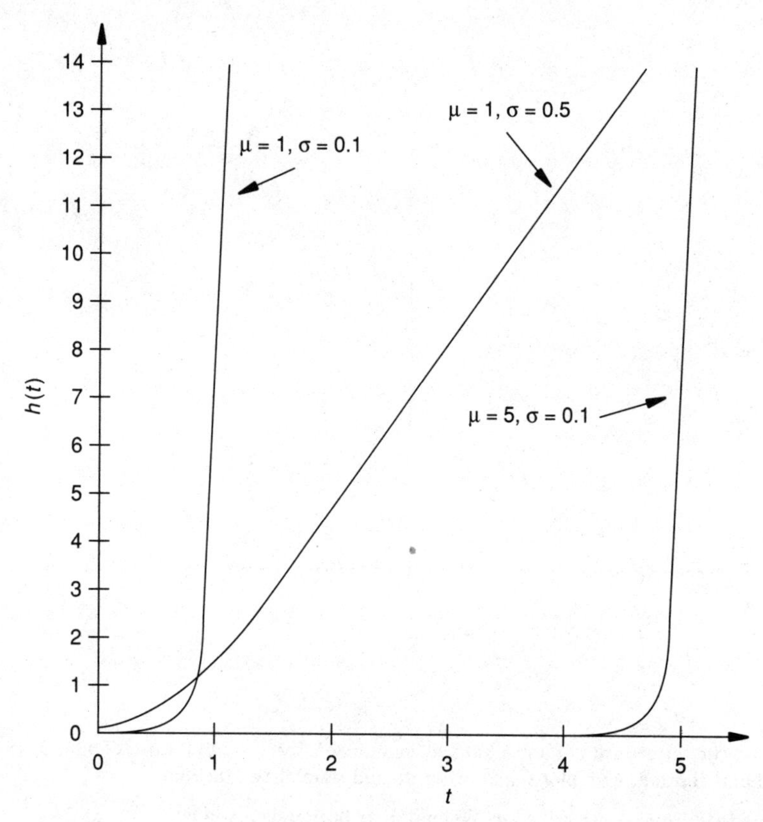

Figure 6.3*b*

and

$$R(t) = 1 - F_T(t) = \exp\left[-\left(\frac{t-l}{\eta - l}\right)^{\beta}\right] \tag{E4}$$

The failure rate can be found as

$$h(t) = \frac{f_T(t)}{R(t)} = \frac{\beta}{\eta - l}\left(\frac{t-l}{\eta - l}\right)^{\beta - 1} \tag{E5}$$

The reliability- and hazard-functions are shown in Figs. 6.4*a* and *b* for different values of β. It can be seen that the hazard function decreases for $\beta < 1$, increases for $\beta > 1$, and remains constant for $\beta = 1$. Thus Weibull distribution with different values of β can be used to model the failure rates during the infant mortality, operating, and wearout periods of a component or a system.

Example 6.5 The failure time of a mechanical component follows Weibull distribution with $\beta = 3, \eta = 2{,}500$, and $l = 1{,}000$. Find the reliability of the component and the failure rate for an operating time of 2,000 hours.

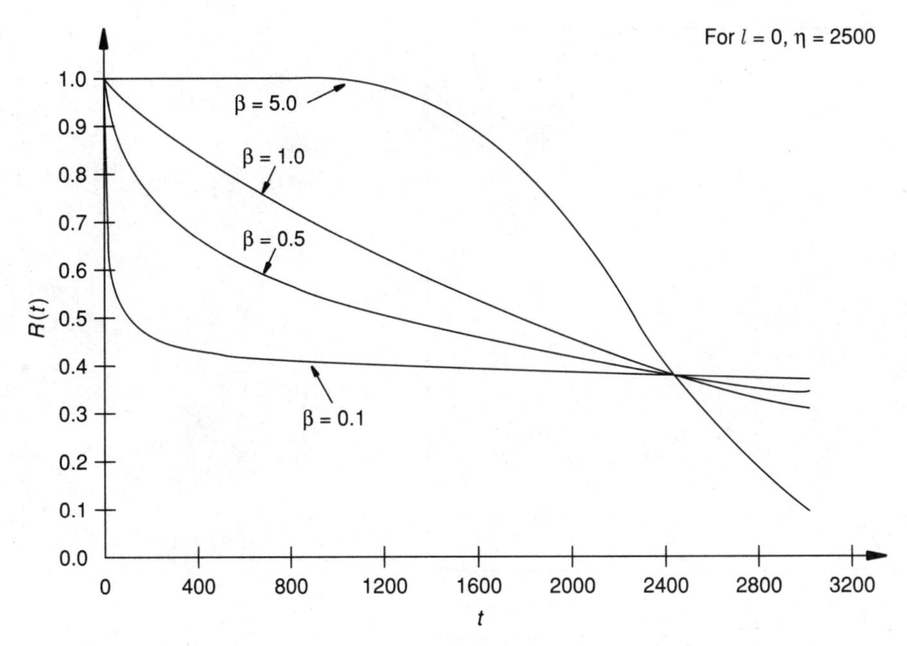

Figure 6.4*a*

solution By substituting the given data into Eqs. (E4) and (E5) of Example 6.4 we get

$$R(2000) = \exp\left[-\left(\frac{2000-1000}{2500-1000}\right)^3\right] = e^{-0.2963} = 0.7436$$

and

$$h(2000) = \frac{3(2000-1000)^{3-1}}{(2500-1000)^3} = \frac{3(1000)^2}{(1500)^3} = 0.00089 \text{ failures per hour}$$

6.7 Series Systems

A series system is one that fails whenever any one of its components fails. A set of electrical switches connected in series as shown in Fig. 6.5 can be considered to be a series system, since the failure of any switch in making contact will prevent the flow of current from one end to the other (i.e., from A to B) in Fig. 6.5. The components need not be physically connected in series for the system to be called a series system. For example, the four tires of an automobile are in series for the purpose of reliability calculation because the failure of any one makes the car inoperative. In Fig. 6.5, a switch is considered to be in operating (failed) condition if it is in a closed (open) position. If R_i denotes the reliability and $(1 - R_i)$ the probability of failure of switch C_i, then the various possibilities at any time t are shown in Table 6.2.

The reliability of the system can be seen to be $R_s = R_1 R_2 R_3$. It is to be noted that the switches C_1, C_2, and C_3 were assumed to be independent in this analysis.

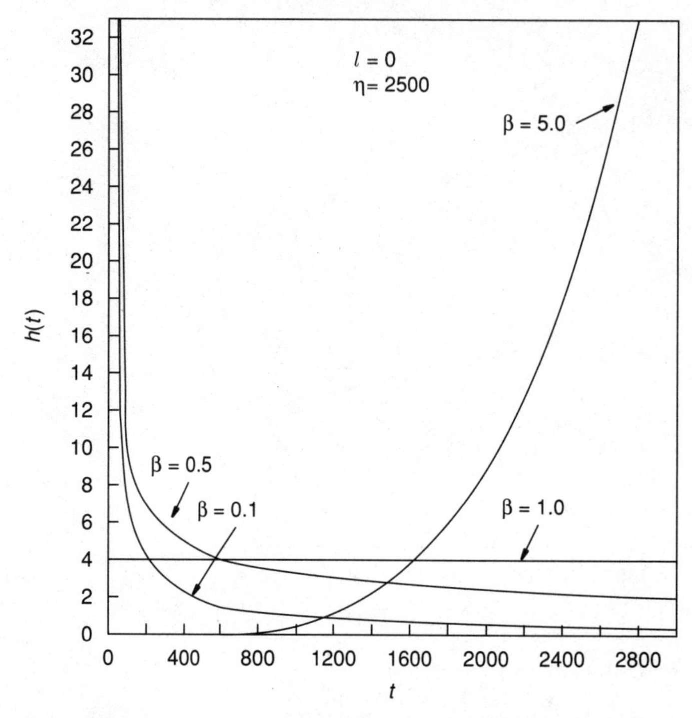

Figure 6.4*b*

TABLE 6.2

Event	Condition of switch C_1	C_2	C_3	Probability of occurrence of the event	Condition of the system
1	Closed	Closed	Closed	$R_1 R_2 R_3$	Operating
2	Closed	Closed	Open	$R_1 R_2 (1 - R_3)$	Failed
3	Closed	Open	Closed	$R_1 (1 - R_2) R_3$	Failed
4	Open	Closed	Closed	$(1 - R_1) R_2 R_3$	Failed
5	Closed	Open	Open	$R_1 (1 - R_2)(1 - R_3)$	Failed
6	Open	Closed	Open	$(1 - R_1) R_2 (1 - R_3)$	Failed
7	Open	Open	Closed	$(1 - R_1)(1 - R_2) R_3$	Failed
8	Open	Open	Open	$(1 - R_1)(1 - R_2)(1 - R_3)$	Failed

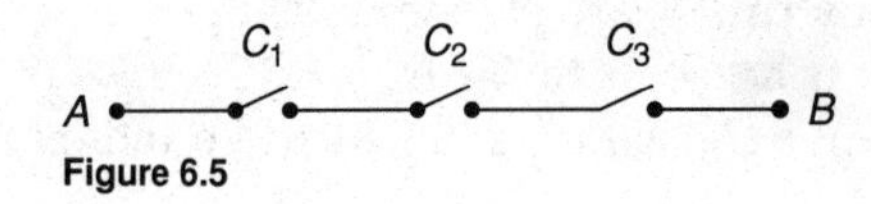

Figure 6.5

If we consider n components $C_1, C_2, \ldots,$ and C_n in series, the reliability of the series system (R_s), assuming the components to be independent, is given by

$$R_s = R_1 R_2 \cdots R_n = \prod_{i=1}^{n} R_i \tag{6.27}$$

where R_i denotes the reliability of component C_i. If the failure time of the ith component is given by t_i, $i = 1, 2, \ldots, n$, the failure time of the series system t_s is given by

$$t_s = \min_{1 \le i \le n} t_i \tag{6.28}$$

since the system fails as soon as one of its components fails. If $F_i(t)$ denotes the probability distribution function of the failure time of the ith component, the failure time distribution of the series system can be found as

$$F_s(t) = P(t_s \le t) = 1 - P(t_s > t) = 1 - \prod_{i=1}^{n} P(t_i > t) = 1 - \prod_{i=1}^{n} \left[1 - F_i(t)\right] \tag{6.29}$$

Thus the probability density function of the failure time of the series system is given by

$$f_s(t) = \frac{dF_s(t)}{dt} = \sum_{j=1}^{n} \frac{\partial F_s}{\partial F_j} \frac{\partial F_j}{\partial t} = \sum_{j=1}^{n} f_j(t) \prod_{i=1, i \ne j}^{n} \left[1 - F_i(t)\right] \tag{6.30}$$

6.7.1 Failure rate of the system

If the failure time of the ith component follows exponential distribution, with a constant failure rate $\lambda_i (i = 1, 2, \ldots, n)$, then

$$R_i(t) = e^{-\lambda_i t} \tag{6.31}$$

and

$$R_s(t) = \prod_{i=1}^{n} R_i(t) = e^{-\left(\sum_{i=1}^{n} \lambda_i\right) t} = e^{-\lambda_s t} \tag{6.32}$$

where the failure rate of the system λ_s can be identified as

$$\lambda_s = \sum_{i=1}^{n} \lambda_i \tag{6.33}$$

Thus, the failure time of the system also follows exponential distribution with the failure rate of the system λ_s given by the sum of the component failure rates. In fact, Eq. (6.33) can be shown to be valid for any failure time distribution. If the failure rate of component i is given by $h_i(t)$, its reliability is given by Eq. (6.7) as

$$R_i(t) = \exp\left[-\int_0^t h_i(x)\,dx\right] \tag{6.34}$$

Equations (6.27) and (6.34) lead to

$$R_s(t) = \prod_{i=1}^{n} R_i(t) = \prod_{i=1}^{n} e^{-\int_0^t h_i(x)\,dx}$$

$$= e^{-\sum_{i=1}^{n}\int_0^t h_i(x)\,dx} = e^{-\int_0^t \left(\sum_{i=1}^{n} h_i(x)\right)dx} \tag{6.35}$$

This shows that the system failure rate, $h_s(t)$, is given by

$$h_s(t) = \sum_{i=1}^{n} h_i(t) \tag{6.36}$$

If $h_i(t)$ is constant, as in the case of exponential failure time distribution, then

$$h_s(t) = \sum_{i=1}^{n} h_i \tag{6.37}$$

which shows that the system failure rate is also a constant.

6.7.2 MTBF of the system

The MTBF of a series system can be computed as

$$\text{MTBF} = \int_0^\infty R(t)\,dt = \int_0^\infty \left(\prod_{i=1}^{n} R_i(t)\right) dt \tag{6.38}$$

If the failure rates of the components follow exponential distribution, Eq. (6.38) gives

$$\text{MTBF} = \int_0^\infty e^{-\left(\sum_{i=1}^{n}\lambda_i\right)t}\,dt$$

$$= \left(-\frac{1}{\sum_{i=1}^{n}\lambda_i}\right)\left[e^{-\left(\sum_{i=1}^{n}\lambda_i\right)t}\right]\Bigg|_0^\infty = \frac{1}{\sum_{i=1}^{n}\lambda_i} \tag{6.39}$$

Example 6.6 A system consists of five components in series, each having a reliability of 0.97. Find the reliability of the system.

solution The system reliability is given by

$$R_s = \prod_{i=1}^{5} R_i = (0.97)^5 = 0.8587$$

It can be seen that the greater the number of components, the less reliable the system becomes. Thus a series system will have a reliability smaller than the reliability of any of its components.

6.8 Parallel Systems

A parallel system is one that fails only when all of its components fail. For example, a set of n switches connected in parallel will constitute a parallel system since all the switches must fail before the system is rendered useless for the flow of current from one end to the other. Consider a parallel system consisting of three independent switches as shown in Fig. 6.6. The eight possible outcomes of the system at any time t are given in Table 6.3.

The reliability of the parallel system is given by the sum of the probabilities of realizing the first seven events, that is,

$$R_p = R_1 R_2 R_3 + R_1 R_2 (1-R_3) + R_1 R_3 (1-R_2) + R_2 R_3 (1-R_1)$$
$$+ R_1 (1-R_2)(1-R_3) + R_2 (1-R_1)(1-R_3) + R_3 (1-R_1)(1-R_2) \quad (6.40)$$

In this case, the system reliability can be expressed more conveniently as

$$R_p = 1 - P_f = 1 - (1-R_1)(1-R_2)(1-R_3) = 1 - \prod_{i=1}^{3}(1-R_i) \quad (6.41)$$

TABLE 6.3

	Condition of switch			Probability of occurrence of the event	Condition of system
Event	C_1	C_2	C_3		
1	Closed	Closed	Closed	$R_1 R_2 R_3$	Operating
2	Closed	Closed	Open	$R_1 R_2 (1-R_3)$	Operating
3	Closed	Open	Closed	$R_1 (1-R_2) R_3$	Operating
4	Open	Closed	Closed	$(1-R_1) R_2 R_3$	Operating
5	Closed	Open	Open	$R_1 (1-R_2)(1-R_3)$	Operating
6	Open	Closed	Open	$(1-R_1) R_2 (1-R_3)$	Operating
7	Open	Open	Closed	$(1-R_1)(1-R_2) R_3$	Operating
8	Open	Open	Open	$(1-R_1)(1-R_2)(1-R_3)$	Failed

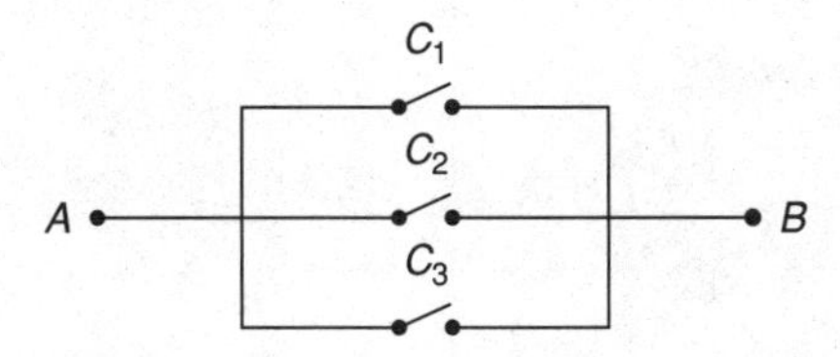

Figure 6.6

This relation can be generalized as follows. If n components $C_1, C_2, \ldots,$ and C_n are connected in parallel, the reliability of the system is given by

$$R_p = 1 - P_f = 1 - P \text{ (all components failing simultaneously)}$$

$$= 1 - \prod_{i=1}^{n}(1 - R_i) = 1 - \prod_{i=1}^{n} F_i \tag{6.42}$$

where $F_i = 1 - R_i$ denotes the probability of failure of the ith component at time t.

Example 6.7 A parallel system consists of 3 components A, B, and C. If the reliabilities of the three components are $R_A = 0.9$, $R_B = 0.8$ and $R_C = 0.7$, find the system reliability.

solution Equation (6.42) gives

$$R_p = 1 - (1 - R_A)(1 - R_B)(1 - R_C) = 1 - (0.1)(0.2)(0.3) = 1 - 0.006 = 0.994$$

This indicates that a larger number of components gives a higher reliability. It can also be seen that the reliability of a parallel system will always be larger than the reliability of any of its components.

If the failure time of the ith component is denoted as t_i, $i = 1, 2, \ldots, n$, the failure time of the parallel system t_p is given by

$$t_p = \max_{1 \le i \le n} t_i \tag{6.43}$$

since the system fails only when all the components fail. The probability distribution function of the failure time of the parallel system $F_p(t)$ can be expressed in terms of the probability distribution functions of the failure times of the individual components $F_i(t)$ as

$$F_p(t) = P(t_p \le t) = \prod_{i=1}^{n} F_i(t) \tag{6.44}$$

Thus the probability density function of the failure time of the parallel system is given by

$$f_p(t) = \frac{dF_p(t)}{dt} = \sum_{j=1}^{n} f_j(t) \prod_{i=1, i \ne j}^{n} F_i(t) \tag{6.45}$$

6.8.1 Failure rate of the system

The system failure rate $h_p(t)$ is given by

$$h_p(t) = \frac{f(t)}{1 - F(t)} = \frac{\sum_{j=1}^{n} f_j(t) \prod_{i=1, i \ne j}^{n} F_i(t)}{1 - \prod_{i=1}^{n} F_i(t)} \tag{6.46}$$

It can be seen that the system failure rate $h_p(t)$ is not equal to the sum of the component failure rates.

6.8.2 MTBF of the system

Consider a parallel system with n components. Let the probability density function of the ith component be exponential with a failure rate of λ_i, $i = 1, 2, \ldots, n$. Then the MTBF of the system can be computed as

$$\text{MTBF} = \int_0^\infty R(t)\,dt = \int_0^\infty \left[1 - \prod_{i=1}^{n}\{1 - R_i(t)\}\right] dt$$

$$= \int_0^\infty \left[1 - \prod_{i=1}^{n}\left\{1 - e^{-\lambda_i t}\right\}\right] dt$$

$$= \int_0^\infty \left[1 - (1 - e^{-\lambda_1 t})(1 - e^{-\lambda_2 t}) \cdots (1 - e^{-\lambda_n t})\right] dt \tag{6.47}$$

Since Eq. (6.47) cannot be simplified any further, we consider the special case of $n = 2$. For this case, Eq. (6.47) gives

$$\text{MTBF} = \int_0^\infty \left[1 - (1 - e^{-\lambda_1 t} - e^{-\lambda_2 t} + e^{-(\lambda_1+\lambda_2)t}\right] dt$$

$$= \frac{1}{\lambda_1} + \frac{1}{\lambda_2} - \frac{1}{\lambda_1+\lambda_2} \tag{6.48}$$

where the relation

$$\int_0^\infty e^{-\lambda t}\,dt = \left[-\frac{1}{\lambda}e^{-\lambda t}\right]\Big|_0^\infty = \frac{1}{\lambda} \tag{6.49}$$

has been used in deriving Eq. (6.48). Similarly, for $n = 3$, Eq. (6.47) yields

$$\text{MTBF} = \frac{1}{\lambda_1} + \frac{1}{\lambda_2} + \frac{1}{\lambda_3} - \frac{1}{\lambda_1+\lambda_2} - \frac{1}{\lambda_2+\lambda_3} - \frac{1}{\lambda_1+\lambda_3} + \frac{1}{\lambda_1+\lambda_2+\lambda_3} \tag{6.50}$$

6.9 (k, n) Systems

A system consisting of n components is called a (k, n) system if it can operate with any k components in operating condition. For example, in a two-engine aircraft, it may be possible to fly the aircraft (probably with less efficiency, but without any loss to property and people) if any one engine is in operating condition. In this case, the aircraft is considered to function as a (1, 2) system. Similarly, in a six-cylinder automobile, it may be possible to drive the car with only five spark plugs firing. In this case, it can be called a (5, 6) system.

If the failure times of all the components of a (k, n) system have the same probability distribution function $F(t)$ then the reliability of the system can be determined as

$$R_k(t) = \sum_{i=k}^{n} \binom{n}{i} [1 - F(t)]^i [F(t)]^{n-i} \tag{6.51}$$

and hence the probability distribution function of the system is given by

$$F_k(t) = 1 - R_k(t) = \sum_{i=0}^{k-1} \binom{n}{i} [1 - F(t)]^i [F(t)]^{n-i} \tag{6.52}$$

Equation (6.52) can be differentiated to find the probability density function of the failure time of the system as

$$f_k(t) = \frac{dF_k(t)}{dt} = \frac{n!}{(n-k)!\,(k-1)!} [F(t)]^{n-k} [1 - F(t)]^{k-1} f(t) \tag{6.53}$$

where $f(t) = \dfrac{dF(t)}{dt}$ denotes the probability density function of the failure time of the components.

6.9.1 MTBF of the system

If the failure times of individual components follow the same exponential distribution, the reliability of a (k, n) system is given by Eq. (6.51) so that the MTBF becomes

$$\text{MTBF} = \int_0^\infty R_k(t)\, dt = \int_0^\infty \left[\sum_{i=k}^{n} \binom{n}{i} \{e^{-\lambda t}\}^i \{1 - e^{-\lambda t}\}^{n-i} \right] dt \tag{6.54}$$

For example, for a (1, 2) system, Eq. (6.54) gives

$$\text{MTBF} = \int_0^\infty \left[\binom{2}{1} e^{-\lambda t} (1 - e^{-\lambda t}) + \binom{2}{2} (e^{-\lambda t})^2 \right] dt$$

$$= \int_0^\infty \left(2 e^{-\lambda t} - e^{-2\lambda t} \right) dt = \frac{3}{2\lambda} \tag{6.55}$$

6.10 Mixed Series and Parallel Systems

It can be noted from Eqs. (6.27) and (6.42) that the reliability of a series system is always smaller than the component reliability while that of a parallel system is always larger than that of the component reliability, that is,

$$R_s < R_i; \quad i = 1, 2, \ldots, n$$

and

$$R_p > R_i; \quad i = 1, 2, \ldots, n \tag{6.56}$$

This indicates that the reliability of a system can be improved by connecting the components in parallel. Other methods of improving the system reliability include placing the components in a combination of series and parallel manner as shown in Figs. 6.7 and 6.8. For the system shown in Fig. 6.7, called the parallel–series system, the reliability R_0 is given by

$$\begin{aligned} R_0 &= (R_{E_1} + R_{E_2} - R_{E_1} R_{E_2}) \\ &= (R_1 R_2 R_3 + R_4 R_5 R_6 - R_1 R_2 R_3 R_4 R_5 R_6) \end{aligned} \tag{6.57}$$

where

$$R_{E_1} = R_1 R_2 R_3$$

$$R_{E_2} = R_4 R_5 R_6$$

and $R_i = R_{C_i}$; $i = 1, 2, \ldots, 6$. If all the components have identical exponential failure time distributions as $F_i(t) = F(t) = 1 - e^{-\lambda t}$, Eq. (6.57) reduces to

$$R_0 = 2R^3 - R^6 = 2(1-F)^3 - (1-F)^6 = 2e^{-3\lambda t} - e^{-6\lambda t} \tag{6.58}$$

For this case, the MTBF of the system can be determined as

$$\text{MTBF} = \int_0^\infty R_0(t)\,dt = \int_0^\infty \left[2e^{-3\lambda t} - e^{-6\lambda t}\right] dt = \frac{2}{3\lambda} - \frac{1}{6\lambda} = \frac{1}{2\lambda} \tag{6.59}$$

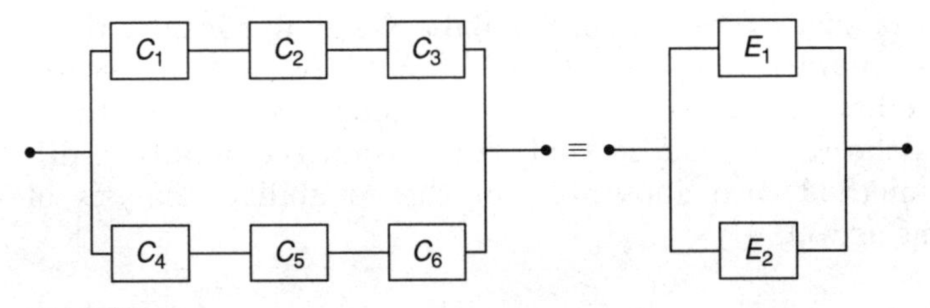

Figure 6.7

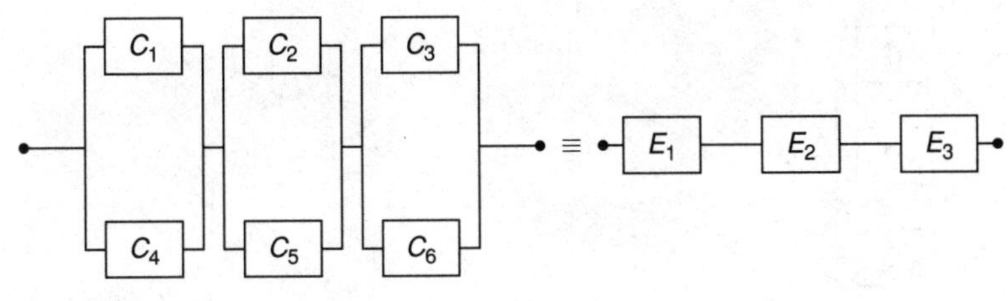

Figure 6.8

The reliability R_0 of the system shown in Fig 6.8, known as the series–parallel system, can be expressed as

$$R_0 = R_{E_1} R_{E_2} R_{E_3} \tag{6.60}$$

$$= [(R_1 + R_4 - R_1 R_4)(R_2 + R_5 - R_2 R_5)(R_3 + R_6 - R_3 R_6)]$$

where $R_i = R_{C_i}$, $i = 1, 2, \ldots, 6$. If all the components have identical exponential distributions as $F_i(t) = F(t) = 1 - e^{-\lambda t}$, Eq. (6.60) reduces to

$$R_0 = (2R - R^2)^3 = (2e^{-\lambda t} - e^{-2\lambda t})^3$$

$$= 8e^{-3\lambda t} - 12e^{-4\lambda t} + 6e^{-5\lambda t} - e^{-6\lambda t} \tag{6.61}$$

The MTBF of the system can be found as

$$\text{MTBF} = \int_0^\infty R_0(t)\,dt = \int_0^\infty \left[8e^{-3\lambda t} - 12e^{-4\lambda t} + 6e^{-5\lambda t} - e^{-6\lambda t}\right] dt$$

$$= \frac{7}{10\lambda} \tag{6.62}$$

6.11 Complex Systems

If a system has a more complex configuration than the series/parallel structures discussed so far, additional evaluation techniques are needed to determine its reliability. A typical system that does not have a series/parallel configuration is shown in Fig. 6.9. There are several methods available for finding the reliability of such systems [6.3, 6.5, 6.6]. We shall consider three methods, namely, the *enumeration*, *conditional probability*, and the *cut-set* approaches, in the following sections. It is to be noted that all of these methods make use of the same basic concept, but present the details in different ways. These methods are applicable for the reliability analysis of series/parallel systems as well.

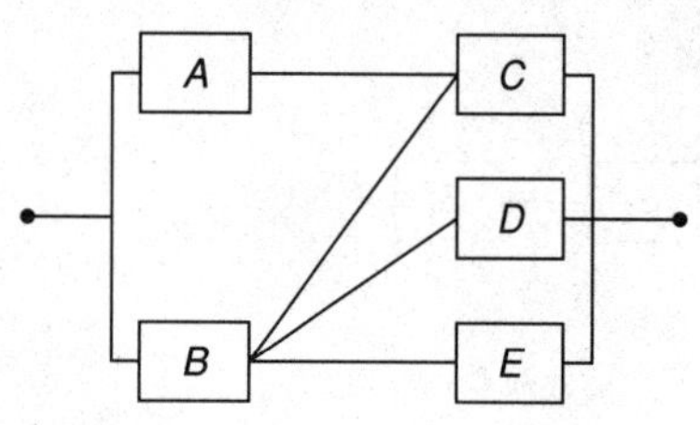

Figure 6.9

6.11.1 Enumeration method

The basis of the enumeration method is the observation that each component of the system can be either in operating condition or in failed condition. We enumerate the various possible events by considering all combinations of the conditions of different components of the system. The events which lead to the operation of the complete system are then identified by inspection and the reliability of the system is computed by finding the probability of realizing all the mutually exclusive events which lead to the operation of the complete system. Example 6.8 illustrates the procedure.

Example 6.8 Find the reliability of the system shown in Fig. 6.9 using the enumeration method.

solution Denoting the operating condition of a component X as X and its failed condition as $\overline{X}$, the possible events of the system are identified as shown in Table 6.4. In this table, the events corresponding to the operation of the system are identified by the letter "O" and those corresponding to the failed condition are indicated by "F." If the reliability of component X is R_X and the probability of failure is $(1 - R_X)$, the reliability of the system can be found as

$$R_0 = \text{probability of realizing the events } 1\text{–}6,\ 8\text{–}10,\ 12\text{–}16,\ 18, \text{ and } 21\text{–}23$$

This can be expressed as

$$\begin{aligned} R_0 = {} & R_A R_B R_C R_D R_E + (1 - R_A) R_B R_C R_D R_E \\ & + (1 - R_B) R_A R_C R_D R_E + (1 - R_C) R_A R_B R_D R_E \\ & + (1 - R_D) R_A R_B R_C R_E + (1 - R_E) R_A R_B R_C R_D \\ & + (1 - R_A)(1 - R_C) R_B R_D R_E + (1 - R_A)(1 - R_D) R_B R_C R_E \\ & + (1 - R_A)(1 - R_E) R_B R_C R_D + (1 - R_B)(1 - R_D) R_A R_C R_E \\ & + (1 - R_B)(1 - R_E) R_A R_C R_D + (1 - R_C)(1 - R_D) R_A R_B R_E \\ & + (1 - R_C)(1 - R_E) R_A R_B R_D + (1 - R_D)(1 - R_E) R_A R_B R_C \\ & + (1 - R_B)(1 - R_D)(1 - R_E) R_A R_C + (1 - R_A)(1 - R_D)(1 - R_E) R_B R_C \\ & + (1 - R_A)(1 - R_C)(1 - R_E) R_B R_D + (1 - R_A)(1 - R_C)(1 - R_D) R_B R_E \end{aligned} \tag{E1}$$

This equation can be simplified to obtain

$$\begin{aligned} R_0 = {} & R_B R_C R_D R_E - R_A R_B R_C \\ & - R_B R_C R_D - R_B R_C R_E - R_B R_D R_E + R_A R_C + R_B R_C \\ & + R_B R_D + R_B R_E \end{aligned} \tag{E2}$$

It can be noted that the method becomes quite tedious as the number of components increases.

TABLE 6.4 Possible Events for the System of Fig. 6.9

Description of the system	Event and condition of various components	Status of the system
No component fails	1. $ABCDE$	O
One component fails	2. $\bar{A}BCDE$	O
	3. $A\bar{B}CDE$	O
	4. $AB\bar{C}DE$	O
	5. $ABC\bar{D}E$	O
	6. $ABCD\bar{E}$	O
Two components fail	7. $\bar{A}\bar{B}CDE$	F
	8. $\bar{A}B\bar{C}DE$	O
	9. $\bar{A}BC\bar{D}E$	O
	10. $\bar{A}BCD\bar{E}$	O
	11. $A\bar{B}\bar{C}DE$	F
	12. $A\bar{B}C\bar{D}E$	O
	13. $A\bar{B}CD\bar{E}$	O
	14. $AB\bar{C}\bar{D}E$	O
	15. $AB\bar{C}D\bar{E}$	O
	16. $ABC\bar{D}\bar{E}$	O
Three components fail	17. $AB\bar{C}\bar{D}\bar{E}$	F
	18. $A\bar{B}C\bar{D}\bar{E}$	O
	19. $A\bar{B}\bar{C}D\bar{E}$	F
	20. $A\bar{B}\bar{C}\bar{D}E$	F
	21. $\bar{A}BC\bar{D}\bar{E}$	O
	22. $\bar{A}B\bar{C}D\bar{E}$	O
	23. $\bar{A}B\bar{C}\bar{D}E$	O
	24. $\bar{A}\bar{B}CD\bar{E}$	F
	25. $\bar{A}\bar{B}C\bar{D}E$	F
	26. $\bar{A}\bar{B}\bar{C}DE$	F
Four components fail	27. $A\bar{B}\bar{C}\bar{D}\bar{E}$	F
	28. $\bar{A}B\bar{C}\bar{D}\bar{E}$	F
	29. $\bar{A}\bar{B}C\bar{D}\bar{E}$	F
	30. $\bar{A}\bar{B}\bar{C}D\bar{E}$	F
	31. $\bar{A}\bar{B}\bar{C}\bar{D}E$	F
Five components fail	32. $\bar{A}\bar{B}\bar{C}\bar{D}\bar{E}$	F

6.11.2 Conditional probability method

In this method, a critical component of the system C_{cri} which possibly prevents the decomposition of the system into a series–parallel system, is identified and the reliability of the system is computed as follows:

$$\begin{pmatrix}\text{Reliability of}\\ \text{the system}\end{pmatrix} = \begin{pmatrix}\text{Reliability of the}\\ \text{system with component } C_{\text{cri}}\\ \text{in operating condition}\end{pmatrix}\begin{pmatrix}\text{Probability of}\\ \text{component } C_{\text{cri}} \text{ in}\\ \text{operating condition}\end{pmatrix}$$

$$+\begin{pmatrix}\text{Reliability of the}\\ \text{system with component } C_{\text{cri}} \text{ in}\\ \text{failed condition}\end{pmatrix}\begin{pmatrix}\text{Probability of}\\ \text{component } C_{\text{cri}} \text{ in}\\ \text{failed condition}\end{pmatrix} \tag{6.63}$$

The method is illustrated through the following example.

Example 6.9 Find the reliability of the system shown in Fig. 6.9 using the conditional probability approach.

solution For this system, we choose the component C as the critical component. Then the system can be divided into two subsystems; one with component C in operating condition and the other with component C in failed condition. This subdivision is indicated in Figs. 6.10*a* and *b*. When C is in operating condition, the subsystem can be modeled as a series-parallel configuration as indicated in Fig. 6.10*a*. The reliability of this system (R_1) can be evaluated as

$$R_1 = R_{AB}\,R_{DE} = [1-(1-R_A)(1-R_B)](1) = R_A + R_B - R_A\,R_B \tag{E1}$$

since the reliability R_{DE} is always equal to unity. Similarly, when C is in failed condition, the subsystem can be modeled as a series-parallel configuration as indicated in Figs. 6.10*b*. The reliability of this subsystem (R_2) can be computed as

$$R_2 = R_B\,R_{DE} = R_B\,[1-(1-R_D)(1-R_E)] = R_B\,R_D + R_B\,R_E - R_B\,R_D\,R_E \tag{E2}$$

Finally the system reliability can be found, using Eq. (6.63), as

$$\begin{aligned} R_0 &= R_1\,R_C + R_2\,(1-R_C)\\ &= (R_A + R_B - R_AR_B)R_C + (R_B\,R_D + R_B\,R_E - R_B\,R_D\,R_E)(1-R_C)\\ &= R_BR_CR_DR_E - R_AR_BR_C - R_BR_CR_D - R_BR_CR_E\\ &\quad - R_BR_DR_E + R_AR_C + R_BR_C + R_BR_D + R_BR_E \end{aligned} \tag{E3}$$

which can be seen to be identical to Eq. (E2) of Example 6.8.

6.11.3 Cut-set method

The cut-set method is more useful in computing the reliability of a system compared to the enumeration and conditional probability methods previously discussed. The cut-set method is more convenient for computer implementation and can identify the various distinct failure modes of the system in a simple manner. A cut set is defined as the set of components that, when failed, will cause the system to fail. A minimal cut set is defined as the set of minimum number of components that, when failed, guarantees the failure of the system. This also implies that if any component of the minimal cut set has not failed, then the system will not fail. The following procedure is adopted in finding the reliability of the system using the cut set method.

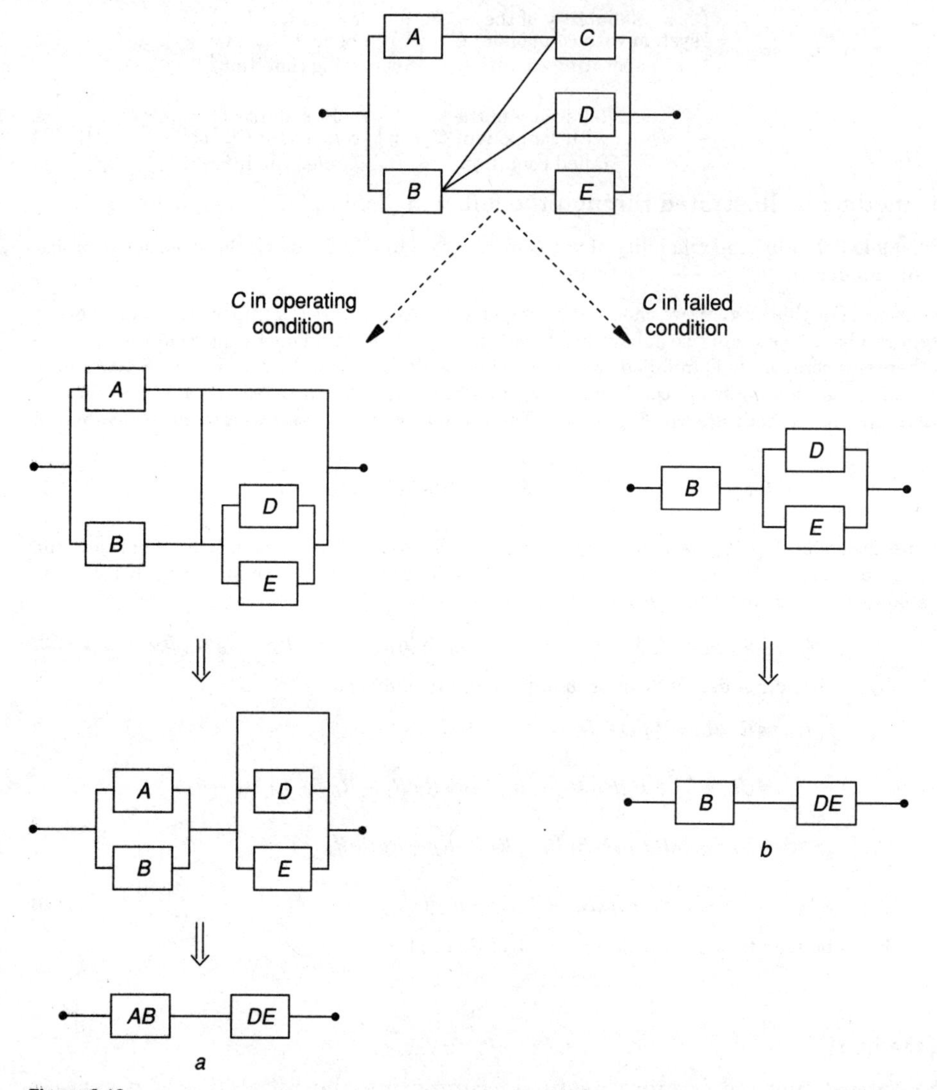

Figure 6.10

1. Identify the minimal cut sets of the system.
2. Model the components of each minimal cut set to be in parallel because, by definition, every component of the cut set must fail for the failure of the system.
3. Assume that the various cut sets are in series since the occurrence of any cut set implies the failure of the system.

4. Find the reliability of the system using the parallel–series model developed in steps 3 and 4. It is to be noted that the concept of series system cannot be used directly because the same component may appear in more than one cut set.

The procedure to be adopted is illustrated through the following example.

Example 6.10 Find the reliability of the system shown in Fig. 6.9 using the cut-set method.

solution By inspection, the independent minimal cut sets of Fig. 6.9 can be identified as *AB*, *BC*, and *CDE*. These cut sets are denoted as C_1, C_2, and C_3 and are assumed to be connected in series as indicated in Fig. 6.11 since the occurrence of any one cut set implies the failure of the system. It will be more convenient to find the probability of failure of the system P_{fs} as:

$$P_{fs} = P(C_1 \cup C_2 \cup C_3) = P(C_1) + P(C_2) + P(C_3) - P(C_1 \cap C_2)$$

$$- P(C_1 \cap C_3) - P(C_2 \cap C_3) + P(C_1 \cap C_2 \cap C_3) \tag{E1}$$

where the probabilities of occurrence of the various events can be expressed as

$$P(C_1) = P_f(A)\, P_f(B) \tag{E2}$$

$$P(C_2) = P_f(B)\, P_f(C) \tag{E3}$$

$$P(C_3) = P_f(C)\, P_f(D)\, P_f(E) \tag{E4}$$

$$P(C_1 \cap C_2) = P(C_1)\, P(C_2) = P_f(A)\, P_f(B)\, P_f(C) \tag{E5}$$

$$P(C_1 \cap C_3) = P(C_1)\, P(C_3) = P_f(A)\, P_f(B)\, P_f(C)\, P_f(D)\, P_f(E) \tag{E6}$$

$$P(C_2 \cap C_3) = P(C_2)\, P(C_3) = P_f(B)\, P_f(C)\, P_f(D)\, P_f(E) \tag{E7}$$

$$P(C_1 \cap C_2 \cap C_3) = P(C_1)\, P(C_2) P(C_3) = P_f(A) P_f(B) P_f(C) P_f(D) P_f(E) \tag{E8}$$

and $P_f(X)$, for example, denotes the probability of failure of component *X*. Also note that if a component appears in more than one cut set, its probability of failure is represented only once. For example, in Eq. (E5), the component *B* appears in both the minimal cut sets C_1 and C_2; however, its probability of failure $P_f(B)$ is represented only once on the right hand side of

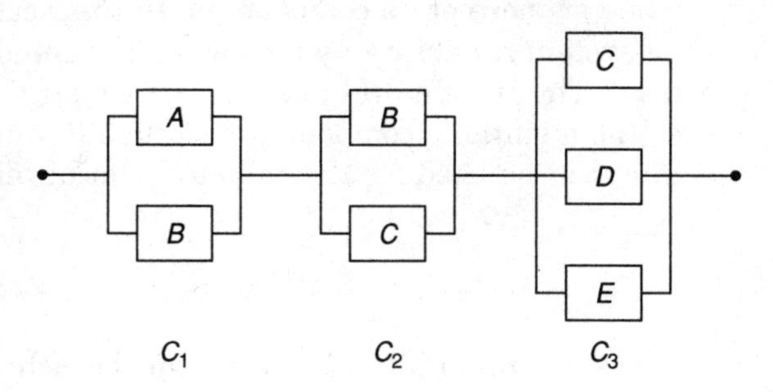

Figure 6.11

Eq. (E5). Thus the probability of failure of the system P_{fs} can be expressed using Eqs. (E1) to (E8) as

$$P_{fs} = P_f(A)\,P_f(B) + P_f(B)\,P_f(C) + P_f(C)P_f(D)(P_fE)$$

$$- P_f(A)P_f(B)P_f(C) - P_f(B)\,P_f(C)\,P_f(D)\,P_f(E) \tag{E9}$$

By using the relation

$$P_f(X) = 1 - R_X \tag{E10}$$

where R_X denotes the reliability of component X, Eq. (E9) can be expressed as

$$P_{fs} = (1 - R_A)(1 - R_B) + (1 - R_B)(1 - R_C) + (1 - R_C)(1 - R_D)(1 - R_E)$$

$$- (1 - R_A)(1 - R_B)(1 - R_C) - (1 - R_B)(1 - R_C)(1 - R_D)(1 - R_E)$$

$$= 1 - R_A R_C + R_A R_B R_C - R_B R_D - R_B R_E + R_B R_D R_E$$

$$- R_B R_C + R_B R_C R_D + R_B R_C R_E - R_B R_C R_D R_E \tag{E11}$$

Finally the reliability of the system can be computed as

$$R_0 = 1 - P_{fs} = R_A R_C + R_B R_C + R_B R_D + R_B R_E - R_A R_B R_C$$

$$- R_B R_D R_E - R_B R_C R_D - R_B R_C R_E + R_B R_C R_D R_E \tag{E12}$$

which can be seen to be identical to Eq. (E2) of Example 6.8.

6.12 Reliability Enhancement

In general, the problem of "reliability allocation" deals with the determination of reliabilities of individual components such that the overall system reliability is equal to a specified value while satisfying any restrictions present on weight, volume, or cost [6.4]. This problem is usually considered at the design stage. The problem of "reliability enhancement" arises when the reliability of an existing system is found to be inedequate. The reliability of a system can be enhanced from a current value R_0 to a new value $R_0 + dR_0$ by improving the reliability of one or more of its components. In this section, we consider the improvement of system reliability by increasing the reliability of the most sensitive component. The weight, volume, or cost constraints also are to be considered in selecting a suitable component for reliability enhancement. The following procedure can be used for the enhancement of reliability of a system.

6.12.1 Series system

Assuming that the desired system reliability $R_0 + dR_0$ can be achieved by improving the reliability of any one component, we have

$$R_0 + dR_0 = R_1 R_2 \cdots R_{i-1} (R_i + dR_i) R_{i+1} \cdots R_n$$

$$= \prod_{j=1}^{n} R_j + \left(\prod_{j=1, j \neq i}^{n} R_j \right) dR_i; \quad i = 1, 2, \ldots, n \tag{6.64}$$

where R_j is the reliability of jth component and dR_i is the improvement made in the reliability of ith component. Since

$$R_0 = \prod_{j=1}^{n} R_j \tag{6.65}$$

Eq. (6.64) gives

$$dR_i = \frac{dR_0}{\left(\sum_{j=1, j \neq i}^{n} R_j \right)}; \quad i = 1, 2, \ldots, n \tag{6.66}$$

With no constraint. If there are no constraints, we select the component which is most sensitive to system reliability for reliability enhancement. The rate of change of the system reliability R_0 with respect to the reliability of kth component R_k can be obtained, from Eq. (6.65), as

$$\frac{\partial R_0}{\partial R_k} = \prod_{j=1, j \neq k}^{n} R_j = \frac{R_0}{R_k}; \quad k = 1, 2, \ldots, n \tag{6.67}$$

Thus the most sensitive component i which can be used for reliability enhancement is given by

$$\frac{R_0}{R_i} = \max_{k=1,2,\ldots,n} \left(\frac{R_0}{R_k} \right) \tag{6.68}$$

The reliability increment dR_i necessary to achieve the overall system reliability improvement of dR_0 is given by Eq. (6.66).

With constraint. Usually, the improvement in the reliability of a component involves extra weight, volume or cost. Assuming, for example, the cost per unit reliability of ith component as c_i, the problem can be stated as follows:

> Find the improvement in the reliability of ith component dR_i necessary to increase the system reliability by dR_0 so as to minimize the cost, $c_i \, dR_i$.

Assume that the reliability of the system can be improved from R_0 to $R_0 + dR_0$ either by increasing the reliability of component i from R_i to $R_i + dR_i$, or of component j from R_j to $R_j + dR_j$.

Thus, Eq. (6.66) gives, in the absence of any cost constraint,

$$dR_0 = dR_i \left(\sum_{k=1, k \neq i}^{n} R_k \right) = dR_j \left(\sum_{k=1, k \neq j}^{n} R_k \right)$$

that is,

$$dR_i \frac{R_0}{R_i} = dR_j \frac{R_0}{R_j}$$

that is,

$$dR_i = \frac{R_i}{R_j} dR_j \tag{6.69}$$

The cost involved in achieving the new system reliability $R_0 + dR_0$ can be expressed as

$$c_i \, dR_i = \left(\frac{c_i \, R_i}{c_j \, R_j} \right) c_j \, dR_j \tag{6.70}$$

This shows that the cost involved in improving the reliability of ith component will be less than that of jth component if $c_i \, R_i$ is less than $c_j \, R_j$. Thus the component i to be selected for system reliability enhancement, for minimum cost, is given by

$$c_i \, R_i = \min_{j=1,2,\ldots,n} \; c_j \, R_j \tag{6.71}$$

with dR_i given by Eq. (6.66).

6.12.2 Parallel system

The reliability of a parallel system, R_0, can be expressed in terms of the component reliabilities R_i as

$$R_0 = 1 - \prod_{i=1}^{n} (1 - R_i) \tag{6.72}$$

Assuming that the desired system reliability $R_0 + dR_0$ can be achieved by increasing the reliability of any one component, we obtain

$$\begin{aligned} R_0 + dR_0 &= 1 - (1 - R_1)(1 - R_2) \\ &\quad \cdots (1 - R_{i-1})(1 - R_i - dR_i)(1 - R_{i+1}) \; \cdots \; (1 - R_n) \\ &= 1 - \left[\prod_{k=1, k \neq i}^{n} (1 - R_k) \right] (1 - R_i - dR_i) \\ &= 1 - \prod_{k=1}^{n} (1 - R_k) + dR_i \left[\prod_{k=1, k \neq i}^{n} (1 - R_k) \right]; \quad i = 1, 2, \ldots, n \end{aligned} \tag{6.73}$$

In view of Eq. (6.72), Eq. (6.73) gives

$$dR_i = \frac{dR_0}{\left[\prod_{k=1,k\neq i}^{n} (1-R_k)\right]}\;; \quad i = 1, 2, \ldots, n \tag{6.74}$$

With no constraint. If there are no constraints, we select the component which is most sensitive to system reliability for reliability enhancement. The sensitivity of the system reliability to changes in component reliability can be found as

$$\frac{\partial R_0}{\partial R_k} = \prod_{j=1,j\neq k}^{n} (1-R_j) = \frac{1-R_0}{1-R_k}\;; \quad k = 1, 2, \ldots, n \tag{6.75}$$

Thus the most sensitive component i which can be used for reliability enhancement is given by

$$\frac{1-R_0}{1-R_i} = \max_{k=1,2,\ldots,n} \left(\frac{1-R_0}{1-R_k}\right) \tag{6.76}$$

The incremental reliability dR_i necessary to achieve the overall system reliability improvement of dR_0 is given by Eq. (6.74).

With constraint. As before, let the cost per unit reliability of ith component be c_i and the problem be to find the value of dR_i necessary to achieve the incremental system reliability dR_0 so as to minimize the cost, $c_i\, dR_i$. If the reliability of the system can be improved from R_0 to $R_0 + dR_0$ either by increasing the reliability of component i from R_i to $R_i + dR_i$ or of component j from R_j to $R_j + dR_j$, Eq. (6.74) gives

$$dR_0 = dR_i \left[\prod_{k=1,k\neq i}^{n} (1-R_k)\right] = dR_j \left[\prod_{k=1,k\neq j}^{n} (1-R_k)\right]$$

that is,

$$\frac{dR_i}{(1-R_i)} \prod_{k=1}^{n}(1-R_k) = \frac{dR_j}{1-R_j} \prod_{k=1}^{n}(1-R_k)$$

that is,

$$dR_i = \left(\frac{1-R_i}{1-R_j}\right) dR_j \tag{6.77}$$

The cost involved in achieving the new system reliability $R_0 + dR_0$ can be expressed as

$$c_i\, dR_i = \frac{c_i\,(1-R_i)}{c_j\,(1-R_j)}\, c_j\, dR_j \tag{6.78}$$

This shows that the cost involved in improving the reliability of ith component will be less than that of jth component if $c_i\,(1-R_i)$ is less than $c_j\,(1-R_j)$. Thus the component i to be selected for system reliability enhancement, for minimum cost, is given by

$$c_i\,(1-R_i) = \min_{j=1,2,\ldots,n} c_j\,(1-R_j) \tag{6.79}$$

with dR_i given by Eq. (6.74).

6.13 Reliability Allocation—Agree Method

Several procedures can be adopted for allocating reliability to individual components in a complex system. The reliability allocation methods based on an optimization criterion are presented in Chapter 11. A simpler method, suggested by AGREE, is presented in this section. The AGREE (Advisory Group on Reliability of Electronic Equipment) method of allocating the component failure rates was given for aircraft electronic systems [6.1, 6.2]. It is also applicable to other systems that can be decomposed into a series of independent subsystems.

The reliability of the system $R_0(t)$, which consists of k subsystems (in series), is given by

$$R_0(t) = \prod_{i=1}^{k} R_i(t) \tag{6.80}$$

where $R_i(t)$ is the reliability of the ith subsystem. If the ith subsystem has a mean life of m_i (with a constant failure rate $\lambda_i = 1/m_i$) and is required to operate over a time period t_i ($t_i < m_i$), then the reliability of the ith subsystem at time t_i is given by

$$R_i = e^{-t_i/m_i} \tag{6.81}$$

and its probability of failure by $(1-R_i)$. If w_i denotes the probability of failure of the total system due to the failure of the ith subsystem, then the probability of simultaneous survival of the system and the failure of the ith subsystem is given by

$$1 - w_i\,[1 - R_i(t_i)]$$

Thus the reliability of the total system can be expressed as

$$R_0 = \prod_{i=1}^{k} \left\{ 1 - w_i[1 - R_i(t_i)] \right\} \tag{6.82}$$

Equations (6.81) and (6.82) yield

$$R_0 = \prod_{i=1}^{k} \left\{ 1 - w_i \left[1 - e^{-t_i/m_i} \right] \right\} \tag{6.83}$$

Since the expected operating time of a subsystem (t_i) is much shorter than its mean life (m_i), the term in square brackets in Eq. (6.83) can be approximated as

$$1 - e^{-t_i/m_i} \approx 1 - \left[1 - \frac{t_i}{m_i}\right] = \frac{t_i}{m_i}$$

and hence Eq. (6.83) can be replaced by

$$R_0 \approx \prod_{i=1}^{k} \left\{1 - \frac{w_i\, t_i}{m_i}\right\} \approx \prod_{i=1}^{k} e^{-w_i\, t_i/m_i}$$

$$= \exp\left[-\sum_{i=1}^{k} \frac{w_i\, t_i}{m_i}\right] \tag{6.84}$$

where the equivalence

$$e^{-w_i\, t_i/m_i} \approx 1 - \frac{w_i\, t_i}{m_i}$$

has been used in Eq. (6.84). Assuming that the ith subsystem consists of n_i identical components in series, each with a mean life of T_i or failure rate of $1/T_i$, the failure rate of the ith subsystem (λ_i) can be expressed as

$$\lambda_i = \frac{1}{m_i} = \frac{n_i}{T_i}$$

or

$$T_i = m_i\, n_i \tag{6.85}$$

By noting that

$$\frac{w_i\, t_i}{m_i} = \frac{w_i\, t_i\, n_i}{T_i} = n_i \left(\frac{w_i\, t_i}{T_i}\right)$$

$$= \frac{w_i\, t_i}{T_i} + \frac{w_i\, t_i}{T_i} + \cdots \quad \text{(total } n_i \text{ terms in the sum)}$$

$$= \sum_{j=1}^{n_i} \frac{W_i t_i}{T_i}$$

and $\sum_{i=1}^{k} n_i = N =$ total number of components in the system, Eq. (6.84) can be rewritten as

$$R_0 \approx \exp\left[-\sum_{i=1}^{k} \frac{w_i\, t_i}{m_i}\right]$$

$$\approx \exp\left[-\sum_{i=1}^{k}\sum_{j=1}^{n_i} \frac{w_i\, t_i}{T_i}\right] \tag{6.86}$$

If each component contributes equally to the system reliability, we have

$$\frac{w_i\, t_i}{T_i} = c = \text{constant}$$

and Eq. (6.86) becomes

$$R_0 \approx e^{-c\,N} = \left[e^{-c}\right]^N = \left[e^{-w_i\, t_i / T_i}\right]^N$$

or

$$\ln R_0 = N\left(-\frac{w_i\, t_i}{T_i}\right) = -\frac{w_i t_i N}{n_i m_i}$$

or

$$m_i = -\frac{N\, w_i\, t_i}{n_i \ln R_0} \tag{6.87}$$

Example 6.11 A system is composed of five subsystems with details as indicated in the following table:

Subsystem (i)	Number of components (n_i)	Operating period in hours (t_i)	Probability of system failure due to failure of subsystem i (w_i)
1	5	10	0.15
2	2	25	0.10
3	8	5	0.20
4	6	20	0.05
5	4	15	0.25

Determine the mean lives of the components of various subsystems (m_i) so as to have a system reliability of 0.99 using the AGREE method.

solution The total number of components is $N = \sum_{i=1}^{5} n_i = 5 + 2 + 8 + 6 + 4 = 25$. The application of Eq. (6.87) yields

$$m_1 = -\frac{N\, w_1\, t_1}{n_1 \ln R_0} = -\frac{25(0.15)(10)}{5 \ln 0.99} = 746.2687 \text{ hours}$$

$$m_2 = -\frac{N\, w_2\, t_2}{n_2 \ln R_0} = -\frac{25(0.10)(25)}{2 \ln 0.99} = 3{,}109.4527 \text{ hours}$$

$$m_3 = -\frac{N\, w_3\, t_3}{n_3 \ln R_0} = -\frac{25(0.20)(5)}{8 \ln 0.99} = 310.9453 \text{ hours}$$

$$m_4 = -\frac{N w_4 t_4}{n_4 \ln R_0} = -\frac{25(0.05)(20)}{6 \ln 0.99} = 414.5937 \text{ hours}$$

$$m_5 = -\frac{N w_5 t_5}{n_5 \ln R_0} = -\frac{25(0.25)(15)}{4 \ln 0.99} = 2{,}332.0896 \text{ hours}$$

References and Bibliography

6.1. D. L. Grosh, *A Primer of Reliability Theory,* John Wiley, New York, 1989.
6.2. K. C. Kapur and L. R. Lamberson, *Reliability in Engineering Design,* John Wiley, New York, 1977.
6.3. E. E. Lewis, *Introduction to Reliability Engineering,* John Wiley, New York, 1987.
6.4. J. G. Rau, *Optimization and Probability in Systems Engineering,* Van Nostrand Reinhold, New York, 1970.
6.5. L. A. Doty, *Reliability for the Technologies,* 2nd ed., Industrial Press, New York, 1989.
6.6. R. Billinton and R. N. Allan, *Reliability Evaluation of Engineering Systems: Concepts and Techniques,* Plenum Press, New York, 1983.
6.7. A. C. King and C. B. Read, *Pathways to Probability,* Holt, Rinehart & Winston, Inc., New York, 1963.
6.8. I. Asimov, *Asimov's Biographical Encyclopedia of Science and Technology,* Doubleday & Co. Inc., Garden City, New York, 1972.

Review Questions

6.1 What is a bathtub curve?

6.2 What is infant mortality period?

6.3 What are the causes for failure of components during the wear-out period?

6.4 What are the causes of failure during the operating period of components?

6.5 Define the hazard function.

6.6 How is the reliability function related to the failure rate?

6.7 Which distribution function is commonly used to model failure time during the operating period of a component?

6.8 Suggest a method for estimating the failure rate of a component from experimental data.

6.9 Define MTBF.

6.10 What is the relationship between MTBF and the probability density function of failure time of a component?

6.11 Express the reliability of a series system in terms of its component reliabilities.

6.12 How is the system-failure rate related to the component-failure rates for a series system?

6.13 Find the reliability of a parallel system in terms of its component reliabilities.

6.14 What is a (k,n) system?

6.15 Give two practical examples each for a series, a parallel, and a (k,n) system.

6.16 What are series–parallel and parallel–series systems?

6.17 What is meant by a complex system? List three methods for finding the reliability of a complex system.

6.18 Define a cut-set. What is the difference between a cut-set and a minimal cut-set?

6.19 What is the AGREE method?

6.20 State some methods of enhancing the reliability of a multicomponent system.

Problems

6.1 Find the reliability of the system shown in Fig. 6.12 when $R_A = 0.95$, $R_B = 0.98$, $R_C = 0.75$, $R_D = 0.85$ and $R_E = 0.90$.

6.2 The reliability function of a system is given by $R(t) = \exp(-t^2)$. Find the corresponding
(a) failure-rate function
(b) probability density function of the failure time.

6.3 Find the reliability of an engine for an operating time of 500 hours if the failure rate is 4 per 10^6 hours.

6.4 If the failure rate of a component is 1 in 10^4 hours, determine the following:
1. reliability function of the component;

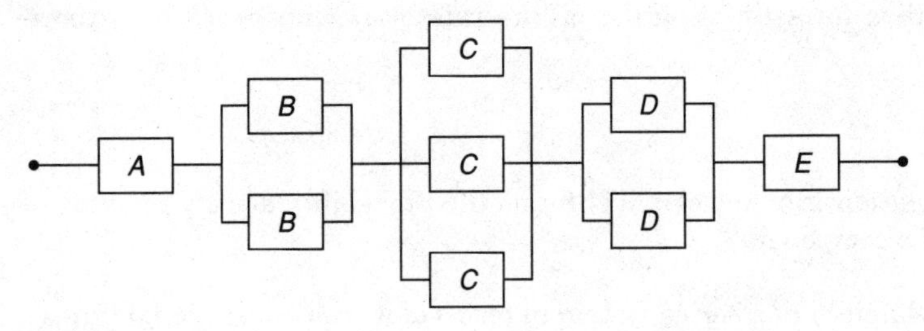

Figure 6.12

2. reliability function of a system with four components in series; and
3. reliability function of a system with four components in parallel.

6.5 Find the MTBF of a component which has a reliability of 0.99 for an operating time of 1000 hours.

6.6 Find the reliability of the system shown in Fig. 6.13 using the enumeration method.

6.7 Find the reliability of the system shown in Fig. 6.13 using the conditional probability approach.

6.8 Find the reliability of the system shown in Fig. 6.13 using the cut-set method.

6.9 Two types of components, one with a failure rate of 10^{-4} per hour (type A) and the other with a failure rate of 10^{-5} per hour (type B) are arranged to form parallel systems. Five components of type A are used to form a parallel system X. Find the minimum number of components of type B needed to form a parallel system Y, which matches with the system X in terms of reliability.

6.10 The data on the lives of 1000 identical components is given in the following table:

Time interval (100 hours)	Number of failures in the interval
1	223
2	68
3	57
4	53
5	50
6	46
7	44
8	46
9	77
10	142

Plot the failure-rate time curve.

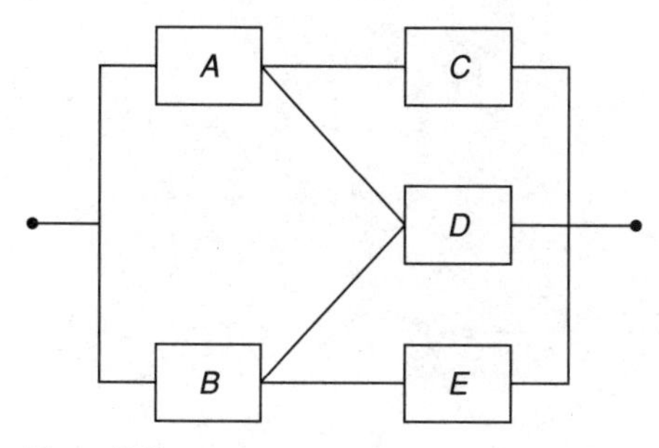

Figure 6.13

6.11 The probability density function of failure time of a component is shown in Fig. 6.14. Find the corresponding hazard- and survivorship-functions.

6.12 The probability density function of failure time of a component is shown in Fig. 6.15. Determine the corresponding failure-rate and reliability functions.

6.13 Ten identical components are connected in parallel to achieve a system reliability of 0.95. Determine the reliability of each component.

6.14 Ten identical components are connected in parallel to achieve a system reliability of 0.9. Determine the additional number of components to be added in parallel to increase the reliability to 0.95.

6.15 Six identical components are connected in two different ways as shown in Figs. 6.16*a* and *b*. Identify the configuration which yields a higher value of reliability.

6.16 A simplified model of a thermal power plant is shown in Fig. 6.17 where the failure rate of each component is also indicated in parentheses. Determine the reliability of the power plant after (a) 1000 hours, and (b) 2000 hours.

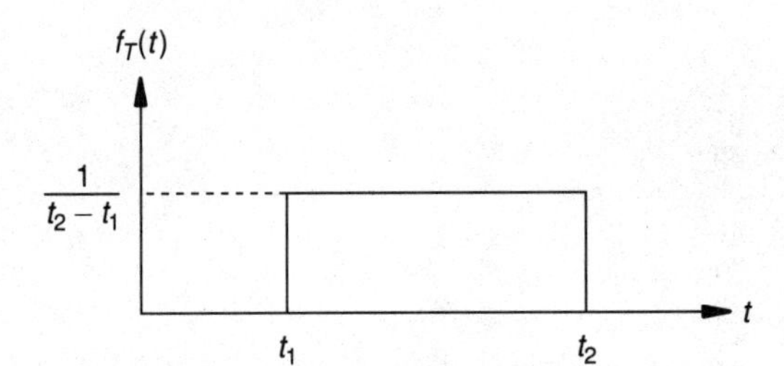

Figure 6.14

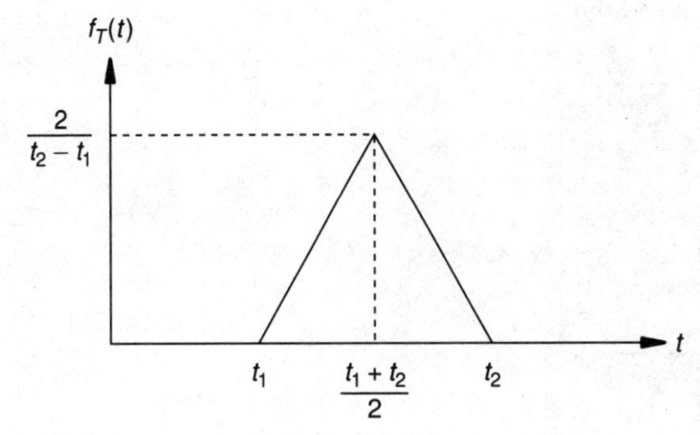

Figure 6.15

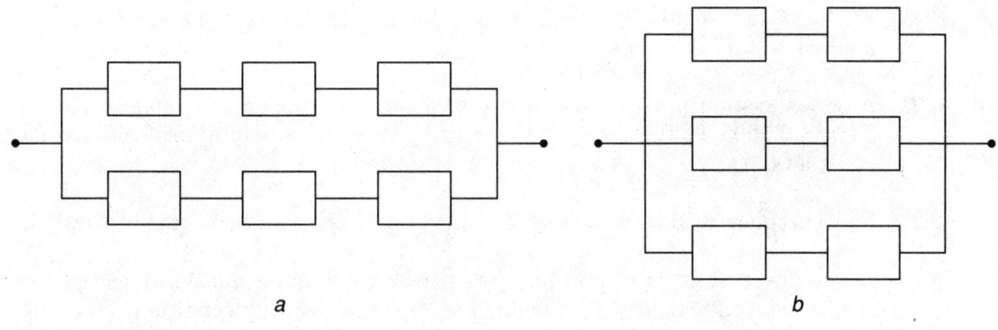

Figure 6.16

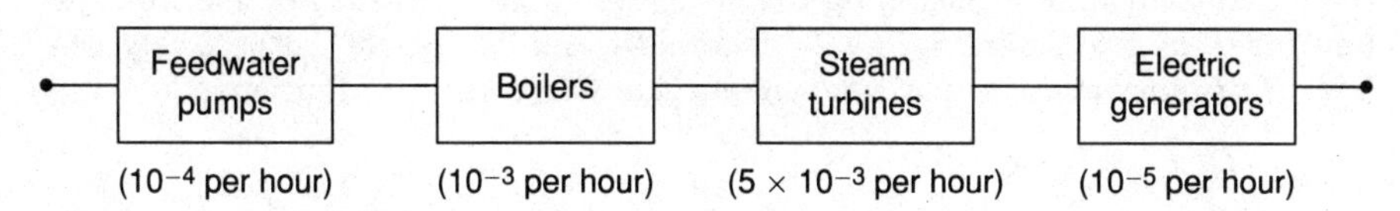

Figure 6.17

6.17 A series system is composed of n identical components, each having a failure rate of 10^{-3} per hour. Find the reliability of the system at time t. Also, plot the reliability function $R(t)$ for $n = 1, 2, 5$ and 10.

6.18 A parallel system is composed of n identical components, each having a failure rate of 10^{-3} per hour. Determine the reliability of the system at time t. Also, plot the reliability function, $R(t)$, for $n = 1, 2, 5$ and 10.

6.19 Find the reliability of the system shown in Fig. 6.18 using the enumeration method.

6.20 Determine the reliability of the system shown in Fig. 6.18 using the conditional probability approach.

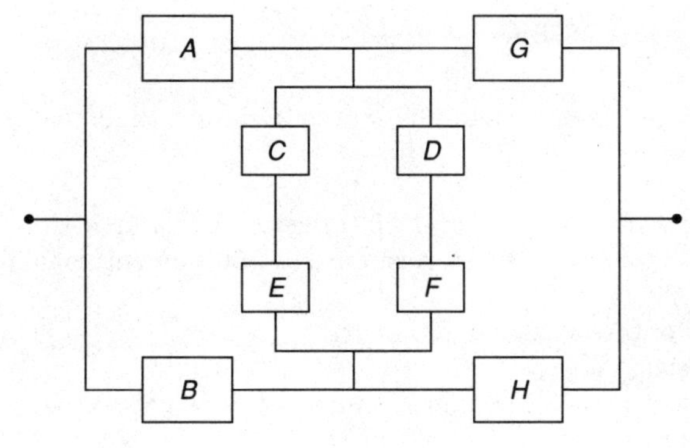

Figure 6.18

6.21 Evaluate the reliability of the system shown in Fig. 6.18 using the cut-set method.

6.22 A series system is composed of five components with failure rates of 0.0005, 0.0001, 0.0008, 0.004, and 0.0002 per hour. Find the reliability of the system at 200 hours.

6.23 Solve problem 6.22 by assuming that the components are connected in parallel.

6.24 A standby system is a parallel system where all the components except one (known as the primary component) are kept in a waiting condition. When the primary component fails, the standby component is switched into use in place of the failed component. Assuming the switch to be 100 percent reliable, the system life T is given by $T = T_1 + T_2$ where T_1 and T_2 denote, respectively, the lives of primary and standby components that follow the distributions:

$$f_{T_1}(t_1) = \lambda_1 \exp(-\lambda_1 t_1) \qquad \text{and} \qquad f_{T_2}(t_2) = \lambda_2 \exp(-\lambda_2 t_2)$$

Derive the following functions for the life of the standby system (T):

1. probability density function;
2. reliability function;
3. probability distribution function; and
4. hazard function.

6.25 The reliability function of a standby system with three components can be expressed as

$$R(t) = \frac{\lambda_2\,\lambda_3 \exp(-\lambda_1\,t)}{(\lambda_2 - \lambda_1)\,(\lambda_3 - \lambda_1)} + \frac{\lambda_1\,\lambda_3 \exp(-\lambda_2\,t)}{(\lambda_1 - \lambda_2)\,(\lambda_3 - \lambda_2)} + \frac{\lambda_1\,\lambda_2 \exp(-\lambda_3\,t)}{(\lambda_1 - \lambda_3)\,(\lambda_2 - \lambda_3)}$$

where the failure time density function of the components is given by

$$f_{T_i}(t_i) = \lambda_i \exp(-\lambda_i t_i); \quad i = 1, 2, 3$$

Show that the MTBF of the system is given by

$$\text{MTBF} = \frac{1}{\lambda_1} + \frac{1}{\lambda_2} + \frac{1}{\lambda_3}$$

Also, show that the MTBF of a standby system is larger than that of a simple parallel system.

6.26 A standby system consists of a total of three components with $\lambda_1 = 0.0005$, $\lambda_2 = 0.001$, and $\lambda_3 = 0.005$ per hour. Assuming a 100 percent reliability of the switch, determine the following:

1. the reliability of the system at 100 hours
2. MTBF of the system
3. MTBF of the system if the same components are connected as a simple parallel system.

6.27 A complex system consists of six subsystems with details as indicated in the following table:

Subsystem (i)	Number of components (n_i)	Operating time (t_i, hours)	Probability of system failure due to failure of subsystem i (w_i)
1	4	10	0.10
2	12	5	0.05
3	8	20	0.25
4	20	15	0.20
5	16	25	0.30
6	6	30	0.15

Find the required failure rates of the components of various subsystems in order to have an overall reliability of 0.995 using the AGREE method.

6.28 A series system consists of five components with failure rates of 0.0001, 0.0003, 0.0002, 0.0005, and 0.0008 per hour. Find the reliability of the system at 100 hours. If it is desired to increase the 100-hour reliability of the system by 5 percent, select a single component and its new failure rate to accomplish the task under the following conditions:

1. When there are no cost constraints; and
2. When a unit reliability improvement costs \$500, \$400, \$600, \$1000, and \$800 for components 1, 2, 3, 4, and 5, respectively.

6.29 A parallel system is composed of five components with failure rates of 0.0001, 0.0003, 0.0002, 0.0005, and 0.0008 per hour. Find the reliability of the system at 100 hours. If it is desired to increase the 100-hour reliability of the system by 5 percent, select a single component and its new failure rate to accomplish the task under the following conditions:

1. When there are no cost constraints; and
2. When a unit reliability improvement costs \$500, \$400, \$600, \$1000, and \$800 for components 1, 2, 3, 4, and 5, respectively.

6.30 The probability density function of failure time of a component is given by Rayleigh distribution

$$f_T(t) = a\, t \exp\left[-\frac{a\, t^2}{2}\right]; \qquad t \geq 0$$

Show that the failure rate and reliability functions of the component are given by

$$h(t) = a\, t \qquad \text{and} \qquad R(t) = \exp\left[-\frac{a\, t^2}{2}\right]$$

6.31 A system is composed of five identical components each having a failure rate of 0.0001 per hour. Find the reliability of the system at 100 hours and 200 hours if at least two components must operate for the survival of the system.

Chapter

7

Modeling of Geometry, Material Strength and Loads

Biographical Note

Ernst Abbe

Ernst Abbe was a German physicist who was born on January 23, 1840, and died on January 14, 1905. He was the son of a master spinner who worked on his feet 16 hours everyday with no break for meals. Abbe won scholarships and was also helped by his father's employer in getting education. He received his doctorate at the University of Jena at the age of 21 for a dissertation on thermodynamics. He became a professor at the University of Jena in 1870 and the director of astronomical and meteorological observatories in 1878. From 1866 onwards Abbe devoted his time to theoretical and practical research in optics and astronomy. He was closely associated with the Carl Zeiss' company. He was an outstanding scientific genius combined with practical inventiveness. The paper he wrote in 1863, "On the law of distribution of errors in observation series," formed a basis for his discovery of the χ^2*-distribution, which can be used to represent a random variable that is the sum of squares of several independent normally distributed random variables. The* χ^2*-distribution also plays an important role in hypothesis testing (i.e., goodness-of-fit tests) [7.51].*

7.1 Introduction

The analysis and design of any mechanical or structural component is influenced by the geometry or dimensions of the component, the strength of the material used, and the loads acting on the component. Although the nominal or design dimensions are known precisely, the actual dimensions will be random in nature. When dimensions are produced on a machine tool, the limitations of the machine and the operator lead to variations in the dimensions in the form of tolerances. Similarly, when structural members are rolled in a mill, the shaping rollers will gradually wear, resulting in variations in the cross-sectional dimensions of the rolled member. Since it is not possible to change the rollers frequently, some variations in the dimensions of the rolled members are unavoidable. The errors in fabrication and assembly also cause variations in the overall dimensions of multicomponent mechanical and structural systems.

Variations occur in the mechanical properties of materials due to uncertainties in the chemistry and geometric shapes, as well as the manufacturing and fabrication procedures used. The mechanical strength is also influenced by the loading and strain rates applied to the component. The load acting on a machine or structure is considered to be static if the magnitude of the load is constant, or dynamic if the load is variable, with respect to time. The static loads include dead loads due to self weights of components and live loads acting on the structure. These loads are subject to variability due to fluctuations in the specific weights of materials as well as changes in the working loads acting on the structure. The dynamic loads include wind, snow, wave, and earthquake induced loads. These loads are to be treated as random processes for the reliability analysis of structures.

This chapter deals with the probabilistic description of geometric dimensions of components, strengths of materials, and loads acting on structures.

7.2 Modeling of Geometry

In mechanical design, the geometric parameters such as shaft diameters, lengths and thicknesses of parts, hole diameters, profiles of gear teeth and cams, and center distances between shafts are of prime importance. Similarly, in structural design the cross-sectional dimensions of beams and lengths, widths, and thicknesses of plates and shells play an important role. Although nominal values of the geometric variables are used in design, physical realization of these values during manufacturing is based on the specification of suitable tolerances on these values. This is due to the fact that all the machining processes such as turning, drilling, and milling and the production processes such as casting, forging, and rolling are subject to change over time. By collecting and statistically analyzing the data from the manufactured components, suitable probability density function can be determined for a particular class of dimensions like length of parts, angular measurements, center distances, diameters, and so on.

In the absence of directly measured data, approximate distributions, based on previous related studies, can be used for the dimension of interest. It has been observed that many manufacturing operations generate normally distributed dimensions if they are controlled. The tolerance limits on dimensions of components are set so as to conform to the capabilities of the manufacturing process except for a small fraction α of the items that will have to be rejected. If a dimension is approximately normally distributed, a good design will have the sample mean approximately equal to the nominal value and the sample standard deviation such that only a small allowable fraction α of the parts manufactured fall outside the tolerance specification limits. Thus the tolerance limits of a dimension, on the average, should permit acceptance of at least $(1 - \alpha)$ of the parts manufactured.

7.2.1 Tolerances on finished metal products

The tolerances commonly obtainable on finished metal products are documented in the literature. The best tolerances that can be achieved in several common manufacturing processes are given in Table 7.1. The rolling and cutting tolerances on I-beams are indicated in Table 7.2. The histograms of geometrical cross-sectional properties of I-beams are shown in Fig. 7.1.

TABLE 7.1 Tolerances Obtainable in Common Manufacturing Processes

Manufacturing process	Tolerance obtainable (inch)
Turning, Shaping, Milling	±0.001
Drilling	+0.002 −0.000
Grinding	±0.0002
Lapping	±0.00005
Sawing	±0.005
Flame cutting	±0.02
Stamping	±0.001
Forging	±0.03 in/in
Cold rolling	±0.001
Drawing	±0.002
Hot extrusion	±0.005
Sand casting	$\pm\frac{1}{32}$ in/ft
Die casting	±0.002 in/in
Spot welding, Fusion welding	±0.01
Heat treatment	±0.01

SOURCE: Joseph Datsko, *Material Properties and Manufacturing Processes*, John Wiley, New York, 1966.

TABLE 7.2 Tolerances on I-Beams

Rolling Tolerances					
Nominal size of section	Depth (A)	Width (B)	Out of square of flanges ($T_1 + T_2$)	Off center of web (E)	Total depth (C)
≤ 12 in	$\pm\frac{1}{8}$ in	$+\frac{1}{4}$ in	$-\frac{3}{16}$ in	$\frac{1}{4}$ in	$\frac{3}{16}$ in
> 12 in	$\pm\frac{1}{8}$ in	$+\frac{1}{4}$ in	$-\frac{3}{16}$ in	$\frac{5}{16}$ in	$\frac{3}{16}$ in

Cutting Tolerances on Length

Beams with a nominal	$\pm\frac{3}{8}$ in · · ·	when length is ≤ 30 ft
depth of ≤ 24 in	$+\frac{3}{8}$ in $+\frac{1}{16}$ in for each additional 5 ft or less · · ·	when length is > 30 ft
	$-\frac{3}{8}$ in	
Beams with a nominal	$\pm\frac{1}{2}$ in · · ·	when length is ≤ 30 ft
depth of > 24 in	$+\frac{1}{2}$ in $+\frac{1}{16}$ in for each additional 5 ft or less · · ·	when length is > 30 ft
	$-\frac{1}{2}$ in	

Other Tolerances

Camber and sweep for most sizes: $\frac{1}{8}$ in $\times \left(\frac{\text{total length in ft}}{10}\right)$

Area and weight ± 2.5 %

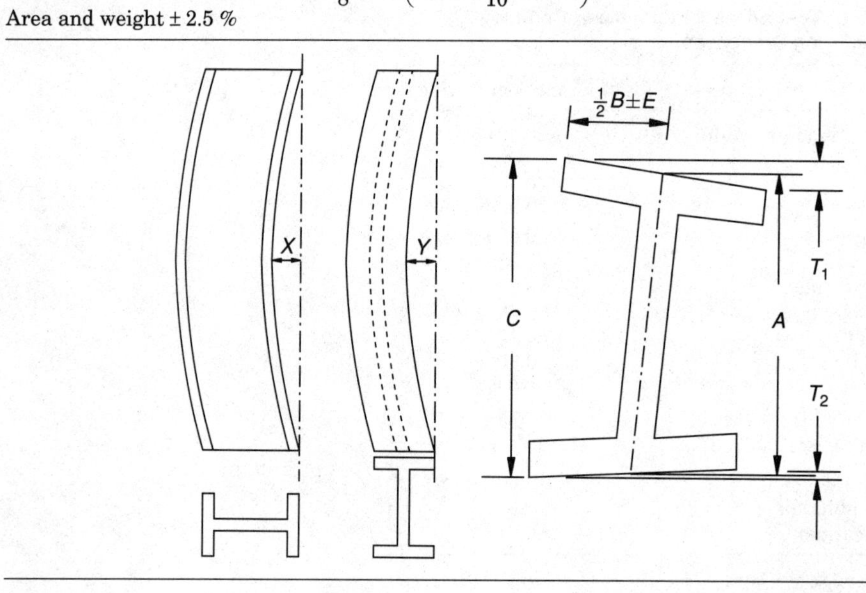

SOURCE: AISC *Manual of Steel Construction*, 9th ed., American Institute of Steel Construction, Chicago, IL, 1989.

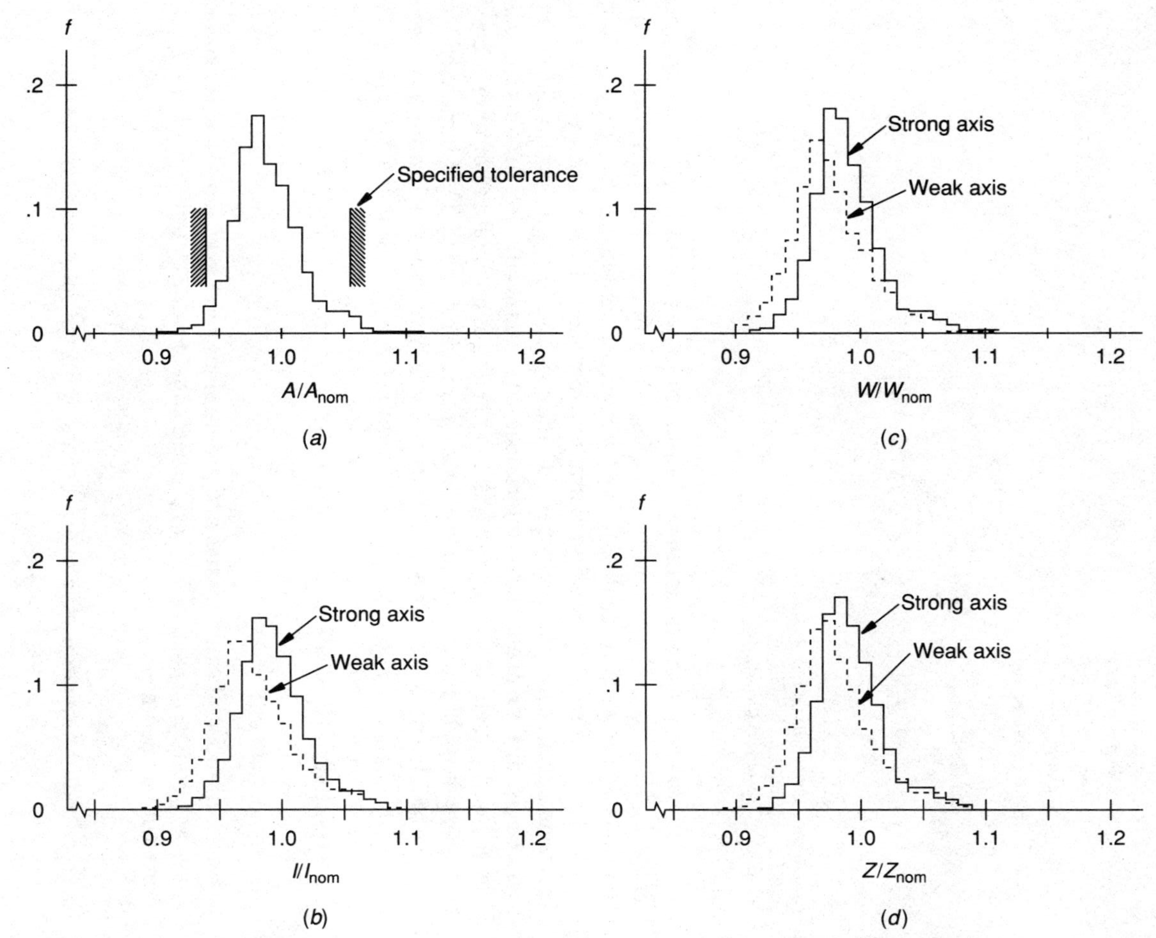

Figure 7.1 *Source: G. A. Alpsten, "Variations in Mechanical and Cross-Sectional Properties of Steel,"* Proceedings of the International Conference on Planning and Design of Tall Buildings, vol. 1*b*, *Lehigh University, Bethlehem, PA, August 1972, ASCE, New York.*

TABLE 7.3 Tolerances on Diameters of Aluminum Wire and Rod

Diameter of wire or rod (inch)	Tolerance (inch)
Drawn wire	
≤ 0.035	±0.0005
0.036–0.064	±0.0010
Cold finished rod:	
0.375–0.500	±0.0015
0.501–1.000	±0.0020
2.001–3.000	±0.0060
Rolled rod:	
1.501–2.000	±0.0060
2.001–3.000	±0.0080

SOURCE: *1990 Aluminum Standards and Data,* Table 10.5, Aluminum Association, Inc., Washington, D.C., 1990.

The tolerances on the diameters of aluminum wire and rod are indicated in Table 7.3. Additional data on tolerances can be found in [7.2, 7.4–7.7]. Most tolerances can be assumed to follow normal distribution for computational convenience. From the design point of view, zero tolerances are desirable. However, from the point of view of manufacturing and production, large tolerances are desirable. As tolerances are reduced, the production costs increase. As such, the designer has to specify an optimal set of tolerances by compromising between accuracy and production cost. The increased cost associated with decreasing tolerances for a general machining process is shown in Fig. 7.2. The tolerances obtainable in different machining processes and the corresponding relative costs are indicated in Fig. 7.3. Additional data related to the costs of tolerances can be found in [7.9–7.11].

7.2.2 Assembly of components

When a number of components are assembled, the variations in their dimensions or geometric parameters are to be considered. Usually, a linear sum of dimensions such as

$$Z = X_1 + X_2 + \cdots + X_N \tag{7.1}$$

is required to determine the overall sizes, a linear combination like

$$Z = X_1 \pm X_2 \pm \cdots \pm X_N \tag{7.2}$$

is required to find the locations, and a difference of dimensions such as

$$Z = X_1 - X_2 \tag{7.3}$$

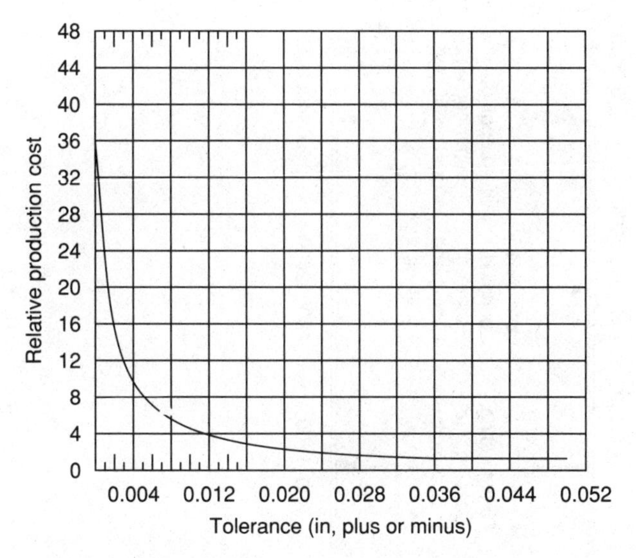

Figure 7.2 *Source: R. E. Bolz, "Design Considerations for Manufacturing Economy,"* Mechanical Engineering, *ASME, vol. 71, fig. 9, December 1949.*

is needed to find the clearance between two parts. If the individual dimensions $X_1, X_2, \ldots, X_N$ are treated as normally distributed independent random variables, the distribution of the dimension Z will also be normal. The mean and standard deviations of Z can be computed once the means and standard deviations of X_i, $i = 1, 2, \ldots, N$ are known. Even if the individual random variables X_i follow different distributions, Z can be considered to follow normal distribution according to the central-limit theorem.

If the overall size of an assembly of N components is given by Eq. (7.1) with the mean value and standard deviation of the dimension of the ith component (X_i) given by $\overline{X}_i$ and σ_{X_i}, respectively, the mean and standard deviations of Z can be determined as

$$\overline{Z} = \overline{X}_1 + \overline{X}_2 + \cdots + \overline{X}_N \tag{7.4}$$

$$\sigma_Z = (\sigma_{X_1}^2 + \sigma_{X_2}^2 + \cdots + \sigma_{X_N}^2)^{1/2} \tag{7.5}$$

If the dimension X_i is specified as $\overline{X}_i \pm t_i$ with $\pm t_i$ representing the 3-sigma tolerance band of X_i, the standard deviation of X_i will be $t_i / 3$. In this case, the standard deviation of Z can be expressed as

$$\sigma_Z = \frac{1}{3}(t_1^2 + t_2^2 + \ldots + t_N^2)^{1/2} \tag{7.6}$$

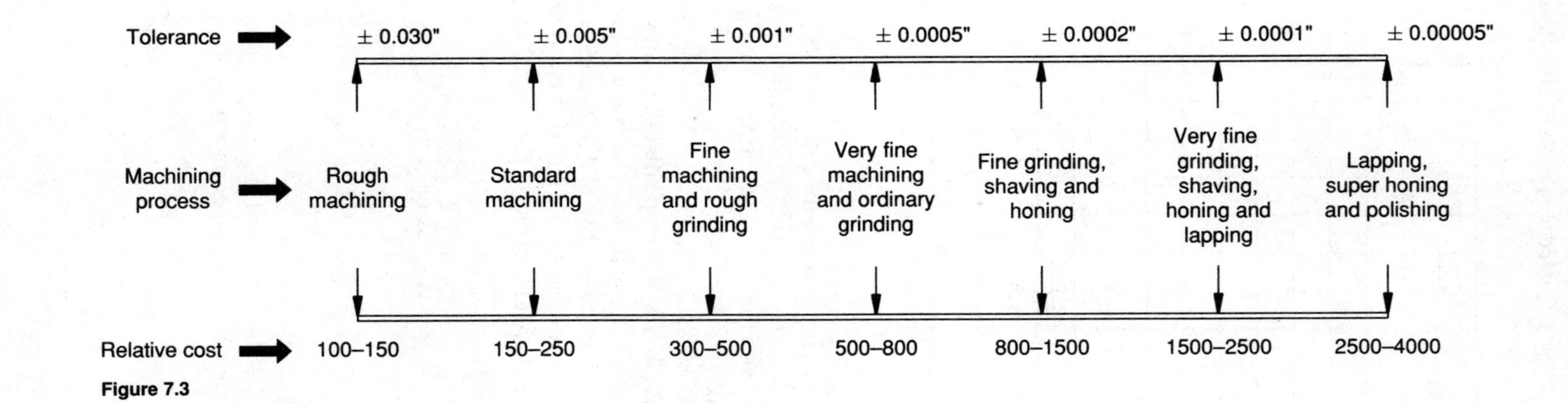
Tolerance
± 0.030"
± 0.005"
± 0.001"
± 0.0005"
± 0.0002"
± 0.0001"
± 0.00005"
Machining process
Rough machining
Standard machining
Fine machining and rough grinding
Very fine machining and ordinary grinding
Fine grinding, shaving and honing
Very fine grinding, shaving, honing and lapping
Lapping, super honing and polishing
Relative cost
100–150
150–250
300–500
500–800
800–1500
1500–2500
2500–4000

Figure 7.3

Example 7.1 Find the statistics of the length of a chain consisting of four links when the length of each link is given by 12 ± 0.2 in.

solution The length of the chain can be expressed as

$$l = l_1 + l_2 + l_3 + l_4$$

with $l_i = 12 \pm 0.2$ in ($i = 1, 2, 3, 4$). The mean and standard deviations of l are given by

$$\bar{l} = \sum_{i=1}^{4} \bar{l}_i = 4\,\bar{l}_i = 4\,(12) = 48 \text{ in}$$

$$\sigma_{\bar{l}}^2 = \sum_{i=1}^{4} \sigma_{\bar{l}_i}^2 = \frac{1}{9} \sum_{i=1}^{4} t_i^2 = \frac{4}{9}\, t_i^2 = \frac{4}{9}\,(0.2)^2 = \frac{0.16}{9}$$

$$\therefore\ \sigma_l = \frac{0.4}{3} = 0.1333 \text{ in}$$

Thus the length of the chain is given by

$$l = \bar{l} \pm 3\,\sigma_l = 48 \pm 0.4 \text{ in}$$

Example 7.2 A mechanical assembly involves inserting the pins of part 1 in the holes A and B of part 2 (see Fig. 7.4). The relevant nominal dimensions and the associated tolerances are also indicated in Fig. 7.4. What is the probability of realizing the distance between the pins

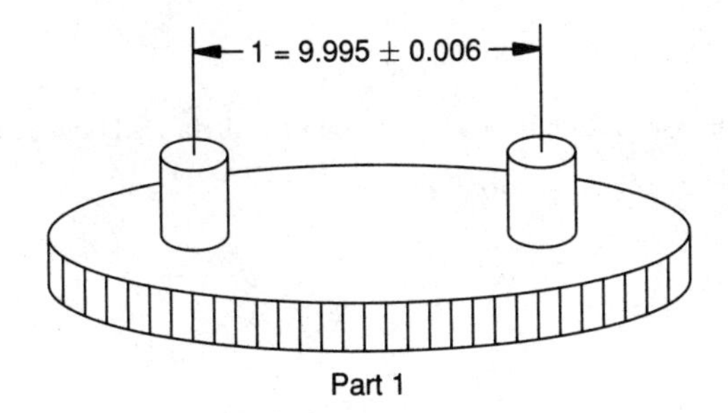

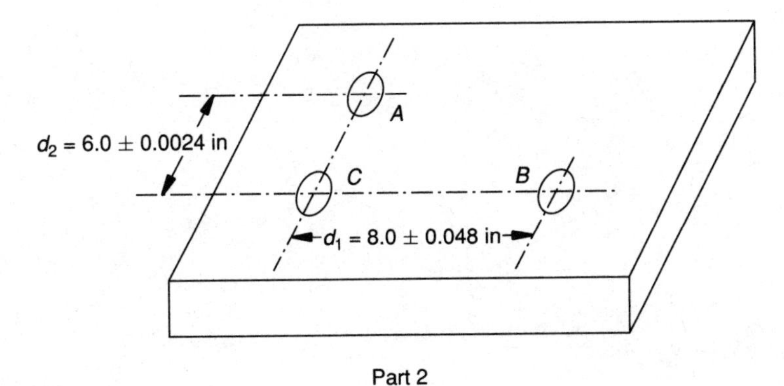

Figure 7.4

larger than the distance between the holes A and B in an assembly? Assume that the dimensions follow normal distribution.

solution The distance between the holes A and B can be determined from the relation (see Fig. 7.5)

$$d = \sqrt{d_1^2 + d_2^2} \tag{E1}$$

where $d_1 = (8.0 \pm 0.0048)$ in and $d_2 = (6.0 \pm 0.0024)$ in. The mean value of d can be found as

$$\bar{d} = \sqrt{\bar{d}_1^2 + \bar{d}_2^2} \tag{E2}$$

and the standard deviation of d can be determined, using the partial derivative rule, as

$$\sigma_d = \left[\frac{\bar{d}_1^2\, \sigma_{d_1}^2 + \bar{d}_2^2\, \sigma_{d_2}^2}{\bar{d}_1^2 + \bar{d}_2^2} \right]^{1/2} \tag{E3}$$

Since the dimensions follow normal distribution, the tolerances can be assumed to correspond to three-sigma values and hence the standard deviations of d_1 and d_2 can be found as

$$\sigma_{d_1} = \frac{0.0048}{3} = 0.0016; \qquad \sigma_{d_2} = \frac{0.0024}{3} = 0.0008 \tag{E4}$$

Equations (E2) and (E3) yield

$$\bar{d} = \sqrt{8.0^2 + 6.0^2} = 10.0 \text{ in} \tag{E5}$$

$$\sigma_d = \left[\frac{(8.0)^2\,(0.0016)^2 + (6.0)^2\,(0.0008)^2}{(8.0)^2 + (6.0)^2} \right]^{1/2}$$

$$= (0.000001869)^{1/2} = 0.001367 \text{ in} \tag{E6}$$

The probability of the distance between the pins (l) larger than the distance between the holes A and B (d) is given by

$$P[l \geq d] = P[l - d \geq 0] \tag{E7}$$

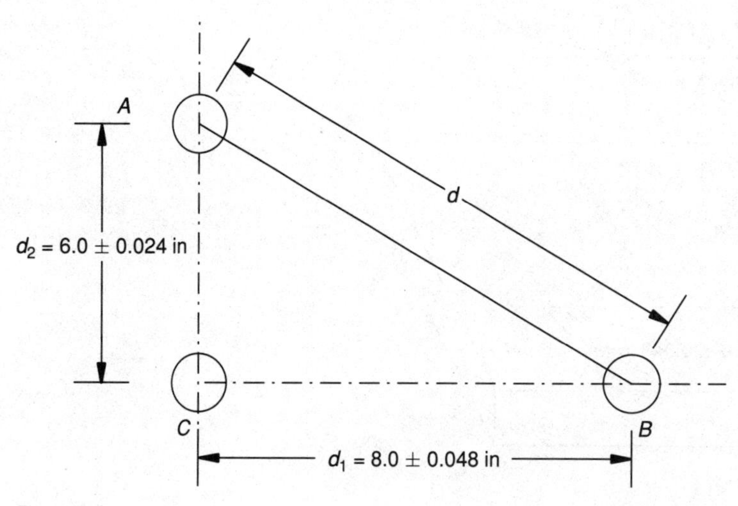

Figure 7.5

Defining $X = l - d$, Eq. (E7) can be rewritten, in terms of the standard normal variate $z = (X - \bar{X}) / \sigma_X$, as

$$P[l \geq d] = P[X \geq 0] = P\left[z = \frac{X - \bar{X}}{\sigma_X} \geq -\frac{\bar{X}}{\sigma_X}\right] \tag{E8}$$

where

$$-\frac{\bar{X}}{\sigma_X} = -\frac{(\bar{l} - \bar{d})}{\sqrt{\sigma_l^2 + \sigma_d^2}} \tag{E9}$$

Since $\bar{l} = 9.995$ in and $\sigma_l = 0.0060 / 3 = 0.0020$ in, Eq. (E9) gives

$$-\frac{\bar{X}}{\sigma_X} = -\frac{(9.995 - 10.000)}{\sqrt{(0.002)^2 + (0.001367)^2}} = \frac{0.005}{0.002422} = 2.0639 \tag{E10}$$

and hence Eq. (E9) gives the desired probability as

$$P\,[l \geq d] = P\,[z \geq 2.0639] = 0.0197 = 1.97\% \tag{E11}$$

7.3 Modeling of Material Strength

7.3.1 Statistics of elastic properties

Most metals and other engineering materials exhibit variable properties due to random differences in chemistry, heat treatment, and mechanical processing. Static strength is the strength displayed by a material subjected to static loading. The nominal elastic properties of structural steel commonly used in standards, codes, and specifications are:

Young's modulus in tension and compression (E): 29×10^6 lb/in^2

Poisson's ratio (ν): 0.03, and shear modulus (G): $1/\{2[E/(1+\nu)]\} = 0.385\,E$.

The statistical characteristics of these properties observed in experiments are indicated in Table 7.4. The yield stress denotes the principal property which influences structural and mechanical designs. The designer selects the grade of steel used for the component to be designed, which gives the corresponding yield stress. The yield stress of steel depends on the chemical composition of the alloy and the method of manufacture [7.12]. The grade of steel reflects these variables. The data on ASTM A7 and A36 steels, for rolled structural shapes, collected over a time span of 40 years prior to 1957, from several United States steel mills are given in Table 7.5. Similar data, collected on steel manufactured by British mills, are given in Table 7.6. The histograms of ultimate and yield strengths of U.S. steel are shown in Figs. 7.6 and 7.7. The statistics of yield and ultimate strengths of some other materials used in mechanical and structural design are given in Table 7.7.

TABLE 7.4 Elastic Properties of Structural Steel

Elastic property	Mean value	Coefficient of variation	Number of tests conducted
E (tension)	29,540 k(lb/in^2)	0.01	67
E (compression)	29,550 k(lb/in^2)	0.01	67
ν (tension)	0.296	0.026	57
ν (compression)	0.298	0.021	48
G (torsion)	12,000 k(lb/in^2)	0.042	5

SOURCE: T. V. Galambos, and M. K. Ravindra, "Properties of Steel for Use in LRFD," *Journal of the Structural Divsion, Proceedings of the ASCE,* vol. 104, no. ST 9, Sept. 1978.

TABLE 7.5 Yield Stress of U.S. Steel (s_y) [7.13, 7.14]

Reference	Mean mill value, s_y (10^3 (lb/in^2))	Specified value, s_y (10^3 (lb/in^2))	Mean mill s_y/ Specified s_y	Coefficient of variation	Number of samples
Julian [7.13]	40.0	33.0	1.21	0.09	3,794
Tall and Alpsten [7.14]	39.4	33.0	1.21	0.08	3,124

SOURCE: O. C. Julian, "Synopsis of First Progress Report of Committee on Factors of Safety," *Journal of the Structural Division, Proceedings of the ASCE,* paper no. 1316, vol. 83, no. ST 4, July 1957.

7.3.2 Statistical models for material strength

Three statistical models can be used for describing the strength of different materials: the *brittle-material model*, the *plastic model* and the *fiber-bundle model* [7.15, 7.16]. An ideal brittle material is one that fails whenever a single particle fails. Thus the strength of the material is governed by the smallest strength of a particle. In an ideal plastic material, yielding starts when the load on a particle reaches its yield capacity. The particle continues to carry the yield load and any additional load is transferred to other particles of the cross section. Thus the cross section carries its maximum load when all the particles yield. In the fiber-bundle model, the failure of a particle implies that it is no longer capable of carrying the load. Hence the load will be redistributed to other particles. This process continues until all the particles of the body fail.

7.3.3 Model for brittle materials

In this model, the material is considered as a chain composed of several (n) elements in series [7.15, 7.16]. The strengths of the individual elements of a series chain are assumed statistically to be independent and follow a common

TABLE 7.6 Yield Stress of British Steel

Type of steel	Plate thickness (inch)	Mill	Coefficient of variation	Mean mill s_y/ Specified s_y*
Structural carbon steel plates	0.375–0.5	Y	0.09	1.15
	0.375–0.5	W	0.05	1.14
	1.5–2	Y	0.12	1.03
	1.5–2	W	0.05	1.07
High-strength steel plates	0.375–0.5	M	0.08	1.11
	0.375–0.5	K	0.04	1.11
	1.5–2	M	0.06	1.06
	1.5–2	L	0.05	1.15
Structural carbon steel webs of shapes	0.375–0.5	Q	0.05	1.20
	0.625–0.75	L	0.12	1.19
High-strength steel webs of shapes	0.25–0.375	N	0.06	1.19
	1.5–2	L	0.05	1.06
Structural carbon steel tubes	0.144	—	0.05	1.27
	0.250	—	0.08	1.32
High strength steel tubes	0.232	—	0.05	1.18
	0.250	—	0.08	1.15

*Nominal $s_y = 36$ k(lb/in²) for structural steel; 50 k(lb/in²) for high strength steel.

SOURCE: T. V. Galambos, and M. K. Ravindra, "Properties of Steel for Use in LRFD," *Journal of the Structural Division, Proceedings of the ASCE*, vol. 104, no. ST 9, Sept. 1978.

distribution function $F(x)$. The distribution function of the smallest strength in n elements $F_n(x)$ is given by (see Eq. 4.2):

$$F_n(x) = 1 - [1 - F(x)]^n \tag{7.7}$$

which can be rewritten as

$$F_n(x) = 1 - \exp[n \ln\{1 - F(x)\}] \tag{7.8}$$

As n tends to infinity, the distribution function of Eq. (7.8) tends to one of the three asymptotic distributions discussed in Chapter 4. Since the strength is limited by a lower bound, the type III extremal distribution for the smallest value will be applicable.

Now consider a one-dimensional structure (bar) of length l made of an ideal brittle material subjected to an axial load. Assume that the structure is divided into n equal parts so that the length of each part is $l_0 = l / n$. Let the

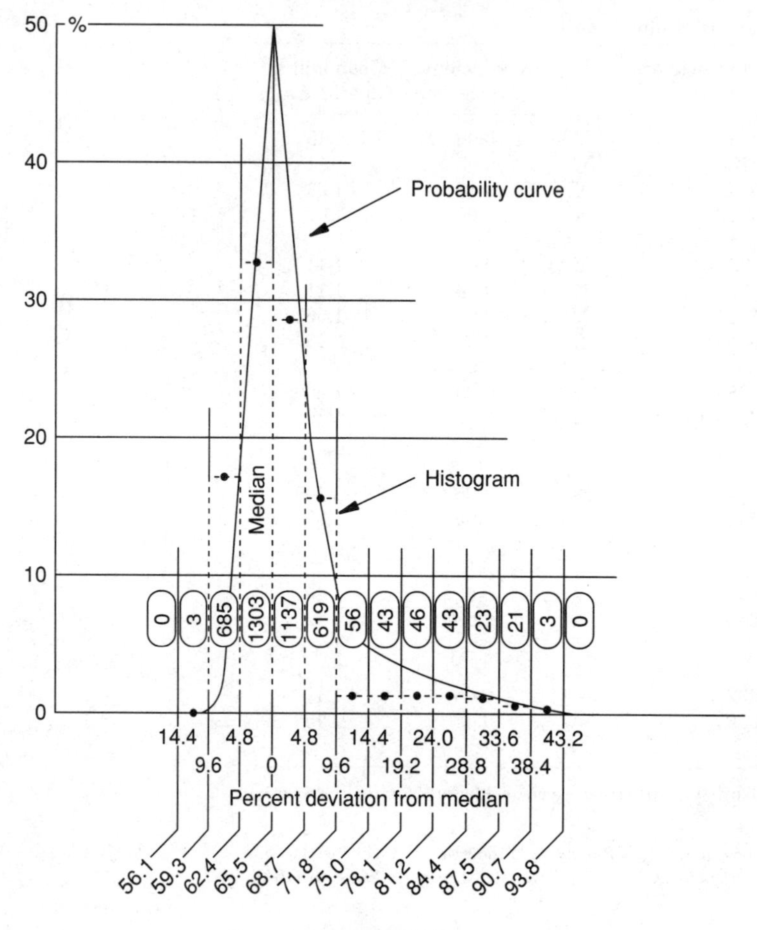

Figure 7.6 *Source: O. G. Julian, "Synopsis of First Progress Report of Committee on Factors of Safety,"* Journal of the Structural Division, Proceedings of the ASCE, *vol. 83, no. ST4, paper no. 1316, July 1957.*

strengths of these n parts be independent statistically and identically distributed with distribution function $F_0(x)$. Then the distribution function of the strength of the structure can be expressed, using Eqs. (7.7) and (7.8), as

$$F_S(x) = 1 - [1 - F_0(x)]^n = 1 - \exp\left\{\frac{l}{l_0}\ln[1 - F_0(x)]\right\} \tag{7.9}$$

It can be seen from Eq. (7.9) that the size of the structure ($l\,/\,l_0$) appears as a parameter in the distribution function. Equation (7.9) can be extended to a brittle solid subjected to a homogeneous stress state. In this case, the volume of the solid V is divided into several elementary volumes V_0 such as the

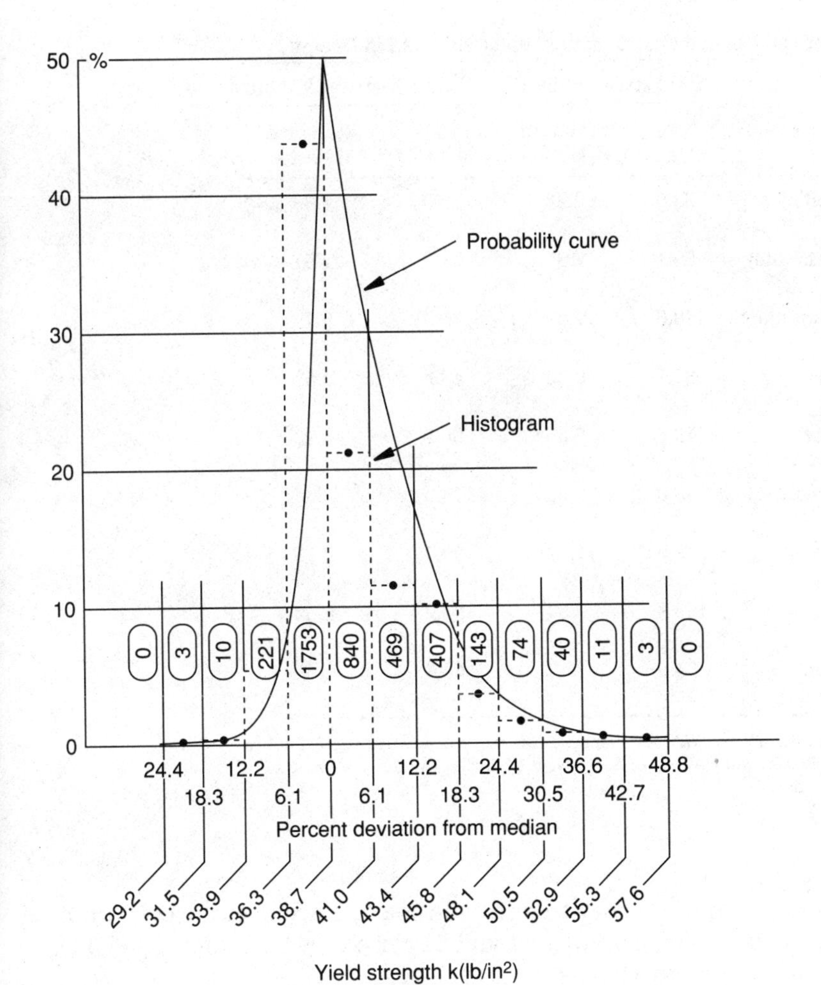

Figure 7.7 *Source: O. G. Julian, "Synopsis of First Progress Report of Committee on Factors of Safety,"* Journal of the Structural Division, Proceedings of the ASCE, *vol. 83, no. ST4, paper no. 1316, July 1957.*

volume of a standard test specimen. Let the distribution function of the strength of elementary volumes be given by $F_0(x)$. Then the distribution of the strength of the solid can be expressed as

$$F_S(x) = 1 - \exp\left\{\frac{V}{V_0}\ln[1 - F_0(x)]\right\} \tag{7.10}$$

The quantity $\ln[1 - F_0(x)]$ is often approximated as

$$\ln[1 - F_0(x)] \approx -\left(\frac{x-a}{b}\right)^c; \qquad x \geq a \tag{7.11}$$

TABLE 7.7 Statistical Properties of Common Materials Used in Design

Material	Yield stress, k(lb/in^2)		Ultimate stress, k(lb/in^2)	
	Mean	Standard deviation	Mean	Standard deviation
2014 Aluminum alloy (Forgings)	63.0	2.23	70.0	1.89
2024-T6 Aluminum alloy (Sheet)	50.1	2.85	63.6	2.51
Ti-6Al-4V Titanium alloy (Sheet and Bar)	130.6	7.2	135.5	6.7
C1006 Carbon steel (Hot rolled sheet)	35.7	0.80	48.3	0.52
C1035 Carbon steel (Hot rolled round bars)	49.5	5.36	86.2	3.92
High strength structural steel	49.6	3.69	76.9	2.06
Type 202 Stainless steel	49.9	1.32	99.7	2.71
Type 301 Stainless steel	166.8	9.37	191.2	5.82
Malleable cast iron (ferritic)	34.9	1.47	53.4	2.68

where a is the lower limit on the strength and b and c are positive constants. Typical values of c are 58 for mild steel and 12 for concrete. Using Eq. (7.11), $F_S(x)$ of Eq. (7.10) can be written as

$$F_S(x) = 1 - \exp\left\{-\frac{V}{V_0}\left(\frac{x-a}{b}\right)^c\right\}; \qquad x \geq a \tag{7.12}$$

which can be identified as a type-III extreme value distribution for the smallest value. The mean and standard deviations of strength S can be expressed as [7.16]

$$\mu_S = \overline{S} = a + b\,\Gamma\left(1 + \frac{1}{c}\right)\left(\frac{V}{V_0}\right)^{-1/c} \tag{7.13}$$

$$\sigma_S^2 = b^2\left\{\Gamma\left(1 + \frac{2}{c}\right) - \Gamma^2\left(1 + \frac{1}{c}\right)\right\}\left(\frac{V}{V_0}\right)^{-2/c} \tag{7.14}$$

The foregoing analysis was given by Weibull for $a = 0$ [7.17]. It can be observed that the mean value of the strength (μ_S) decreases with the volume and for $a = 0$ the coefficient of variation is independent of the volume. The size effect is found to be valid for many brittle materials. The decrease in tensile strength with increasing volume is shown qualitatively in Fig. 7.8 for glass threads [7.16].

7.3.4 Model for plastic materials

Consider a bar under axial loading. Assume the force-deformation relation for a typical (ith) particle in the cross section of the bar to be ideal plastic or elastic-ideal plastic with a yield strength (force) S_i. The strength of the bar (S) can be expressed as [7.16]

$$S = \sum_{i=1}^{n} S_i \tag{7.15}$$

where the summation is over all the n particles in the cross section. The mean and standard deviation of S are given by

$$\overline{S} = \sum_{i=1}^{n} \overline{S}_i \tag{7.16}$$

$$\sigma_S^2 = \text{var}(S) = \sum_{i=1}^{n} \sum_{j=1}^{n} \text{cov}\,(S_i, S_j) \tag{7.17}$$

If the number of particles n is large, the strength of the bar can be assumed to follow normal distribution according to the central limit theorem. If the yield stress s_y is known to vary as a continuous function within the cross

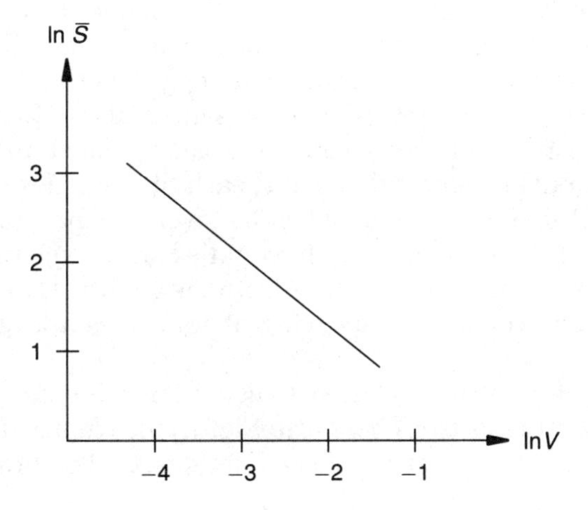

Figure 7.8

section, Eqs. (7.15) to (7.17) can be rewritten, by replacing summations by integrals, as

$$S = \iint_A s_y(x,y)\, dA \tag{7.18}$$

$$\mu_S = \overline{S} = \iint_A E[s_y(x,y)]\, dA \tag{7.19}$$

$$\sigma_S^2 = \iint_A \iint_A \operatorname{cov}[s_y(x_i,\, y_i),\, s_y(x_j,\, y_j)]\, dA_i dA_j \tag{7.20}$$

If the yield stress is homogeneous, Eqs. (7.19) and (7.20) reduce to [7.16]

$$\mu_S = \overline{S} = \mu_{s_y}\, A = \overline{s}_y\, A \tag{7.21}$$

$$\sigma_S^2 = \iint_A \iint_A \operatorname{cov}(x_j - x_i,\, y_j - y_i)\, dA_i\, dA_j \tag{7.22}$$

where $\mu_{s_y} = \overline{s}_y$ denotes the mean value of strength. It can be seen from Eq. (7.21) that the mean strength per unit area is independent of the area.

7.3.5 Model for fiber bundles

The first contribution to fiber-bundle theory was made by Daniels [7.18] in the context of predicting the statistical properties of the combined strength of bundles of threads in textile yarns, given the statistics of the individual threads. In the fiber-bundle model, the cross section is assumed to be composed of ideal fibers in a parallel arrangement. The force-displacement relation for a fiber is such that when a fiber breaks, it will no longer be capable of carrying any load. Thus the failure of a fiber does not necessarily imply failure of the system, but the load is redistributed among the other unbroken fibers which may then be able to carry the increased loading. Consider a bundle of n fibers, each with a unit cross-sectional area, as shown in Fig. 7.9 [7.19]. Since the entire bundle supports the load s, each fiber carries a load of s/n. The strengths of all fibers are assumed to be given by the probability distribution, $F(x)$. The bundle is assumed to have failed whenever the fibers $1, 2, \ldots, n$ fail successively in some order. Also, it is known that when j fibers fail $(0 < j < n)$, each of the $(n - j)$ surviving fibers carrys a larger load, $s/(n - j)$.

For simplicity, we first derive the probability distribution function for the strength of a bundle consisting of two fibers (see Fig. 7.10). We assume that, as a typical sequence, fiber 1 fails first and fiber 2 fails next. The probability

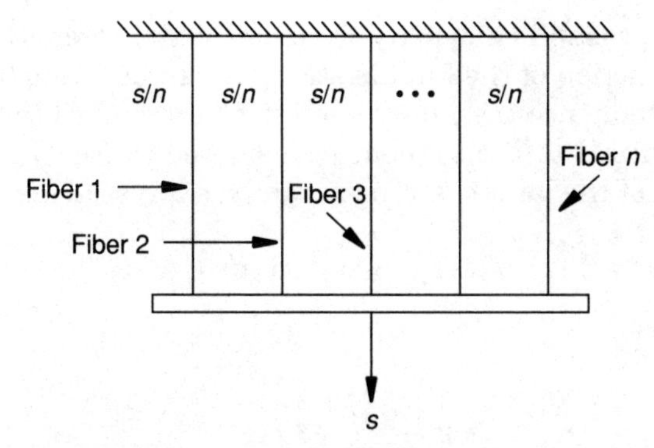

Figure 7.9 *Source: J. F. McCarthy, Jr. and O. Orringer, "Some Approaches to Assessing Failure Probabilities of Redundant Structures," in* Composite Reliability, ASTM Special Publication 580, *ASTM, Philadelphia, 1974. Copyright ASTM. Reprinted with permission.*

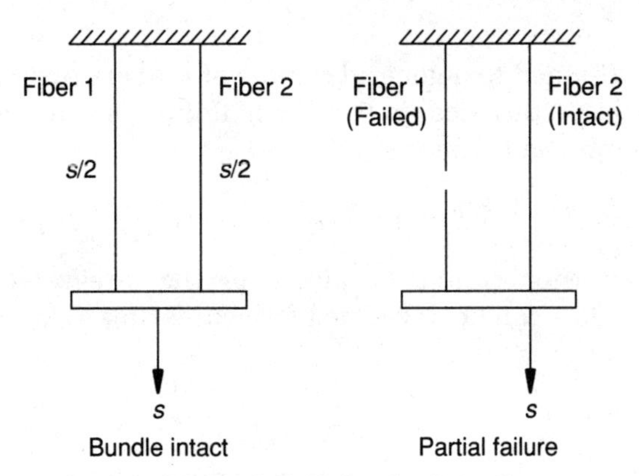

Figure 7.10 *Source: J. F. McCarthy, Jr. and O. Orringer, "Some Approaches to Assessing Failure Probabilities of Redundant Structures," in* Composite Reliability, ASTM Special Publication 580, *ASTM, Philadelphia, 1974. Copyright ASTM. Reprinted with permission.*

of occurrence of this sequence is given by

$$P(\text{fiber 1 failing first and fiber 2 failing next at a total load} \le s)$$

$$= P(\text{strength of fiber 1 } (x_1) \le \frac{s}{2})\, P(x_1 \le \text{strength of fiber 2 } (x_2) \le s)$$

$$= \int_0^{s/2} f(x_1) \int_{x_1}^{s} f(x_2)\, dx_2\, dx_1 \qquad (7.23)$$

where $f(x_i)$ denotes the probability density function of the strength of fiber $i(x_i)$. The distribution function of the bundle strength can be found by noting that the other possible sequence (i.e., fiber 2 failing first and fiber 1 failing next) is statistically identical to the sequence considered in Eq. (7.23). Thus the distribution function of the bundle strength can be expressed as

$$F_B(s) = 2\int_0^{s/2} f(x_1)\int_{x_1}^{s} f(x_2)\,dx_2\,dx_1 \tag{7.24}$$

which can be rewritten as

$$F_B(s) = 2\,F(s)\,F\left(\frac{s}{2}\right) - \left\{F\left(\frac{s}{2}\right)\right\}^2 \tag{7.25}$$

This result can be compared with the corresponding simple parallel system for which

$$F_P(s) = \left\{F\left(\frac{s}{2}\right)\right\}^2 \tag{7.26}$$

Equation (7.23) can be extended to a general bundle of n fibers by recognizing that there are $n!$ possible, statistically identical, failure sequences [7.19]. Thus, we obtain, for a bundle of n fibers

$$F_B(s) = n!\int_0^{s/n} f(x_1)\int_{x_1}^{s/(n-1)} f(x_2)\cdots\int_{x_{n-1}}^{s} f(x_n)\,dx_n\cdots dx_2\,dx_1 \tag{7.27}$$

Although a general expression cannot be given for the integrated form of Eq. (7.27), recursive formulas can be developed for evaluating $F_B(s)$.

7.4 Fatigue Strength

It is well-known that mechanical and structural components fail under the action of fully reversed, repeated, fluctuating, or random loads, even when the peak stress levels are less than the static ultimate or the static yield strength of the material. Such failures are known as *fatigue failures*. There are two stages of fatigue failure: stage one, known as the *crack-initiation stage*, and stage two, known as the *crack-propagation stage*. If N denotes the number of cycles to failure, 60 to 90 percent of cycles cause the initiation of the crack and 10 to 40 percent of the cycles cause the propagation of the crack. A fatigue crack is usually initiated at the surface of a component where stresses are often the highest, imperfections may be present, and the grains are least supported. Surface scratches, inclusions, gas pockets, machining irregularities, changes in cross section, fillets, keyways, and holes are common sites for the crack initiation. The initial crack is too small for detection by visual or standard inspection methods. Once the crack is initiated, local geometric-stress concentration effects aid in its growth. At the macro level, as the cross-sectional area is reduced, the stress intensity grows in magnitude until sudden fracture over the remaining area occurs.

7.4.1 Constant amplitude fatigue strength

In a standard fatigue test such as the rotating-beam test [7.20], a constant-amplitude bending load, which causes a constant stress at the extreme fiber of the beam, is applied to each specimen. As the load varies cyclically, the stress also varies as shown in Fig. 7.11. The number of stress cycles required to cause failure N and the corresponding stress amplitude $S = s_{\max}$ are noted. The test is repeated with different stress amplitudes and the corresponding number of cycles to failure in each case is noted. The data are plotted as cycle life N versus applied tensile stress S on a log-log scale through which a best-fit line is drawn. The resulting graph is known as the *S-N* diagram (see Fig. 7.12). In this diagram, any ordinate value S_f, just below the failure line is called the fatigue strength corresponding to the number of cycles n of interest. The objective of any design is to assure that the applied load does not cause a stress amplitude larger than the fatigue strength for the material specified for the number of load cycles expected during the life of the component.

It has been observed that failure would not occur at stress levels below the horizontal part of the *S-N* curve irrespective of the number of cycles applied. The stress corresponding to the horizontal line in Fig. 7.12 is called the endurance limit (S_e). As with most experimental data, a certain amount of scatter of the test-data points about the best-fit *S-N* line has been observed. The statistical fatigue data needed for probabilistic design can be collected in two possible ways as follows.

1. The statistical *S-N* envelope can be developed for a material by plotting the distributions of cycles to failure for a number of stress amplitudes. A

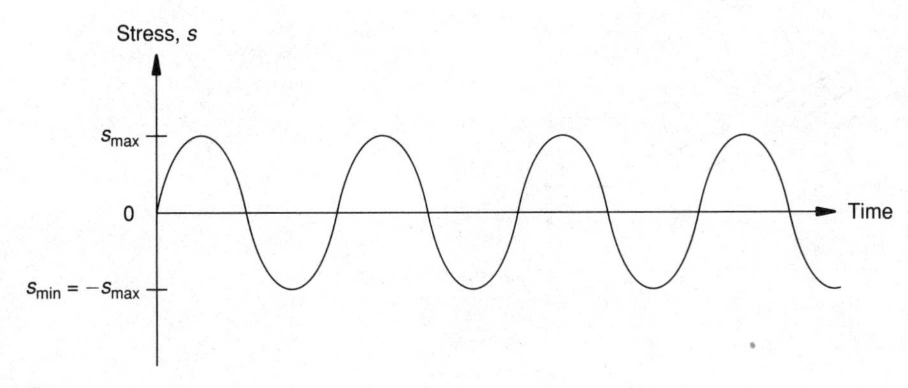

Figure 7.11

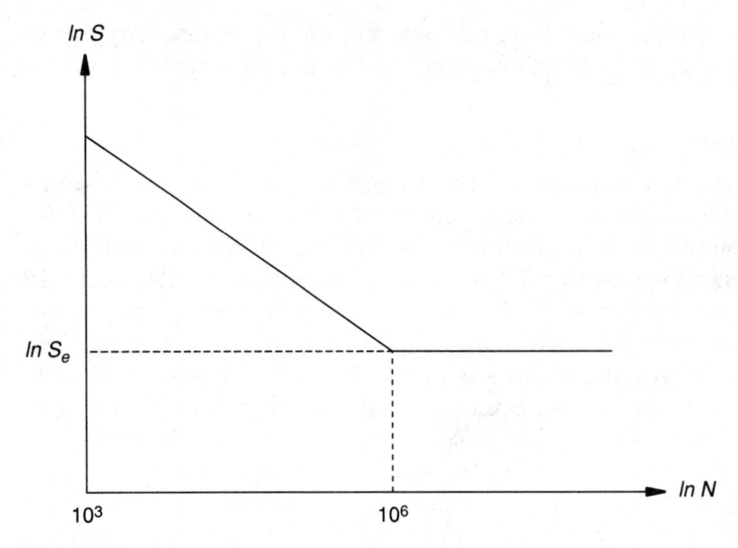

Figure 7.12

typical result, including the scatter in the experimental results, is shown in Fig. 7.13. The number of cycles to failure at a fixed level of maximum bending stress has been found to follow approximately lognormal distribution. The three-parameter Weibull distribution was also observed to be fitting the fatigue-life data of titanium and steel very closely [7.7, 7.21]. Since the mean value ($\overline{S}_f$) and standard deviation (σ_{S_f}) are frequently

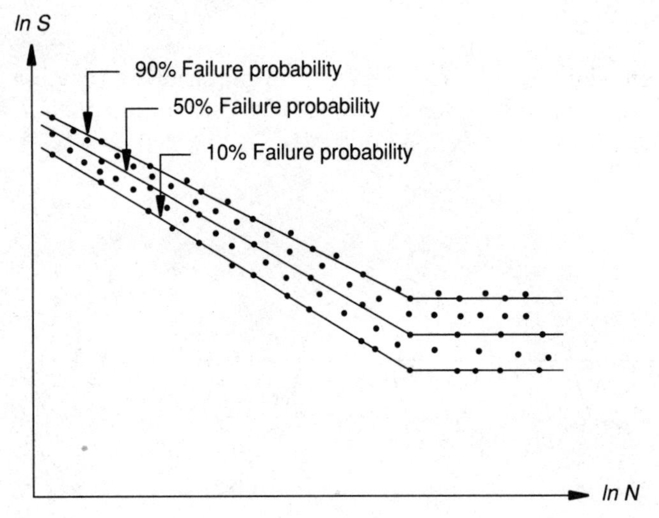

Figure 7.13

used in engineering design, the values of $\overline{S}_f$ and σ_{S_f} corresponding to Weibull distribution have been expressed in the following form:

$$\overline{S}_f = x_0 + (\theta - x_0)\Gamma\left(1 + \frac{1}{b}\right) \tag{7.28}$$

$$\sigma_{S_f} = \frac{(\overline{S}_f - x_0)}{b^{0.926}}; \qquad 0.7 < b < 10.0 \tag{7.29}$$

where x_0, θ, and b are the Weibull parameters. Table 7.8 gives typical fatigue data for steel and aluminum subjected to constant-amplitude fatigue loading.

2. The statistical *S-N* envelope can be developed by plotting the distributions of stress at failure for a number of fixed-cycle levels. Figure 7.13 shows typical results, including the scatter in the experimental results. The fatigue strength at fixed cycle life was found to follow approximately normal distribution.

From practical point of view, the three distributions shown in Fig. 7.14 are important. The distribution of the endurance limit (curve *A*) is useful in the design of mechanical and structural components for infinite life (more than a million cycles of life). The distribution of fatigue life (curve *B*) is useful in the design of components such as bearings where L_{10}-life is important. The distribution *C* becomes important when the failure criterion involves fatigue strength at a finite life.

7.4.2 Variable amplitude fatigue strength

In many engineering applications, the amplitude of fatigue stress varies during the life of a component. First we shall consider a deterministic fatigue

TABLE 7.8 Statistical Fatigue Data from Rotating Bending Tests [7.45, 7.46]

	At	10^4	cycles	At	10^5	cycles	At	10^6	cycles
Material	x_0 k(lb/in^2)	b	θ k(lb/in^2)	x_0 k(lb/in^2)	b	θ k(lb/in^2)	x_0 k(lb/in^2)	b	θ k(lb/in^2)
AISI 1045 Steel (S_u = 105 k(lb/in^2))	79.0	2.60	86.2	67.0	2.75	73.0	56.7	2.85	61.65
M 10 Tool Steel (S_u = 330 k(lb/in^2))	152.0	2.67	185.7	133.0	2.70	163.7	117.0	2.73	144.0
AISI 2340 Steel (S_u = 122 k(lb/in^2))	87.0	4.4	94.4	75.0	4.4	81.55	64.0	4.9	70.5
2014 Aluminum (S_u = 68-78 k(lb/in^2))	57.85	1.781	63.92	48.94	1.636	53.81	41.34	1.522	45.31
1100 Aluminum (S_u = 10 k(lb/in^2))	3.077	3.47	7.318	2.006	3.46	4.757	1.304	3.46	3.092

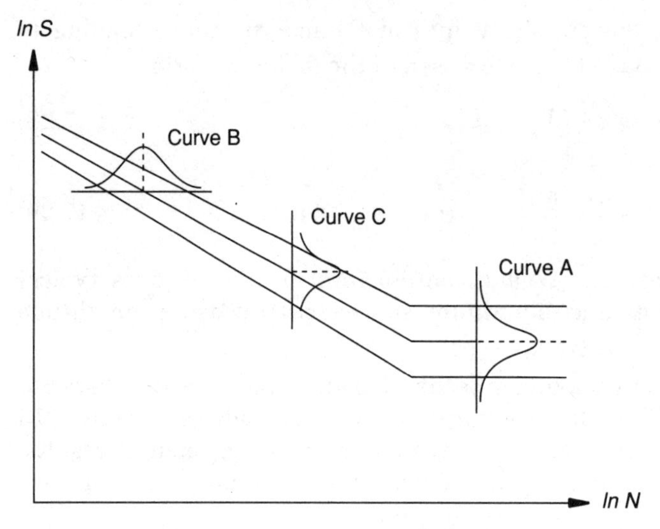

Figure 7.14

strength model according to Palmgren and Miner's hypothesis [7.22]. Let $N(S)$ denote the number of load cycles with constant stress amplitude S to cause fatigue failure. Then a linear approximation can be obtained from experimental data for typical materials used in structural and machine design between $N(S)$ and S on a log-log scale as shown in Fig. 7.12:

$$\log N = b - \alpha \log S \tag{7.30}$$

or,

$$\log (N\, S^{\alpha}) = b$$

that is,

$$N\, S^{\alpha} = \beta \tag{7.31}$$

where the constants α and $\beta = 10^b$ depend only on the material. Actually wide scatter prevails in this fatigue law and Eq. (7.31) should be interpreted as representing a mean fatigue curve. Based on this idealized experimental law, Palmgren and Miner's hypothesis states that the percentage or fraction of damage D due to the application of stress amplitude S_i for n_i cycles is accumulated linearly, that is,

$$D = \frac{n_i}{N(S_i)} \tag{7.32}$$

where $N(S_i)$ is the fatigue life with cyclic load of constant amplitude S_i. When stress amplitudes $S_1, S_2, \ldots,$ are applied for $n_1, n_2, \ldots,$ cycles, respectively, the total percentage (fractional) damage is given by

$$D = \frac{n_1}{N(S_1)} + \frac{n_2}{N(S_2)} + \cdots = \sum_i \frac{n_i}{N(S_i)} \tag{7.33}$$

Clearly, fatigue failure is defined by $D = 1$ (100 percent damage). Equation (7.33) can be extended to continuous variations of loading cycles as

$$D = \int_{S=0}^{\infty} \frac{n(S)dS}{N(S)} \tag{7.34}$$

where $n(S)dS$ represents the number of cyclic loadings with amplitude in the interval $(S, S + dS)$ and $N(S)$ is the fatigue life at stress amplitude S.

We now consider the cyclic loading as a stationary narrow band random process and derive an expression for the expected fatigue damage [7.23, 7.47]. Now $n(S)dS$ represents the random number of load cycles with amplitudes in the interval $(S, S + dS)$. From Eq. (7.33), we see that the percentage damage D is also a random variable. The mean or expected value of D is given by

$$E(D) = \int_{0}^{\infty} \frac{E[n(S)dS]}{N(S)} \tag{7.35}$$

by assuming fatigue life $N(S)$ to be a deterministic or nonrandom function of the amplitude S. In Eq. (7.35), $E(n(S)dS)$ denotes the mean number of load cycles with stress amplitude or peak in the interval $(S, S + dS)$. It can be shown [7.24, 7.47] that the mean number of load cycles is equal to the difference between the mean numbers of up-crossings of the two peaks S and $S + dS$:

$$E[n(S)\,dS] = (\nu_S^+ - \nu_{S+dS}^+)T = (-\,d\nu_S^+)T \tag{7.36}$$

where ν_S^+ is the mean up-crossing rate of the stress level S and T is the time duration of load application. The probability density function of the amplitude of the load (S) [7.23] can be derived as

$$f(S)dS = -\frac{d\nu_S^+}{\nu_0^+} \tag{7.37}$$

where ν_0^+ is the zero crossing rate. Substitution of Eqs. (7.36) and (7.37) into Eq. (7.35) leads to

$$E(D) = \int_{0}^{\infty} \nu_0^+ \, T \, \frac{f(S)dS}{N(S)} = \nu_0^+ \, T \int_{0}^{\infty} \frac{f(S)dS}{N(S)} \tag{7.38}$$

If the stress history $S(t)$ follows normal distribution, then the stress peaks follow Rayleigh's distribution as [7.47]

$$f(S) = \frac{S}{\sigma_S^2} \exp\left(-\frac{S^2}{2\,\sigma_S^2}\right) \tag{7.39}$$

where σ_S is the standard deviation of the random stress, S. The Palmgren-Miner fatigue law can be written as

$$N(S)S^{\alpha} = \beta \tag{7.40}$$

By using Eqs. (7.39) and (7.40), Eq. (7.38) can be expressed as [7.23, 7.47]

$$E(D) = \frac{\nu_0^+ T}{\sigma_S^2 \beta} \int_0^\infty S^{\alpha+1} \exp\left(-\frac{S^2}{2\sigma_S^2} \right) dS$$

$$= \frac{\nu_0^+ T}{\beta} \left(\sqrt{2}\, \sigma_S \right)^\alpha \Gamma\left(1 + \frac{\alpha}{2} \right) \tag{7.41}$$

where

$$\Gamma(1+x) = \int_0^\infty e^{-y} y^x \, dy \tag{7.42}$$

is the gamma function. Noting that

$$\nu_0^+ = \frac{\omega_0}{2\pi} \tag{7.43}$$

where ω_0 is the frequency of cyclic load variation, the expected fatigue damage in time T can be expressed as

$$E(D) = \frac{\omega_0 T}{2\pi\beta} \left(\sqrt{2}\, \sigma_S \right)^\alpha \Gamma\left(1 + \frac{\alpha}{2} \right) \tag{7.44}$$

7.5 Modeling of Loads

7.5.1 Introduction

The analysis and design of any machine or structure requires the identification of all important environmental factors that induce loads on the machine or structure, as well as the quantification of those loads. The following types of loads are important in the design of machines and structures:

1. dead loads;
2. live loads;
3. wind loads; and
4. earthquake loads.

7.5.2 Dead loads

Dead load or gravity load is due to the weight of the machine or structure including all the permanent attachments. The mean values of dead loads can be determined from the specific weights of materials and components. The variability in the dead load is due to the variations in the specific weight of the material and permanent attachments. The dead load is usually described by normal distribution with a coefficient of variation of 0.1. Although normal distribution is often used to describe the dead loads, it is desirable to use a lognormal distribution due to its ability to consider positive loads only.

7.5.3 Live loads

In structural design, the live load consists of weights of people, furniture and portable fixtures, and equipment. Since these loads vary in time and space, it is very difficult to establish the random variability of these loads. However, several comprehensive statistical surveys have been conducted on live loads [7.25–7.27]. Typical values of the minimum live loads recommended by the 1979 Uniform Building Code (UBC) are given in Table 7.9. Culver [7.25, 7.27] conducted a survey of live loads of 625 general and clerical offices in private office buildings and observed the mean value and standard deviation of the live load to be 9.0 lb/ft^2 and 5.5 lb/ft^2 respectively. The histogram shows a skew shape and hence a lognormal distribution will be more appropriate than normal distribution to describe the live load.

Example 7.3 If the mean value and the standard deviation of the live load in a heavy manufacturing plant are 125 lb/ft^2 and 62.5 lb/ ft^2, respectively, determine the probability of the live load exceeding a value of 150 lb/ft^2. Assume lognormal distribution for the magnitude of the live load.

solution The probability density function of the live load L can be expressed as

$$f_L(l) = \frac{1}{l\,\sigma_X\sqrt{2\,\pi}}\exp\left\{-\frac{1}{2}\left(\frac{\ln l - \overline{X}}{\sigma_X}\right)^2\right\}$$

where

$$\gamma_L = \frac{\sigma_L}{\overline{L}} = \frac{62.5}{125.0} = 0.5$$

$$\sigma_X = \ln(\gamma_L^2 + 1) = \ln(0.5^2 + 1) = 0.223143$$

$$\overline{X} = \ln\overline{L} - \frac{1}{2}\,\sigma_X^2 = \ln 125.0 - \frac{1}{2}\,(0.223143)^2 = 4.80341$$

TABLE 7.9 Minimum Live Loads Recommended by UBC†

Nature of occupancy	Concentrated load (lb)	Uniform load†† (lb/ft^2)
Wholesale stores	3000	100
Heavy storage	—	250
School classrooms	1000	40
Residential buildings	0	40
Offices	2000	50
Heavy manufacturing	3000	125
Libraries:		
Reading rooms	1000	60
Stack rooms	1500	125
Hospitals	1000	40
Printing plants (press rooms)	2500	150
Auditoriums (stage areas)	0	125

† Loads specified in Table 7.9 are from Table 23-A of the 1985 Uniform Building Code Copyright©1985, and are reproduced with the written permission of the publishers, the International Conference of Building Officials.

†† See Table 23-A of [7.29] for more details.

Defining the standard normal variate z_1 corresponding to the specified live load of 150 lb/ft² as

$$z_1 = \frac{\ln l - \overline{X}}{\sigma_X} = \frac{\ln 150 - 4.80341}{0.223143} = 0.92864$$

the desired probability can be found as

$$P[L > 150] = P[z > z_1] = P[z > 0.92864]$$

$$= 1 - P[z \le 0.92864] = 1 - 0.8238 = 0.1762$$

7.5.4 Wind loads

The statistical analysis of the maximum annual wind velocities at several locations in the United States was conducted by Thom [7.30]. He found that type-II extreme value distribution for largest values, also known as the *Frechet distribution*, is the most appropriate one to describe the yearly maximum wind velocity at any location. The corresponding probability density and distribution functions are given by

$$f_V(v) = \left(\frac{\gamma}{\beta}\right)\left(\frac{\beta}{v}\right)^{\gamma+1} \exp\left(-\left\{\frac{\beta}{v}\right\}^{\gamma}\right); \qquad v \ge 0,\ \gamma > 0,\ \beta > 0 \tag{7.45}$$

$$F_V(v) = P(V \le v) = \exp\left\{-\left(\frac{\beta}{v}\right)^{\gamma}\right\} \tag{7.46}$$

where V is the maximum wind velocity, and γ and β are the shape and scale parameters of the distribution. Thom determined the numerical values of the parameters γ and β for 141 locations in the U.S. The mean and standard deviation of the random variable V are given by

$$\overline{V} = \beta\,\Gamma\left(1 - \frac{1}{\gamma}\right) \tag{7.47}$$

$$\sigma_V^2 = \beta^2 \left[\Gamma\left(1 - \frac{2}{\gamma}\right) - \Gamma^2\left(1 - \frac{1}{\gamma}\right)\right] \tag{7.48}$$

The probability that the wind velocity will exceed a specified value v in any single year can be found as

$$P(V > v) = 1 - F_V(v) = r(say) \tag{7.49}$$

The return period corresponding to a wind velocity of magnitude v is defined as the number of years elapsed before the recurrence of the wind velocity of magnitude v (see Section 4.7). By denoting the return period as $R(v)$, the probability of the return period exactly equal to i years can be determined as

$$P[R(v) = i] = (1 - r)^{i-1} r \tag{7.50}$$

where r denotes the probability that the wind velocity will exceed v in the ith year and $(1 - r)^{i-1}$ represents the probability that the wind velocity will not exceed v in each of the first $(i-1)$ years. The mean return period can be found as

$$E[R(v)] = \bar{R} = \sum_{i=1}^{\infty} i\, P[R(v) = i] = \sum_{i=1}^{\infty} i\,(1-r)^{i-1}\, r \tag{7.51}$$

From Eqs. (7.49) and (7.51), we obtain

$$\bar{R} = \frac{1}{1 - F_V(v)}$$

or

$$F_V(v) = 1 - \frac{1}{\bar{R}} \tag{7.52}$$

For any specified mean return period $\bar{R}$ at any particular location, the value of the probability distribution function $F_V(v)$ and the corresponding value of the wind velocity v can be determined. The mean values and coefficients of variation of maximum wind velocities at seven sites are given in Table 7.10.

Example 7.4 The annual maximum wind velocities measured at a particular location are given in the following table:

Year	Maximum wind velocity (mi/h)
1978	82.4
1979	76.9
1980	70.2
1981	91.8
1982	76.1
1983	72.7
1984	58.6
1985	85.3
1986	92.5
1987	68.5

TABLE 7.10 Wind Velocity Data [7.31]

	Maximum annual value, V		50-year maximum value, V_{50}	
Site	$\bar{V}$ (mi/h)	γ_V	$\bar{V}_{50}$ (mi/h)	$\gamma_{V_{50}}$
Baltimore, MD	55.9	0.12	76.9	0.09
Rochester, NY	53.5	0.10	69.3	0.08
Detroit, MI	48.9	0.14	69.8	0.10
St. Louis, MO	47.4	0.16	70.0	0.11
Austin, TX	45.1	0.12	61.9	0.09
Tucson, AZ	51.4	0.17	77.6	0.11
Sacramento, CA	46.0	0.22	77.3	0.13

Determine the parameters of the probability density function assuming type-II extreme value distribution for the largest values to be valid for the maximum wind velocity.

solution The mean and standard deviation of the measured annual maximum wind velocity are given by

$$\overline{V} = \frac{1}{10} \sum_{i=1}^{10} v_i$$

$$= \frac{1}{10} (82.4 + 76.9 + 70.2 + 91.8 + 76.1 + 72.7 + 58.6 + 85.3 + 92.5 + 68.5)$$

$$= 77.5 \text{ mi/h} \tag{E1}$$

$$\sigma_V^2 = \frac{1}{10} \sum_{i=1}^{10} (v_i - \overline{V})^2$$

$$= \frac{1}{10} \left[4.9^2 + 0.6^2 + 7.3^2 + + 14.3^2 + 1.4^2 + 4.8^2 + 18.9^2 + 7.8^2 + 15.0^2 + 9.0^2 \right]$$

$$= 103.12 \text{ (mi/h)}^2 \tag{E2}$$

Equation (E2) gives

$$\sigma_V = 10.1548 \text{ mi/h}$$

and

$$\gamma_V = \frac{\sigma_V}{\overline{V}} = \frac{10.1548}{77.5} = 0.13103 \tag{E3}$$

For type-II extremal distribution for the maximum values, Eqs. (7.47) and (7.48) give

$$\gamma_V = \frac{\sigma_V}{\overline{V}} = \frac{\left\{ \Gamma\left(1 - \frac{2}{\gamma}\right) - \Gamma^2\left(1 - \frac{1}{\gamma}\right) \right\}^{1/2}}{\Gamma\left(1 - \frac{1}{\gamma}\right)} = 0.13103 \tag{E4}$$

Squaring of Eq. (E4) and rearranging the terms gives

$$1 + \gamma_V^2 = 1 + (0.13102)^2 = 1.01717 = \frac{\Gamma\left(1 - \frac{2}{\gamma}\right)}{\Gamma^2\left(1 - \frac{1}{\gamma}\right)} \tag{E5}$$

Using tables of gamma function [7.50], the solution of Eq. (E5) can be found by trial and error as $\gamma \approx 10$. Equation (7.47) can be used to find β as

$$\beta = \frac{\overline{V}}{\Gamma\left(1 - \frac{1}{\gamma}\right)} = \frac{77.5}{\Gamma(0.9)} = \frac{77.5}{1.0686} = 72.5248$$

Example 7.5 Find the probability of the maximum wind velocity exceeding a value of 85 mi/h in any year for the data given in Example 7.4. Also find the return period for maximum velocities greater than 85 mi/h.

solution Assuming type-II distribution for the largest values to be valid for the maximum wind velocity V the probability of V, exceeding 85 mi/h is given by

$$P[V > 85 \text{ mi/h}] = 1 - F_V(85) = 1 - \exp\left\{-\left(\frac{\beta}{85}\right)^{\gamma}\right\}$$

$$= 1 - \exp\left\{-\left(\frac{72.5248}{85}\right)^{10}\right\} = 0.18493$$

The return period for maximum wind velocity greater than 85 mi/h can be found from Eq. (7.52) as

$$\overline{R} = \frac{1}{P[V > 85]} = \frac{1}{1 - F_V(85)} = \frac{1}{0.18493} = 5.40745 \text{ years}$$

Variation with height. Wind velocity is found to vary with height. The mean wind velocity decreases at ground level due to the friction forces created by the roughness and other obstructions on the ground level. In addition, the fluctuations around the mean wind velocity will also be larger at ground level because of the same surface roughness. The height at which the frictional effect of the surface roughness is negligible is known as the gradient height H_g. The wind velocity at this height is called the gradient wind velocity V_g. At heights greater than the gradient height, the wind velocity is assumed to be same as the gradient wind velocity. The variation of mean wind velocity with height is described by the equation [7.28, 7.48]

$$V_H = \left(\frac{H}{H_g}\right)^{\frac{1}{n}} V_g \tag{7.53}$$

where V_H is the mean wind velocity at height H above the ground and n is constant. The values of H_g and n can be taken as 950 ft and 7 for level ground with some obstructions (e.g., airports and farm land with scattered trees and buildings), 1150 ft and 5 for level ground with numerous obstructions of varying sizes (e.g., suburbs with lot sizes of half acre or more), and 1800 ft and 2 for grounds with large obstructions (e.g., centers of large cities).

Example 7.6 If the annual maximum wind velocity data of Example 7.4 was obtained at a height of 100 ft in the middle of a city, find the annual maximum wind velocity corresponding to a 25 year return period. Also find the corresponding annual maximum wind velocity at the gradient height.

solution The 25-year return period $\overline{R}$ is given by (see Eq. (7.52))

$$1 - F_V(v) = \frac{1}{\overline{R}} = \frac{1}{25} = 0.04$$

or

$$F_V(v) = 0.96$$

Using Eq. (7.46) and the solution of Example 7.4, we can obtain

$$F_V(v) = \exp\left\{-\left(\frac{72.5248}{v}\right)^{10}\right\} = 0.96$$

This equation gives $v = V_H = 99.8613$ mi/h. Since the gradient height and constant n in Eq. (7.53) for middle of city are given by $H_g = 1800$ ft and $n = 2$, the gradient wind velocity can be determined as

$$V_g = V_H\left(\frac{H_g}{H}\right)^{\frac{1}{n}} = (99.8613)\left(\frac{1800}{100}\right)^{0.5} = 423.6756 \text{ mi/h}$$

Determination of wind load. The wind velocity can be related to the wind pressure acting on any given structure once the geometry of the structure, the vibrational characteristics of the structure, and the density of air are known. In general, the wind pressure load W acting at a particular point on a structure can be expressed as

$$W = cEGC_pV^2 \tag{7.54}$$

where c is a constant, E is an exposure coefficient, G is the gust factor, C_p is the pressure coefficient and V is the wind velocity. The pressure coefficient, C_p is used to account for both the building geometry and the site exposure. The gust factor G is used to account for the gustiness of the wind (i.e., the dynamic variations of the wind velocity) and the vibrational characteristics of the structure. Methods, based on the time series analysis of the wind velocity data and the theory of structural dynamics, have been developed to compute the gust factor [7.32]. Once the wind pressure is computed using Eq. (7.54), it must be integrated over the surface of the structure to find the total lateral force and moment at any section of the structure. These values can be used directly in the structural design. The mean values of c, E, G and C_p are given by the codes [7.33]. The coefficients of variation of E, G, and C_p are on the order of 0.16, 0.11 and 0.12, respectively. Normal distribution can be used to describe these parameters [7.34]. As stated earlier, the yearly maximum wind velocity V is described by the type-II extreme value distribution. Thus the probability distribution of the wind load, given by Eq. (7.54), cannot be expressed in closed form. The Monte Carlo approach can be used to determine the probability characteristics of the wind load.

7.5.5 Earthquake loads

The discontinuities present in earth's crust are always subjected to stress due to the natural forces acting on them. A rapid release of this stress produces seismic waves which cause the ground to shake, leading to an earthquake. An earthquake causes vibration of the buildings and the machinery located on the ground. Since the structural/machinery mass times the earthquake-induced acceleration represents the inertia forces acting on the system, the structure/machine must be designed to withstand the stresses created by these inertia forces. Figure 7.15 gives the common terminology used in the study of earthquakes. The focus or hypocenter represents the point of origin (center of initial rupture) of an earthquake. The epicenter denotes a point on the surface of the ground directly above the focus. The focal depth is the depth of the focus beneath the surface of the ground. The intensity of an earthquake is commonly measured on the *Richter scale*. A magnitude of M on the Richter scale [7.28, 7.35] is defined as

$$M = \log_{10}\left(\frac{a}{a_0}\right) \tag{7.55}$$

where a denotes the maximum amplitude of the earthquake recorded by an instrument called Wood-Anderson seismograph at a distance of 100 km from the epicenter and a_0 is the reference amplitude (one thousandth of a millimeter). The Richter magnitude [7.28] is related to the energy released by the earthquake W as

$$\log W = 11.8 + 1.5\,M \tag{7.56}$$

where the energy W is in ergs. A magnitude of M greater than 5 on the Richter scale indicates potential damage to structures. The value of M also gives the length of the slipped fault which, in turn, reflects the size of the area affected by the earthquake. A magnitude of 4 corresponds to a fault

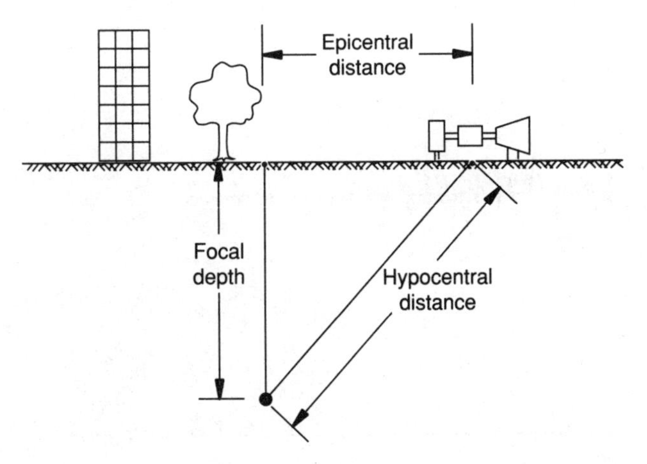

Figure 7.15

length of approximately 1 mile while a magnitude of 9 denotes a 1000 mile slip.

The ground motion during an earthquake is measured by an instrument called a *strong motion accelerograph* placed on the ground at a particular location. A typical graph of the acceleration versus time (also known as an *accelerogram*) recorded by an accelerograph is shown in Fig. 7.16. The accelerograms are generally recorded on photographic paper or film and are digitized for engineering applications. The peak ground acceleration, duration, and frequency content of the earthquake can be obtained from an accelerogram. An accelerogram can be integrated to obtain the time variations of the ground velocity and ground displacement.

Response spectrum. The response spectrum is a graph between the maximum response of a single degree of freedom oscillator under a base excitation equal to the earthquake acceleration and the natural period (or frequency) of the oscillator. The response spectrum provides the most descriptive representation of the influence of a given earthquake on a structure. The maximum relative displacement of the oscillator for any specified values of the natural period of vibration (τ_n) and damping (ζ) is called the spectral displacement and is represented as $S_d(\tau_n,\zeta)$. Similarly, the maximum relative velocity of the oscillator for any specified values of τ_n and ζ is called the spectral velocity and is denoted as $S_v(\tau_n,\zeta)$. The spectral velocity S_v and the

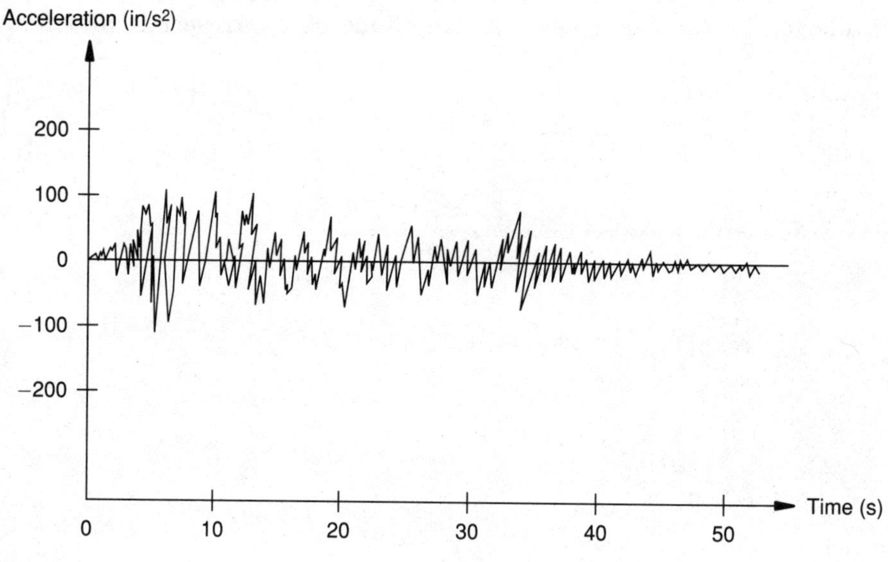

Figure 7.16

spectral acceleration S_a can be computed from the spectral displacement S_d as

$$S_v(\tau_n, \zeta) = \frac{2\pi}{\tau_n} S_d(\tau_n, \zeta) \tag{7.57}$$

$$S_a(\tau_n, \zeta) = \left(\frac{2\pi}{\tau_n}\right)^2 S_d(\tau_n, \zeta) \tag{7.58}$$

The response spectra are plotted on a four-way logarithmic paper as shown in Fig. 7.17. In this figure, the vertical axis denotes the spectral velocity, the horizontal axis represents the natural time period, the 45° inclined axis going up from left (right) indicates the spectral acceleration (displacement). The characteristics of the peak ground motions or the spectral response parameters can be related to the Richter magnitude and the hypocentral distance as [7.28, 7.36]

$$x = \frac{c_1\, 10^{c_2 M}}{(d + 25)^{c_3}} \tag{7.59}$$

where x = characteristic of peak ground motion; M = Richter magnitude; d = hypocentral distance (km); and c_1, c_2, and c_3 = constants. The values of c_1, c_2, and c_3 are given respectively, by 472.3, 0.278, and 1.301 for peak ground acceleration; 5.64, 0.401, and 1.202 for peak ground velocity; 0.393, 0.434, and 0.885 for peak ground displacement; 11.0, 0.278, and 1.346 for spectral velocity with 0 percent damping; and 10.09, 0.233, and 1.341 for spectral velocity with 5 percent damping [7.42]. For reliability studies, the peak ground acceleration (or velocity or displacement) can be considered as a random variable. The coefficient of variation is given by 0.548 for peak ground acceleration, 0.969 for peak ground velocity, 0.883 for peak ground displacement, 0.941 for spectral velocity with 0 percent damping, and 0.651 for spectral velocity with 5 percent damping [7.42].

Example 7.7 Determine the following for an earthquake with $M = 7$ and $d = 50$ km:

1. The mean and standard deviation of S_v for $\tau_n = 1$ s and $\zeta = 0.05$ using Eq. (7.59).
2. The mean and standard deviation of the corresponding S_a assuming τ_n to be deterministic.
3. The probability of the spectral acceleration S_a exceeding a value of 0.15 g by assuming normal distribution for S_a.

Assume the values of c_1, c_2, and c_3 in Eq. (7.59) for $\tau_n = 1$ s as 0.432, 0.399 and 0.704, respectively.

solution

1. The mean value of the spectral velocity S_v is given by Eq. (7.59) as

$$\bar{x} = \bar{S}_v = \frac{c_1\, 10^{c_2 M}}{(d + 25)^{c_3}}$$

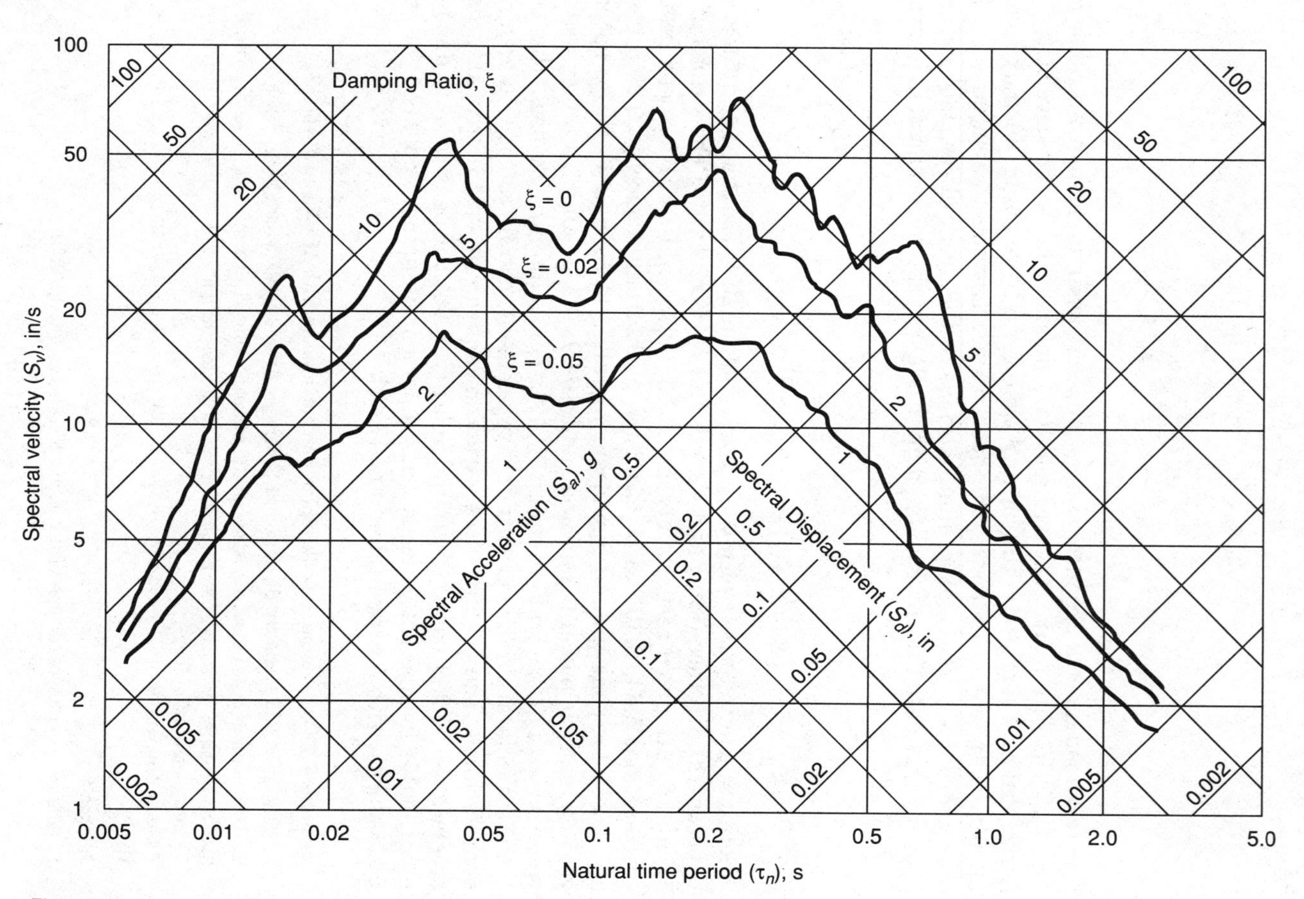

Damping Ratio, ξ
ξ = 0
ξ = 0.02
ξ = 0.05
Spectral Acceleration (S_a), g
Spectral Displacement (S_d), in
Spectral velocity (S_v), in/s
Natural time period (τ_n), s
100
50
20
10
5
2
1
0.005
0.01
0.02
0.05
0.1
0.2
0.5
1.0
2.0
5.0

Figure 7.17

Using the given values of c_1, c_2, and c_3, we obtain

$$\bar{x} = \bar{S}_v = \frac{0.432\,(10)^{0.399 \times 7}}{(50 + 25)^{0.704}} = \frac{268.2154}{20.8950} = 12.8363 \text{ cm/s}$$

The standard deviation of spectral velocity is given by

$$\sigma_x = \sigma_{S_v} = \gamma_x\,\bar{x} = 0.703\,(12.8363) = 9.0239 \text{ cm/s}$$

2. The spectral acceleration S_a is given by

$$S_a = \left(\frac{2\pi}{\tau_n}\right) S_v$$

and hence the mean and standard deviation of S_a can be obtained as

$$\bar{S}_a = \left(\frac{2\pi}{\tau_n}\right) \bar{S}_v = \left(\frac{2\pi}{1}\right) 12.8363 = 80.6529 \text{ cm/s}^2$$

$$\sigma_{S_a} = \left(\frac{2\pi}{\tau_n}\right) \sigma_{S_v} = \left(\frac{2\pi}{1}\right) 9.0239 = 56.6988 \text{ cm/s}^2$$

3. If the spectral acceleration S_a follows normal distribution, the required probability can be determined as

$$P(S_a > 0.15\text{ g}) = P(S_a > 147.15 \text{ cm/s}^2) = P\left(\frac{S_a - \bar{S}_a}{\sigma_{S_a}} > \frac{147.15 - 80.6529}{56.6988}\right)$$

$$= P(z > 1.1728) = 1 - P(z \le 1.1728) = 1 - 0.8796 = 0.1204$$

Earthquake loads. Whenever any structure or machine is subjected to ground acceleration due to an earthquake, all the components of the structure or machine will vibrate. Since the acceleration times the mass gives the inertia load or force, the structure must be designed to withstand the stresses produced by the inertia loads. The equation of motion of the structure, idealized as an undamped single-degree-of-freedom oscillator, subjected to base excitation [7.37], can be expressed as

$$m\ddot{y}(t) + kx(t) = 0 \tag{7.60}$$

where $m = W/g$ = mass of the structure or oscillator, k = spring stiffness or elasticity of the structure, $x(t)$ = displacement of the mass of the oscillator relative to the base, and $\ddot{y}(t)$ = absolute acceleration of the mass of the oscillator due to the earthquake support motion. From Eq. (7.60), the earthquake inertia force can be expressed as

$$m\,\ddot{y}(t) = -k\,x(t) \tag{7.61}$$

Thus the maximum force F is given by k times the maximum relative displacement induced by the earthquake excitation (S_d):

$$F = kS_d(\tau_n, \zeta) = k\left(\frac{\tau_n}{2\pi}\right)^2 S_a(\tau_n, \zeta) \tag{7.62}$$

where S_a is the spectral acceleration and $\tau_n = (2\pi)/\omega_n$ is the natural period of vibration and $\omega_n = (k/m)^{1/2}$ is the natural frequency of vibration of the oscillator. Equation (7.62) can be rewritten as

$$F = \frac{k}{\omega_n^2} S_a = m S_a$$

or

$$\frac{F}{W} = \frac{S_a}{g} \tag{7.63}$$

This shows that the ratio F/W is equal to the spectral acceleration expressed in units of g.

Power spectrum of ground acceleration. As seen in the last section, the response spectrum and its coefficient of variation can be used to predict the mean value and standard deviation of the earthquake induced loads in structures. An alternative procedure involves the use of the power spectrum of the ground acceleration. Although the random process techniques are not as well developed, it is possible to apply the power spectral analysis to earthquake analysis.

It has been suggested that an earthquake can be adequately represented by a time history, which consists of a portion of a stationary random process of finite duration, say 15–50 seconds. One form of the power spectrum of ground acceleration, $S_a(\omega)$, is given by Tajimi [7.38] as

$$\frac{\omega S_a(\omega)}{\sigma_a^2} = \frac{4}{\pi}\left(\frac{H}{1 + 4H^2}\right) \frac{\xi[1 + 4H^2 \xi^2]}{[(1 - \xi^2)^2 + 4H^2 \xi^2]} \tag{7.64}$$

where a = ground acceleration, σ_a = standard deviation of ground acceleration, $\xi = \omega/\omega_g = \omega/(2\pi f_g)$ = nondimensional frequency, ω = frequency, f_g = characteristic ground frequency (2–3 Hz) and H = nondimensional characteristic damping property of the ground (0.6–0.7). The graph of Eq. (7.64) is shown in Fig. 7.18. The power spectrum can be used to find the variance of the response once the modal characteristics of the structure are known [7.36].

References and Bibliography

7.1. J. Datsko, *Material Properties and Manufacturing Processes*, John Wiley, New York, 1966.
7.2. *Manual of Steel Construction*, 9th ed., American Institute of Steel Construction, Inc., Chicago, IL, 1989.

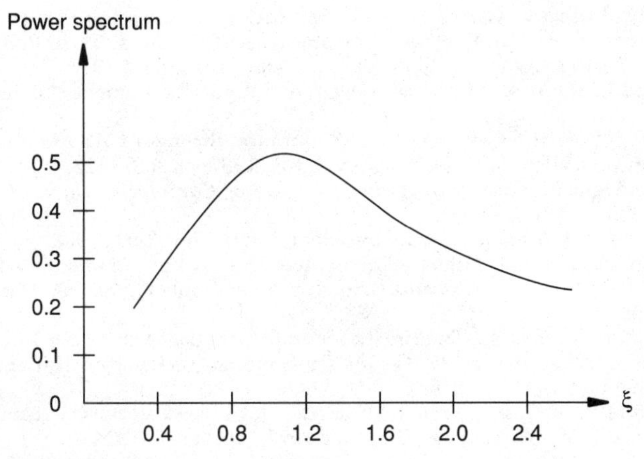

Figure 7.18

7.3. G. A. Alpsten, "Variations in Mechanical and Cross-Sectional Properties of Steel," in *Proceedings of International Conference on Planning and Design of Tall Buildings*, Vol. 1b, *Tall Building Criteria and Loading*, T. C. Kavanagh (Group Coordinator), Lehigh University, Bethlehem, PA, August 1972, pp. 755–805.

7.4. J. C. Bralla (ed.), *Handbook of Product Design for Manufacturing*, McGraw-Hill, New York, 1986.

7.5. J. E. Shigley and C. R. Mischke (eds.), *Standard Handbook of Machine Design*, McGraw-Hill, New York, 1986.

7.6. *Metals Handbook*, vol. 1, T. Lyman (ed.), *Properties and Selection of Metals*, 8th ed., American Society for Metals, Metals Park, Ohio, 1961.

7.7. E. B. Haugen, *Probabilistic Mechanical Design*, John Wiley, New York, 1980.

7.8. F. E. Bolz, "Design Considerations for Manufacturing Economy," *Mechanical Engineering*, ASME, vol. 71, Dec. 1949.

7.9. L. J. Bayer, "Analysis of Manufacturing Costs Relative to Product Design," *ASME* paper no. 56–SA–9, 1956.

7.10. M. F. Spotts,"Allocation of Tolerances to Minimize Cost of Assembly," *ASME J. of Eng. for Industry*, vol. 95, 1973, p. 762.

7.11. M. F. Spotts, *Dimensioning and Tolerancing for Quality Production*, Prentice-Hall, Englewood Cliffs, NJ, 1983.

7.12. T. V. Galambos and M. K. Ravindra, "Properties of Steel for Use in LRFD," *J. of the Structural Div., Proc. of the ASCE*, vol. 104, no. ST9, Sept. 1978, pp. 1459–1468.

7.13. O. C. Julian, "Synopsis of First Progress Report of Committee on Factor of Safety," *J. of the Structural Div., ASCE*, vol. 83, No. ST4, July 1957, pp. 1316.1–1316.22.

7.14. L. Tall and G. A. Alpsten, "On the Scatter of Yield Strength and Residual Stresses in Steel Members," in *Symposium on Concepts of Safety of Structures and Methods of Design, IABSE*, London, 1969, pp. 151–163.

7.15. B. Epstein, "Statistical Aspects of Fracture Problems," *J. of Applied Physics*, vol. 19, 1948, pp. 140–147.

7.16. H. O. Madsen, S. Krenk, and N. C. Lind, *Methods of Structural Safety*, Prentice-Hall, Englewood Cliffs, NJ, 1986.

7.17. W. Weibull, "A Statistical Distribution Function of Wide Applicability," *J. of Applied Mechanics*, vol. 18, 1951, pp. 293–297.

7.18. H. E. Daniels, "The Statistical Theory of the Strength of Bundles of Threads," *Proceedings of the Royal Society*, London, vol. A183, 1945, pp. 405–435.

7.19. J. F. McCarthy, Jr. and O. Orringer, "Some Approaches to Assessing Failure Probabilities of Redundant Structures," in *Composite Reliability*, ASTM Special Tech. Pub. 580, American Society for Testing Materials, Philadelphia, Penn., 1975, pp. 5–31.
7.20. R. E. Little and E. H. Jebe, "Statistical Design of Fatigue Experiments," John Wiley, New York, 1975.
7.21. C. R. Mischke, "Prediction of Stochastic Endurance Strength," *ASME J. of Vibration, Acoustics, Stress, and Reliability in Design*, vol. 109, 1987, pp. 113–122.
7.22. J. L. Bogdanoff and F. Kozin, *Probabilistic Models of Cumulative Damage*, John Wiley, New York, 1985.
7.23. T. Y. Yang, *Random Vibration of Structures*, John Wiley, New York, 1986.
7.24. A. Powell, "On the Fatigue Failure of Structures Due to Vibrations Excited by Random Pressure Fields," *J. of the Acoustical Soc. of Amer.*, vol. 30, no. 12, December 1958, pp. 1130–1135.
7.25. C. G. Culver, "Survey Results for Fire Loads and Live Loads in Office Buildings," *NBS Building Science Series Report 85*, Center for Building Technology, National Bureau of Standards, Washington, D.C., 1976.
7.26. R. B. Corotis and V. A. Doshi, "Probability Models for Live Load Survey Results," *J. of the Structural Div., ASCE*, vol. 103, no. ST6, June 1977, pp. 1257–1274.
7.27. B. Ellingwood and C. Culver, "Analysis of Live Loads in Office Buildings," *J. of the Structural Div., ASCE*, vol. 103, no. ST8, August 1977, pp. 1551–1560.
7.28. G. C. Hart, *Uncertainty Analysis, Loads, and Safety in Structural Engineering*, Prentice-Hall, Englewood Cliffs, NJ, 1982.
7.29. "Uniform Building Code," International Conference on Building Officials, Whittier, Calif., 1985.
7.30. H. C. S. Thom, "New Distributions of Extreme Winds in the United States," *J. of the Structural Div., ASCE*, vol. 94, no. ST7, July 1968, pp. 1787–1802.
7.31. B. Ellingwood, T. V. Galambos, J. G. MacGregor, and C. A. Cornell, "Development of a Probability Based Load Criterion for American National Standard A58," *NBS Special Publication 577*, National Bureau of Standards, Washington, D.C., 1980.
7.32. E. Simiu and R. H. Scanlan, *Wind Effects on Structures*, John Wiley, New York, 1978.
7.33. "Building Code Requirements for Minimum Design Loads in Buildings and Other Structures," *ANSI* A58.1, American National Standards Institution, New York, 1972.
7.34. G. I. Schueller, H. Hirtz, and G. Booz, "The Effects of Uncertainties in Wind Load Estimation on Reliability Assessments," *J. of Wind Eng. and Indust. Aerodynamics*, vol. 14, 1983, pp. 15–26.
7.35. F. Naeim (ed.), *The Seismic Design Handbook*, Van Nostrand Reinhold, New York, 1989.
7.36. P. L. Gould and S. H. Abu-Sitta, *Dynamic Response of Structures to Wind and Earthquake Loading,* John Wiley, New York, 1980.
7.37. S. S. Rao, "Mechanical Vibrations," 2d ed., Addison-Wesley Publishing Co., Reading, MA, 1990.
7.38. H. Tajimi, "A Statistical Method of Determining the Maximum Response of a Building Structure During an Earthquake," *Proceedings of the Second World Conference on Earthquake Engineering*, Tokyo, Japan, vol. 2, 1960, pp. 781–796.
7.39. R. E. Melchers, "Structural Reliability Analysis and Prediction," Ellis Horwood Ltd., Chichester, England, 1987.
7.40. E. Simiu and J. J. Filliben, "Weibull Distributions and Extreme Wind Speeds," *J. of the Structural Div., ASCE*, vol. 106, no. ST12, 1980, pp. 491–501.
7.41. W. H. Melbourne, "Probability Distributions Associated with the Wind Loading of Structures," *Civil Eng. Trans. of Inst. of Engs.*, Australia, CE, vol. 19, no. 1, 1977, pp. 58–67.
7.42. R. K. McGuire, "Seismic Structural Response Risk Analysis, Incorporating Peak Response Regressions on Earthquake Magnitude and Distance," *Department of Civil Engineering Research Report R74-51*, MIT, Cambridge, MA, 1974.
7.43. M. Vorlicek and M. Holicky, "Analysis of Dimensional Accuracy of Building Structures," Elsevier Science Publishing Co. Inc., Amsterdam, The Netherlands, 1989.
7.44. *Aluminum Standards and Data*, Aluminum Association, Inc., Washington, D.C., 1990.
7.45. C. Lipson, N. J. Sheth, and R. Disney, "Reliability Prediction—Mechanical Stress/Strength Interference (Ferrous)," Rome Air Development Center, Technical Report No. RADC-TR-66-710, AD 813574, March 1967.

7.46. C. Lipson, N. J. Sheth, R. Disney, and M. Alton, "Reliability Prediction—Mechanical Stress/Strength Interference (Nonferrous)," Rome Air Development Center, Technical Report No. RADC-TR-68-403, AD 856021, Feb. 1969.
7.47. S. H. Crandall and W. D. Mark, *Random Vibration in Mechanical Systems*, Academic Press, New York, 1963.
7.48. A. G. Davenport, "Gust Loading Factors," *J. of the Structural Div., ASCE*, vol. 93, no. ST3, June 1967, pp. 11–34.
7.49. R. L. Wiegel (ed.), *Earthquake Engineering*, Prentice-Hall, Englewood Cliffs, New Jersey, 1970.
7.50. H. T. Davis, *Tables of the Mathematical Functions*, vol. I, The Principia Press of Trinity University, San Antonio, Texas, 1963.
7.51. H. Volkmann, "Ernst Abbe and His Work," *Applied Optics*, vol. 5, Nov. 1966, pp. 1720–1731.

Review Questions

7.1 What are the primary reasons for the randomness of geometry of machine parts?

7.2 What are the causes of variation in the strength of a material?

7.3 What type of loads are of interest to a mechanical designer?

7.4 Indicate qualitatively the variation of cost with tolerance on a machine component.

7.5 How are the dimensions of an assembly influenced by the dimensions of individual parts?

7.6 Name three statistical models that can be used to describe the strength of different types of materials.

7.7 How is the strength of a brittle material described probabilistically?

7.8 How can we model the strength of a fiber-reinforced plastic material?

7.9 How does a fatigue failure occur?

7.10 What is the significance of *S-N* diagram?

7.11 What is endurance limit?

7.12 How can we estimate the damage of a component subjected to a variable amplitude cyclic load?

7.13 What is the difference between a dead load and a live load?

7.14 Which probability distribution is commonly used to describe the annual maximum wind velocities?

7.15 What is the significance of return period in the wind loading on a structure?

7.16 Describe the variation of wind velocity with height.

7.17 How is the magnitude of an earthquake defined on the Richter scale?

7.18 What is an accelerogram and what is its use?

7.19 What is a response spectrum?

7.20 Write the equation of motion of a structure, idealized as a single degree of freedom system, subjected to a base excitation.

Problems

7.1 The link lengths of a robot manipulator (Fig. 7.19) are given by

$$l_1 = 10 \pm 0.1 \text{ in}, \qquad l_2 = 15 \pm 0.3 \text{ in}, \qquad l_3 = 12 \pm 0.2 \text{ in}$$

Find the probability of the end effector P reaching a point located at a distance of 37 in from the base point O.

7.2 The battery compartment (see Fig. 7.20) in an electronic instrument has a length of $d = 2.0 \pm 0.1$ in. The free length of the spring is given by $d_1 = 0.3 \pm 0.05$ in. If the length of the battery is specified as $d_2 = 0.8 \pm 0.05$ in, find the probability of compressing the spring more than 0.15 in. Assume the tolerances to correspond to 3-sigma values.

7.3 A packaging box has a length of 20 ± 0.2 in. If the box is to be packed with components of length 4 ± 0.05 in each, determine the probability of accommodating five components in the box without interference. Assume normal distribution for all the dimensions.

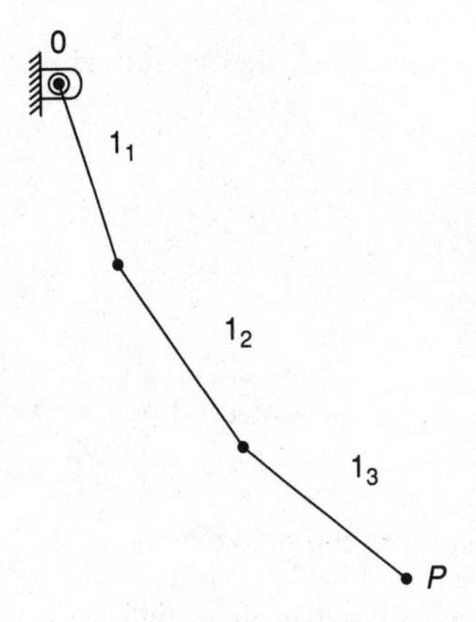

Figure 7.19

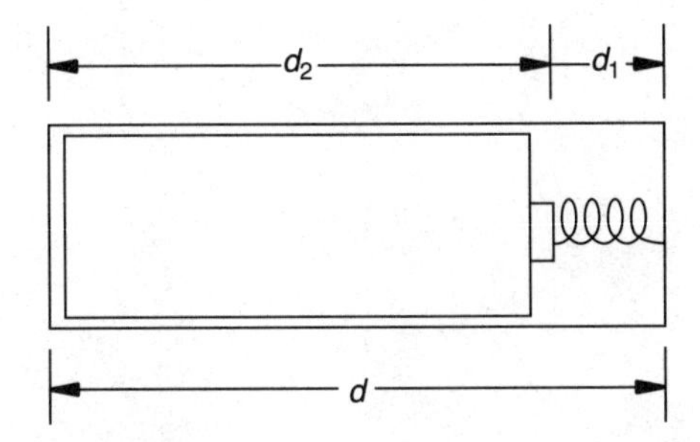

Figure 7.20

7.4 The radial clearance between the rotor and stator of a Francis water turbine is given by 5 ± 1 mm (see Fig. 7.21). The deflection of the shaft, on which the rotor is mounted, can cause a radial motion of 4 ± 1 mm. Determine the probability of rotor touching the stator during the operation of the turbine.

7.5 The joint between two links is formed by first aligning the holes of links 1 and 2 (Fig. 7.22*a* and *b*) and then inserting the pin shown in Fig. 7.22*c*. If the dimension A and a are specified as $A = 1.2 \pm 0.1$ in and $a = 1.1 \pm 0.05$ in, determine the probability of interference between the links 1 and 2. Assume the tolerances to correspond to 3-sigma values.

7.6 A load is supported by three identical cables as shown in Fig. 7.23. If the load and strength of each cable follow normal distribution with the parameters $\mu_L = 200$, $\sigma_L = 50$, $\mu_{S_i} = 90$ and $\sigma_{S_i} = 10$, determine the probability of failure of the system.

7.7 The mean value and the standard deviation of live load in a heavy-manufacturing building are given by 125 lb/ft^2 and 25 lb/ft^2, respectively. Find

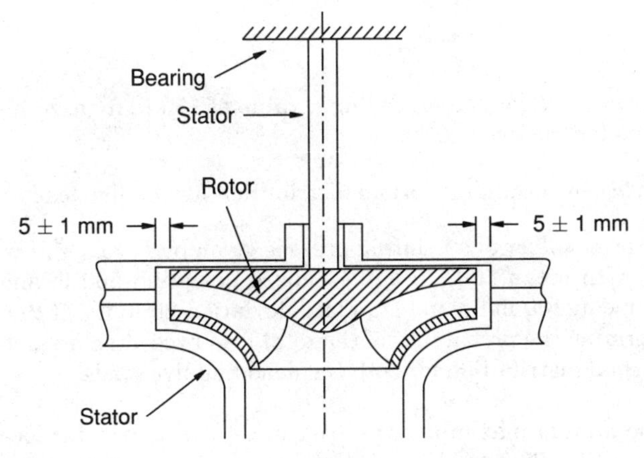

Figure 7.21

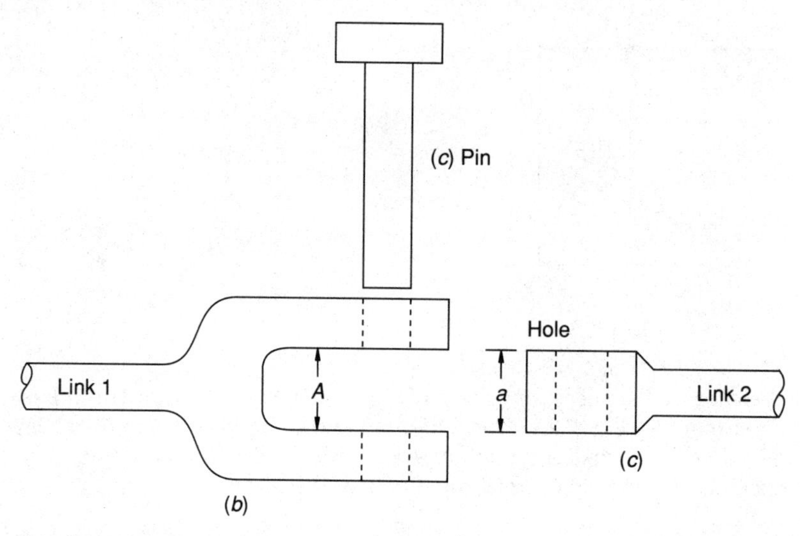

Figure 7.22

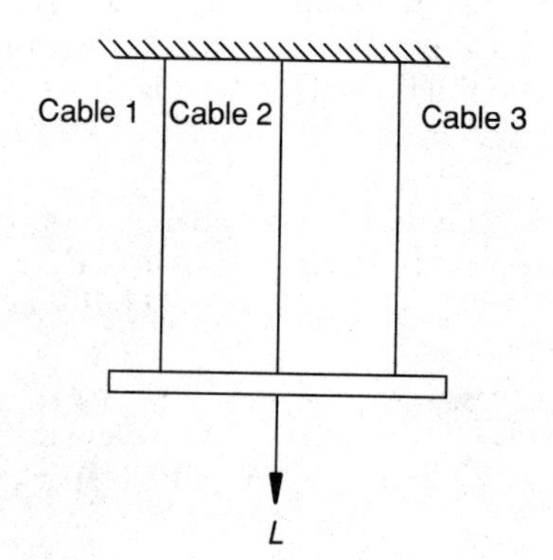

Figure 7.23

the probability of the live load exceeding a value of 150 lb/ft^2 assuming lognormal distribution for the live load.

7.8 Solve Problem 7.7 by assuming normal distribution for the live load.

7.9 The total load on the floor of a printing press is given by $L = L_d + L_l$ where L_d is the dead load with mean 5000 lb and standard deviation 500 lb and L_l is the live load with mean 100 lb/ft^2 and standard deviation 20 lb/ft^2. If the floor area is 400 ft^2, determine the probability of the total load exceeding a value of 50,000 lb. Assume normal distribution for both the dead and live loads.

7.10 A record of the annual maximum wind velocities at a particular location gave the mean value as 79.5 mi/h and the standard deviation as 41.2 mi/h.

Determine the parameters of the distribution using a type-II extreme value distribution for the maximum wind velocity.

7.11 Using the solution of Problem 7.10, find the probability of the maximum wind velocity exceeding a value of 90 mi/h in any single year. Compare your result with the one obtained by assuming normal distribution for the maximum wind velocity.

7.12 Determine the mean return period for annual maximum wind velocities greater than 90 mi/h using the data of Problem 7.10.

7.13 Determine the following quantities corresponding to an earthquake with $M = 8$ and $d = 100$ km:

1. The mean and standard deviation of S_v for $\tau_n = 1.0$ sec and $\zeta = 0.05$ using Eq. (7.59).
2. The mean and standard deviation of the corresponding S_a assuming τ_n to be deterministic.
3. The probability of S_a exceeding a value of 0.1 g by assuming normal distribution for S_a.

7.14 Plot the power spectrum given by Eq. (7.64) for the following data: $\sigma_a = 10$ in/s^2, $H = 0.6$, $f_g = 2$ Hz.

7.15 A structure has a natural frequency of vibration of 2 Hz and a damping ratio of $\zeta = 0.02$. Find the spectral displacement and spectral acceleration of the structure corresponding to a spectral velocity of 40 in/s^2.

7.16 The response spectrum of an earthquake is shown in Fig. 7.17. Find the spectral displacement, spectral velocity, and spectral acceleration of a machine structure whose natural frequency of vibration is 10 Hz and the damping ratio is $\zeta = 0.05$.

Chapter

8

Strength Based Reliability and Interference Theory

Biographical Note

Siméon Denis Poisson

Siméon Denis Poisson was a French mathematician who was born on June 21, 1781, and died on April 25, 1840. During his childhood, he once suspended himself from a ceiling and started swinging as a pendulum to avoid being bothered by the small animals and bugs that roamed around the floor. The study of the pendulum occupied much of his time in later years. His family wanted him to become a physician. However, during his medical studies, when his first patient died while under his ministrations, he developed a dislike of the profession; he changed to the study of mathematics, joined the Ecole Polytechnic in 1798, and succeeded Fourier as professor there in 1808. He made many contributions to the study of mathematical physics, elasticity of materials, Fourier series, heat, acoustics, and probability. The Poisson's ratio in elasticity and Poisson's density and distribution functions in probability are associated with his name [8.15, 8.16].

8.1 Introduction

This chapter deals with the reliability analysis of structural and mechanical components. There is a major difference between electronic and electrical systems and mechanical and structural systems. Most electronic and electrical components and systems deteriorate during use as a result of elevated operating temperature or other similar reasons. A particular component may fail as a direct result of one of these reasons, or it may fail as an indirect result of the failure of some other component of the system. Thus the reliability of an electronic or electrical system is closely associated with the life of the system. The time at which failure occurs will be the primary random variable in these cases. The analysis methods presented in Chapter 6 are applicable to these systems. On the other hand, many mechanical and structural components fail due to extremely small values of strength and/or large values of load. Another difference between the systems is that most electronic and electrical systems are produced in large numbers and hence can be assumed to be nominally identical. The existence of a large sample of nominally identical systems enables us to interpret failure probabilities in terms of relative frequencies. On the other hand, many mechanical and structural systems are not mass produced, hence their failure data is very limited. Thus the reliability analysis of machines and structures may have to be carried in the face of lack of knowledge relating to the properties of the machine or structure and to the uncertain nature of the loads to which it is subjected during its life time.

A mechanical or structural component is considered to be safe and reliable when the strength or resistance of the component exceeds the value of the load acting on it. Thus, the computation of the reliability of the component requires a knowledge of the random nature of the strength (S) and the load (L). If the probability density functions of S and L are known to be $f_S(s)$ and $f_L(l)$ as shown in Fig. 8.1, the reliability of the component can be evaluated by constructing integral equations [8.1]. In certain cases, such as the cases in which S and L follow normal, lognormal, exponential, or Weibull distributions, the integral equations can be reduced to a simple form. However, in a more general case, the reliability of the component can be found only by evaluating the integrals numerically. The methods of evaluating the reliability of structural and mechanical components are discussed in this chapter.

8.2 General Expression for Reliability

The reliability of a component R is given by

$$R = P(S > L) = P(S - L > 0)$$

$$= \int\int f_{S,L}(s,l)\, ds\, dl \tag{8.1}$$

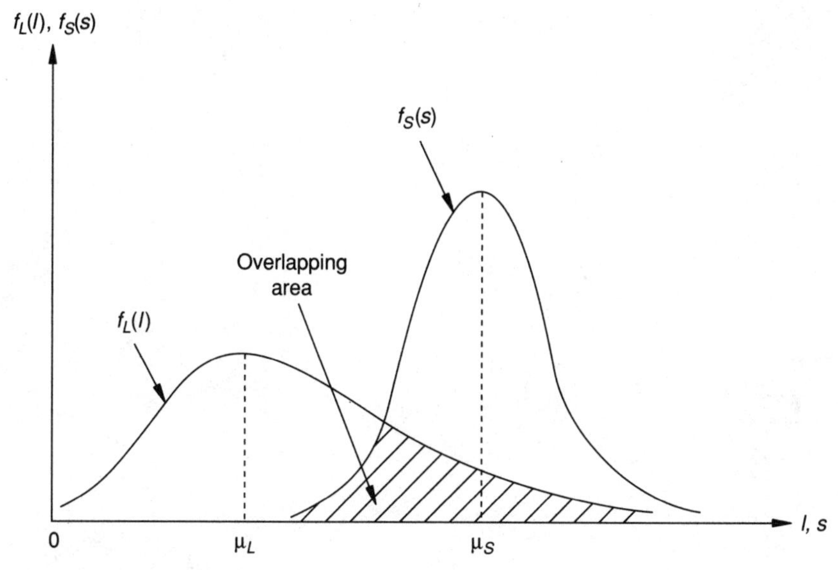

Figure 8.1

where $f_{S,L}(s,l)$ is the joint density function of S and L, and the double integral has to be evaluated over the region $\{(s,l)|_{s-l>0}\}$. If S and L are independent,[1] then the interference area shown in Fig. 8.1, between the probability density functions of S and L, gives a measure of the probability of failure. With reference to Fig. 8.2, the probability of the load (L) falling in a small interval dl around l is equal to the area A_l [8.2, 8.17]:

$$P\left(l - \frac{dl}{2} \leq L \leq l + \frac{dl}{2}\right) = \text{area } A_l = f_L(l)\, dl \tag{8.2}$$

The probability of strength (S) assuming a value larger than the load l is equal to the shaded area under the density function of the strength (i.e., area A_s)

$$P(S > l) = \text{area } A_s = \int_l^\infty f_S(s)\, ds = 1 - F_S(l) \tag{8.3}$$

If the load and the strength are independent, the probability of the load value falling in the interval dl around l and the strength value exceeding the value l happening simultaneously gives the elemental reliability, dR, as

$$dR = f_L(l)\,dl \int_l^\infty f_S(s)\,ds = f_L(l)\,dl\,[1 - F_S(l)] \tag{8.4}$$

[1]The terms strength (S) and load (L) are used in a general sense and are measured in the same units. For example, in most structural and mechanical designs, S represents the allowable or permissible stress and L denotes the induced or applied stress.

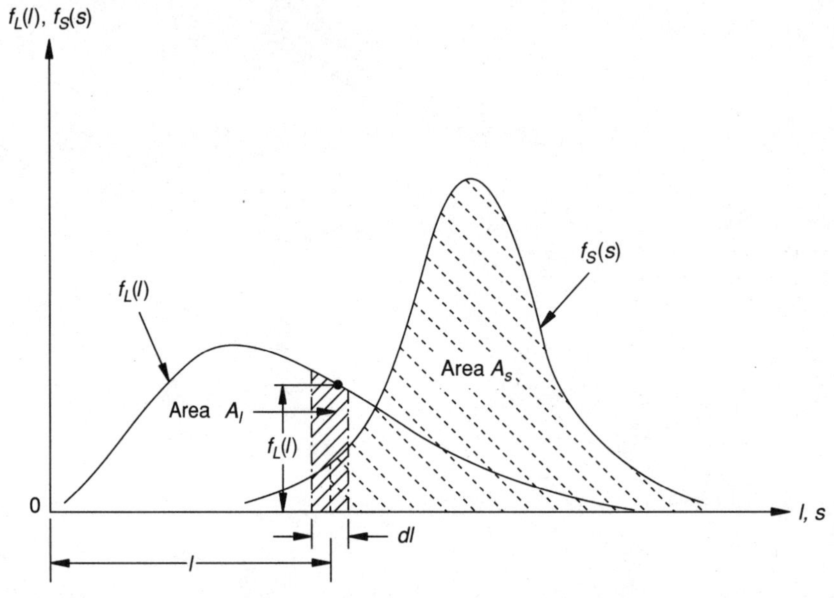

Figure 8.2

Since the reliability of the component is given by the probability of strength exceeding the load for all possible values of the load, we have

$$R = \int dR = \int_{-\infty}^{\infty} f_L(l) \left[\int_l^{\infty} f_S(s)\, ds\right] dl = \int_{-\infty}^{\infty} f_L(l)[1 - F_S(l)]\,dl \tag{8.5}$$

8.2.1 Alternate expression for reliability

The expression for reliability can also be derived by finding the probability of the load assuming a smaller value than the value of the strength. As before, the probability of strength falling in the interval ds around s is given by (see Fig. 8.3)

$$P\left(s - \frac{ds}{2} \le S \le s + \frac{ds}{2}\right) = \text{area } A_s' = f_S(s)\, ds \tag{8.6}$$

The probability of the load assuming a smaller value than s can be obtained as

$$P(L < s) = \text{area } A_l' = \int_{-\infty}^{s} f_L(l)\,dl = F_L(s) \tag{8.7}$$

The probability of strength value falling in the interval ds and the load assuming a value smaller than s occurring simultaneously gives the elemental reliability, dR, as

$$dR = f_S(s)\, ds \int_{-\infty}^{s} f_L(l)\,dl = f_S(s)\, ds\, F_L(s) \tag{8.8}$$

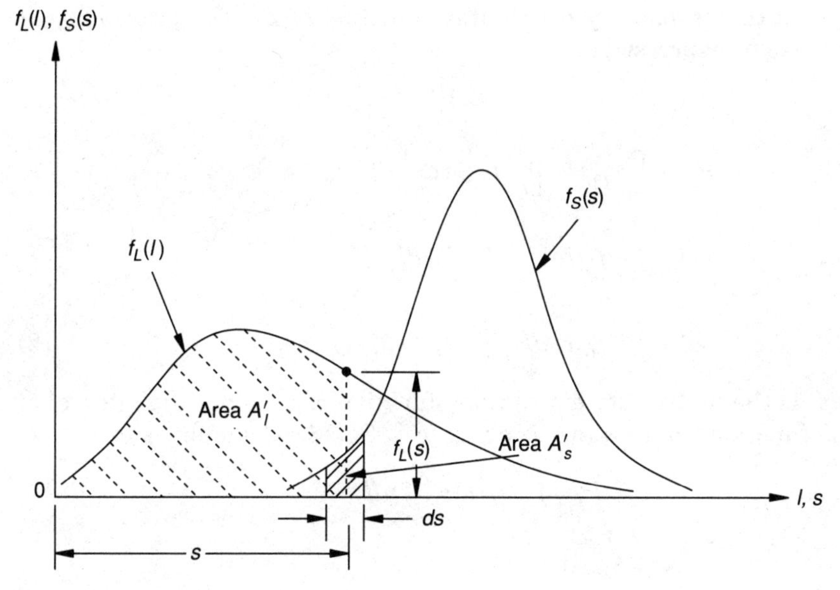

Figure 8.3

Hence, the reliability of the component can be determined as the probability of load assuming a smaller value than strength for all possible values of strength

$$R = \int dR = \int_{-\infty}^{\infty} f_S(s) \left[\int_{-\infty}^{s} f_L(l)\, dl\right] ds = \int_{-\infty}^{\infty} f_S(s)\, F_L(s)\, ds \tag{8.9}$$

8.3 Expression for Probability of Failure

Just as we can derive an expression for reliability, we can derive an expression for the probability of failure or unreliability of the component. The probability of failure (P_f) is the probability of strength falling to a value smaller than the load:

$$P_f = P(S \le L) = 1 - P(L \le S) = 1 - R \tag{8.10}$$

which can be rewritten as

$$\begin{aligned} P_f &= 1 - \int_{-\infty}^{\infty} f_S(s) \left[\int_{-\infty}^{s} f_L(l)\, dl\right] ds \\ &= 1 - \int_{-\infty}^{\infty} f_S(s)\, F_L(s)\, ds \\ &= \int_{-\infty}^{\infty} f_S(s)\, ds - \int_{-\infty}^{\infty} f_S(s)\, F_L(s)\, ds \\ &= \int_{-\infty}^{\infty} \left[1 - F_L(s)\right] f_S(s)\, ds \end{aligned} \tag{8.11}$$

where $F_L(l)$ is the probability distribution function of L. The probability of failure can also be expressed as

$$P_f = 1 - R = 1 - P(S \geq L)$$

$$= 1 - \int_{-\infty}^{\infty} f_L(l) \left[\int_l^{\infty} f_S(s)\, ds\right] dl$$

$$= 1 - \int_{-\infty}^{\infty} f_L(l)\, [1 - F_S(l)]\, dl$$

$$= 1 - \int_{-\infty}^{\infty} f_L(l)\, dl + \int_{-\infty}^{\infty} f_L(l)\, F_S(l)\, dl \tag{8.12}$$

where $F_S(s)$ is the probability distribution function of S. Since the integral of the density function from $-\infty$ to ∞ is equal to 1, Eq. (8.12) simplifies to

$$P_f = \int_{-\infty}^{\infty} f_L(l) F_S(l) dl \tag{8.13}$$

8.4 Reliability when S and L Follow Normal Distribution

For certain distributions of strength and load, simple expressions can be derived for the reliability. We shall consider normal distribution in this section. Let the strength (S) and load (L) be known to follow normal distribution with density functions

$$f_S(s) = \frac{1}{\sigma_S \sqrt{2\pi}}\, e^{-1/2\{(s - \mu_S)/\sigma_S\}^2}; \qquad -\infty < s < \infty \tag{8.14}$$

and

$$f_L(l) = \frac{1}{\sigma_L \sqrt{2\pi}}\, e^{-1/2\{(l - \mu_L)/\sigma_L\}^2}; \qquad -\infty < l < \infty \tag{8.15}$$

were μ_S and μ_L denote the mean values, and σ_S and σ_L represent the standard deviations of the variables S and L, respectively. Then the reliability of the component can be expressed as

$$R = P(S - L \geq 0) = P(X \geq 0) \tag{8.16}$$

where $X = S - L$ is a new random variable. Since X is a linear function of the normally distributed random variables S and L, it also follows normal distribution. The probability density function of X (see Fig. 8.4) is given by

$$f_X(x) = \frac{1}{\sigma_X \sqrt{2\pi}}\, e^{-1/2\{(x - \mu_X)/\sigma_X\}^2}; \qquad -\infty < x < \infty \tag{8.17}$$

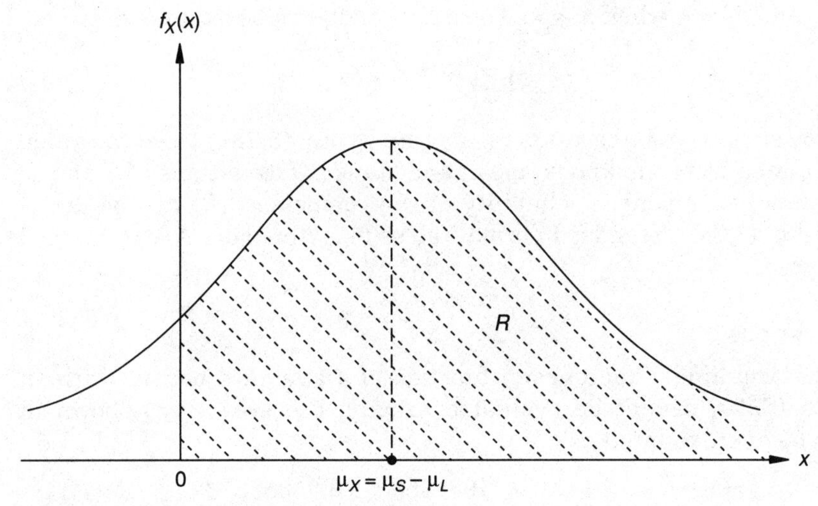

Figure 8.4 Probability density function of $X = S - L$.

If S and L are assumed to be independent, the mean and standard deviations of X can be expressed as

$$\mu_X = \mu_S - \mu_L \tag{8.18}$$

$$\sigma_X = (\sigma_S^2 + \sigma_L^2)^{1/2} \tag{8.19}$$

Equation (8.16) can be rewritten as

$$R = P(X \geq 0) = \int_0^\infty f_X(x)\, dx$$

$$= \frac{1}{\sigma_X \sqrt{2\pi}} \int_0^\infty e^{-1/2\{(x - \mu_X)/\sigma_X\}^2}\, dx \tag{8.20}$$

By defining the standard normal variate z as

$$z = \frac{x - \mu_X}{\sigma_X} \tag{8.21}$$

we can express

$$dz = \frac{dx}{\sigma_X} \quad \text{or} \quad \mathrm{dx} = \sigma_X\, dz \tag{8.22}$$

The values of z corresponding to the lower and upper limits of the integration of X are given by

$$z_1 = \frac{0 - \mu_X}{\sigma_X} = -\frac{\mu_S - \mu_L}{\sqrt{\sigma_S^2 + \sigma_L^2}} \tag{8.23}$$

when $X = 0$, and $z_2 = \infty$ when $X = \infty$. Thus, Eq. (8.20) can be expressed as

$$R = \frac{1}{\sqrt{2\pi}} \int_{z=z_1}^{\infty} e^{-1/2 z^2} \, dz \qquad (8.24)$$

where the lower limit of integration z_1 is given by Eq. (8.23). Once the value of z_1 is calculated from the known mean and standard deviations of S and L, the corresponding component reliability of the component (R) can be determined from Eq. (8.24). Standard normal tables (see Appendix A) can be used for this purpose.

8.4.1 Approximate expressions

Although the area under the density function of a standard normal variate, given by Eq. (8.24), cannot be evaluated exactly, the area $P_f(z)$ shown in Fig. 8.5 can be approximated as

$$P_f(z) = 1 - R(z) = f(z)\,\{0.31938153\,t - 0.356563782\,t^2 + 1.781477937\,t^3 - 1.821255978\,t^4 + 1.330274429\,t^5\} + \text{error} \qquad (8.25)$$

where

$$f(z) = \frac{1}{\sqrt{2\pi}} e^{-1/2 z^2} \qquad (8.26)$$

$$t = \frac{1}{(1 + 0.2316419z)} \qquad (8.27)$$

and the error is less than 7.5×10^{-8}. Similarly, if $P_f(z) = P_o$ is given, we can find z approximately as

$$z = t - \frac{2.515517 + 0.802853\,t + 0.010328\,t^2}{1 + 1.432788\,t + 0.189269\,t^2 + 0.001308\,t^3} + \text{error} \qquad (8.28)$$

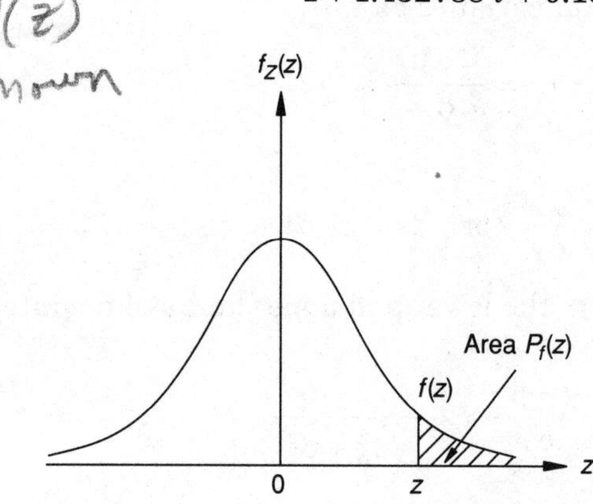

Figure 8.5

where

$$t = \sqrt{\log_e \left(\frac{1}{P_o^2} \right)} \tag{8.29}$$

and the error is less than 4.5×10^{-5}.

Example 8.1 The torque transmitting capacity (T) of a cone clutch under uniform pressure [8.8, 8.11] is given by

$$T = \frac{F\,f}{3 \sin \alpha} \left[\frac{D^3 - d^3}{D^2 - d^2} \right]$$

where D = outer diameter of the cone, d = inner diameter of the cone, α = semi-cone angle, f = coefficient of friction between cup and cone, and F = axial force applied (see Fig. 8.6). For a specific clutch, which has molded asbestos on a cast iron cone, $D = 10$ in, $d = 8$ in, $\alpha = 20^o$, and $F = 100$ lb. The coefficient of friction can be assumed to be normally distributed with a mean value of 0.3 and a standard deviation of 0.02. If the torque required to be transmitted by the cone clutch is also normally distributed with mean and standard deviations of 400 lb·in and 40 lb·in, respectively, estimate the reliability of the cone clutch.

solution Since the coefficient of friction (f) is the only random variable, the mean and standard deviations of the torque transmitting capacity (T) of the clutch can be found as

$$\overline{T} = \frac{F\,\overline{f}}{3 \sin \alpha} \left[\frac{D^3 - d^3}{D^2 - d^2} \right] = \frac{100(0.3)}{3 \sin 20^o} \left[\frac{10^3 - 8^3}{10^2 - 8^2} \right] = 396.33 \text{ lb·in}$$

$$\sigma_T = \frac{F\,\sigma_f}{3 \sin \alpha} \left[\frac{D^3 - d^3}{D^2 - d^2} \right] = \frac{100(0.02)}{3 \sin 20^o} \left[\frac{10^3 - 8^3}{10^2 - 8^2} \right] = 26.4225 \text{ lb·in}$$

The required torque to be transmitted (T_r) is given by

$$N(T_r, \sigma_{T_r}) = N(400, 40) \text{ lb·in}$$

By defining a random variable X as $X = T - T_r$, we find

$$\overline{X} = \overline{T} - \overline{T}_r = 396.33 - 400 = -3.67 \text{ lb·in}$$

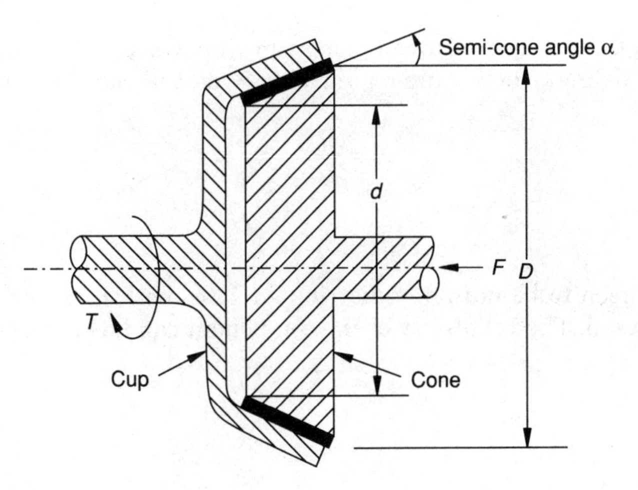

Figure 8.6

and

$$\sigma_X = \sqrt{\sigma_T^2 + \sigma_{T_r}^2} = \sqrt{26.4225^2 + 40^2} = 47.9390 \text{ lb·in}$$

Thus, the reliability of the cone clutch can be determined as

$$R = 1 - \Phi\left(-\frac{\bar{X}}{\sigma_X}\right) = 1 - \Phi\left(\frac{3.67}{47.9390}\right) = 1 - \Phi(0.07639)$$

From standard normal tables, we find that $R = 1 - 0.530446 = 0.469554$.

Example 8.2 A mechanical fuse has a specially designed screw head that is required to shear off when the applied torque exceeds a value of 400 in·lb. The fuse is permitted to have a maximum of 100 failures per 50,000 units. Find the torque at which the fuse is to be designed to fail if the standard deviation of the shear strength of the material of the fuse, expressed in terms of the applied torque, is 15 in·lb. The standard deviation of the torque wrenches used to tighten the screw head of the fuse is 10 in·lb of torque. Assume that the applied torque and the torsional strength of the fuse follow normal distribution.

solution The permissible probability of failure of the fuse is

$$P_f = \frac{100}{50{,}000} = 0.002$$

If S and L denote strength and load, we have, in the present case,

$$\mu_S = 400 \text{ in·lb}, \qquad \sigma_S = 10 \text{ in·lb}, \qquad \sigma_L = 15 \text{ in·lb}$$

If $X = S - L$, $\mu_X = 400 - \mu_L$ and $\sigma_X = \sqrt{\sigma_S^2 + \sigma_L^2} = \sqrt{10^2 + 15^2} = 18.0278$ in·lb. Since $1 - \Phi(z_1) = 0.002$, we find from the standard normal tables (Appendix A) that $z_1 = 2.88$, and hence

$$2.88 = \frac{400 - \mu_L}{18.0278}, \qquad \text{or} \qquad \mu_L = 348.0799 \text{ in·lb}$$

Thus, the mechanical fuse must be designed to fail at a mean torque of $\mu_L = 348.0799$ in·lb.

8.5 Reliability when S and L Follow Lognormal Distribution

When the strength and load of a component are known to follow lognormal distribution, we define a new random variable X as the ratio of strength to load:

$$X = \frac{S}{L} \tag{8.30}$$

This gives

$$\ln X = \ln S - \ln L \tag{8.31}$$

and $\ln X$ can be seen to be normally distributed since both $\ln S$ and $\ln L$ are normally distributed. The reliability of the component can be expressed as

$$R = P\left(\frac{S}{L} > 1\right) = P(X > 1)$$

$$= \int_1^\infty f_X(x)\, dx \tag{8.32}$$

Since $\ln X$ is normally distributed, X follows lognormal distribution. We define a standard normal variate z as

$$z = \frac{\ln x - \mu_{\ln X}}{\sigma_{\ln X}} \tag{8.33}$$

where $\mu_{\ln X}$ and $\sigma_{\ln X}$ are the mean and standard deviation of $\ln X$, respectively. If Eq. (8.32) is rewritten in terms of z, the new limits of integration can be found as follows:

When

$$x = 1,\ z = z_1 = \frac{\ln 1 - \mu_{\ln X}}{\sigma_{\ln X}} = \frac{0 - (\mu_{\ln S} - \mu_{\ln L})}{\sqrt{\sigma_{\ln S}^2 + \sigma_{\ln L}^2}}$$

When

$$x = \infty,\ z = z_2 = \infty.$$

Thus Eq. (8.32) can be rewritten in terms of z as

$$R = \int_{z_1}^{\infty} \frac{1}{\sqrt{2\pi}} e^{-1/2 z^2}\, dz \tag{8.34}$$

where

$$z_1 = -\frac{(\mu_{\ln S} - \mu_{\ln L})}{\sqrt{\sigma_{\ln S}^2 + \sigma_{\ln L}^2}} \tag{8.35}$$

8.5.1 Mean and standard deviation of $\ln X$ in terms of mean and standard deviation of X [8.2]

If X is lognormally distributed, its density function is given by

$$f_X(x) = \frac{1}{\sqrt{2\pi}\,\sigma x} \exp\left\{-\frac{1}{2}\left(\frac{\ln x - \mu}{\sigma}\right)^2\right\}; \qquad x > 0 \tag{8.36}$$

where $\mu = E(\ln X) = \mu_{\ln X}$, $\sigma = \sigma_{\ln X}$ and the variable $\ln x$ is normally distributed. Defining $Y = \ln X$, we can find the expected value and the standard deviation of X as follows:

$$E(X) = \int_{-\infty}^{\infty} \frac{1}{\sqrt{2\pi}\,\sigma} \exp\left\{-\frac{1}{2}\left(\frac{\ln x - \mu}{\sigma}\right)^2\right\} dx \tag{8.37}$$

Since $x = e^y$ and $dy = \frac{1}{x} dx$ or $dx = x\, dy$, Eq. (8.37) can be rewritten as

$$E(X) = E(e^Y) = \int_{-\infty}^{\infty} \frac{1}{\sqrt{2\pi}\,\sigma} e^y \exp\left\{-\frac{1}{2}\left(\frac{\ln x - \mu}{\sigma}\right)^2\right\} dy \tag{8.38}$$

This equation can be simplified (see Problem 8.31) as:

$$E(X) = \exp\left(\mu + \frac{\sigma^2}{2}\right) \tag{8.39}$$

The computation of σ_X requires the evaluation of $E(X^2)$

$$E(X^2) = E(e^{2y}) = \int_{-\infty}^{\infty} \frac{e^{2y}}{\sqrt{2\pi}\ \sigma x} \exp\left\{-\frac{1}{2}\left(\frac{\ln x - \mu}{\sigma}\right)^2\right\} dx \tag{8.40}$$

This equation can be simplified (see Problem 8.31) as:

$$E(X^2) = \exp\{2(\mu + \sigma^2)\} \tag{8.41}$$

Thus the variance of X is given by

$$\sigma_X^2 = E[\{X - E(X)\}^2] = E(X^2) - [E(X)]^2$$

$$= \exp\{2(\mu + \sigma^2)\} - \exp\left\{2\left(\mu + \frac{\sigma^2}{2}\right)\right\}$$

$$= \exp(2\mu + \sigma^2)(\exp(\sigma^2) - 1) \tag{8.42}$$

Equations (8.39) and (8.42) can be solved to find (see Eqs. 3.61 and 3.62):

$$\mu = \ln E(X) - \frac{\sigma^2}{2} = \ln \mu_X - \frac{\sigma^2}{2} \tag{8.43}$$

and

$$\sigma^2 = \ln\left[\frac{\sigma_X^2}{\{E(X)\}^2} + 1\right] = \ln\left[\frac{\sigma_X^2}{\mu_X^2} + 1\right] \tag{8.44}$$

Example 8.3 The bending stress developed in a gear tooth is given by the Lewis equation [8.7]

$$s = \frac{F_t\, p_d}{w\, Y}$$

where F_t = tangential load acting on the gear tooth, p_d = diametral pitch $= \pi / p$, p = circular pitch, w = face width, and Y = Lewis form factor (see Fig. 8.7). In an automobile transmission, a gear has full-depth teeth with a pressure angle of 20°, a diametral pitch of unity, a face width of 0.5 in, and a Lewis form factor of 0.3. The tangential load on the gear tooth is lognormally distributed with a mean value of 3720 lb and a coefficient of variation of 0.080645. The yield strength of the material is known to follow lognormal distribution with a mean value of 32,000 (lb/in^2). Find the maximum permissible value of the coefficient of variation of yield strength required in order to have a reliability of 0.9999 for the gear tooth.

solution Since the bending stress induced in the gear tooth ($s \equiv L$) is given by a constant $[p_d / \{w\ Y\} = 1/\{0.5(0.3)\} = 1/0.15]$ times the tangential load (F_t), L can also be assumed to

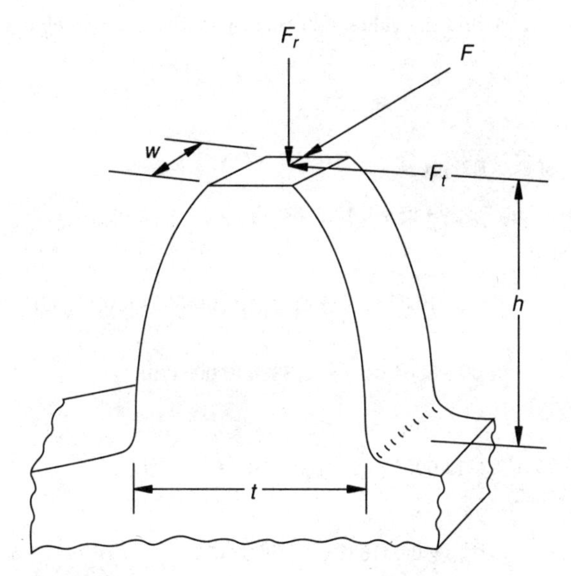

Figure 8.7

follow lognormal distribution with a mean value

$$\overline{L} = \overline{s} = \frac{\overline{F}_t}{0.15} = \frac{3720}{0.15} = 24{,}800 \text{ lb/in}^2$$

and a standard deviation

$$\sigma_L = \sigma_s = \frac{1}{0.15}\sigma_{F_t} = \frac{1}{0.15}\gamma_{F_t}\,\overline{F}_t = \frac{1}{0.15}\,(0.080645)(3720)$$

$$= 2000 \text{ lb/in}^2$$

Thus $E(S) = 32{,}000$ lb/in^2, $E(L) = 24{,}800$ lb/in^2, $\sigma_L = 2000$ lb/in^2, and hence

$$\sigma^2_{\ln L} = \ln\left[\frac{2000^2}{24{,}800^2} + 1\right] = \ln 1.0065 = 0.006479$$

$$\mu_{\ln L} = \ln E(L) - \left\{\frac{\sigma^2_{\ln L}}{2}\right\}$$

$$= 10.118599 - \frac{1}{2}\,(0.006479) = 10.115360$$

$$\mu_{\ln S} = \ln E(S) - \left\{\frac{\sigma^2_{\ln S}}{2}\right\}$$

$$= 10.3735 - \frac{1}{2}\,\sigma^2_{\ln S} \qquad \text{(E1)}$$

From standard normal tables, we find the value of z_1 corresponding to a realiability of 0.9999 as

$$z_1 = -\frac{\mu_{\ln S} - \mu_{\ln L}}{\sqrt{\sigma^2_{\ln S} + \sigma^2_{\ln L}}} = -3.72 \tag{E2}$$

Squaring and rearranging of Eq. (E2) gives

$$\mu^2_{\ln S} - 2\,\mu_{\ln S}\,\mu_{\ln L} + \mu^2_{\ln L} = 13.8384\,\sigma^2_{\ln S} + 13.8384\,\sigma^2_{\ln L} \tag{E3}$$

Equations (E3) and (E1) yield

$$\left(10.3735 - \frac{1}{2}\,\sigma^2_{\ln S}\right)^2 - 2\left(10.3735 - \frac{1}{2}\,\sigma^2_{\ln S}\right)(10.11536) + (10.11536)^2$$

$$= 13.8384\,\sigma^2_{\ln S} + 13.8384\,(0.006479)$$

that is,

$$0.25\,\sigma^4_{\ln S} - 14.0965\,\sigma^2_{\ln S} - 0.02306 = 0 \tag{E4}$$

The roots of Eq. (E4) are

$$\sigma^2_{\ln S} = 0.001636\,; \qquad 56.3876$$

Considering the smaller value, we have

$$\sigma^2_{\ln S} = 0.001636$$

By definition,

$$\sigma^2_{\ln S} = \ln\left[\frac{\sigma^2_S}{\{E(S)\}^2} + 1\right]$$

that is,

$$0.001636 = \ln\left[\frac{\sigma^2_S}{32000^2} + 1\right]$$

that is,

$$\sigma^2_S = (e^{0.001636} - 1)(32000)^2 = (0.001637)(32000)^2 = (1294.8495)^2$$

Thus, the permissible value of the coefficient of variation of yield strength is

$$\gamma_S = \frac{\sigma_S}{E(S)} = \frac{1294.8495}{32000} = 0.040464$$

8.6 Reliability when S and L Follow Exponential Distribution

When S and L follow exponential distribution, their probability density functions can be expressed as

$$f_S(s) = \frac{1}{\mu_S}\,e^{-s/\mu_S}; \qquad 0 \le s < \infty \tag{8.45}$$

$$f_L(l) = \frac{1}{\mu_L} e^{-l/\mu_L}; \qquad 0 \le l < \infty \tag{8.46}$$

where μ_S and μ_L denote the mean values of strength and load, respectively. The reliability of the component can be computed by substituting Eqs. (8.45) and (8.46) into Eq. (8.9) as

$$R = \int_o^\infty f_S(s) \cdot \left[\int_o^s f_L(l) \cdot dl \right] \cdot ds$$

$$= \int_o^\infty \frac{1}{\mu_S} e^{-s/\mu_S} \left[\int_o^s \frac{1}{\mu_L} e^{-l/\mu_L}\, dl \right] ds$$

$$= \frac{1}{\mu_S \, \mu_L} \int_o^\infty e^{-s/\mu_S} \left[-\mu_L \, e^{-l/\mu_L} \right] \Big|_o^s \, ds$$

$$= -1/\mu_S \int_o^\infty e^{-s/\mu_S} \left[e^{-s/\mu_L} - 1 \right] ds$$

$$= -\frac{1}{\mu_S} \left[\int_o^\infty e^{-(1/\mu_S + 1/\mu_L)s} \, ds - \int_o^\infty e^{-s/\mu_S} \, ds \right]$$

$$= \frac{1}{\mu_S \left(\dfrac{1}{\mu_S} + \dfrac{1}{\mu_L} \right)} \left[e^{-(1/\mu_S + 1/\mu_L)s} \right] \Big|_o^\infty - \frac{1}{\mu_S} \frac{1}{\left(\dfrac{1}{\mu_S} \right)} \left[e^{-s/\mu_S} \right] \Big|_o^\infty$$

$$= \frac{1}{\mu_S \left(\dfrac{1}{\mu_S} + \dfrac{1}{\mu_L} \right)} [0 - 1] - [0 - 1]$$

$$= \frac{\mu_S}{\mu_L + \mu_S} \tag{8.47}$$

Equation (8.47) can also be written as

$$R = \frac{\lambda_L}{\lambda_L + \lambda_S} \tag{8.48}$$

where

$$\lambda_S = \frac{1}{\mu_S} \tag{8.49}$$

and

$$\lambda_L = \frac{1}{\mu_L} \tag{8.50}$$

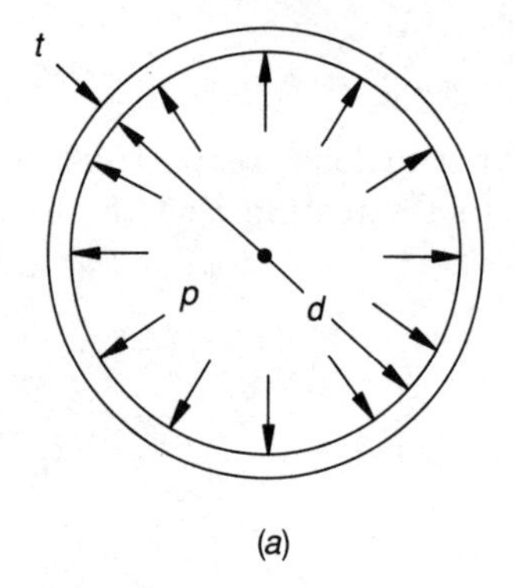

(a)

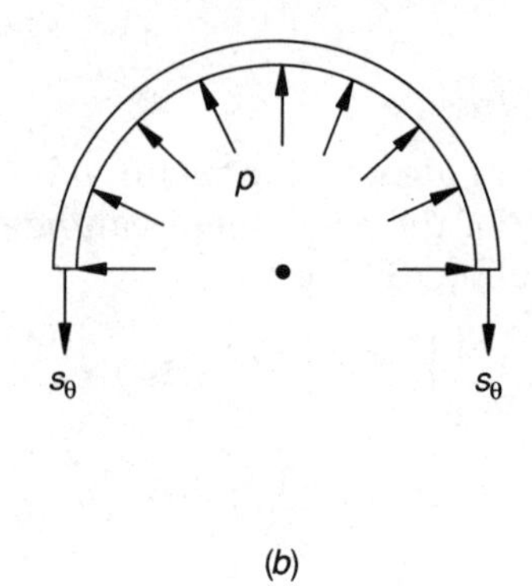

(b)

Figure 8.8

Example 8.4 The hoop (circumferential) stress developed in a thin-walled cylindrical pressure vessel is given by

$$s_\theta = \frac{pd}{2t}$$

where $d =$ inside diameter, $p =$ internal pressure and $t =$ wall thickness (see Fig. 8.8). In a particular application, the vessel is subjected to an exponentially distributed pressure with a mean value of 400 lb/in^2. Assuming exponential distribution for the yield strength of the vessel (s_y) with a mean value of 30,000 lb/in^2, determine the reliability of the pressure vessel. Assume the values of d and t to be 72 in and 0.5 in, respectively.

solution The mean value of the hoop stress induced in the pressure vessel is given by

$$\bar{s}_\theta = \frac{\bar{p}d}{2t} = \frac{(400)(72)}{(2)(0.5)} = 28{,}800 \text{ lb/in}^2$$

Thus we have $\bar{S} = \bar{s}_y = 30{,}000$ lb/in^2 and $\bar{L} = \bar{s}_\theta = 28{,}800$ lb/in^2 and hence the reliability of the pressure vessel is given by Eq. (8.47)

$$R = \frac{\bar{S}}{\bar{L} + \bar{S}} = \frac{30{,}000}{28{,}800 + 30{,}000} = 0.5102$$

8.7 Reliability when S and L Follow Extreme Value Distributions

As discussed in Chapters 4 and 7, the strength of most materials can be described, based on the weakest link concept, using extreme value distributions for the smallest value. Similarly, if the load is caused by earthquakes, wind velocities, gusts, landing loads, or ocean waves, it can be described by the extreme value distributions for the largest value. The computation of reliability of a component is considered in this section with extreme value distributions for strength and load. The density functions of extreme value distributions for the smallest value of strength S (see Chapter 4) can be expressed as follows:

Type-I

$$f_S(s) = a \exp[a\,(s - w) - \exp\{a\,(s - w)\}]; \qquad -\infty < s < \infty \tag{8.51}$$

where a and w are the parameters of the distribution.

Type-II

$$f_S(s) = \frac{m}{s}\left|\frac{w}{s}\right|^{m-1} \exp\left[-\left|\frac{w}{s}\right|^{m}\right] \tag{8.52}$$

where $s \leq 0$ and $m > 0$.

Type-III

$$f_S(s) = \frac{m}{v - \varepsilon}\left(\frac{s - \varepsilon}{v - \varepsilon}\right)^{m-1} \exp\left[-\left(\frac{s - \varepsilon}{v - \varepsilon}\right)^{m}\right] \tag{8.53}$$

where $s \geq \varepsilon$, $v > \varepsilon$, and $m > 0$. The density functions of extremal distributions corresponding to the largest value of load L (see Chapter 4) are given by:

Type-I

$$f_L(l) = a \exp\{-a\,(l - w) - \exp[-a\,(l - w)]\} \tag{8.54}$$

where $-\infty < l < \infty$ and a and w are the parameters of the distribution.

Type-II

$$f_L(l) = \frac{m}{w}\left(\frac{w}{l}\right)^{m-1} \exp\left[-\left(\frac{w}{l}\right)^{m}\right] \tag{8.55}$$

where $l \geq 0$, $w > 0$, and $m > 0$.

Type-III

$$f_L(l) = \frac{m}{w - v}\left(\frac{w - l}{w - v}\right)^{m-1} \exp\left[-\left(\frac{w - l}{w - v}\right)^{m}\right] \tag{8.56}$$

where $l \leq w$, $v < w$, and $m > 0$.

The reliability of a component can be determined using Eq. (8.5) for any specified $f_L(l)$ and $f_S(s)$:

$$R = \int_{-\infty}^{\infty} f_L(l) \left[\int_{l}^{\infty} f_S(s) \cdot ds\right] dl \tag{8.57}$$

Equation (8.9) can also be used for this purpose. The procedure is illustrated by considering the typical case given in the next section.

8.7.1 When S and L follow type-III extremal distributions

When S and L follow type-III extremal distributions that correspond to the smallest and the largest values, respectively, we have

$$f_L(l) = \frac{m_1}{w_1 - v_1}\left(\frac{w_1 - l}{w_1 - v_1}\right)^{m_1 - 1} \exp\left[-\left(\frac{w_1 - l}{w_1 - v_1}\right)^{m_1}\right] \tag{8.58}$$

$$f_S(s) = \frac{m_2}{v_2 - \varepsilon_2}\left(\frac{s - \varepsilon_2}{v_2 - \varepsilon_2}\right)^{m_2 - 1} \exp\left[-\left(\frac{s - \varepsilon_2}{v_2 - \varepsilon_2}\right)^{m_2}\right] \tag{8.59}$$

with $l \le w_1$, $v_1 < w_1$, $m_1 > 0$, $s \ge \varepsilon_2$, $v_2 > \varepsilon_2$, and $m_2 > 0$. Thus the reliability of the component can be evaluated as

$$R = \int_{-\infty}^{\infty} \frac{m_1}{w_1 - v_1}\left(\frac{w_1 - l}{w_1 - v_1}\right)^{m_1 - 1} \exp\left[-\left(\frac{w_1 - l}{w_1 - v_1}\right)^{m_1}\right] \left\{\int_l^{\infty} \frac{m_2}{v_2 - \varepsilon_2}\left(\frac{s - \varepsilon_2}{v_2 - \varepsilon_2}\right)^{m_2 - 1} \exp\left[-\left(\frac{s - \varepsilon_2}{v_2 - \varepsilon_2}\right)^{m_2}\right] ds\right\} dl \tag{8.60}$$

that is,

$$R = \frac{m_1 m_2}{(w_1 - v_1)^{m_1}(v_2 - \varepsilon_2)^{m_2}} \int_{-\infty}^{\infty} (w_1 - l)^{m_1 - 1} \exp\left[-\left(\frac{w_1 - l}{w_1 - v_1}\right)^{m_1}\right] \left\{\int_l^{\infty} (s - \varepsilon_2)^{m_2 - 1} \exp\left[-\left(\frac{s - \varepsilon_2}{v_2 - \varepsilon_2}\right)^{m_2}\right] ds\right\} dl \tag{8.61}$$

The integral in Eq. (8.61) has to be evaluated numerically using a suitable procedure [8.5].

8.8 Reliability in Terms of Experimentally Determined Distributions of S and L

When experiments are conducted to find the values of the random variables S and L, several values of S and L can be observed. From these observed values, empirical distribution functions can be determined for describing the random variables S and L. In this case, a graphical procedure can be used for finding the reliability of the component as described by the following steps [8.2, 8.17].

Step 1 Let n experiments be conducted to find the values of the random variable S and let $s_1, s_2, \ldots, s_n$ be the observed values of S. From these values, we can plot the approximate distribution function of S, that is, $F_S(s)$.

Step 2 Similarly, let m experiments be conducted to find the values of the random variable L and let $l_1, l_2, \ldots, l_m$ be the observed values of L. From these values, plot the approximate distribution function of L, that is, $F_L(l)$.

Step 3 Define two variables X and Y as

$$X = \int_0^l f_L(l')\,dl' = F_L(l) \tag{8.62}$$

$$Y = \int_l^\infty f_S(s)\,ds = 1 - F_S(l) \tag{8.63}$$

and find the values of the pairs (X_1,Y_1), (X_2,Y_2), . . . , corresponding to different discrete values of the load l_1, l_2, . . . , respectively, from the results of Steps 1 and 2. Plot a graph of X versus Y from the computed pairs of values (X_i,Y_i); i = 1, 2, . . . , m.

Step 4 According to Eq. (8.5), the reliability of the component is given by

$$R = \int_{-\infty}^\infty f_L(l)\left[\int_l^\infty f_S(s)\,ds\right]dl \tag{8.64}$$

Noting that

$$\int_l^\infty f_S(s)\,ds = Y,$$

$$f_L(l)\,dl = dX,$$

and the range of X is from 0 to 1, Eq. (8.64) can be expressed in terms of X and Y as

$$R = \int_0^1 Y\,dX \tag{8.65}$$

This equation shows that reliability of the component is the same as the area under X versus Y curve. Since the curve X versus Y was plotted in Step 3, the area under the curve (and hence the reliability R) can be found graphically. The procedure is illustrated with the help of the following example.

Example 8.5 Several machine tool belt drives were tested experimentally under identical operating (cutting) conditions. The stresses induced in ten belts were found to be 22,300, 11,600, 15,850, 19,900, 13,650, 16,950, 26,750, 12,700, 18,400 and 15,100 lb/in^2. The values of strengths found by testing twelve of the belts used in the machine tool drives were 21,100, 26,950, 19,200, 30,150, 22,050, 24,350, 18,250, 25,700, 23,400, 19,950, 21,600 and 20,500 lb/in^2. Find the reliability of the belt drives.

solution First, the estimated cumulative distribution functions for the strength and load (induced stress) are to be determined. For finding the distribution functions of strength, the strength data are arranged in an increasing order and a cumulative probability of (i/n) is assigned to the ith value with N denoting the total number of values observed. A similar procedure is adopted for finding the distribution function of the induced stress. The results are shown in Table 8.1. These results are used to plot the approximate probability distributions shown in Figs. 8.9a and b. Once the approximate $F_S(s)$ and $F_L(l)$ are determined, the values of X and Y for different values of the induced stress (l) can be found as indicated in Table 8.2. These results can be used to plot a graph between X and Y as shown in Fig. 8.10. The area under the $X - Y$ curve gives the reliability of the belt drive. This area and, hence, the reliability of the belt drive is found to be approximately 0.8525.

TABLE 8.1 Distribution Functions of Strength and Stress

Strength data			Stress (load) data		
Number	Strength (lb/in^2)	Cumulative probability	Number	Stress (load) (lb/in^2)	Cumulative Probability
1	18,250	0.083	1	11,600	0.1
2	19,200	0.167	2	12,700	0.2
3	19,950	0.250	3	13,650	0.3
4	20,500	0.333	4	15,100	0.4
5	21,100	0.417	5	15,850	0.5
6	21,600	0.500	6	16,950	0.6
7	22,050	0.583	7	18,400	0.7
8	23,400	0.667	8	19,900	0.8
9	24,350	0.750	9	22,300	0.9
10	25,700	0.833	10	26,750	1.0
11	26,950	0.917			
12	30,150	1.000			

8.9 Factor of Safety Corresponding to a Given Reliability

The factor of safety (n) corresponding to any specified reliability can be derived once the probability distribution functions of strength and load are known [8.3]. To illustrate the procedure, we assume that S and L follow normal distribution. Let $X = S - L$ so that the mean and standard deviation of X are given by

$$\mu_X = \mu_S - \mu_L \tag{8.66}$$

$$\sigma_X = \sqrt{\sigma_S^2 + \sigma_L^2} \tag{8.67}$$

where S and L have been assumed to be independent. By defining a standard normal variate z as

$$z = \frac{x - \mu_X}{\sigma_X} \tag{8.68}$$

the reliability can be expressed as

$$R = P(X \geq 0) = P\left[\left(z_1 = -\frac{\mu_X}{\sigma_X}\right) \geq 0\right]$$

$$= P\left[\frac{\mu_L - \mu_S}{\sqrt{\sigma_S^2 + \sigma_L^2}} \geq 0\right] \tag{8.69}$$

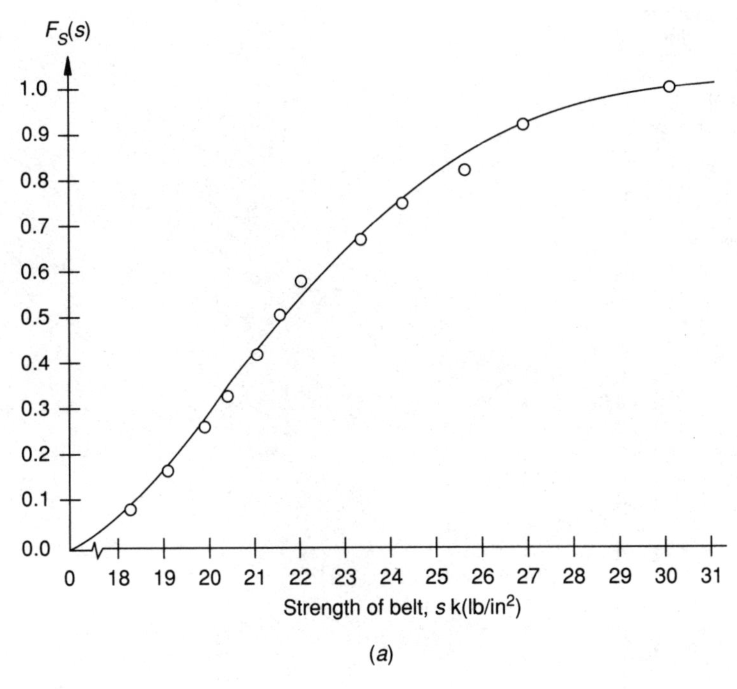

(a)

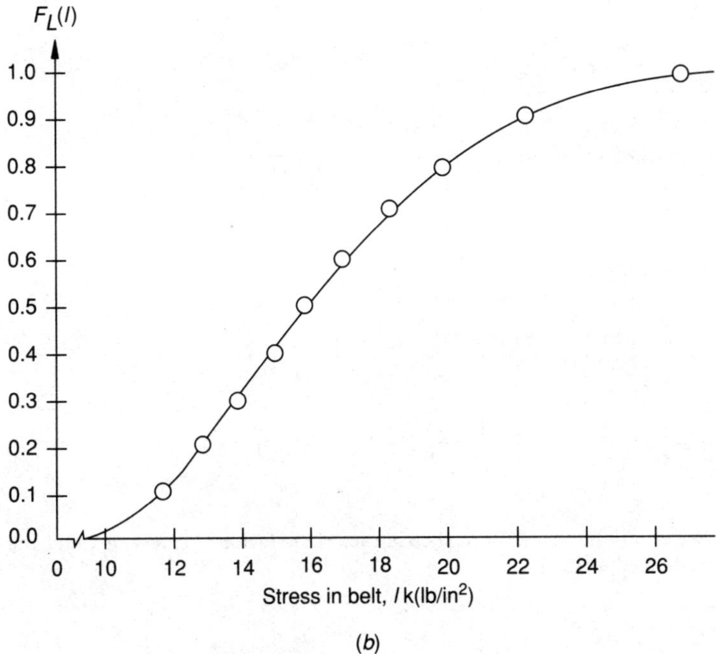

(b)

Figure 8.9 (*a*) Approximate distribution function of strength s; (*b*) approximate distribution function of induced stress l.

TABLE 8.2 Values of X and Y for Different Settings of the Induced Stress

Stress, l (lb/in^2)	$X = F_L(l)$ (from Fig. 8.9b)	$Y = 1 - F_S(l)$ (from Fig. 8.9a)
0	0.0	1 − 0 = 1
2000	0.0	1 − 0 = 1
4000	0.0	1 − 0 = 1
6000	0.0	1 − 0 = 1
8000	0.0	1 − 0 = 1
10,000	0.02	1 − 0 = 1
12,000	0.13	1 − 0 = 1
14,000	0.31	1 − 0 = 1
16,000	0.50	1 − 0 = 1
17,000	0.60	1 − 0.02 = 0.98
18,000	0.67	1 − 0.06 = 0.94
19,000	0.74	1 − 0.22 = 0.78
20,000	0.80	1 − 0.28 = 0.72
21,000	0.85	1 − 0.42 = 0.58
22,000	0.89	1 − 0.54 = 0.46
23,000	0.93	1 − 0.65 = 0.35
24,000	0.95	1 − 0.73 = 0.27
25,000	0.97	1 − 0.81 = 0.19
26,000	0.98	1 − 0.86 = 0.14
27,000	1.00	1 − 0.91 = 0.09
28,000	1.00	1 − 0.95 = 0.05
29,000	1.00	1 − 0.97 = 0.03
30,000	1.00	1 − 0.99 = 0.01
31,000	1.00	1 − 1.00 = 0.00
32,000	1.00	1 − 1.00 = 0.00

By defining the central factor of safety[2] as

$$n = \frac{\mu_S}{\mu_L} \tag{8.70}$$

Eq. (8.69) can be rewritten as

$$R = P\left[\frac{1-n}{\sqrt{\gamma_S^2 n^2 + \gamma_L^2}} \geq 0\right]$$

$$= \int_{z_1}^{\infty} \frac{1}{\sqrt{2\pi}} \exp\left(-\frac{1}{2} z^2\right) dz = 1 - \Phi(z_1) \tag{8.71}$$

[2]Other factors of safety that are often used are the characteristic factor of safety, n_k, and the design factor of safety, n^*, defined as [8.4]

$$n_k = \frac{S_{0.05}}{L_{0.95}}$$

$$n^* = \frac{S_{0.005}}{L_{0.95}} \quad \text{(see Fig. 8.11)}$$

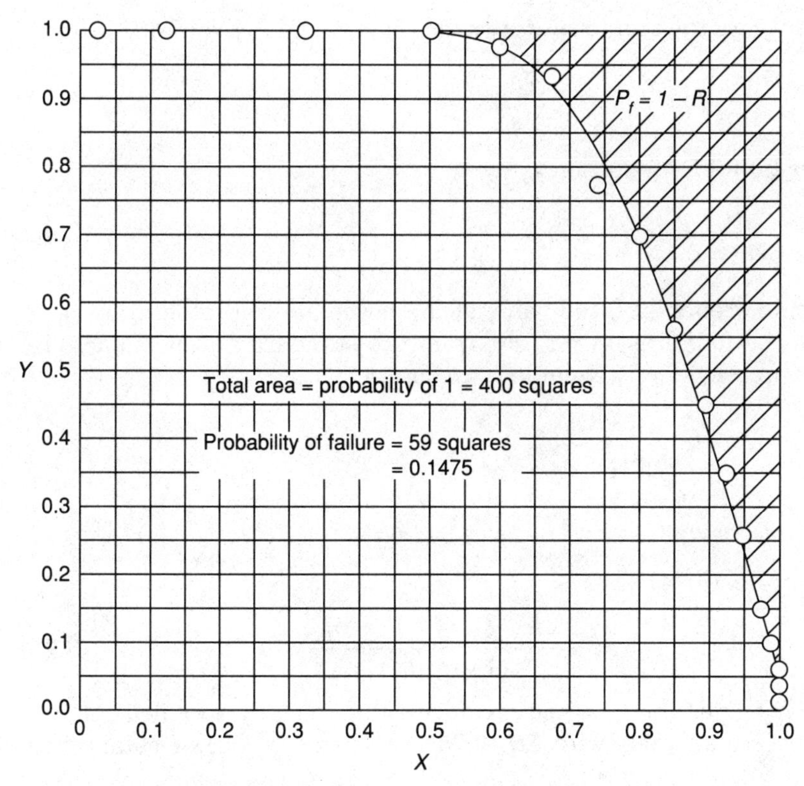

Figure 8.10

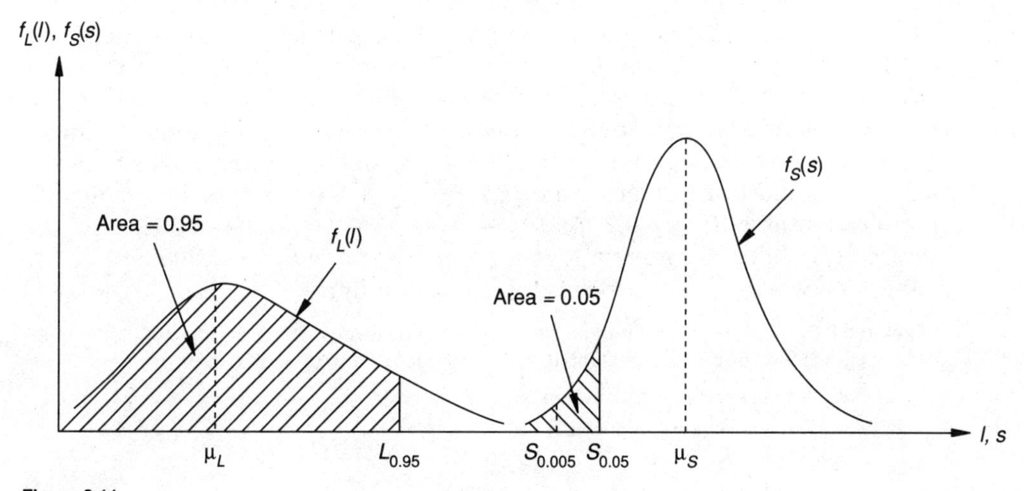

Figure 8.11

where γ_S and γ_L are the coefficients of variation of S and L given by

$$\gamma_S = \frac{\sigma_S}{\mu_S}; \qquad \gamma_L = \frac{\sigma_L}{\mu_L} \tag{8.72}$$

and the lower limit of integration z_1 by

$$z_1 = \frac{1-n}{\sqrt{\gamma_S^2 n^2 + \gamma_L^2}} \tag{8.73}$$

For any specified reliability, the value of the lower limit of integration z_1 in Eq. (8.73) can be found as, say $z_1 = t$, from the standard normal tables. By setting the expression for z_1 equal to t, we obtain

$$t = \frac{1-n}{\sqrt{n^2\gamma_S^2 + \gamma_L^2}}$$

or

$$0 = n^2(1 - t^2\gamma_S^2) - 2n + (1 - t^2\gamma_L^2) \tag{8.74}$$

The solution of Eq. (8.74) gives

$$n = \frac{1 \pm t\sqrt{\gamma_S^2 + \gamma_L^2 - t^2\gamma_S^2\,\gamma_L^2}}{(1 - t^2\,\gamma_S^2)} \tag{8.75}$$

Thus, Eq. (8.75) gives the required factor of safety for any specified reliability. We can see an anomaly with Eq. (8.75). By letting $\gamma_L = 0$, we obtain from Eq. (8.75),

$$n = \frac{1}{1 \pm t\gamma_S} \tag{8.76}$$

If we choose the negative sign on the right hand side of Eq. (8.76) and plot n as a function of γ_S (for any fixed value of reliability), we obtain a curve as shown in Fig. 8.12. It can be seen that the value of n required for achieving the specified reliability R tends to infinity as the product $t\gamma_S$ tends to unity. The reason for this apparent anomaly can be seen from Fig. 8.13 where the load is assumed to be deterministic ($\gamma_L = 0$), and the strength is assumed to be probabilistic with $\gamma_S = 0.2$. As $t\,\gamma_S \to 1$ or $|z_1| = |t| \to 5.0$ (i.e., as the required reliability of the component approaches a value of 0.99999971), the distance between μ_L and μ_S needs to approach infinity.

Example 8.6 Find the reliability and the corresponding central factor of safety of a system for which $\mu_S = 15000$ lb/in^2, $\mu_L = 10000$ lb/in^2, $\sigma_S = 3000$ lb/in^2 and $\sigma_L = 1000$ lb/in^2 if

1. S and L follow normal distribution, and
2. S and L follow lognormal distribution.

solution

1. When S and L follow normal distribution,

$$R = P(z_1 \geq 0) = P\left[\frac{\mu_L - \mu_S}{(\sigma_L^2 + \sigma_S^2)^{1/2}} \geq 0\right]$$

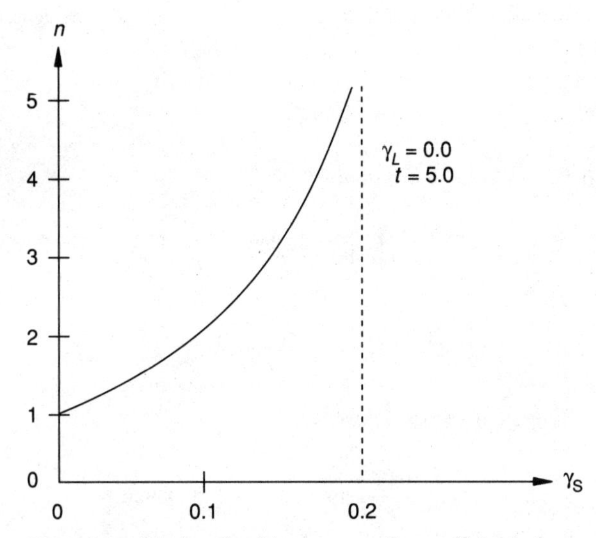

Figure 8.12 Factor of safety n versus the coefficient of variation of strength.

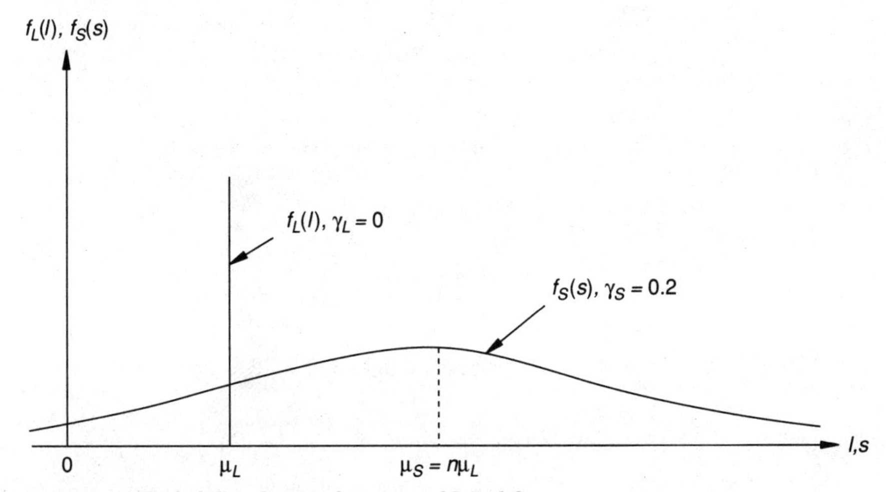

Figure 8.13 Probability density functions of L and S.

Here

$$z_1 = \frac{\mu_L - \mu_S}{(\sigma_L^2 + \sigma_S^2)^{1/2}} = \frac{10000 - 15000}{(1000^2 + 3000^2)^{1/2}}$$

$$= -\frac{5000}{3162} = -1.5811$$

From standard normal tables, we find that $R = 1 - \Phi(z_1) = 0.94305$. The central factor of safety is given by

$$n = \frac{\mu_S}{\mu_L} = \frac{15{,}000}{10{,}000} = 1.5$$

2. When S and L follow lognormal distribution,

$$\sigma_{\ln S}^2 = \ln\left(\frac{\sigma_S^2}{\mu_S^2} + 1\right) = \ln\left(\frac{3000^2}{15{,}000^2} + 1\right)$$

$$= \ln\left(\frac{9}{225} + 1\right) = \ln 1.04 = 0.03922$$

$$\therefore \sigma_{\ln S} = 0.19804$$

$$\mu_{\ln S} = \ln \mu_S - \frac{\sigma_{\ln S}^2}{2} = \ln 15{,}000 - \frac{1}{2}(0.03922)$$

$$= 9.61581 - 0.01961 = 9.59620$$

$$\sigma_{\ln L}^2 = \ln\left(\frac{\sigma_L^2}{\mu_L^2} + 1\right) = \ln\left(\frac{1000^2}{10000^2} + 1\right)$$

$$= \ln\left(\frac{1}{100} + 1\right) = \ln 1.01 = 0.00995$$

$$\therefore \sigma_{\ln L} = 0.09975$$

$$\mu_{\ln L} = \ln \mu_L - \frac{1}{2}\sigma_{\ln L}^2 = \ln 10000 - \frac{1}{2}(0.00995)$$

$$= 9.21034 - 0.00497 = 9.20537$$

The reliability of the system can be determined as

$$R = P(z_1 \geq 0)$$

where

$$z_1 = \frac{\mu_{\ln L} - \mu_{\ln S}}{\sqrt{\sigma_{\ln L}^2 + \sigma_{\ln S}^2}} = \frac{9.20537 - 9.59620}{\sqrt{0.09975^2 + 0.19804^2}} = -1.762536$$

From standard normal tables, the reliability corresponding to this value of z_1, can be found as

$$R = 1 - \Phi(z_1) = 1 - \Phi(-1.762536) = 1 - 0.0391 = 0.9609$$

$n = \frac{\mu_S}{\mu_L} = \frac{15{,}000}{10{,}000} = 1.5$ (Same as for normal dist)

References and Bibliography

8.1. P. T. Christensen and M. J. Baker, *Structural Reliability Theory and Its Applications,* Springer-Verlag, Berlin, 1986.
8.2. K. C. Kapur and L. R. Lamberson, *Reliability in Engineering Design,* John Wiley, New York, 1977.
8.3. C. Mischke, "A Method of Relating Factor of Safety and Reliability," *J. of Eng. for Industry,* vol. 92, Aug. 1970, pp. 537–542.
8.4. D. Kececioglu and E. B. Haugen, "A Unified Look at Design Safety Factors, Safety Margins and Measures of Reliability," *Annals of Assurance Sciences—Seventh Reliability and Maintainability Conference* 1968, pp. 520–530.
8.5. M. C. Kohn, *Practical Numerical Methods: Algorithms and Programs,* Macmillan Publishing Co., New York, 1987.
8.6. G. C. Hart, *Uncertainty Analysis, Loads, and Safety in Structural Engineering,* Prentice-Hall, Englewood Cliffs, New Jersey, 1982.
8.7. R. C. Juvinall, *Fundamentals of Machine Component Design,* John Wiley, New York, 1983.
8.8. J. E. Shigley and C. R. Mischke, *Mechanical Engineering Design,* 5th ed., McGraw-Hill, New York, 1989.
8.9. E. Sandgren, G. Gim and K. M. Ragsdell, "Optimal Design of a Class of Welded Structures Based on Design for Latitude," *ASME J. of Mechanisms, Transmissions, and Automation in Design,* vol. 107, 1985, pp. 482–487, 1985.
8.10. J. H. Faupel and F. E. Fisher, *Engineering Design,* 2d ed., John Wiley, New York, 1981.
8.11. S. S. Rao, "Application of Complementary Geometric Programming in Mechanical Design," *International J. of Mech. Eng. Educ.*, vol. 13, 1985, pp. 19–29.
8.12. E. B. Haugen, *Probabilistic Approaches to Design,* John Wiley, New York, 1968.
8.13. E. B. Haugen, *Probabilistic Mechanical Design,* John Wiley, New York, 1980.
8.14. R. L. Disney, C. Lipson, and N. J. Sheth, "The Determination of the Probability of Failure by Stress/Strength Interference Theory," *Proceedings of 1968 Annual Symposium on Reliability*, IEEE, New York, 1968.
8.15. A. C. King and C. B. Read, *Pathways to Probability,* Holt, Rinehart & Winston, Inc., New York, 1963.
8.16. C. C. Gillispie (ed.-in-chief), *Dictionary of Scientific Biography,* vol. 15, suppl. I, Charles Scribner's Sons, New York, 1980.
8.17. D. Kececioglu and D. Cormier, "Designing a Specified Reliability Directly into a Component," *Third Annual Aerospace Reliability and Maintainability Conference*, Society of Automotive Engineers, Inc., New York, June 29–July 1, 1964, pp. 546–565.

Review Questions

8.1 What is the major difference between the failure of electronic components and mechanical components?

8.2 How is the strength-based reliability of a component defined?

8.3 Derive an expression for the reliability of a component when $f_L(l)$ and $F_S(s)$ are known.

8.4 Find an expression for the reliability of a component when $f_S(s)$ and $F_L(l)$ are known.

8.5 What is a central factor of safety?

8.6 Define the characteristic factor of safety.

8.7 How is design factor of safety defined?

8.8 What type of anomaly is observed in relating the factor of safety to a specified value of reliability?

8.9 Why is reliability defined as $P(S - L > 0)$ for normally distributed S and L and as $P(S / L > 1)$ for lognormally distributed S and L?

8.10 What is the justification for assuming S and L as statistically independent?

Problems

8.1 The maximum stress induced near the hole in a rectangular plate of width w and thickness t subjected to an axial load P (see Fig. 8.14) is given by

$$s_{\max} = k_t s_o$$

where $s_o = P / \{(w-d)t\}$ is the nominal stress induced in the cross-section A-A and k_t is the stress concentration factor. The mean and standard deviations are given by (for $d / w = 0.2$) $\bar{k}_t = 2.5$ and $\sigma_{k_t} = 0.1$. The yield stress of the plate material, s_y, has mean and standard deviations of $\bar{s}_y = 30{,}000$ lb/in^2 and $\sigma_{s_y} = 1{,}000$ lb/in^2. If $P = 10{,}000$ lb, $w = 10$ in, $d = 2$ in, $t = 0.1$ in, and both k_t and s_y follow normal distribution, find the reliability of the plate.

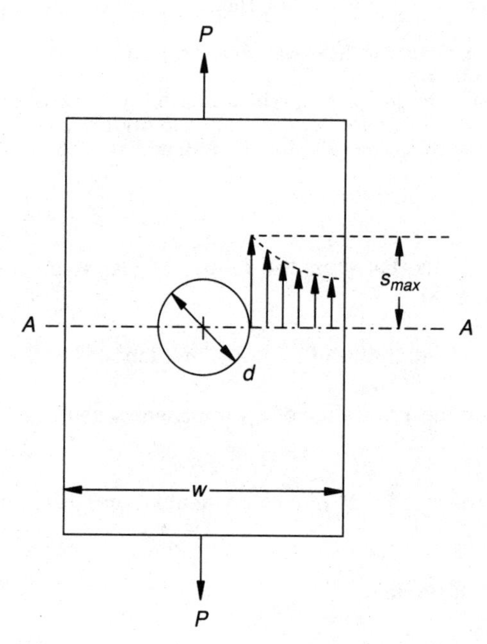

Figure 8.14

8.2 Find the reliability of the plate described in Problem 8.1 if k_t and s_y are uniformly distributed over the ranges 2.2–2.8 and 27,000–33,000 lb/in^2, respectively.

8.3 The mean and standard deviation of shear strength of a bolted joint are 45 k(lb/in^2) and 4.2 k(lb/in^2), respectively. The joint is loaded such that the stress induced has a mean value of 38 k(lb/in^2) and a standard deviation of 6.4 k(lb/in^2). Assuming that the shear strength and the induced stress are independent and normally distributed, find the probability of failure of the bolted joint.

8.4 If the internal pressure and the yield strength of the vessel described in Example 8.4 are lognormally distributed, determine the reliability of the pressure vessel. Assume the standard deviation of internal pressure as 50 lb/in^2 and of yield strength as 1,000 lb/in^2.

8.5 Two metal strips are clamped at the ends and are welded as shown in Fig. 8.15. The thermal stress (compressive) that developed in these metal strips due to a uniform rise in temperature during welding is given by

$$s = \alpha E \, \Delta T$$

where α = coefficient of thermal expansion, E = Young's modulus and ΔT = temperature rise. For steel, $\alpha = 12 \times 10^{-6} / {}^\circ C$ and $E = 207$ GPa. If the temperature rise during welding is normally distributed with $N(\Delta\overline{T}, \sigma_{\Delta T}) = N(40, 5)\,{}^\circ C$, and the permissible compressive stress in steel is also normally distributed with $N(\overline{s}_y, \sigma_{s_y}) = N(29, 1)$ lb/in^2, determine the reliability of the steel strips.

8.6 Solve Problem 8.5 if the temperature rise, ΔT, and the permissible compressive stress, s_y, follow exponential distributions

$$f(\Delta T) = 0.025 \exp(-0.025\,\Delta T); \qquad \Delta T \geq 0$$

$$f(s_y) = 0.035 \exp(-0.035\, s_y); \qquad s_y \geq 0$$

8.7 The knuckle joint shown in Fig. 8.16*a* is subjected to a tensile load F. Although the actual load distribution on the pin AB is unknown, the loads on AB can be approximated as shown in Fig. 8.16*b*. For $x = 4$ in, $y = 8$ in, $d = 4$

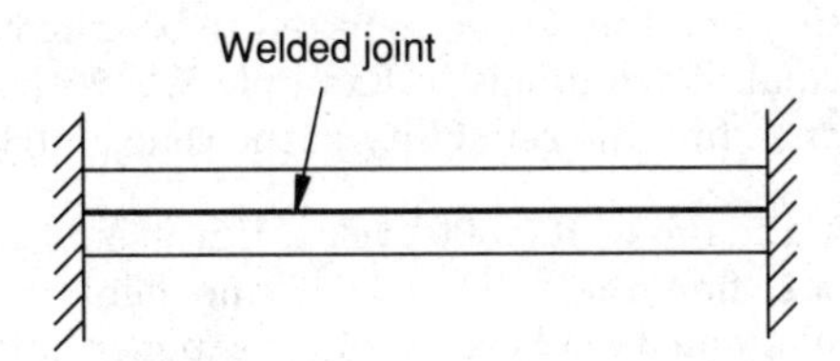

Figure 8.15

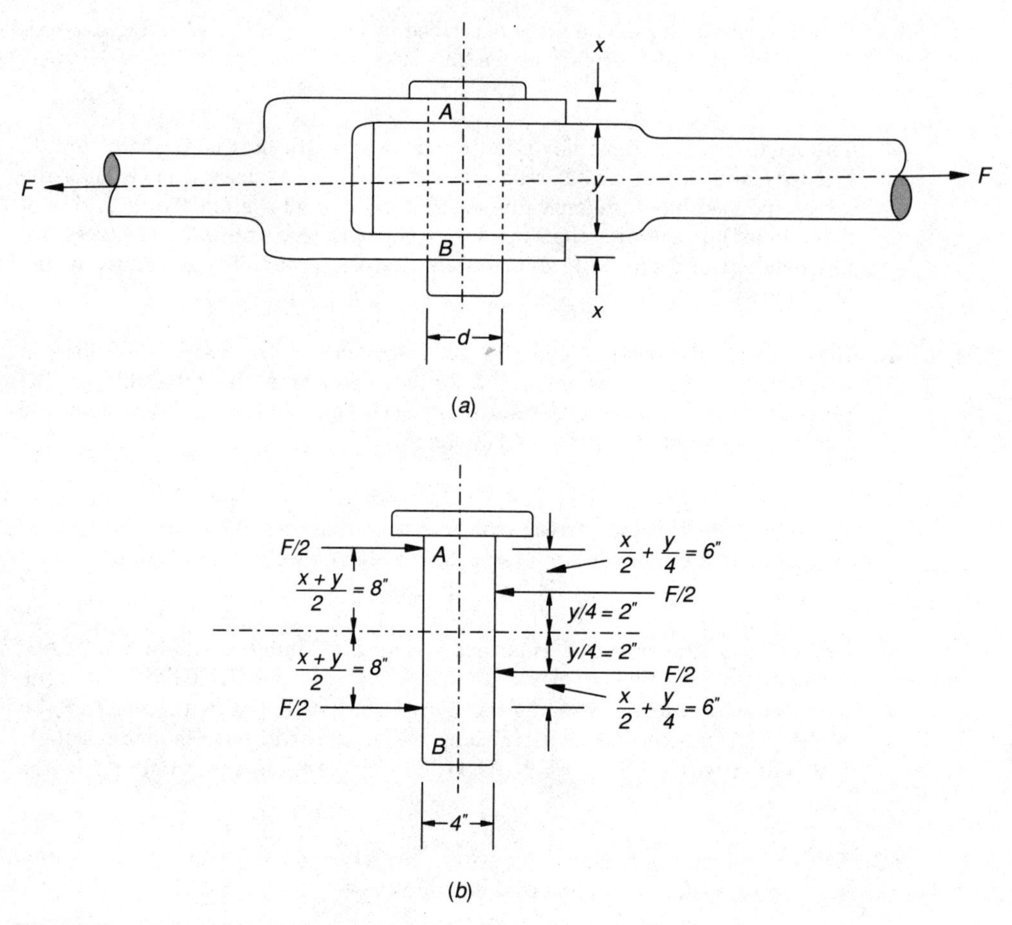

Figure 8.16

in, $F = N(60{,}000, 6000)$ lb, and $s_y = N(30{,}000, 2000)$ lb/in^2, determine the reliability of the knuckle joint.

8.8 If the tensile load and the yield stress are uniformly distributed as $F = U(40{,}000, 80{,}000)$ lb and $s_y = U(20{,}000, 40{,}000)$ lb/in^2, find the reliability of the knuckle joint described in Problem 8.7.

8.9 If both the coefficient of friction and the torque required to be transmitted by the cone clutch follow lognormal distributions in Example 8.1 (with the same means and standard deviations), find the reliability of the cone clutch.

8.10 A bolted joint has a mean fatigue life of 100,000 cycles. If it is subjected to an average of 80,000 load cycles, find the reliability of the joint if both the fatigue life and the number of applied-load cycles follow exponential distribution.

8.11 If the tangential load on the gear tooth is uniformly distributed in the range 4000–6000 lb, and the yield strength of the material is uniformly distributed in the range 28000–32000 psi, determine the reliability of the gear tooth described in Example 8.3.

8.12 When a shaft rotates in a guide bearing as shown in Fig. 8.17, the shear stress developed in the lubricant is given by [8.8]

$$\tau = \frac{\pi d \mu n}{\varepsilon}$$

and the pressure exerted on the bearing by

$$p = \frac{\pi^2 d \mu n}{\varepsilon}$$

where d = diameter of the shaft, μ = viscosity of the lubricant, n = rotational speed of the shaft, and ε = radial clearance. The data are $d = 2$ in, $n = 50$ r/s, $\varepsilon = 0.002$ in, and the viscosity of the lubricant is a random variable given by $N(4, 0.4)$ μ reyn. If it is desirable to limit the pressure on the bearing to 10 ± 0.2 lb/in^2, find the reliability of the bearing.

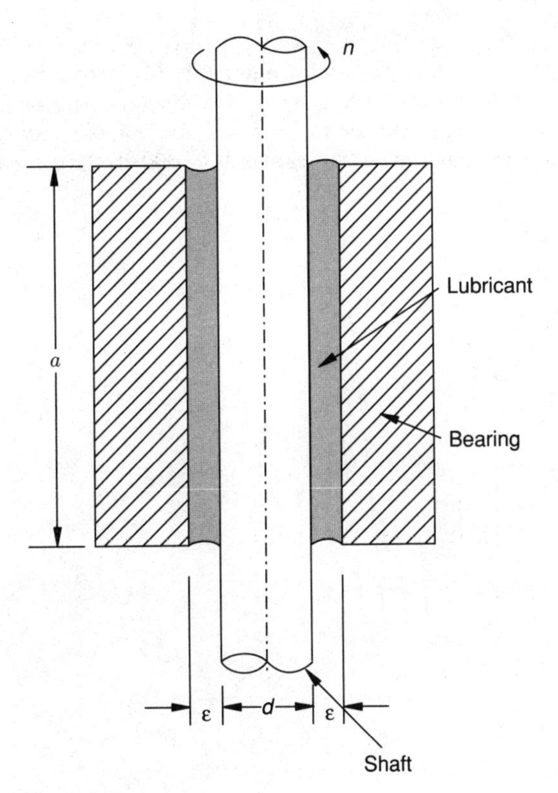

Figure 8.17

8.13 The spring rate of a helical torsional spring (Fig. 8.18) is given by [8.7]

$$k = \frac{T}{\theta} = \frac{d^4 E}{64 D n}$$

where T = torsional moment applied, θ = angular deflection of the spring, d = wire diameter, D = mean spring diameter, n = number of turns (coils), and E = Young's modulus of the material. A torsional spring used in a door hinge is made up of music wire and has $D = 1$ in, $d = 0.1$ in, and $n = 5.5$. The Young's modulus of music wire is normally distributed as $N(\overline{E}, \sigma_E) = N(30 \times 10^6, 3 \times 10^6)$ lb/in². If it is required to achieve a spring rate of 50 ± 5 lb·in per turn, determine the reliability of the helical spring.

8.14 If the Young's modulus is uniformly distributed in the range 25×10^6 lb/in²–35×10^6 lb/in², and the spring rate required is also uniformly distributed in the range 40 lb·in/turn–60 lb·in/turn, determine the reliability of the torsional spring discussed in Problem 8.13.

8.15 The shear stress induced in a helical spring τ under compression (see Fig. 8.19) can be expressed as [8.8]

$$\tau = \left(\frac{2C + 1}{2C}\right) \frac{8FD}{\pi d^3}$$

where C = spring index $= D/d$, d = wire diameter, D = mean coil diameter, and F = compressive load. The spring used in a machine suspension (isolation) system has $d = 0.5$ in, D = 10 in, and the axial force induced in the spring is uniformly distributed in the range 400–600 lb. If the permissible

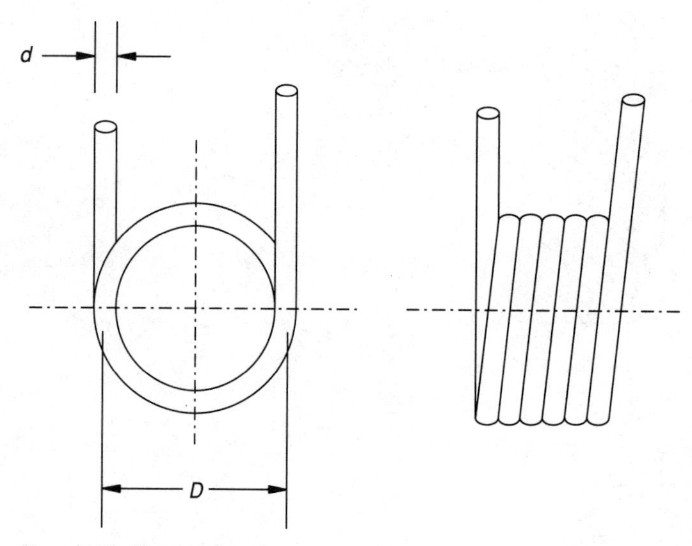

Figure 8.18 Torsional spring.

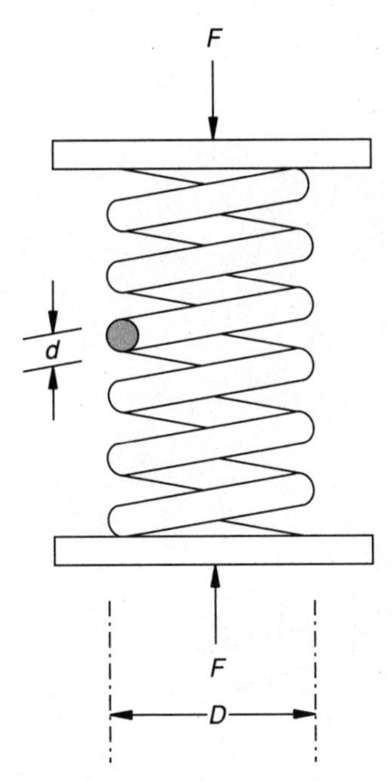

Figure 8.19 Compression spring.

yield strength in shear of the spring material τ_o is also uniformly distributed in the range 90–110 k(lb/in^2), find the reliability of the compression spring.

8.16 If the probability density functions of the compressive force on the spring and the permissible yield strength in shear of the spring material are given by

$$f(F) = 0.002 \exp(-0.002 F); \qquad F \geq 0$$

and

$$f(\tau_o) = 10^{-5} \exp(-10^{-5} \tau_o); \qquad \tau_o \geq 0$$

find the reliability of the compression spring described in Problem 8.15.

8.17 A beam, having length L and cross-sectional dimensions t and b, is welded to a fixed support as shown in Fig. 8.20. The weld length is l on both top and bottom surfaces and the beam is required to support a load P. The weld is in the form of a triangle of depth h. The maximum shear stress developed in the weld, τ, is given by [8.9]

$$\tau = \sqrt{(\tau')^2 + 2\,\tau'\,\tau''\cos\theta + (\tau'')^2}$$

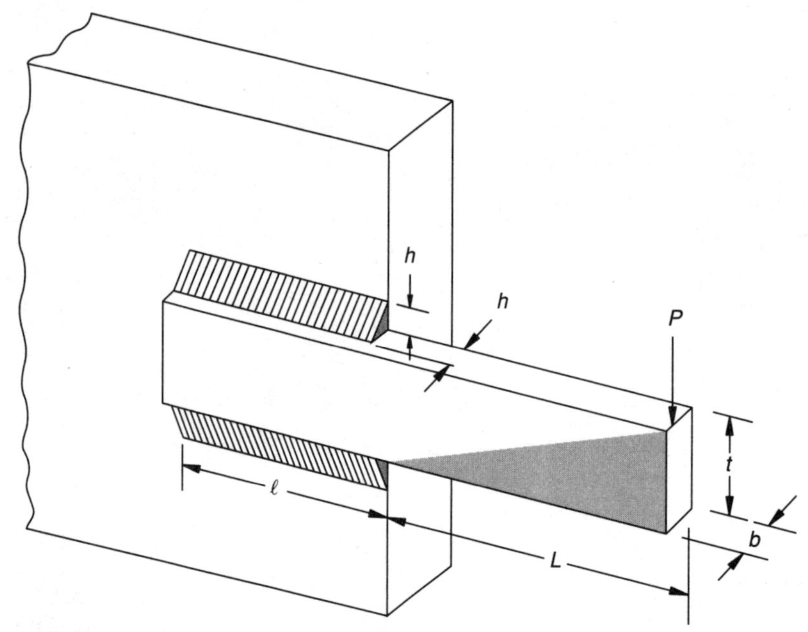

Figure 8.20 Welded beam.

where

$$\tau' = \frac{P}{\sqrt{2}\, h\, l}$$

$$\tau'' = \frac{M R}{J} = \left[P\left(L + \frac{1}{2}\right)\right] \frac{\sqrt{\left[\frac{l^2}{4} + \left(\frac{h+t}{2}\right)^2\right]}}{2\left(\sqrt{2}\, h\, l\left[\frac{l^2}{12} + \left(\frac{h+t}{2}\right)^2\right]\right)}$$

and

$$\cos\theta = \frac{l}{2\sqrt{\left[\frac{l^2}{4} + \left(\frac{h+t}{2}\right)^2\right]}}$$

The dimensions of the weld and the beam are $L = 25$ in, $h = 2$ in, $l = 10$ in, $t = 10$ in, $b = 2$ in and the load is normally distributed with $P = N(500, 50)$ lb. If the design shear stress, τ_o, in the weld is given by $\tau_o = N(\bar{\tau}_o, \sigma_{\tau_o}) = N(3500, 500)$ lb/in^2, find the reliability of the welded joint.

8.18 Solve Problem 8.17 by assuming that the load, P, and the design shear stress in the weld, τ_o, are exponentially distributed as

$$f(P) = 0.002 \exp(-0.002\, P); \quad P \geq 0$$

and

$$f(\tau_o) = 0.0003 \exp(-0.0003\, \tau_o); \qquad \tau_o \geq 0$$

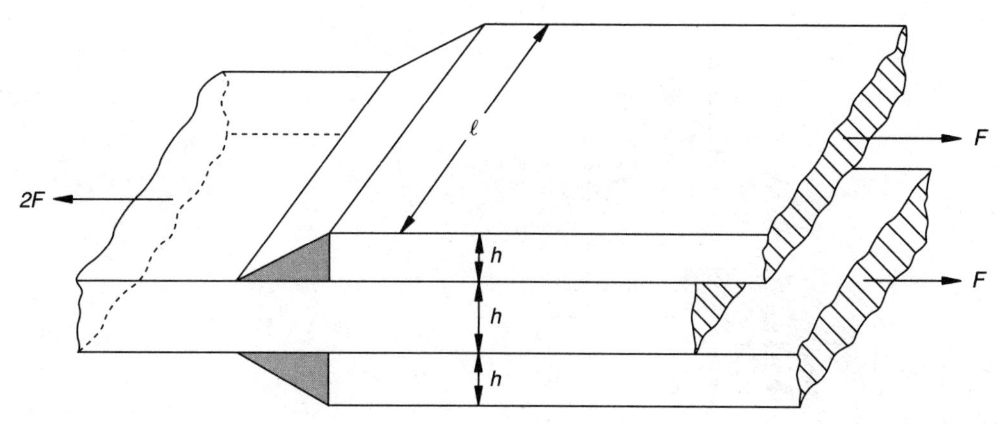

Figure 8.21

8.19 Three plates are jointed by a fillet weld as shown in Fig. 8.21. The maximum principal stress, s_1, and the maximum shear stress, τ, induced in the weld are given by [8.8]

$$s_1 = 1.61803 \frac{F}{hl}$$

$$\tau = 1.11803 \frac{F}{hl}$$

In a specific application, $h = 0.1$ in, $l = 10$ in and $F = N(10{,}000,\ 2500)$ lb. If the tensile and shear strengths of the weld material are normally distributed with $s_o = N(\bar{s}_o, \sigma_{s_o}) = N(18{,}000,\ 2500)$ lb/in^2, and $\tau_o = N(\bar{\tau}_o, \sigma_{\tau_o}) = N(12{,}000,\ 2000)$ lb/in^2, find the reliability of the welded joint.

8.20 Solve Problem 8.19 by assuming that the load applied, tensile strength, and shear strength of the weld material follow lognormal distribution with the same values of mean and standard deviation given in Problem 8.19.

8.21 Solve Problem 8.19 by assuming that the load applied, F, the tensile strength of the weld material, s_o, and the shear strength of the weld material, τ_o, follow exponential distributions as

$$f(F) = 0.0001 \exp(-0.0001\, F); \qquad F \geq 0$$

$$f(s_o) = 0.000056 \exp(-0.000056\, s_o); \qquad s_o \geq 0$$

$$f(\tau_o) = 0.000083 \exp(-0.000083\, \tau_o); \qquad \tau_o \geq 0$$

8.22 The torque required, T, to lower a load by a screw jack (see Fig. 8.22) is given by [8.7]

$$T = \frac{Fd_m}{2}\left(\frac{\pi \mu d_m - l}{\pi d_m + \mu l}\right)$$

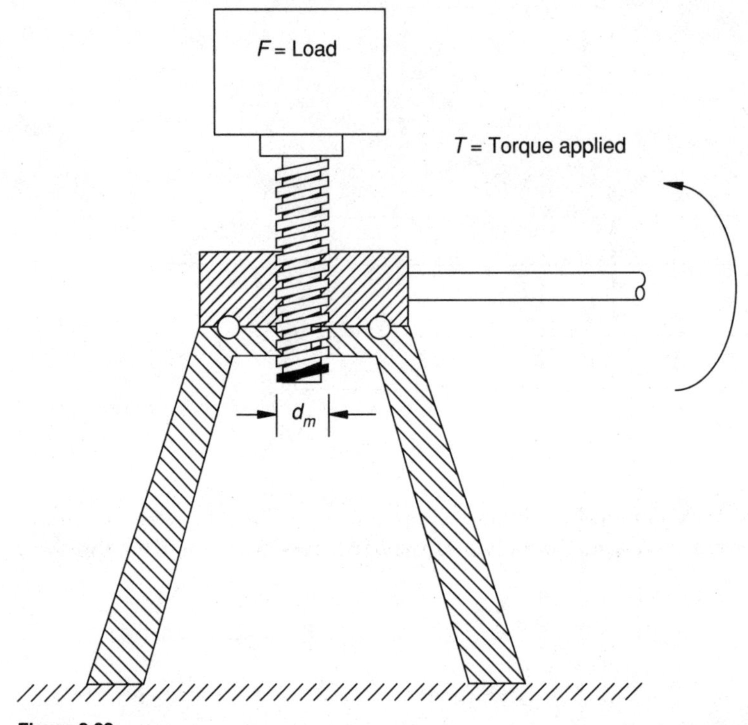

Figure 8.22

where F = load, d_m = mean diameter, l = lead (same as the pitch for a single threaded screw jack) and μ = coefficient of friction. It is desirable to have a positive value of T (self-locking condition) to avoid the load coming down by itself without any external effort. The condition for self-locking is

$$T > 0 \quad \text{or} \quad \pi \mu d_m > l$$

The manufacturing tolerances make d_m and l as random variables with $d_m = N(30, 0.05)$ mm and $l = N(8, 0.01)$ mm. Find the value of μ necessary in order to achieve the self-locking condition with a reliability of 0.999.

8.23 The split piston ring, shown in Fig. 8.23*a*, is mounted on a piston. The tangential deflection of the split ring under a force F is given by [8.10]

$$\delta = \frac{3\pi F R^3}{EI} + \frac{\pi F R}{EA} + \frac{6\pi F R}{5GA}$$

where the three terms on the right-hand side represent the contributions from bending moment, axial load, and transverse shear. A particular cast iron piston ring has $R = 2$ in, $b = 0.2$ in, $h = 0.3$ in, $E = 18 \times 10^6$ lb/in^2, and $G = 7 \times 10^6$ lb/in^2. The tangential force applied is a normally distributed random variable with a mean value of 50 lb and a standard deviation of 10 lb. Find the probability of the two ends X and Y overlapping.

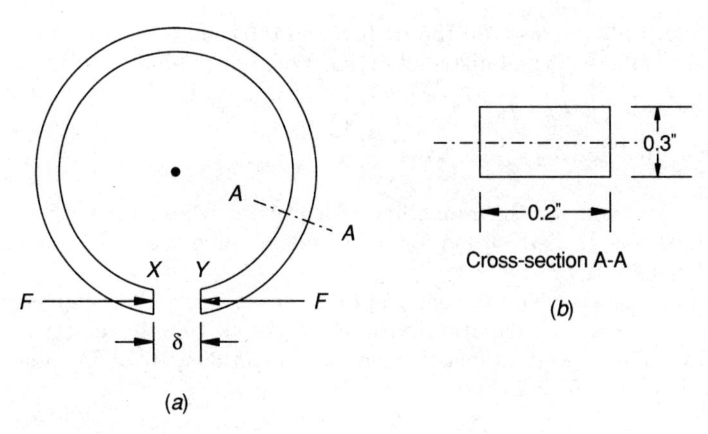

Figure 8.23

8.24 A connecting rod of length $l = 20$ in and a diameter of 0.5 in is subjected to an axial compressive force as shown in Fig. 8.24*a*. Assuming the ends as pin joints, the Euler buckling load, P_{cri} [8.10], is given by

$$P_{\text{cri}} = \frac{\pi^2 EI}{l^2}$$

where E = Young's modulus and I = moment of inertia of the cross section. The Young's modulus and the compressive load can be assumed to be random variables with $E = N(30 \times 10^6, 3 \times 10^5)$ (lb/in^2) and $P = N(2000, 400)$ lb. Find the reliability of the connecting rod in buckling.

8.25 Solve Problem 8.24 by assuming that the Young's modulus and the compressive load on the connecting rod follow uniform distributions over the range $28 \times 10^6 – 32 \times 10^6$ lb/in^2 and 1500–2500 lb, respectively.

8.26 Solve Problem 8.24 for the case where the Young's modulus follows Type-I distribution for the smallest value:

$$f(E) = a_1 \exp\left[a_1 (E - w_1) - e^{a_1(E - w_1)}\right]; \qquad -\infty < E < \infty$$

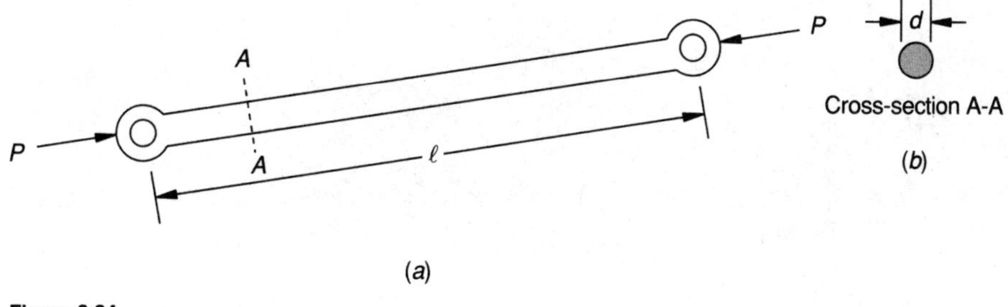

Figure 8.24

with $a_1 = 4.2752 \times 10^{-6}$ and $w_1 = 30.1350 \times 10^6$, and the compressive load on the connecting rod follows Type-I distribution for the largest value:

$$f(P) = a_2 \exp\left[a_2 (P - w_2) - e^{-a_2 (P - w_2)}\right]; \qquad -\infty < P < \infty$$

with $a_2 = 3.2064{\times}10^{-3}$ and $w_2 = 1820.0464$.

8.27 (a) Derive an expression for the reliability of a system whose strength is described by the Type-III distribution for the smallest value, Eq. (8.55), and load by normal distribution.
(b) Using the result of (a), solve Problem 8.24 for the case where the compressive load follows normal distribution with $P = N(2000, 400)$ lb, and the Young's modulus follows Weibull distribution with mean 30×10^6 lb/in^2 and standard deviation 30×10^5 lb/in^2.

8.28 (a) Derive an expression for the reliability of a system whose strength is described by the Type-III distribution for the smallest value, Eq. (8.55), and the load by the Type-I distribution for the largest value, Eq. (8.56).
(b) Using the result of (a), solve Problem 8.24 for the case where the compressive load follows Type-I distribution for the largest value with mean 2000 lb and has a standard deviation of 400 lb, and the Young's modulus follows Type-III distribution for the smallest value with mean 30×10^6 lb/in^2 and standard deviation 30×10^5 lb/in^2.

8.29 A steel gear with a central hole of diameter 1.500 in is shrink-fitted over a solid steel shaft of diameter 1.505 in as shown in Fig. 8.25. The gear has a pitch circle diameter of 6 in. The pressure developed at the interface p due to shrink-fit is given by [8.8]

$$p = \frac{E\delta}{2R}\left[\frac{(r_o^2 - R^2)}{r_o^2}\right]$$

where E = Young's modulus, δ = radial interference, r_o = outer radius of the outer part, and R = transition (nominal) radius at the interface. If the

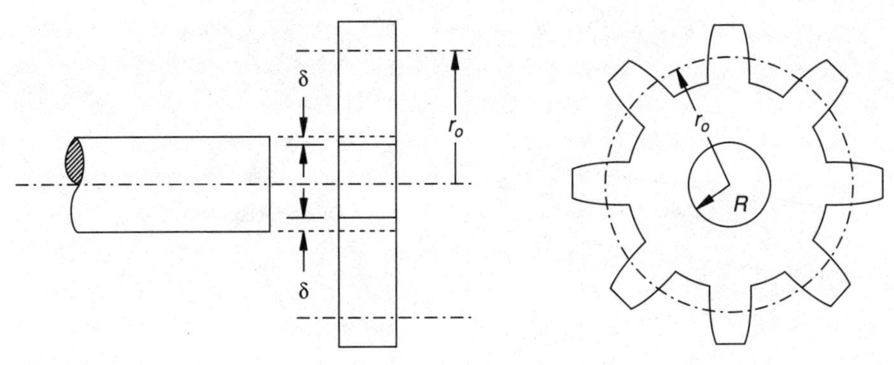

Figure 8.25

Young's modulus is normally distributed with $E = N(30 \times 10^6,\ 3 \times 10^6)$ lb/in^2 and the permissible shrink-fit pressure, p_o, can be taken as $p_o = N(45{,}000,\ 5{,}000)$ lb/in^2, determine the failure probability of the assembly.

8.30 Experiments conducted to find the stresses induced in power transmitting shafts gave the values 23.4, 28.6, 23.9, 25.2, 25.5, 26.7, 20.3, 27.7, 19.2, 22.8, 23.1, 24.6, 25.9, 26.8, 27.5, 21.7, 24.1, 24.9, 24.8, and 22.6 k(lb/in^2). The experimental values of the strength of the shaft material yielded the values 31.8, 25.2, 28.3, 29.6, 30.1, 32.7, 26.5, 28.9, 30.8, 33.5, 27.4, 29.1, 30.7, 32.2, 29.9, 30.6, 31.7, 31.3, 33.9, and 34.3 k(lb/in^2). Find the reliability of the shafts.

8.31 Derive Eqs. (8.39) and (8.41).

8.32 The following values of compressive load, in lb, have been observed to act on the spring described in Problem 8.15:

400, 420, 440, 460, 470, 480, 490, 500, 510, 520, 540, 550, 570, 580, 590, 610

The yield strength of the spring material in shear, in k(lb/in^2), for different specimens has been observed to be:

99, 100, 101, 103, 105, 106, 108, 110, 95, 97, 98, 91

Determine reliability of the spring.

8.33 Plot a graph between reliability, R, and the lower limit of integration, z_1, when strength and load follow normal distribution (Eq. 8.24).

8.34 For normally distributed strength and load, plot the variations of n with respect to t for different values of γ_S and γ_L given by $\gamma_S = 0.1, 0.2$ and $\gamma_L = 0.1, 0.2$.

8.35 For the case of normally distributed strength and load, plot the variation of the probability of failure, P_f, with respect to the factor of safety, n, for the following cases:

1. $\gamma_S = 0.1,\ \gamma_L = 0.1$
2. $\gamma_S = 0.1,\ \gamma_L = 0.2$
3. $\gamma_S = 0.2,\ \gamma_L = 0.1$
4. $\gamma_S = 0.2,\ \gamma_L = 0.2$

Chapter

9

Design of Mechanical Components and Systems

Biographical Note

William Sealy Gosset

William Sealy Gosset was an Englishman who is more popularly known under the pen name of "Student." He was born in Canterbury on 13 June 1876 and died in Beaconsfield on 16 October 1937. He worked as a consulting statistician for a famous Irish brewery. Originally the brewery did not permit him to publish his research work fearing that the stockholders might not like to approve spending money on research. Later when the company permitted him to publish his research work on the condition that he use a pen name, Gosset used the name, "Student." He was one of the first who used the theory of probability to statistical data relating the brewing methods, characteristics of the raw materials used, and the quality of the finished product. The experiments he conducted using small samples have led to the t-distribution (or Student's t-distribution) and t-test. His works include such diverse topics as testing of cereals, agronomy, and mathematics. He toured the barley fields before harvest and suggested methods of improving the crops using principles of statistics [9.14, 9.15].

9.1 Introduction

The reliability-based design of mechanical components and systems is considered in this chapter. The geometry, material properties, and external loads are treated as random variables. In most cases, the random variables are assumed to follow normal distribution for computational simplicity. The assumption of normal distribution for the behavior or response parameters can be justified on the basis of central-limit theorem. When the induced stress (or load) and the strength of a component follow normal distribution, the reliability can be determined, in terms of the standard normal variate, as

$$R_0 = \frac{1}{\sqrt{2\pi}} \int_{z_1}^{\infty} e^{-1/2z^2} dz \tag{9.1}$$

where the lower limit of integration is given by

$$z_1 = -\left(\frac{\mu_S - \mu_L}{\sqrt{\sigma_S^2 + \sigma_L^2}} \right) \tag{9.2}$$

with μ_S and σ_S (μ_L and σ_L) denoting, respectively, the mean value and standard deviation of the strength (load). If the induced stress (load) and the strength follow lognormal distribution, the lower limit of integration in Eq. (9.1) can be computed as

$$z_1 = -\left(\frac{\mu_{\ln S} - \mu_{\ln L}}{\sqrt{\sigma_{\ln S}^2 + \sigma_{\ln L}^2}} \right) \tag{9.3}$$

The designs of a connecting rod, a pressure vessel, a helical spring, fatigue design of machine components, and a compound gear train, and the reliability analysis of cam-follower systems and function generating mechanisms are considered as specific examples.

9.2 Design of Mechanical Components

Example 9.1 A connecting rod of length l and diameter d is subjected to an axial compressive force P. The Euler buckling load of the connecting rod, P_c, assuming the ends to be pin-connected, is given by [9.1]

$$P_c = \frac{\pi^2 E I}{l^2}$$

where E = Young's modulus and I = moment of inertia of the cross section. The parameters are given by $l = N(20, 0.5)$ in, $d = N(\bar{d}, 0.1\bar{d})$ in, $P = N(2000, 200)$ lb, and $E = N(30 \times 10^6, 3 \times 10^6)$ lb/in^2. Design the connecting rod to achieve a reliability of 0.99 against buckling.

solution The specified reliability of 0.99 corresponds to a lower limit of $z_1 = -2.32635$ in Eq. (9.1) so that

$$z_1 = -2.32635 = -\left(\frac{\bar{P}_c - \bar{P}}{\sqrt{\sigma_{P_c}^2 + \sigma_P^2}} \right) \tag{E1}$$

where $\bar{P} = 2000$ lb and $\sigma_P = 200$ lb. Using the relation

$$I = \frac{\pi d^4}{64} \tag{E2}$$

the buckling load can be expressed as

$$P_c = \frac{\pi^3 E d^4}{64 l^2} \tag{E3}$$

so that the mean value of P_c is given by

$$\bar{P}_c = \frac{\pi^3 \bar{E} \bar{d}^4}{64 \bar{l}^2} = \frac{\pi^3 (30 \times 10^6) \bar{d}^4}{64 (20)^2} = 36{,}331.1091 \, \bar{d}^4 \tag{E4}$$

The partial derivative rule can be used to find the standard deviation of P_c as

$$\sigma_{P_c}^2 = \left[\left(\frac{\partial P_c}{\partial E} \right)^2 \sigma_E^2 + \left(\frac{\partial P_c}{\partial d} \right)^2 \sigma_d^2 + \left(\frac{\partial P_c}{\partial l} \right)^2 \sigma_l^2 \right] \tag{E5}$$

where

$$\frac{\partial P_c}{\partial E} = \frac{P_c}{E}, \quad \frac{\partial P_c}{\partial d} = 4 \frac{P_c}{d}, \quad \text{and} \quad \frac{\partial P_c}{\partial l} = -2 \frac{P_c}{l}$$

Thus Eq. (E5) gives

$$\gamma_{P_c}^2 = \gamma_E^2 + 16 \, \gamma_d^2 + 4 \, \gamma_l^2$$

$$= (0.1)^2 + 16 \, (0.1)^2 + 4 \, (0.1)^2 = 21 \, (0.1)^2 = (0.45825757)^2$$

that is,

$$\sigma_{P_c} = \gamma_{P_c} \bar{P}_c = 16{,}649.0058 \, \bar{d}^4 \tag{E6}$$

Substitution of Eqs. (E5) and (E6), along with the known values of $\bar{P}$ and σ_P, into Eq. (E1) gives

$$-2.32635 = -\left(\frac{36331.1091 \, \bar{d}^4 - 2000}{\sqrt{(16649.0058 \, \bar{d}^4)^2 + (200)^2}} \right) \tag{E7}$$

Squaring both sides of Eq. (E7) leads to a quadratic equation in $\bar{d}^4$

$$1.80365827 \times 10^8 \, \bar{d}^8 + 1.45324436 \times 10^8 \, \bar{d}^4 - 378.3496 \times 10^4 = 0 \tag{E8}$$

The solution of Eq. (E8) gives

$$\bar{d}^4 = \frac{-1.45324436 \times 10^8 \pm \sqrt{(1.45324436 \times 10^8)^2 + 4 \, (1.80365827 \times 10^8) \, (0.03783496 \times 10^8)}}{2 \, (1.80365827 \times 10^8)}$$

$$= 0.0252439096; \quad -0.83096436 \tag{E9}$$

which leads to $\bar{d}^2 = 0.15888332$ or $\bar{d} = 0.3986$ in.

Example 9.2 A pressure vessel, in the form of a torus (see Fig. 9.1), is to be designed for a scientific experiment. The cross section of the cylinder has a radius of $r = N(50, 1)$ in, while the ring has a radius of $R = N(200, 2)$ in. The expected pressure in the vessel is $p = N(500, 50)$

lb/in^2. Find the thickness of the cylinder $t = N(\bar{t}, 0.1\bar{t})$ in to achieve a reliability of 0.9999997 when the strength of the material is given by $S = N(30000, 2000)$ lb/in^2.

solution The stress induced around the rim of the cross section of the torus [9.2] is given by

$$s_\phi = \frac{p\,r}{2t}\left(1 + \frac{R}{R \pm r}\right) \tag{E1}$$

where the plus and minus signs in Eq. (E1) correspond to the stresses induced at points A and B, respectively, in Fig. 9.1. The larger stress induced (at point B) will be

$$s_\phi = \frac{p\,r}{2\,t}\left(1 + \frac{R}{R - r}\right) \tag{E2}$$

The stress induced in the circumferential direction of the cross section is given by [9.2]

$$s_\theta = \frac{p\,r}{2\,t} \tag{E3}$$

Since the reliability of 0.9999997 corresponds to $z_1 = -5$, Eq. (9.2) can be written, by considering the larger induced stress, s_ϕ, as

$$z_1 = -5.0 = -\frac{\bar{S} - \bar{s}_\phi}{\sqrt{\sigma_S^2 + \sigma_{s_\phi}^2}} \tag{E4}$$

where $\bar{S} = 30{,}000$ lb/in^2, $\sigma_S = 2000$ lb/in^2 and $\bar{s}_\phi$ is, from Eq. (E2),

$$\bar{s}_\phi = \frac{\bar{p}\,\bar{r}}{2\,\bar{t}}\left(1 + \frac{\bar{R}}{\bar{R} - \bar{r}}\right)$$

$$= \frac{500\,(50)}{2\,\bar{t}}\left(1 + \frac{200}{200 - 50}\right) = \frac{29166.6667}{\bar{t}} \tag{E5}$$

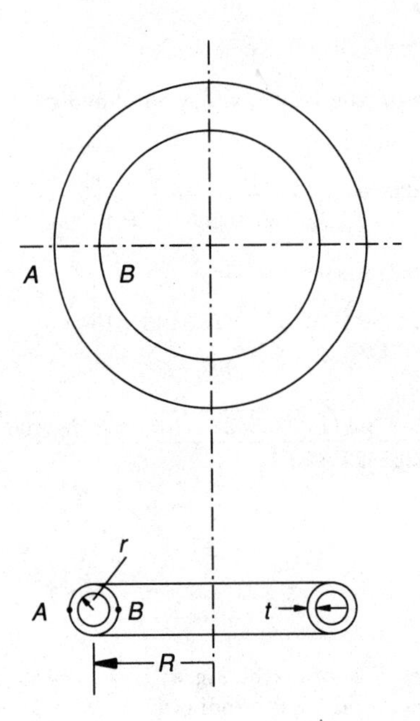

Figure 9.1

The standard deviation of s_ϕ can be found, using the partial derivative rule, as

$$\sigma_{s_\phi}^2 = \left(\frac{\partial s_\phi}{\partial p}\right)^2 \sigma_p^2 + \left(\frac{\partial s_\phi}{\partial r}\right)^2 \sigma_r^2 + \left(\frac{\partial s_\phi}{\partial t}\right)^2 \sigma_t^2 + \left(\frac{\partial s_\phi}{\partial R}\right)^2 \sigma_R^2 \tag{E6}$$

where the partial derivatives, evaluated at the mean values of the random variables, can be determined as

$$\frac{\partial s_\phi}{\partial p} = \frac{s_\phi}{p} = \frac{29,166.6667}{\bar{t}\,(500)} = \frac{58.3333}{\bar{t}}$$

$$\frac{\partial s_\phi}{\partial t} = -\frac{s_\phi}{t} = -\left(\frac{29,166.6667}{\bar{t}^2}\right)$$

$$\frac{\partial s_\phi}{\partial r} = \frac{s_\phi}{r} + \frac{p\,r\,R}{2\,t\,(R-r)^2}$$

$$= \frac{29,166.6667}{\bar{t}\,(50)} + \frac{500\,(50)\,(200)}{2\,\bar{t}\,(200-50)^2} = \frac{694.4444}{\bar{t}}$$

$$\frac{\partial s_\phi}{\partial R} = \frac{p\,r}{2\,t\,(R-r)}\left(1 - \frac{R}{R-r}\right)$$

$$= \frac{500\,(50)}{2\,\bar{t}\,(200-50)}\left(1 - \frac{200}{200-50}\right) = -\frac{27.775}{\bar{t}}$$

Thus Eq. (E6) gives

$$\sigma_{s_\phi}^2 = \left(\frac{58.3333}{\bar{t}}\right)^2 (50)^2 + \left(\frac{694.4444}{\bar{t}}\right)^2 (1)^2 + \left(-\frac{29166.6667}{\bar{t}^2}\right)^2 (0.1\,\bar{t})^2 + \left(\frac{27.775}{\bar{t}}\right)^2 (2)^2$$

$$= \frac{17.499219 \times 10^6}{\bar{t}^2}$$

that is,

$$\sigma_{s_\phi} = \frac{4183.207}{\bar{t}}$$

Substitution of the known quantities into Eq. (E4) yields

$$-5 = -\frac{30,000 - \left(\dfrac{29,166.6667}{\bar{t}}\right)}{\sqrt{(2000)^2 + \left(\dfrac{4183.207}{\bar{t}}\right)^2}} \tag{E7}$$

Squaring and rearranging Eq. (E7) results in

$$\frac{413.213971 \times 10^6}{\bar{t}^2} - \frac{1750.0 \times 10^6}{\bar{t}} + 800 \times 10^6 = 0$$

that is,

$$\bar{t}^2\,(800) - 1750\,\bar{t} + 413.213971 = 0 \tag{E8}$$

The solution of Eq. (E8) gives $\bar{t}$ = 1.9182 in and 0.2693 in. Selecting the smaller value, the required mean thickness of the pressure vessel can be taken as $\bar{t}$ = 0.2693 in.

Example 9.3 The spring rate of a helical torsional spring [9.3] is given by

$$k = \frac{T}{\theta} = \frac{d^4 E}{64\, D\, n}$$

where T = torque applied, θ = angular deflection of the spring, d = wire diameter, D = mean spring coil diameter, n = number of turns, and E = Young's modulus of the material. The torsional spring used for a door hinge is composed of music wire and has the following probabilistic parameters: $D = N(1, 0.1)$ in, $n = N(5.5, 0.55)$, and $E = N(30 \times 10^6, 3 \times 10^6)$ lb/in^2. Find the mean value of the wire diameter if a spring rate of 50 ± 5 lb · in per turn is to be achieved with a reliability of 0.99. Assume the coefficient of variation of the wire diameter as 0.1.

solution If K denotes the required stiffness and k reflects the actual stiffness of the spring, the application of Eq. (9.2) for the specified reliability of R_0 gives

$$z_1 = -2.3265 = \frac{\bar{K} - \bar{k}}{\sqrt{\sigma_K^2 + \sigma_k^2}} \tag{E1}$$

where $\bar{K} = 50$ and $\sigma_K = \Delta K / 3 = 5/3 = 1.6667$. The mean value of k is given by

$$\bar{k} = \frac{\bar{d}^4 \bar{E}}{64 \bar{D} \bar{n}} = \frac{\bar{d}^4 (30 \times 10^6)}{64\,(1)\,(5.5)} = 85{,}227.2727\, \bar{d}^4 \tag{E2}$$

The standard deviation of k can be obtained, from the partial derivative rule, as

$$\sigma_k^2 = \left[\left(\frac{\partial k}{\partial E}\right)^2 \sigma_E^2 + \left(\frac{\partial k}{\partial d}\right)^2 \sigma_d^2 + \left(\frac{\partial k}{\partial D}\right)^2 \sigma_D^2 + \left(\frac{\partial k}{\partial n}\right)^2 \sigma_n^2\right] \tag{E3}$$

where

$$\left(\frac{\partial k}{\partial E}\right)^2 \sigma_E^2 = \left(\frac{k}{E}\right)^2 \sigma_E^2 = k^2\, \gamma_E^2$$

$$\left(\frac{\partial k}{\partial d}\right)^2 \sigma_d^2 = \left(\frac{4\, d^3\, E}{64\, D\, n}\right)^2 \sigma_d^2 = \left[4 \left(\frac{k}{d}\right)\right]^2 \sigma_d^2 = 16\, k^2\, \gamma_d^2$$

$$\left(\frac{\partial k}{\partial D}\right)^2 \sigma_D^2 = \left(\frac{-d^4\, E}{64\, D^2\, n}\right)^2 \sigma_D^2 = \left(-\frac{k}{D}\right)^2 \sigma_D^2 = k^2\, \gamma_D^2$$

and

$$\left(\frac{\partial k}{\partial n}\right)^2 \sigma_n^2 = \left(\frac{-d^4\, E}{64\, D\, n^2}\right)^2 \sigma_n^2 = \left(-\frac{k}{n}\right)^2 \sigma_n^2 = k^2\, \gamma_n^2$$

Thus Eq. (E3) can be rewritten as

$$\gamma_k^2 = \gamma_E^2 + 16\, \gamma_d^2 + \gamma_D^2 + \gamma_n^2 = (0.1)^2 + 16\,(0.1)^2 + (0.1)^2 + (0.1)^2 = 19\,(0.1)^2$$

that is,

$$\gamma_k = 0.4358899$$

and hence

$$\sigma_k = \gamma_k\, \bar{k} = 0.4358899\, \bar{k} = 37{,}149.7074\, \bar{d}^4 \tag{E4}$$

Substitution of Eqs. (E2) and (E4) into (E1) yields

$$-2.3265 = \frac{50 - 85227.2727\,\bar{d}^4}{\sqrt{(1.6667)^2 + (37149.7074\,\bar{d}^4)^2}} \tag{E5}$$

Squaring Eq. (E5) and rearranging the terms gives

$$2.062456 \times 10^8\,\bar{d}^8 + 852.27273 \times 10^4\,\bar{d}^4 - 2484.964338 = 0 \tag{E6}$$

The solution of Eq. (E6) gives

$$\bar{d}^4 = \frac{-852.27273 \times 10^4 \pm \sqrt{72.636881 \times 10^{12} + 2.050052 \times 10^{12}}}{4.124912 \times 10^8}$$

$$= 0.02895405 \times 10^{-2}; \qquad -4.16127362 \times 10^{-2}$$

This gives $\bar{d}^2 = 0.0170158894$ or $\bar{d} = 0.1304$ inch.

9.3 Fatigue Design

A fatigue failure is one in which the component fails gradually when subjected to repeated loads that induce stresses that are often below the yield strength of the material. If the induced repeated stress is below the endurance strength of the material, theoretically the component will not fail no matter how many cycles are repeated. High-cycle fatigue design is concerned with designing the component for infinite life. Thus, in the limiting case (deterministic design procedure), the induced stress is equated to the endurance strength of the component.

The endurance strength of the component is determined using the fatigue-strength modification factors suggested by Marin [9.3–9.5] as follows:

$$S_e = k_a\, k_b\, k_c\, k_d\, k_e\, k_f\, S_e' \tag{9.4}$$

where S_e = fatigue strength at the critical location of the component in the geometry and condition of use, k_a = surface-condition modification factor, k_b = size modification factor, k_c = load modification factor, k_d = temperature modification factor, k_e = stress-concentration and notch-sensitivity modification factor, k_f = miscellaneous-effects modification factor, and S_e' = endurance limit of the laboratory specimen, which is often estimated from the ultimate tensile strength of the material (S_{ut}), as

$$S'_e = \begin{cases} 0.504\, S_{ut} & \text{if } S_{ut} \le 200\ \text{k(lb/in}^2) \\ 100\ \text{k(lb/in}^2) & \text{if } S_{ut} > 200\ \text{k(lb/in}^2) \end{cases} \tag{9.5}$$

9.3.1 Deterministic design procedure

The modification factors k_a, k_b, k_c, k_d, k_e, and k_f and the ultimate tensile strength S_{ut} are treated as constants [9.3–9.5]. The surface modification factor, k_a, is given by

$$k_a = a\,S_{ut}^{b} \tag{9.6}$$

where $a = 16.45$ and $b = -0.7427$ for hot-rolled components, $a = 39.9$ and $b = -0.995$ for as-forged components, $a = 2.7$ and $b = -0.2653$ for machined surfaces, and $a = 1.34$ and $b = -0.0848$ for ground surfaces when S_{ut} is expressed in k(lb/in^2). The size modification factor, k_b, is expressed as

$$k_b = \begin{cases} \left(\dfrac{d}{0.3}\right)^{-0.1133}; & \text{for bending and torsion loads; } 0.11 \text{ in } \le d \le 2 \text{ in} \\ 1; & \text{for axial loads} \end{cases} \tag{9.7}$$

where d is the diameter (or equivalent diameter) of the component in inches. The load modification factor, k_c, is given by

$$k_c = \begin{cases} 0.923; & \text{for axial loads} \quad (\text{when } S_{ut} \le 220 \text{ k(lb/in}^2)) \\ 1; & \text{for bending loads} \\ 0.583; & \text{for torsional loads} \end{cases} \tag{9.8}$$

The temperature modification factor, k_d, can be expressed as

$$k_d = \begin{cases} 1; & \text{when operating temperature is } 70^\circ\text{F} \\ 1.018; & \text{when operating temperature is } 400^\circ\text{F} \end{cases} \tag{9.9}$$

The stress-concentration and notch-sensitivity modification factor, k_e, is given by

$$k_e = \frac{1}{K_F} \tag{9.10}$$

where K_F is the fatigue-strength reduction factor whose value depends on the type and size of the stress concentration present in the component. The value of K_F [9.4] can be computed as

$$K_F = \frac{K_t}{1 + \dfrac{2}{\sqrt{r}}\left(\dfrac{K_t - 1}{K_t}\right)\sqrt{a}} \tag{9.11}$$

where

$$\sqrt{a} = \begin{cases} \dfrac{5}{S_{ut}}; & \text{for a transverse hole} \\ \dfrac{4}{S_{ut}}; & \text{for a shoulder} \\ \dfrac{3}{S_{ut}}; & \text{for a groove} \end{cases} \tag{9.12}$$

K_t is the geometric stress concentration factor [9.6] and r is the notch radius. The value of the miscellaneous-effects modification factor, k_f, depends on factors such as the residual stresses present in the component, and the nature of corrosive atmosphere in which the component operates. The deterministic design procedure is illustrated by working through the following example.

Example 9.4 The machined steel shaft shown in Fig. 9.2*a* is subjected to a force $F = 3{,}000$ lb. The material has an ultimate tensile strength of 80 k(lb/in^2). Determine the factor of safety of the shaft. The geometric stress concentration factor at the shoulder C is known to be 1.75.

solution The reactions at the bearings can be determined as $P_A = (1/5)$ F and $P_B = (4/5)$ F. The bending-moment diagram is shown in Fig. 9.2*b*. The critical section can be seen to be the section C. The bending moment at section C, M_C, is given by

$$M_C = 19\,P_A = \frac{19}{5}F = \frac{19}{5}(3{,}000) = 11{,}400 \text{ lb} \cdot \text{in}$$

The stress induced at C, s_C, is given by

$$s_C = \frac{M_C\, c}{I}$$

where

$$\frac{I}{c} = \frac{\dfrac{\pi}{64}d^4}{\dfrac{d}{2}} = \frac{\pi d^3}{32} = \frac{\pi}{32}(2)^3 = \frac{\pi}{4}\text{ in}^3$$

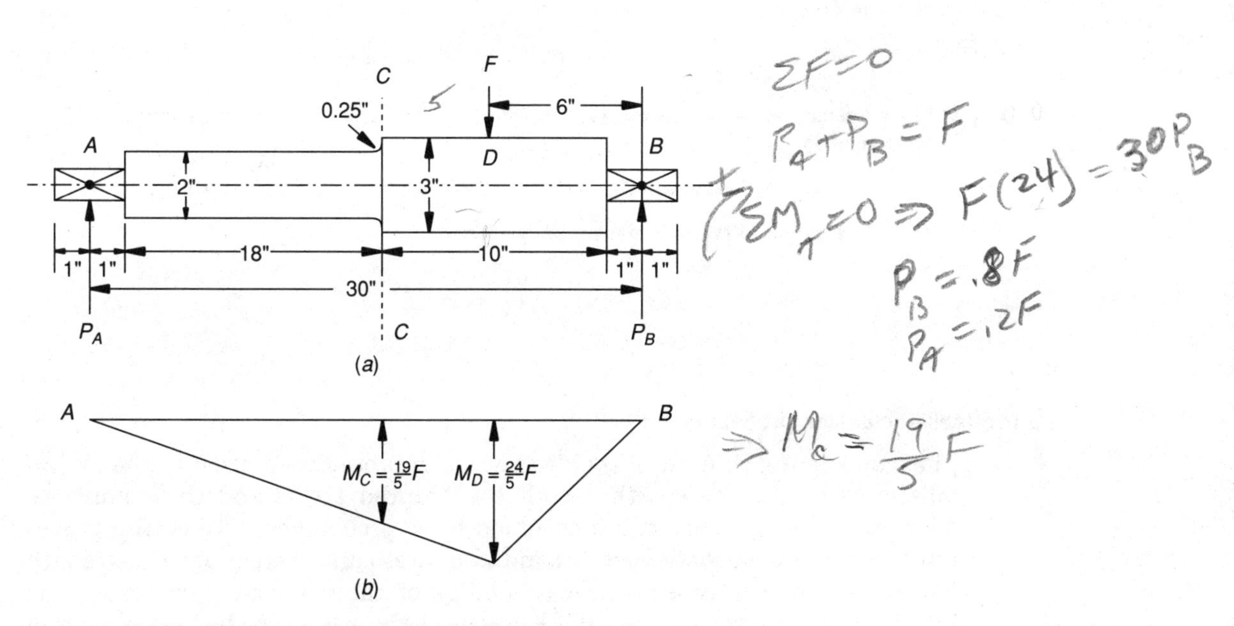

Figure 9.2 (*a*) Machined steel shaft; and (*b*) bending moment diagram.

Thus

$$s_C = \frac{4\,M_C}{\pi} = \frac{4\,(11{,}400)}{\pi}$$

$$= 14{,}514.896868 \text{ lb/in}^2$$

The endurance strength of the component at section C can be computed using Eq. (9.4) with

$$S_e' = 0.504(80) = 40.32 \text{ k(lb/in}^{2)} \qquad \text{(using Eq. 9.5)}$$

$$k_a = 2.7\,(80)^{-0.2653} = 0.844256 \text{ (using Eq. 9.6)}$$

$$k_b = \left(\frac{2}{0.3}\right)^{-0.1133} \qquad \text{(using Eq. 9.7)}$$

$$k_c = 1 \qquad \text{(using Eq. 9.8)}$$

$$k_d = 1 \qquad \text{(using Eq. 9.9)}$$

To compute the fatigue-strength reduction factor, the notch radius is taken as $r = 0.25$ in, $\sqrt{a}$ as $\sqrt{a} = 4/80 = 0.05$ for the shoulder, and the geometric stress concentration factor K_t as $K_t = 1.75$. This gives

$$K_F = \frac{1.75}{1 + \dfrac{2}{\sqrt{0.25}}\left(\dfrac{1.75-1}{1.75}\right)(0.05)} = \frac{1.75}{1.021429} = 1.713287$$

and hence

$$k_e = \frac{1}{K_F} = 0.583673 \qquad \text{(using Eq. 9.10)}$$

The miscellaneous effects factor, k_f, is taken as 1. This gives the endurance strength of the component as

$$S_e = (0.844256)(0.806587)(1)(1)(0.583673)(1)(40.32) = 16.025644 \text{ k(lb/in}^2\text{)}$$

Thus the factor of safety, n, is given by

$$n = \frac{S_e}{s_C} = \frac{16.025644}{14.514897} = 1.104082$$

9.3.2 Probabilistic design procedure

The same approach used in the deterministic procedure is used in the probabilistic case, also, except that both the induced stress and the endurance strength of the component are assumed to be probabilistic. Thus, the probability distribution functions of the induced stress and the endurance strength play a role in determining the reliability of the component by using the interference theory. Mischke [9.4] proposed the use of a stochastic form of Marin factors to express the endurance strength in probabilistic form as

$$S_e = k_a\, k_b\, k_c\, k_d\, k_e\, k_f\, \phi\, \overline{S}_{ut} \tag{9.13}$$

where S_e, k_a, k_b, k_c, k_d, k_e, and k_f are treated as random variables; ϕ is an empirical random variable proportionality factor; and $\overline{S}_{ut}$ is the deterministic, mean ultimate tensile strength of the material. The random surface modification factor is given by:

$$k_a = a\, \overline{S}_{ut}^{\,b}\, (1,\, \gamma_{k_a}) \tag{9.14}$$

where a and b are same as those given for Eq. (9.6) and γ_{k_a} is the coefficient of variation of k_a given by 0.098 for hot-rolled components, 0.078 for as-forged components, 0.131 for ground surfaces, and 0.06 for machined surfaces. The size factor, k_b, is considered as a deterministic quantity so that the value given by Eq. (9.7) is taken as its mean value and its coefficient of variation γ_{k_b} is taken as zero. The loading modification factor k_c is given by

$$k_c = \begin{cases} (1,\, 0); & \text{for bending loads} \\ 0.774\,(1,\, 0.163); & \text{for axial loads} \\ 0.583\,(1,\, 0.123); & \text{for torsional loads} \end{cases} \tag{9.15}$$

The temperature and miscellaneous-effects factors, k_d and k_f, depend on the details of the particular application and cannot be expressed in a general form. The stress concentration and notch-sensitivity modification factor, k_e, can be expressed as

$$k_e = \frac{1}{K_F} \tag{9.16}$$

where

$$K_F = \frac{K_t\,(1,\, \gamma_t)}{1 + \dfrac{2}{\sqrt{r}}\left(\dfrac{K_t - 1}{K_t}\right)\sqrt{a}} = \overline{K}_F\,(1,\, \gamma_F) \tag{9.17}$$

with

$$\sqrt{a} = \begin{cases} \dfrac{5}{\overline{S}_{ut}}; & \text{for a transverse hole} \\ \dfrac{4}{\overline{S}_{ut}}; & \text{for a shoulder} \\ \dfrac{3}{\overline{S}_{ut}}; & \text{for a groove} \end{cases} \tag{9.18}$$

$\gamma_t =$ coefficient of variation of K_t, $\gamma_F =$ coefficient of variation of K_F

$$\gamma_F = \begin{cases} 0.11; & \text{for a transverse hole} \\ 0.08; & \text{for a shoulder} \\ 0.13; & \text{for a groove} \end{cases} \tag{9.19}$$

$r =$ notch radius and $\overline{K}_F =$ mean value of K_F. The random proportionality factor ϕ is given by

$$\phi = \begin{cases} 0.504\ (1,\ 0.146); & \text{for bending loads} \\ 0.390\ (1,\ 0.309); & \text{for axial loads} \\ 0.294\ (1,\ 0.269); & \text{for torsional loads} \end{cases} \tag{9.20}$$

Since the endurance strength, S_e, given by Eq. (9.13), is a product of several random variables, its mean value and standard deviation can be found using the partial derivative rule. The results can be expressed as

$$\overline{S}_e = \overline{k}_a\ \overline{k}_b\ \overline{k}_c\ \overline{k}_d\ \overline{k}_e\ \overline{k}_f\ \overline{\phi}\ \overline{S}_{ut} \tag{9.21}$$

and

$$\gamma^2_{S_e} = \gamma^2_{k_a} + \gamma^2_{k_b} + \gamma^2_{k_c} + \gamma^2_{k_d} + \gamma^2_{k_e} + \gamma^2_{k_f} + \gamma^2_{\phi} \tag{9.22}$$

where a bar over a symbol indicates its mean value and γ denotes the coefficient of variation.

Example 9.5 The machined steel shaft shown in Fig. 9.2*a* is subjected to a random force $F = N(3000,\ 300)$ lb. The material has a mean ultimate tensile strength of 80 k(lb/in^2). The geometric stress concentration factor is known to be 1.75. Determine the reliability of the shaft assuming normal distribution for the induced and permissible stresses.

solution From the bending-moment diagram shown in Fig. 9.2*b*, the bending moment at section C, M_C, can be expressed as

$$M_C = 19\ P_A = \frac{19}{5}\ F = \frac{19}{5}\ (3000,\ 300)\ \text{lb} \cdot \text{in} = (11400,\ 1140)\ \text{lb} \cdot \text{in}$$

The stress induced at C, s_C, is given by

$$s_C = \frac{M_C\ c}{I}$$

where

$$\frac{I}{c} = \frac{\pi\ d^3}{32} = \frac{\pi\ (2)^3}{32} = \frac{\pi}{4}\ \text{in}^3$$

so that

$$s_C = \frac{4\ M_C}{\pi} = \frac{4\ (11400,\ 1140)}{\pi}$$

$$= 14514.896868(1,\ 0.1)\ \text{lb/in}^2$$

The endurance strength of the component at section C can be computed using Eq. (9.13) with:

$$k_a = a\ \overline{S}^b_{ut}\ (1,\ \gamma_{k_a}) = 2.7\ (80)^{-0.2653}\ (1,\ 0.06) = 0.844256\ (1,\ 0.06)$$

$$k_b = \left(\frac{d}{0.3}\right)^{-0.1133}\ (1,\ 0) = \left(\frac{2}{0.3}\right)^{-0.1133}\ (1,\ 0) = 0.806587\ (1,\ 0)$$

$$k_c = (1,0)$$

$$k_d = (1,0)$$

$$k_f = (1,0)$$

since no specific information is known about the operating temperature and other effects. Here $K_t = 1.75$, $r = 0.25$ in and $\sqrt{a} = 4/\overline{S}_{ut} = 4/80 = 0.05$ and hence the mean value of K_F can be determined using Eq. (9.17) as

$$\overline{K}_F = \frac{K_t}{1 + \frac{2}{\sqrt{r}}\left(\frac{K_t - 1}{K_t}\right)\sqrt{a}} = \frac{1.75}{1.75 + \frac{2}{\sqrt{0.25}}\left(\frac{1.75 - 1}{1.75}\right)(0.05)} = 1.713287$$

and hence $K_F = 1.713287(1, 0.08)$ with

$$k_e = \frac{1}{K_F} = \frac{1}{1.713287\,(1, 0.08)}$$

The random proportionality factor ϕ is given by

$$\phi = 0.504(1, 0.146)$$

Thus the endurance strength of the component, given by Eq. (9.13), is

$$S_e = [0.844256\,(1, 0.06)]\,[0.806587\,(1, 0)]\,[(1, 0)]\,[(1, 0)]\left[\frac{1}{1.713287\,(1, 0.08)}\right]$$

$$\times \left[(1, 0)\right]\left[0.504\,(1, 0.146)\right]\left[80\right] \tag{E1}$$

Equation (E1) gives the mean value of S_e as

$$\overline{S}_e = (0.844256)\,(0.806587)\,(1)\,(1)\left(\frac{1}{1.713287}\right)(1)\,(0.504)\,(80) = 16.025644 \text{ k(lb/in}^2\text{)}$$

The coefficient of variation of S_e can be determined using Eq. (9.22) as

$$\gamma_{S_e}^2 = (0.06)^2 + (0)^2 + (0)^2 + (0)^2 + (0.08)^2 + (0)^2 + (0.146)^2$$

$$= 0.0036 + 0.0064 + 0.021316 = 0.031316 = (0.176963)^2$$

from which the standard deviation of S_e can be found as

$$\sigma_{S_e} = 0.176963\,(16.025644) = 2.835946 \text{ k(lb/in}^2\text{)}$$

If s_C and S_e follow normal distribution, the lower limit of integration, z_1, can be found as

$$z_1 = -\frac{\overline{S}_e - \overline{s}_C}{\sqrt{\sigma_{S_e}^2 + \sigma_{sC}^2}}$$

$$= -\frac{16.025644 - 14.514897}{\sqrt{(2.835946)^2 + (1.451490)^2}} = -\frac{1.510747}{\sqrt{10.149413}} = -0.474211$$

From standard normal tables, this value of z_1 can be seen to correspond to a reliability of 0.6822.

$\phi(z_1) = (1 - .6822)$ $R = 1 - \phi(z_1)$

$\Rightarrow R = 1 - (1 - .6822) = .6822$

Example 9.6 Find the reliability of the shaft described in Example 9.5 by assuming lognormal distribution for both the induced stress and the endurance strength of the component.

solution The mean and standard deviations of the induced stress and the endurance strengths have been found, in Example 9.5, to be

$$\bar{s}_C = 14.514897 \text{ k(lb/in}^2), \qquad \sigma_{s_C} = 1.451490 \text{ k(lb/in}^2)$$

$$\bar{S}_e = 16.025644 \text{ k(lb/in}^2), \qquad \sigma_{S_e} = 2.835946 \text{ k(lb/in}^2)$$

Since s_C and S_e follow lognormal distributions, $\ln s_C$ and $\ln S_e$ follow normal distributions with

$$\mu_{\ln s_C} = \ln \bar{s}_C - \frac{1}{2}\left(\frac{\sigma_{s_C}}{\bar{s}_C}\right)^2 = \ln 14.514897 - \frac{1}{2}\left(\frac{1.451490}{14.514897}\right)^2 = 2.670175$$

$$\sigma_{\ln s_C} = \frac{\sigma_{s_C}}{\bar{s}_C} = \frac{1.451490}{14.514897} = 0.1$$

$$\mu_{\ln S_e} = \ln \bar{S}_e - \frac{1}{2}\left(\frac{\sigma_{S_e}}{\bar{S}_e}\right)^2 = \ln 16.025644 - \frac{1}{2}\left(\frac{2.835946}{16.025644}\right)^2 = 2.758532$$

and

$$\sigma_{\ln S_e} = \frac{\sigma_{S_e}}{\bar{S}_e} = \frac{2.835946}{16.025644} = 0.176963$$

The lower limit of integration, z_1, in Eq. (9.3), can be found as

$$z_1 = -\frac{\mu_{\ln S_e} - \mu_{\ln s_C}}{\sqrt{\sigma^2_{\ln S_e} + \sigma^2_{\ln s_C}}} = -\frac{2.758532 - 2.670175}{\sqrt{(0.176963)^2 + (0.1)^2}}$$

$$= -\frac{0.088357}{\sqrt{0.0413159}} = -0.434693$$

From standard normal tables, this value of z_1 can be seen to correspond to a reliability of 0.6679.

9.4 Design of Mechanical Systems

9.4.1 Reliability-based design of gear trains

Introduction. A gear train is composed of several gears mounted on different shafts. Different gears can be brought into mesh by shifting the positions of certain gears. Depending on the number of teeth on the various meshing gear pairs, different speeds can be obtained at the output shaft for a single input speed. Gear trains find application in automobiles, machine tools, helicopters, and a variety of industrial machinery. A typical gear train is shown in Fig. 9.3. The safe performance of a gear train depends on its overall reliability, which in turn depends on the probability of failure of each of the gear pairs in the drive in different failure modes. The problem, thus, is the design of a multicomponent transmission for a specified overall probability of failure [9.8, 9.9].

Assumptions. The following assumptions are made in the analysis:

1. Two failure modes, namely, the bending and surface-wear failure modes, are considered for each gear pair.
2. The gear train is idealized as a weakest-link kinematic chain, a concept analogous to the series system.
3. The layout of the gears, the number of teeth on different gears, the module, and the interconnection of the various gear pairs of the gear train are known. The interconnection of the gear pairs and the output shaft speeds obtainable for the gear train of Fig. 9.3 are shown in Fig. 9.4.
4. The power transmitted by all the gear pairs is the same.
5. The power transmitted, the speeds of the input shaft, the allowable and induced stresses in bending and surface wear, the center distances of the gear pairs, and the face widths of the gear pairs are random variables and are statistically independent.
6. All the random variables follow normal distribution.

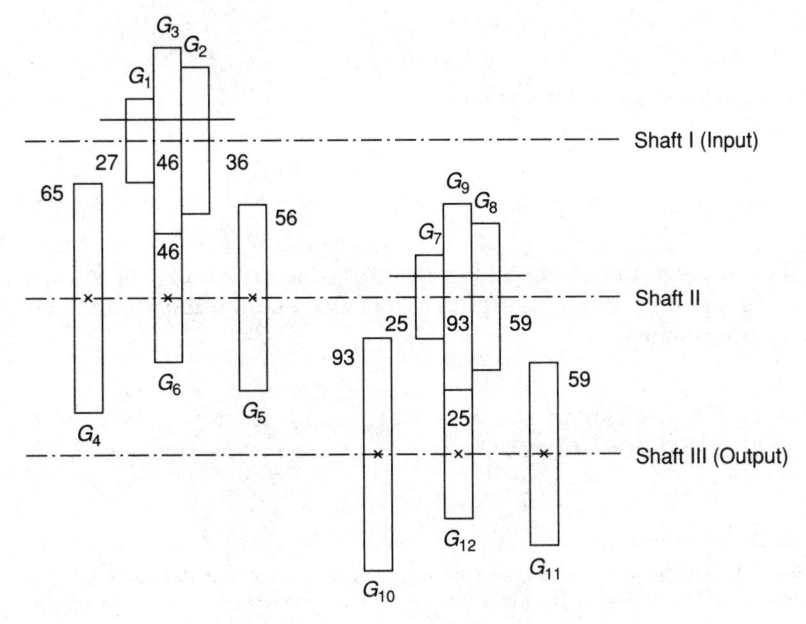

Figure 9.3 A typical gear train (nine-speed gear train).

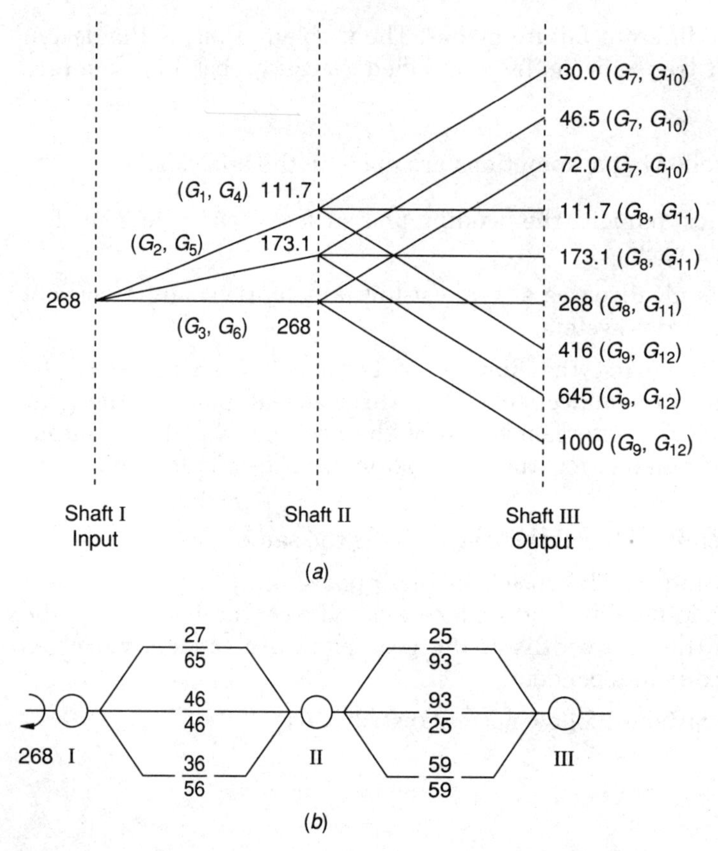

Figure 9.4 (*a*) Interconnection of the gear pairs; and (*b*) the output shaft speeds.

7. The mean values of the face widths of the gear pairs are the design variables.
8. A design is considered to be safe and adequate if the probability of failure of the gear train is less than or equal to a specified small quantity in each of the two failure modes.

Design Equations. If two meshing gears (see Fig. 9.5) transmit a power of p (watts) at a wheel speed[1] of n_w (rpm), the torque acting on the gear teeth, M_t (N/mm), is given by

---

[1]The gear having a larger diameter in a gear pair is called the wheel and the one having a smaller diameter is called the pinion. If both the gears have the same diameter, any one can be called the wheel.

$$M_t = \frac{7.1324 \times 10^9\, p}{n_w} \tag{9.23}$$

The induced bending stress, $s_b(N/\mathrm{mm}^2)$, assuming the tooth to be a cantilever beam with end load [9.8, 9.9], can be computed as

$$s_b = \frac{\beta\, M_t}{A\, t} \tag{9.24}$$

where

$$\beta = \frac{K_c\, K_d(i + 1)}{imy \cos \alpha} \tag{9.25}$$

A is the center distance between the gears (mm), t is the face width (mm), K_c is the stress concentration factor, K_d is the dynamic load factor, $i = T_w / T_p = n_p / n_w$ is the ratio of speeds (transmission ratio), T_w is the number of teeth on the wheel, T_p is the number of teeth on the pinion, n_p is the speed of the pinion (rpm), m is the module (mm), α is the pressure angle, and y is the Lewis form factor given by

$$y = 0.52\left(1 + \frac{20}{T_w}\right) \tag{9.26}$$

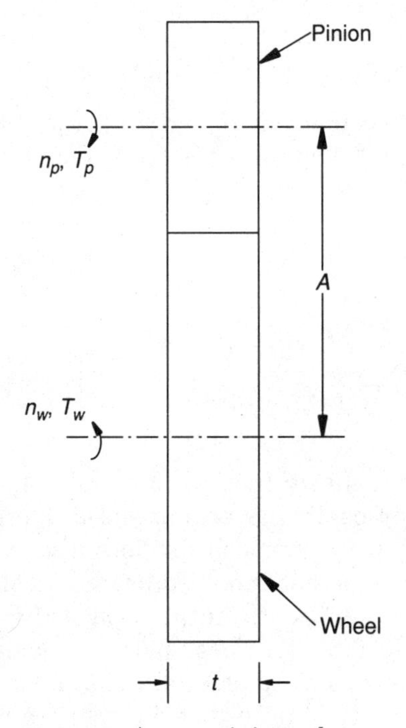

Figure 9.5 A gear pair in mesh.

The induced wear stress s_w (N/mm^2) [9.8, 9.9] is given by

$$s_w = \frac{\gamma}{A}\sqrt{\frac{M_t}{t}} \tag{9.27}$$

where

$$\gamma = 0.1848\left(\frac{i+1}{i}\right)\left\{\frac{(i+1)\,EK_cK_d}{\sin 2\alpha}\right\}^{1/2} \tag{9.28}$$

and E is the Young's modulus of the material of the gear pair.

Mean and standard deviations of induced stresses. The mean and standard deviation of s_b are given by

$$\bar{s}_b = 7.1324 \times 10^9 \frac{\beta\,\bar{p}}{\bar{n}_w\,\bar{A}\,\bar{t}} \tag{9.29}$$

and

$$\sigma_{s_b} = \frac{1}{\bar{A}\,\bar{t}}\left[\frac{\beta^2\,\bar{M}_t^2\left(\bar{t}^2\,\sigma_A^2 + \bar{A}^2\,\sigma_t^2\right) + \beta^2\,\bar{A}^2\,\bar{t}^2\,\sigma_{M_t}^2}{\bar{A}^2\,\bar{t}^2 + \bar{t}^2\,\sigma_A^2 + \bar{A}^2\,\sigma_t^2}\right]^{1/2} \tag{9.30}$$

where

$$\sigma_{M_t} = \frac{7.1324 \times 10^9}{\bar{n}_w}\left[\frac{\bar{p}^2\,\sigma_{n_w}^2 + \bar{n}_w^2\,\sigma_p^2}{\bar{n}_w^2 + \sigma_{n_w}^2}\right]^{1/2} \tag{9.31}$$

and the partial derivative rule has been used in deriving Eqs. (9.30) and (9.31). Similarly, the mean and standard deviation of s_w are given by

$$\bar{s}_w = \frac{\gamma}{\bar{A}}\sqrt{\frac{\bar{M}_t}{\bar{t}}} \tag{9.32}$$

and

$$\sigma_{s_w} = \frac{1}{\bar{A}\sqrt{\bar{t}}}\left[\frac{4\,\bar{M}_t^2\,\gamma^2\,\bar{t}^2\,\sigma_A^2 + \gamma^2\,\bar{A}^2\,\bar{M}_t^2\,\sigma_t^2 + \bar{A}^2\,\bar{t}^2\,\gamma^2\,\sigma_{M_t}^2}{4\,\bar{M}_t\,\bar{A}^2\,\bar{t}^2 + 4\,\bar{M}_t\,\bar{t}\,\sigma_A^2 + \bar{M}_t\,\bar{A}^2\,\sigma_t^2}\right]^{1/2} \tag{9.33}$$

Probabilistic model. For the gear train shown in Figs. 9.3 and 9.4, two gear pairs are in mesh while delivering any particular output speed. For example, the gear pairs (G_1, G_4) and (G_9, G_{12}) are in mesh for obtaining an output speed of 416 rpm when the input speed is 268 rpm. To use the design equations, we consider a typical ith gear pair in jth output speed ($i = 1, 2$ and $j = 1, 2, \ldots, 9$ for the gear train of Fig. 9.3). This gear pair is considered as a single link in a series kinematic chain consisting of a total of k links ($k = 2$ for the gear train of Fig. 9.3). The probability of failure of the gear train, P_f, can be expressed, using the weakest-link model, as

$$P_f = 1 - \prod_{i=1}^{k} (1 - P_{f_i}) \tag{9.34}$$

where P_{f_i} is the probability of failure of the ith link (gear pair). Since two failure modes are considered for each gear pair, the overall probabilities of failure of the gear train while transmitting the jth output speed, can be computed as

$$P_{f_{bj}} = 1 - \prod_{i=1}^{k} (1 - P_{F_{bij}}); \qquad j = 1, 2, \ldots, N \tag{9.35}$$

$$P_{f_{wj}} = 1 - \prod_{i=1}^{k} (1 - P_{f_{wij}}); \qquad j = 1, 2, \ldots, N \tag{9.36}$$

where $P_{f_{bj}}$ ($P_{f_{wj}}$) is the probability of failure of the gear train in bending mode (surface-wear mode), $P_{f_{bij}}$ ($P_{f_{wij}}$) is the probability of failure of the ith gear pair in delivering the jth output shaft speed and N is the total number of speeds obtainable at the output shaft. The probabilities of failure of all the gear pairs, $P_{f_{bij}}$ and $P_{f_{wij}}$, are assumed to be same in the present analysis.

If we consider the ith gear pair in jth output shaft speed as a single link being acted upon by the bending stress, the probability of failure of the gear pair (link) is given by

$$P_{f_b} = P_{f_{bij}} = \int_{-\infty}^{\infty} F_{S_b}(r)\, f_{s_b}(r)\, dr \tag{9.37}$$

where $f_{s_b}(r)$ is the probability density function of the induced stress, s_b, and $F_{S_b}(r)$ is the probability distribution function of the permissible bending stress. Since s_b and S_b are assumed to follow normal distribution with means $(\bar{s}_b, \bar{S}_b)$ and standard deviations $(\sigma_{s_b}, \sigma_{S_b})$, the probability of failure can also be expressed as

$$P_{f_{bij}} = P(\bar{S}_b - \bar{s}_b < 0) \tag{9.38}$$

with the corresponding upper limit of integration given by

$$z_{1b} = -\left[\frac{\bar{S}_b - \bar{s}_b}{\sqrt{\sigma_{S_b}^2 + \sigma_{s_b}^2}}\right] \tag{9.39}$$

The value of z_{1b} can be found from standard normal distribution tables for any specified probability of failure $P_{f_{bij}}$. Similarly, the upper limit of integration corresponding to the probability of failure in surface wear $P_{f_{wij}}$ is given by

$$z_{1w} = -\left[\frac{\bar{S}_w - \bar{s}_w}{\sqrt{\sigma_{S_w}^2 + \sigma_{s_w}^2}}\right] \tag{9.40}$$

Design procedure. Assuming that the speed of the input shaft, the number of teeth on different gears, the power transmitted, the allowable stresses, the allowable probabilities of failure of the gear train, and other factors involved in the various equations are known as design data, the following procedure can be adopted for designing the gear train for bending strength [9.8, 9.9]

1. From the known $P_{f_b} = P_{f_{bj}}$, compute the probability of failure of any gear pair at any speed by assuming equal probabilities of failure for all gear pairs.

2. Find the upper limit of integration in Eq. (9.39) from standard normal tables.

3. Calculate the coefficients of variation of torque and induced bending stress using Eqs. (9.31) and (9.30) as

$$\gamma_{M_t}^2 = \frac{\gamma_p^2 + \gamma_{n_w}^2}{1 + \gamma_{n_w}^2} \tag{9.41}$$

and

$$\gamma_{s_b}^2 = \frac{\gamma_t^2 + \gamma_A^2 + \gamma_{M_t}^2}{1 + \gamma_A^2 + \gamma_t^2} \tag{9.42}$$

4. Find the mean value of the induced stress as

$$\bar{s}_b = \frac{\bar{S}_b \pm [-\bar{S}_b^2 (1 - z_{1b}^2 \gamma_{S_b}^2)(1 - z_{1b}^2 \gamma_{s_b}^2) + \bar{S}_b^2]^{1/2}}{(1 - z_{1b}^2 \gamma_{s_b}^2)} \tag{9.43}$$

5. Using the smaller value of $\bar{s}_b$ given by Eq. (9.43), compute the mean value of the face width of the particular gear pair using Eq. (9.29). From this, find the standard deviation of t as

$$\sigma_t = \gamma_t \bar{t} \tag{9.44}$$

6. Repeat steps 3 through 5 until the thicknesses of different gear pairs are calculated for all the speeds.

7. Determine the design thickness $\bar{t}$ of any gear pair as the largest value out of all the speeds in which the gear pair participates.

A similar procedure can be adopted for the design of the gear train for surface-wear strength. The relevant equations in this case are

$$\gamma_{s_w} = \frac{\gamma_t^2 + \gamma_{M_t}^2 + 4\,\gamma_A^2}{4 + \gamma_t^2 + 4\,\gamma_A^2} \tag{9.45}$$

$$\bar{s}_w = \frac{\bar{S}_w \pm [-\bar{S}_w^2 (1 - z_{1w}^2 \gamma_{S_w}^2)(1 - z_{1w}^2 \gamma_{s_w}^2) + \bar{S}_w^2]^{1/2}}{(1 - z_{1w}^2 \gamma_{s_w}^2)} \tag{9.46}$$

and

$$\bar{t} = \left(\frac{\gamma \sqrt{\bar{M}_t}}{\bar{A}\, \bar{s}_w} \right)^2 \tag{9.47}$$

Numerical results. The reliability-based design of the three shaft, nine-speed lathe gear train shown in Figs. 9.3 and 9.4 is illustrated for the following data

mean power, 7.457 kW
mean input shaft speed, 268 rpm
gear tooth profile, 20° involute type
module of gears, 4 mm
material of gears, hardened steel
mean allowable stress in bending, 2.4525×10^8 N/m^2
mean allowable stress in surface wear, 17.1675×10^8 N/m^2
density of the material, 7.60275×10^4 N/m^3
stress concentration factor, 1.5
dynamic load factor, 1.1

The results obtained for an allowable probability of failure of $P_{f_b} = P_{f_w} = 0.2 \times 10^{-5}$ for different values of the coefficient of variation of the input shaft speed are shown in Table 9.1. The results given by the deterministic procedure are also shown in the table. It can be seen that the thicknesses of all the gear pairs except that of the sixth one are governed by the surface-wear strength. The deterministic design gives the weight of the gear train as 1264.0185 N, whereas the probabilistic design gives a varying weight of 1236.8448 N to 1305.2205 N as the value of γ_{n_w} changes from 0.01 to 0.10. Thus the factor γ_{n_w}, which is not even mentioned in the deterministic procedure, can be seen to influence the weight by 5.33 percent. A similar behavior can be expected in respect of the other coefficients of variation like γ_p, γ_t, γ_{S_b}, and γ_{S_w}. It is to be noted that the weight of the gear train is obtained by adding the weights of the individual gears without considering the weights of the shafts and bearings. The effect of the variation of the probability of failure on the design of the gear train is shown in Table 9.2. The results indicate that the face widths of the gear pairs (hence the weight of the gear train) increase as the probability of failure decreases. A decrease in the probability of failure from 0.2 to 0.2×10^{-6} can be seen to increase the weight of the gear train by 114.14 percent for $\gamma_{n_w} = 0.05$.

TABLE 9.1 Mean Value of the Face Width (cm) for $P_{f_b} = P_{f_w} = 0.2 \times 10^{-5}$

Gear pair number	Speed of wheel n_w	Coefficient of variation of input shaft speed 0.01	0.05	0.10	Deterministic design (safety factor = 2)
		By considering bending strength only			
1 (G_1, G_4)	111.7	2.733	2.786	2.936	2.586
2 (G_3, G_6)	268.0	1.467	1.495	1.576	1.388
3 (G_2, G_5)	173.1	2.053	2.093	2.206	1.943
4 (G_7, G_{10})	30.0	7.654	7.802	8.222	7.242
	46.5	4.938	5.034	5.305	4.672
	72.0	3.189	3.251	3.426	3.018
5 (G_9, G_{12})	416.0	0.552	0.563	0.593	0.522
	643.0	0.357	0.364	0.384	0.338
	1000.0	0.230	0.234	0.247	0.217
6 (G_8, G_{11})	111.7	2.940	2.997	3.159	2.782
	173.1	1.897	1.934	2.038	1.795
	268.0	1.226	1.249	1.316	1.160
Weight of the gear train (N)*		1198.782	1224.288	1289.034	1134.036
		By considering surface wear strength only			
1 (G_1, G_4)	111.7	3.104	3.144	3.266	3.216
2 (G_3, G_6)	268.0	1.516	1.536	1.595	1.571
3 (G_2, G_5)	173.1	2.086	2.113	2.194	2.161
4 (G_7, G_{10})	30.0	7.821	7.922	8.228	8.104
	46.5	5.046	5.111	5.309	5.228
	72.0	3.259	3.301	3.428	3.377
5 (G_9, G_{12})	416.0	0.564	0.571	0.593	0.584
	643.0	0.365	0.370	0.384	0.378
	1000.0	0.235	0.238	0.247	0.243
6 (G_8, G_{11})	111.7	2.211	2.240	2.327	2.291
	173.1	1.427	1.445	1.501	1.479
	268.0	0.922	0.934	0.970	0.955
Weight of the gear train (N)*		1187.991	1202.706	1250.775	1231.155
		By considering both bending and surface wear strengths			
Weight of the gear train (N)†		1236.845	1254.601	1305.221	1264.019

SOURCE: Reprinted with permission from S. S. Rao, "A Probabilistic Approach to the Design of Gear Trains," *International Journal of Machine Tool Design and Research*, vol. 14, copyright 1974, Pergamon Press PLC, pp. 267–278.

*The thickness of the gear pairs 4, 5, and 6 correspond to the speeds 30.0, 416.0, and 111.7 rpm, respectively.

†The thickness of any gear pair is taken as the larger one of the two values given by bending and surface wear.

Data: $\gamma_p = 0.10$, $\gamma_A = \gamma_t = 0.01$, $\gamma_{S_b} = \gamma_{S_w} = 0.10$, $E = 20.601 \times 10^{10}$ N/m^2.

TABLE 9.2 Effect of Variation on the Probability of Failure*

Value of $p_{f_{bj}} = P_{f_{wj}}$	Face width of gear pair (cm) 1	2	3	4	5	6	Weight of gear train (N)†
2×10^{-1}	1.567	0.841	1.177	4.388	0.316	1.686	687.5829
2×10^{-2}	1.836	0.986	1.380	5.144	0.371	1.976	805.8915
2×10^{-3}	2.076	1.115	1.560	5.815	0.419	2.234	912.3300
2×10^{-4}	2.308	1.239	1.734	6.464	0.466	2.483	1012.5882
2×10^{-5}	2.626	1.363	1.908	7.113	0.513	2.733	1119.3210
2×10^{-6}	3.144	1.536	2.113	7.922	0.571	2.997	1254.6009
2×10^{-7}	3.763	1.838	2.528	9.480	0.684	3.282	1472.3829

SOURCE: Reprinted with permission from S. S. Rao, "A Probabilistic Approach to the Design of Gear Trains," *International Journal of Machine Tool Design and Research*, vol. 14, copyright 1974, Pergamon Press PLC, pp. 267–278.

†Note: Weight of the gear train according to deterministic design = 1264.019 N.

*Data: $\gamma_{n_w} = 0.05$, $\gamma_p = 0.10$, $\gamma_A = \gamma_t = 0.01$, $\gamma_{S_b} = \gamma_{S_w} = 0.10$, and $E = 20.601 \times 10^{10}$ N/m^2.

9.5 Reliability Analysis of Mechanical Systems

9.5.1 Cam-follower systems

Introduction. It is well known that the limitations of the manufacturing machines, poor workmanship, inaccuracies in assembly, and clearances in the bearings of kinematic pairs cause errors that affect the kinematic accuracy, dynamic performance, noise level, and life of cam mechanisms. In the case of internal combustion engines with valves, the lift of a valve, which is governed by a cam-follower system, should be accurately known so as to clearly define the closing and opening of the valves. Otherwise it may result in the inefficient use of the fuel. Thus, the analysis and control of errors in cam mechanisms is extremely important to achieve the desired level of accuracy. In this section, the effect of errors in the geometrical parameters of the cam-follower system on the kinematic response is studied. The errors are treated as probabilistic in nature. The disc cam with translating roller-ended follower system shown in Fig. 9.6 [9.10] is considered for illustration.

Kinematic response of the cam-follower system. For the cam-follower mechanism shown in Fig. 9.6, the condition of contact [9.10] can be expressed as:

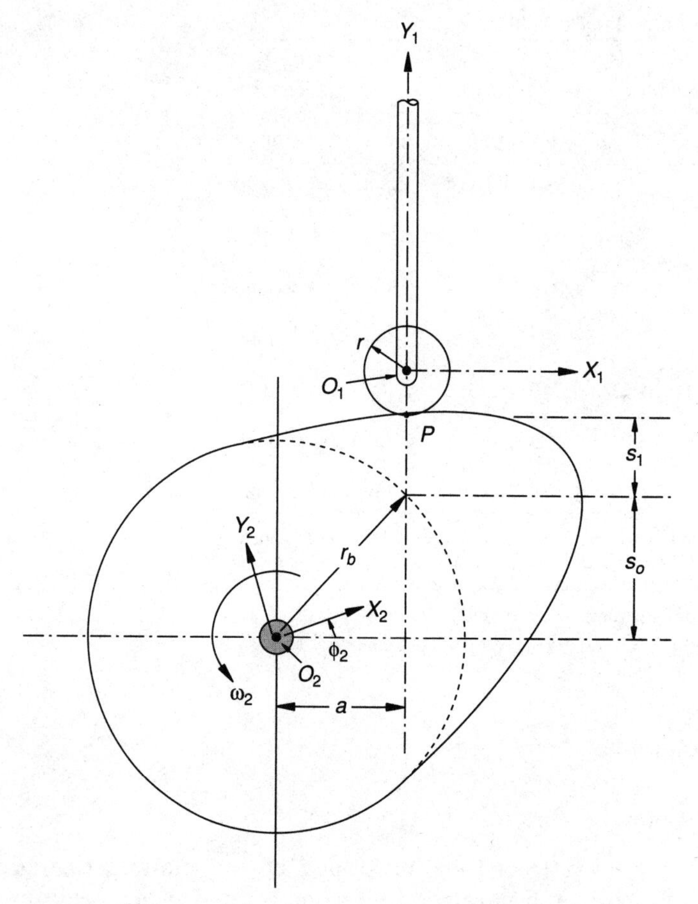

Figure 9.6 Disc cam with translating-roller follower.

$$E \sin \theta + F \cos \theta + G = 0 \tag{9.48}$$

with

$$E = s'_1 - a$$

$$F = s_1 + s_o$$

$$G = 0$$

$$s'_1 = \frac{ds_1}{d\phi_2}$$

$$s_o = \sqrt{r_b^2 - a^2}$$

s_1 = displacement of the follower, a = offset or eccentricity, r_b = radius of the base circle of the cam, ϕ_2 = cam rotation angle, and θ = parameter of the follower surface. Equation (9.48) can be used to express θ as

$$\theta = \tan^{-1}\left(\frac{s_1 + s_o}{-s'_1 + a}\right) \tag{9.49}$$

The cam profile [9.10–9.12] is defined by

$$X_2 = -r\cos(\theta + \phi_2) + a\cos\phi_2 + (s_1 + s_o)\sin\phi_2 \tag{9.50}$$

$$Y_2 = -r\sin(\theta + \phi_2) - a\sin\phi_2 + (s_1 + s_o)\cos\phi_2 \tag{9.51}$$

where X_2 and Y_2 denote the coordinates of the point of contact, P, and r indicates the radius of the roller (see Fig. 9.6). Using Eqs. (9.50) and (9.51), the displacement or response of the follower can be expressed as

$$s_1 = X_2\sin\phi_2 + Y_2\cos\phi_2 - r\sin\theta - s_o \tag{9.52}$$

This equation shows that the follower response is not an explicit function of the geometric parameters, because the variables θ, X_2, and Y_2 themselves are functions of s_1 as given by Eqs. (9.49) to (9.51).

Here the errors in X_2, Y_2, r, r_b, and a are assumed to be independent random variables following normal distribution. The variations in the geometrical parameters r and r_b are treated as random variables because of the nature of manufacturing tolerances. Some error is bound to occur in mounting the cam-follower system, and hence the offset or eccentricity a is assumed to be a random variable. To account for the manufacturing errors, the tolerances in the coordinates of the cam profile are assumed to be random variables. Although these coordinates change with the cam-rotation angle, the tolerances ΔX and ΔY on the coordinates X_2 and Y_2 will be same for all the cam-rotation angles. By defining a vector of random parameters as

$$\mathbf{X} = \begin{Bmatrix} x_1 \\ x_2 \\ x_3 \\ x_4 \\ x_5 \end{Bmatrix} = \begin{Bmatrix} \Delta X \\ \Delta Y \\ \Delta r \\ \Delta r_b \\ \Delta a \end{Bmatrix}$$

the standard deviation of the kinematic response s_1 can be obtained by applying the partial derivative rule as

$$(\sigma_{s_1})_j = \left[\sum_{i=1}^{5}\left(\frac{\partial s_1}{\partial x_i}\right)^2\Bigg|_{\mathbf{X}} \sigma_{x_i}^2\right]^{1/2} \tag{9.53}$$

where $(\sigma_{s_1})_j$ = standard deviation of s_1 at $\phi_2 = (\phi_2)_j$, and $\mathbf{X}$ = mean design vector $= (0.0, 0.0, 0.0, 0.0, 0.0)^T$. The partial derivatives required for the computation of $(\sigma_{s_1})_j$ can be found as

$$\frac{\partial s_1}{\partial x_1} = \sin \phi_2$$

$$\frac{\partial s_1}{\partial x_2} = \cos \phi_2$$

$$\frac{\partial s_1}{\partial x_3} = -\sin \theta$$

$$\frac{\partial s_1}{\partial x_4} = -r \cos \theta \left(\frac{\partial \theta}{\partial r_b}\right) - \left(\frac{\partial s_o}{\partial r_b}\right)$$

$$\frac{\partial s_1}{\partial x_5} = -r \cos \theta \left(\frac{\partial \theta}{\partial a}\right) - \left(\frac{\partial s_o}{\partial a}\right)$$

$$\frac{\partial \theta}{\partial r_b} = \frac{\left(\frac{\partial s_o}{\partial r_b}\right)(s_1' - a)}{[(s_1' - a)^2 + (s_1 + s_o)^2]}$$

$$\frac{\partial \theta}{\partial a} = \frac{\left(\frac{\partial s_o}{\partial a}\right)(s_1' - a)}{[(s_1' - a)^2 + (s_1 + s_o)^2]}$$

$$\frac{\partial s_o}{\partial r_b} = \frac{r_b}{\sqrt{r_b^2 - a^2}}$$

$$\frac{\partial s_o}{\partial a} = \frac{-a}{\sqrt{r_b^2 - a^2}}$$

Mean and standard deviation of the kinematic response. The various steps involved in the computation of the mean and standard deviation of the follower response are now outlined [9.10]:

1. Divide the cam angle for total lift, ϕ_o, into a convenient number of intervals, $n - 1$. The discrete values of ϕ_2 are labeled as $(\phi_2)_j$, $j = 1, 2, \ldots, n$. Set $j = 1$.
2. Assume a suitable lift curve. The cycloidal motion is assumed in this work. The lift s_1 corresponding to any cam angle ϕ_2, according to cycloidal motion, is given by

$$s_1 = \frac{L_o}{\pi}\left[\frac{\pi\,\phi_2}{\phi_o} - \frac{1}{2}\sin\left(\frac{2\,\pi\,\phi_2}{\phi_o}\right)\right] \tag{9.54}$$

and hence

$$s_1' = \frac{ds_1}{d\phi_2} = \left(\frac{L_o}{\phi_o}\right)\left[1 - \cos\left(\frac{2\,\pi\,\phi_2}{\phi_o}\right)\right] \tag{9.55}$$

Calculate the values of s_1 and s_1' for a particular value of $\phi_2 = (\phi_2)_j$ and substitute them into Eq. (9.49) to find the corresponding value of θ at $\phi_2 = (\phi_2)_j$.

3. Determine the mean lift $(s_1)_j$ and standard deviation $(\sigma_{s_1})_j$ using Eqs. (9.52) and (9.53).

4. Set new j = current $j + 1$, and repeat steps 2. through 4. until the angle $(\phi_2)_n$ is considered.

Numerical results. The tolerances for all the geometrical parameters are assumed to be normally distributed with $\pm 3\,\sigma$ band. Hence, if the mean value of one of the parameters is r with a standard deviation σ_r, then the allowable value will be $(r \pm 3\,\sigma_r)$. Thus $\sigma_r = \Delta r / 3 = x_3 / 3$ where Δr is the tolerance given to r. Similarly $\sigma_{r_b} = \Delta r_b / 3 = x_4 / 3$, $\sigma_a = \Delta a / 3 = x_5 / 3$, $\sigma_{X_2} = \Delta X / 3 = x_1 / 3$, and $\sigma_{Y_2} = \Delta Y / 3 = x_2 / 3$. The following data are assumed for the numerical computation: $\overline{r} = 1.0$, $\overline{r}_b = 2.5$, $\overline{a} = 1.2$, and total angle of rise = ϕ_o = 150°. The standard deviation of the response is computed for two different sets of errors—one given by

$$\mathbf{X}_1 = \begin{Bmatrix} 0.010 \\ 0.010 \\ 0.003 \\ 0.006 \\ 0.003 \end{Bmatrix}$$

and the other by

$$\mathbf{X}_2 = \begin{Bmatrix} 0.1847 \\ 0.1843 \\ 0.0299 \\ 0.0731 \\ 0.0357 \end{Bmatrix}$$

The numerical results, along with the mean kinematic response, are shown in Table 9.3. Since normal distribution is assumed for all the random variables, the actual response of the cam-follower system is expected to fall in the range (mean response ± 3 standard deviations) 99.73 percent of the time. This can be interpreted as a measure of reliability of the system.

TABLE 9.3 Standard Deviation of the Kinematic Response of the Cam-Follower System

Angle ϕ_2 (degrees)	Mean kinematic response	Standard deviation corresponding to $\mathbf{X}_1$	Standard deviation corresponding to $\mathbf{X}_2$
0	0.0000	0.004051	0.06521
10	0.00773	0.004074	0.06742
20	0.06023	0.004138	0.06795
30	0.1945	0.004194	0.06844
40	0.4335	0.004209	0.06860
50	0.7820	0.004232	0.06886
60	1.226	0.004237	0.06952
70	1.734	0.004230	0.06890
80	2.266	0.004219	0.06880
90	2.774	0.004209	0.06869
100	3.218	0.004201	0.06860
110	3.566	0.004197	0.06854
120	3.805	0.004196	0.06852
130	3.940	0.004192	0.06847
140	3.992	0.004187	0.06840
150	4.000	0.004184	0.06837

SOURCE: S. S. Rao, S. S. Gavane, "Analysis of Synthesis of Mechanical Error in Cam-Follower Systems," *Journal of Mechanical Design*, ASME, vol. 104, Jan. 1982.

9.5.2 Four-bar mechanisms

A probabilistic model of linkage mechanisms, considering tolerances on the link lengths and clearances in the hinges, is developed in this section. Consider the planar four-bar mechanism shown in Fig. 9.7. In this mechanism, l_1, l_2, l_3, and l_4 are the link lengths, θ is the input angle and ϕ is the output angle. Let this mechanism be used for generating a function, $\phi = \phi(\theta)$. For an arbitrary choice of the link lengths, the output angle ϕ may vary with θ as $\phi = \phi_s(\theta)$, instead of $\phi(\theta)$ as shown in Fig. 9.8. The difference, $|\phi(\theta) - \phi_s(\theta)|$, is known as the structural error. In practical mechanisms, the play or clearances in the joints and tolerances on the link lengths of the mechanism cause an additional error, known as the mechanical error, and the output angle may vary with θ as $\phi = \phi_m(\theta)$. Depending on the values of the clearances in the joints and the tolerances on the link lengths, the mechanical error may be much larger than the structural error. Since the tolerances and clearances are probabilistic in nature, a probabilistic model is necessary for the analysis of the four-bar mechanism.

Deterministic analysis. The input-output relation of the ideal mechanism [9.13] shown in Fig. 9.7 can be expressed as:

$$\phi_a(\theta) = 2\tan^{-1}\left(\frac{A \pm D}{B + C}\right) \tag{9.56}$$

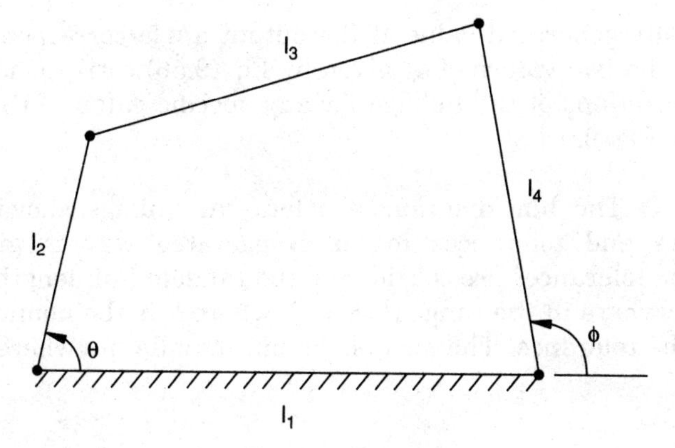

Figure 9.7 A planar four-bar mechanism.

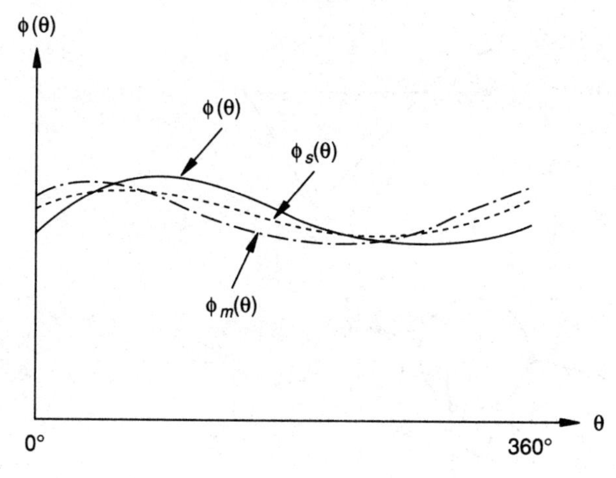

Figure 9.8 Generalized output angle; deterministic analysis.

where

$$A = \sin\theta \tag{9.57}$$

$$B = \cos\theta - \frac{l_1}{l_2} \tag{9.58}$$

$$C = \frac{l_1^2 + l_2^2 - l_3^2 + l_4^2}{2\,l_2\,l_4} - \frac{l_1}{l_4}\cos\theta \tag{9.59}$$

$$D = \sqrt{A^2 + B^2 - C^2} \tag{9.60}$$

and ϕ_a is the actually generated value of the output angle corresponding to the input angle θ. The two values of ϕ_a given by Eq. (9.56) correspond to the two possible configurations of the linkage for any specific value of the input angle θ as shown in Fig. 9.9.

Probabilistic analysis. The line diagram of a four-bar linkage showing the effect of clearances and tolerances in an exaggerated way is given in Fig. 9.10. If only the tolerances are considered, the random link length, R_i, of the ith link lies anywhere in the range ($l_i \pm \Delta l_i$), where l_i is the nominal link length and Δl_i is the tolerance. The axis of the pin may lie anywhere in the

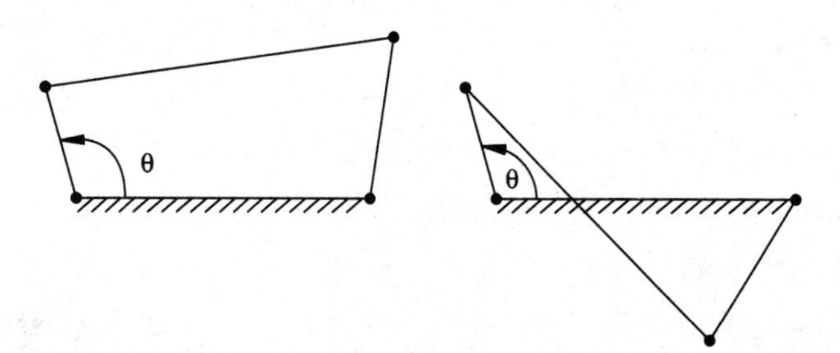

Figure 9.9

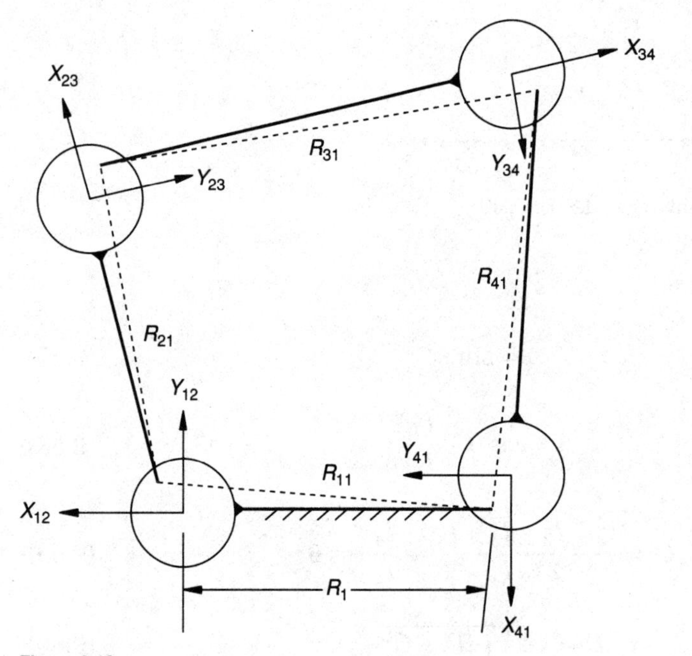

Figure 9.10

clearance circle of radius r_{ij}, where r_{ij} is the radial clearance in the hinge between jth and ith links. Let us fix four sets of rectangular coordinate systems X_{ij} Y_{ij} at the hinges so that the X_{ij} is coincident with the center line of the link ending in the race, Fig. 9.11 [9.13]. When the effects of both tolerances and clearances are considered, the equivalent linkage is shown by the dashed lines in Fig. 9.10 [9.13]. The link lengths of the equivalent mechanism are given by

$$R_{i1}^2 = (R_i + x_{ij})^2 + y_{ij}^2 \approx (R_i + x_{ij})^2;$$

$$j = i + 1 \text{ for } i = 1, 2, 3; \qquad j = 1 \text{ for } i = 4 \tag{9.61}$$

as $R_i + x_{ij} \gg y_{ij}$ and hence

$$R_{i1} \approx R_i + x_{ij} \tag{9.62}$$

The output angle of the equivalent linkage [9.13] can be expressed as:

$$\phi = 2 \tan^{-1}\left(\frac{A' \pm D'}{B' + C'}\right) \tag{9.63}$$

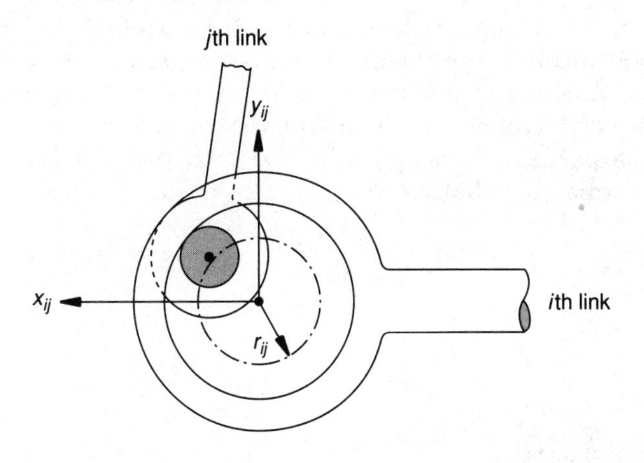

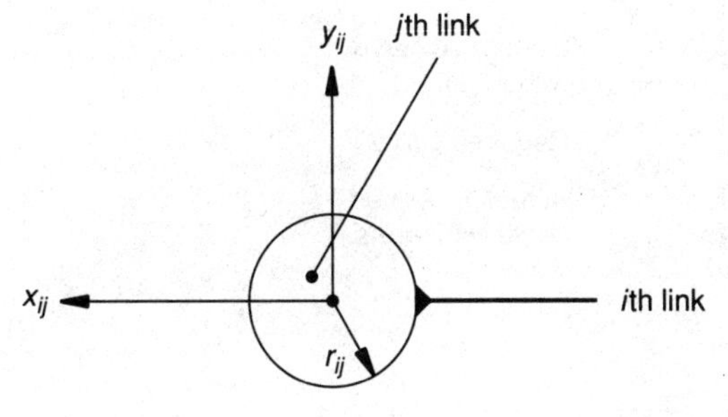

Figure 9.11

where

$$A' = \sin\theta \tag{9.64}$$

$$B' = \cos\theta - \frac{R_{11}}{R_{21}} \tag{9.65}$$

$$C' = \frac{R_{11}^2 + R_{21}^2 - R_{31}^2 + R_{41}^2}{2\,R_{21}\,R_{41}} - \frac{R_{11}}{R_{41}}\cos\theta \tag{9.66}$$

$$D' = \sqrt{A'^2 + B'^2 - C'^2} \tag{9.67}$$

By defining a set of random variables V_i as $V_1 = R_1$, $V_2 = R_2$, $V_3 = R_3$, $V_4 = R_4$, $V_5 = x_{12}$, $V_6 = x_{23}$, $V_7 = x_{34}$, and $V_8 = x_{41}$, Eq. (9.63) can be rewritten as

$$\phi = \phi(\theta; V_i, i = 1, 2, \ldots, 8) \tag{9.68}$$

The actual output angle, in the presence of random tolerances and clearances, will fall in, say, $\pm 3\,\sigma_\phi$ band with a 99.73 percent probability if ϕ follows normal distribution. This is indicated in Fig. 9.12. To find the probability of generating $\phi(\theta)$ within a specified range of ϕ, we need to know the probability distribution function of ϕ which is a difficult task. An approximate probability analysis can be conducted by finding the mean and standard deviation of ϕ using the partial derivative rule. For this, the link lengths are assumed to follow normal distribution with

$$\overline{V}_i = E(V_i) = l_i\,; \qquad \sigma_{V_i} = \left(\frac{\Delta l_i}{3}\right); \qquad i = 1, 2, 3, 4 \tag{9.69}$$

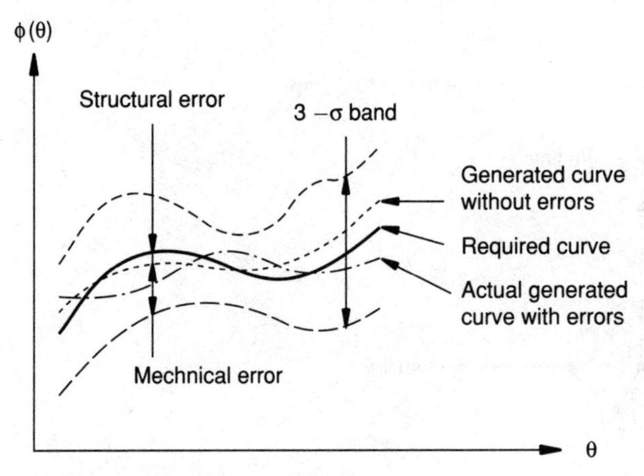

Figure 9.12 Generated-output angle.

For the location of the pin center in the clearance circle, we assume a bivariate normal probability density function with independent random variables (x_{ij}, y_{ij})

$$f(x_{ij}, y_{ij}) = \frac{1}{2\,\pi\,\sigma_{x_{ij}}\,\sigma_{y_{ij}}}\,\exp\left\{-\frac{1}{2}\left(\frac{x_{ij}-\overline{x}_{ij}}{\sigma_{x_{ij}}}\right)^2 - \frac{1}{2}\left(\frac{y_{ij}-\overline{y}_{ij}}{\sigma_{y_{ij}}}\right)^2\right\} \tag{9.70}$$

In the present case, $\overline{x}_{ij} = \overline{y}_{ij} = 0$, $\sigma_{x_{ij}} = \sigma_{y_{ij}} = r_{ij}/3$, and hence Eq. (9.70) reduces to

$$f(x_{ij}, y_{ij}) = \frac{9}{2\,\pi\,r_{ij}^2}\,\exp\left\{-\frac{9\,(x_{ij}^2 + y_{ij}^2)}{2\,r_{ij}^2}\right\} \tag{9.71}$$

By using the partial derivative rule, the mean and standard deviations of ϕ can be determined, from Eq. (9.68), as

$$\overline{\phi} = \phi(\theta;\, \overline{V}_i,\ i = 1, 2, \ldots, 8) \tag{9.72}$$

$$\sigma_\phi = \left\{\sum_{i=1}^{8}\left(\frac{\partial\phi}{\partial V_i}\bigg|_{\overline{V}_1, \overline{V}_2, \ldots, \overline{V}_8}\right)^2 \sigma_{V_i}^2\right\}^{1/2} \tag{9.73}$$

Once $\overline{\phi}$ and σ_ϕ are known, the $\pm 3\,\sigma_\phi$ band of the output angle ϕ about the mean value $\overline{\phi}$ can be constructed. Since all V_i are assumed to follow normal distribution, ϕ also follows normal distribution in view of the central-limit theorem. Thus, the actually generated output angle ϕ is guaranteed to fall in the band 99.73 percent of the time.

Numerical results. The $\pm 3\,\sigma_\phi$ band of the output angle is determined using the following data:

Function generated: $y = \sin x$

Range of x: $0^\circ \le x \le 90^\circ$, x is proportional to θ, y is proportional to ϕ

Range of the input angle: $\Delta\theta = 90^\circ$

Range of the output angle: $\Delta\phi = 90^\circ$

Nominal link lengths: $l_1 = 1$, $l_2 = 1.9$, $l_3 = 2.7$, $l_4 = 0.85$

Tolerances: $\Delta l_i = 0.0002$; $i = 1,2,3,4$

Clearances: $\Delta r_{ij} = 0.0002$; $ij = 12,23,34,41$

The range of the input angle $(\Delta\theta)$ is divided into 10 equal parts for computational convenience so that θ_1 and θ_{11} indicate the starting and the final angular positions of the input link. θ_1 is assumed to be 2.0283 radians. The required and mean-generated angular positions of the output link, the structural error, the $3\,\sigma_\phi$ value of the mechanical error at various positions of the input link are given in Table 9.4. Since ϕ follows normal distribution, the

TABLE 9.4 $3\sigma_\phi$**—Values for the Four-bar Mechanism**

	Angular position of the output link (degrees)			
Input angle (i)	Required value ($\phi_{required}$)	Mean generated value ($\bar{\phi}$)	Structural (degrees)	Mechanical error (degrees) 3–σ value
1	43.4510	43.4510	0.0000	3.2946
2	57.5301	56.5167	– 1.0134	3.1956
3	71.2625	68.3578	– 2.9047	3.1601
4	84.3101	79.0859	– 4.2242	3.1630
5	96.3516	88.7510	– 7.6006	3.1851
6	107.0905	97.3676	– 9.7229	3.2104
7	116.2624	104.9340	– 11.3284	3.2260
8	123.6415	111.4473	– 12.1942	3.2232
9	129.0460	116.9146	– 12.1314	3.1989
10	132.3429	121.3578	– 10.9851	3.1563
11	133.4509	124.8113	– 8.6396	3.1045

SOURCE: Reprinted with permission from S. S. Rao and C. P. Reddy, "Mechanism Design by Chance Constrained Programming," *Mechanism and Machine Theory*, vol. 14, copyright 1979, Pergamon Press PLC, pp. 413–424.

probability of the actual value of ϕ falling in the band (mean generated value $\pm 3\,\sigma_\phi$) at any value of θ will be 99.73 percent.

References and Bibliography

9.1. R. W. Fitzgerald, *Mechanics of Materials,* 2nd ed., Addison-Wesley, Reading, Mass., 1982.
9.2. J. H. Faupel and F. E. Fisher, *Engineering Design,* 2nd ed., John Wiley, New York, 1981.
9.3. J. E. Shigley and C. R. Mischke, *Mechanical Engineering Design,* 5th ed., McGraw-Hill, New York, 1989.
9.4. C. R. Mischke, "Prediction of Stochastic Endurance Strength," *ASME J. of Vibration, Acoustics, Stress, and Reliability in Design*, vol. 109, 1987, pp. 113–122.
9.5. J. Marin, *Mechanical Behavior of Engineering Materials,* Prentice-Hall, Englewood Cliffs, New Jersey, 1962.
9.6. R. E. Peterson, "Design Factors for Stress Concentration," *Machine Design*, vol. 23, no. 2, Feb. 1951, p. 169; no. 3, March 1951, p. 161; no. 5, May 1951, p. 159; no. 6, June 1951, p. 173; no. 7, July 1951, p. 155.
9.7. R. C. Juvinall and K. M. Marshek, *Fundamentals of Machine Component Design,* 2nd ed., John Wiley, New York, 1991.

9.8. S. S. Rao, "A Probabilistic Approach to the Design of Gear Trains," *Intern. J. of Machine Tool Design and Research*, vol. 14, 1974, pp. 267–278.
9.9. S. S. Rao and G. Das, "Reliability Based Optimum Design of Gear Trains," *ASME J. of Mechanisms, Transmissions, and Automation in Design*, vol. 106, 1984, pp. 17–22.
9.10. S. S. Rao and S. S. Gavane, "Analysis and Synthesis of Mechanical Error in Cam-Follower Systems," *J. of Mech. Design, Trans. of ASME*, vol. 104, 1982 pp. 52–62.
9.11. S. S. Rao, "Error Analysis of Cam-Follower Systems: A Probabilistic Approach," *Proceedings of the Institution of Mechanical Engineers, London*, part C, *Mechanical Engineering Science*, vol. 198C, 1984, pp. 155–162.
9.12. H. R. Kim and W. R. Newcombe, "Stochastic Error Analysis in Cam Mechanisms," *Mechanism and Machine Theory*, vol. 13, 1978, pp. 631–641.
9.13. S. S. Rao and C. P. Reddy, "Mechanism Design by Chance Constrained Programming," Mechanism and Machine Theory, vol. 14, 1979, pp. 413–424.
9.14. A. C. King and C. B. Read, "Pathways to Probability," Holt, Rinehart and Winston, Inc., New York, 1963.
9.15. C. C. Gillispie (ed.-in-chief), "Dictionary of Scientific Biography," vol. 5, Charles Scribner's Sons, New York, 1980.

Review Questions

9.1 What is the purpose of Marin factors in fatigue design of machine components?

9.2 What is meant by endurance strength of a material?

9.3 How is the ultimate tensile strength related to the endurance strength of a material?

9.4 What is a fatigue failure?

9.5 What is the significance of three-sigma band of the performance of a mechanism?

9.6 What type of probabilistic model is useful for a multispeed gear train?

9.7 Give two examples of mechanical systems where the weakest-link concept is useful for probabilistic modeling.

9.8 Give two examples of mechanical systems where the fail-safe concept is useful for probabilistic modeling.

9.9 Suggest a method of including the probabilities of failure of shafts and bearings in the reliability analysis of a gear box.

9.10 What is the difference between structural error and mechanical error in a linkage mechanism?

9.11 What are the sources of error in a cam-follower mechanism?

9.12 What is the influence of error in the kinematic response of a cam-follower mechanism?

Problems

9.1 A cold rolled-steel rectangular plate, having a central hole, is subjected to a completely reversed axial load of $F = 1$ kip as shown in Fig. 9.13. If the ultimate tensile strength of the material is 90 k(lb/in)^2 and the geometric stress concentration factor is 2.42, find the factor of safety against fatigue failure.

9.2 For the rectangular steel plate described in Problem 9.1, the reversed axial load is given by $F = N(1.0, 0.08)$ kip. If the mean ultimate tensile strength of the material is 90 k(lb/in^2, determine the reliability of the plate by assuming normal distribution for both the induced stress and the endurance strength of the component.

9.3 For the rectangular steel plate described in Problem 9.1, the reversed axial load is given by $F = N(1.0, 0.08)$ kip. If the mean ultimate tensile strength of the material is 90 k(lb/in^2), estimate the reliability of the plate by assuming lognormal distribution for both the induced stress and the endurance strength of the component.

9.4 A thin-walled steel spherical shell of radius 40 ft is supported along the equator as shown in Fig. 9.14. It is filled with a liquid having a specific weight of $w = N(62.4, 3.12)$ lb/ft^3. The shell material has a yield stress of $s_y = N(20{,}000, 2000)$ lb/in^2. Determine the shell wall thickness for a reliability of 0.99.

9.5 A thin-walled steel cylinder of diameter $D = N(24, 0.5)$ in and thickness $t = N(0.1, 0.01)$ in is reinforced by a layer of closely wound steel wire of diameter $d = N(\bar{d}, 0.1\,\bar{d})$ in. The wire is wound with an initial tension of $p_1 = N(10{,}000, 1000)$ lb/in^2 while the cylinder is empty. The cylinder is then subjected to an internal pressure of $p_2 = N(1000, 100)$ lb/in^2. Find the mean wire diameter to achieve a reliability of 0.999 assuming that the stress in the cylinder in the longitudinal direction is negligible (as in the case of a section of a pipe with expansion joints).

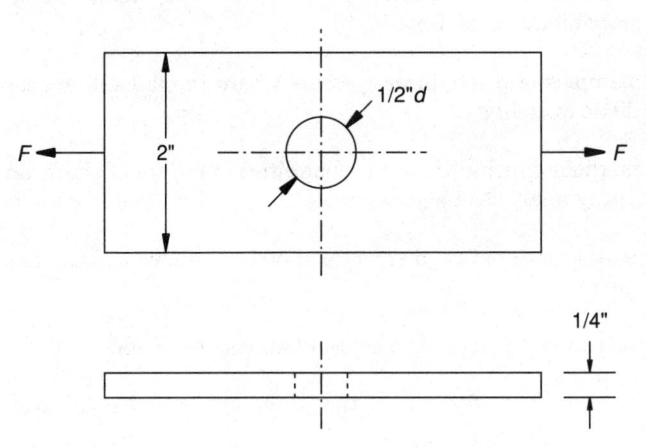

Figure 9.13

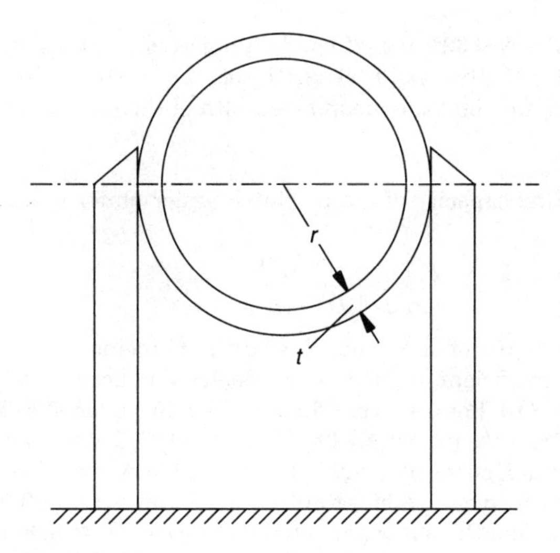

Figure 9.14

9.6 A steel gear with a central hole of radius $R = N(1.5, 0.001)$ in is shrink-fitted over a solid steel shaft of radius $(1.5 + \delta)$ in as shown in Fig. 8.25. The gear has a pitch circle radius of $r_o = N(6, 0.1)$ in. The pressure developed at the interface, p, due to shrink-fit [9.3] is given by

$$p = E\,\frac{\delta}{R}\left[\frac{(r_o^2 - R^2)(R^2 - r_i^2)}{2R^2(r_o^2 - r_i^2)}\right]$$

where $E =$ Young's modulus, $\delta =$ radial interference, $r_o =$ outer radius of the outer part, $r_i =$ inner radius of the inner part, and $R =$ transition (nominal) radius at the interface. If the Young's modulus is normally distributed with $E = N(30 \times 10^6, 3 \times 10^6)$ lb/in^2 and the permissible shrink-fit pressure, p_o, can be taken as $p_o = N(45{,}000, 4500)$ lb/in^2, determine the radial interference δ to achieve a reliability of 0.99.

9.7 The bending stress developed in a gear tooth is given by the Lewis equation [9.7]

$$s = \frac{F_t\, p_d}{wY}$$

where $F_t =$ tangential load acting on the gear tooth, $p_d =$ diametral pitch $= \pi / p$, $p =$ circular pitch, $w =$ face width, and $Y =$ Lewis form factor:

$$Y = \frac{p_d\, t^2}{6\,h}$$

where $t =$ root thickness and $h =$ radial depth of the tooth (see Fig. 8.7). In an automobile transmission, a gear has full-depth teeth with a pressure angle of 20°, a diametral pitch of $p_d = N(1, 0.05)$, a face width of $w = N(\overline{w}, 0.1\,\overline{w})$ in, a

Lewis form factor of $Y = N(0.322, 0.0322)$ and is subjected to a tangential load of $F_t = N(5000,\ 500)$ lb. If the yield strength of the material is given by $s_y = N(30{,}000,\ 3000)$ lb/in^2, find the mean face width of the gear tooth for a reliability of 0.99.

9.8 The torque transmitting capacity of a cone clutch under uniform pressure [9.3] is given by

$$T = \frac{F\,f}{3\sin\alpha}\left[\frac{D^3 - d^3}{D^2 - d^2}\right]$$

where $D =$ outer diameter of the cone, $d =$ inner diameter of the cone, $\alpha =$ semi-cone angle, $f =$ coefficient of friction between cup and cone, and $F =$ axial force applied (see Fig. 8.6). For a specific clutch, which has molded asbestos on a cast iron cone, $d = N(8, 0.08)$ in, $\alpha = N(20°, 2°)$, $F = N(100, 10)$ lb, and $f = N(0.3, 0.03)$. If the torque required to be transmitted by the cone clutch is also normally distributed with mean and standard deviations of 400 lb · in and 40 lb · in, respectively, find the outer diameter of the cone (D) to achieve a reliability of 0.99. Assume that the outer diameter of the cone is given by $D = N(\bar{D},\ 0.05\,\bar{D})$ in.

9.9 The cantilever beam shown in Fig. 9.15 is subjected to an alternating force $F = N(400, 40)$ lb. The beam is machined from AISI 1040 steel and heat treated with a mean ultimate tensile strength of $s_y = N(85, 8.5)$ k(lb/in^2). Find the factor of safety based on a deterministic approach.

9.10 Find the reliability of the beam described in Problem 9.9 by assuming that the induced stress and the strength of the material follow normal distribution.

9.11 Find the reliability of the beam described in Problem 9.9 by assuming that the induced stress and the strength of the material follow lognormal distribution.

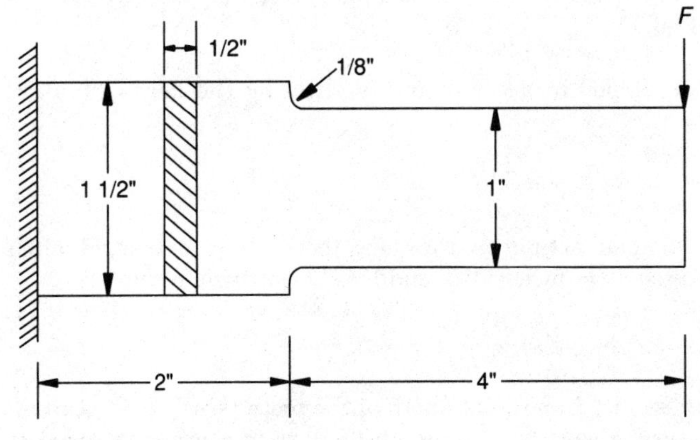

Figure 9.15

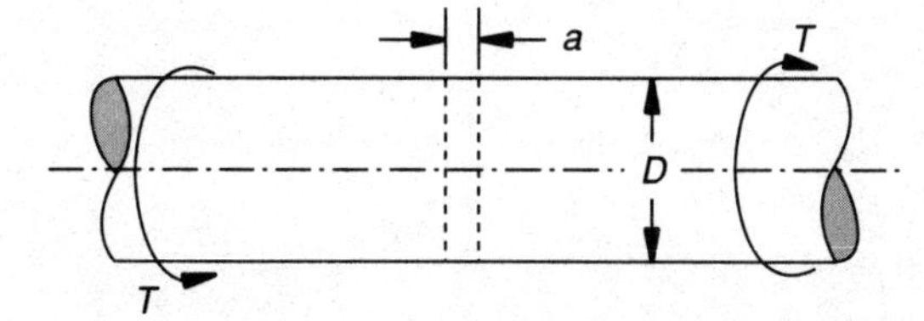

Figure 9.16

9.12 The steel shaft shown in Fig. 9.16 has a transverse hole of diameter a and is subjected to an alternating torque of $T = N(30{,}000, 3000)$ lb · in. The mean ultimate tensile strength of the material is 150 k(lb/in^2). The geometric stress concentration factor at the hole is 1.62. If the diameter of the shaft is $D = N(\bar{D}, 0.1\,\bar{D})$ in, and that of the hole is $a = N(0.25, 0.025)$ in, determine the mean diameter of the shaft, $\bar{D}$, for a factor of safety of 2.

9.13 Find the mean diameter of the shaft, $\bar{D}$, described in Problem 9.12 to achieve a reliability of 0.99 by assuming that the induced stress and the strength of the material follow normal distribution.

9.14 Find the mean diameter of the shaft described in Problem 9.12 to achieve a reliability of 0.99 by assuming that the induced stress and the strength of the material follow lognormal distribution.

Chapter

10

Structural Reliability: Weakest Link and Fail-Safe Systems

Biographical Note

John Venn

John Venn was born on August 4, 1834 in Hull, England and died on April 4, 1923 in Cambridge, England. He studied at Gonville and Caius College in Cambridge and received his degree in mathematics in 1857. He received the Cambridge Sc.D. in 1883 and was elected a fellow of the Royal Society in the same year. He was one of the first persons to explain the basic concepts of probability from the standpoint of logic. Although many of his arguments are now discounted, his book, "The Logic of Chance," is well known for its simplicity. In fact, his books on probability and logic were highly esteemed textbooks in the late nineteenth and early twentieth centuries. Along with other topics, he discussed the subjects of life expectancy and the law of error. According to his theory, a person can expect to live to be of certain age, based on the average age observed for other people, since all persons (not animals, for example) have similarly constructed bodies. Venn is better known for the "Venn diagrams" that are used in set theory as well as in probability theory. He explained the problem of inverse probability through Venn diagrams. In recent years, Venn diagrams have been used in elementary mathematics to stimulate logical thinking beginning in the early stages of a child's education [10.16, 10.17].

10.1 Introduction

The concept of strength-based reliability has been introduced in Chapter 8 by considering a single structural element for which loading is described by the random variable L and strength by the random variable S. The probability of failure of the element is given by

$$P_f = P(L \geq S) = \int\int f_{S,L}(s,l)\, ds\, dl \tag{10.1}$$

where $f_{S,L}(s,l)$ is the joint probability density function of S and L and the double integral in Eq. (10.1) is to be evaluated over the failure region, $\{(s,l) \mid s - l \leq 0\}$. When $f_{S,L}$ is known, the probability of failure can be computed relatively easily either by numerical integration or by simulation.

Most machines and structures consist of more than one element (member) and are subjected to more than one load condition. In addition, a machine or structure may have more than one possible failure mode. Systems with multiple members can be classified into two categories, namely, weakest-link systems and fail-safe systems. A system is called a *weakest link system* (also, known as a series system) if it fails whenever any of its elements fails. An example of a weakest-link system is the statically determinate truss shown in Fig. 10.1. A system is called a *fail-safe system* (also, known as a parallel system) if the failure of a single element will not always result in the failure of the total system, because the remaining elements may be able to sustain the applied loads. An example of a fail-safe system is the statically indeterminate truss shown in Fig. 10.2. In general, there will be more than one way in which a parallel system can fail under a given load. For example, a parallel system may fail due to the failure of different elements or the elements may fail in different orders. All these considerations are important in the reliability analysis of a parallel system.

The material of the elements can be brittle or ductile in a multicomponent structure. If an element is perfectly brittle, it will lose its load-carrying capacity after its failure (see Fig. 10.3*a*). On the other hand, if the element is

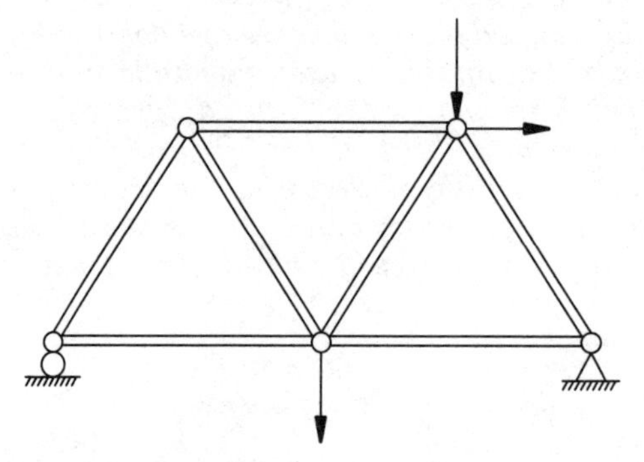

Figure 10.1 A statically determinate truss (weakest-link system).

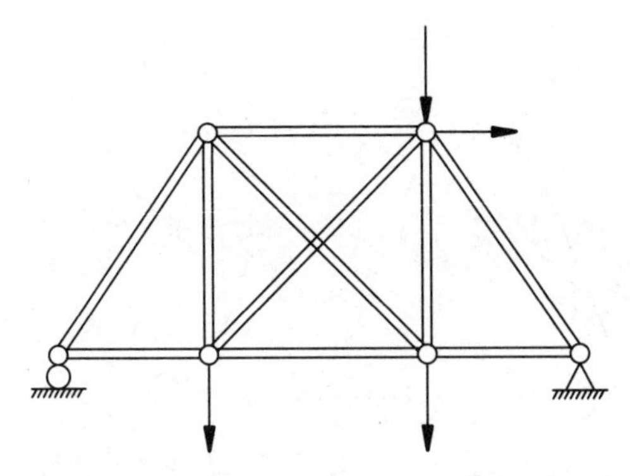

Figure 10.2 A statically indeterminate truss (fail-safe system).

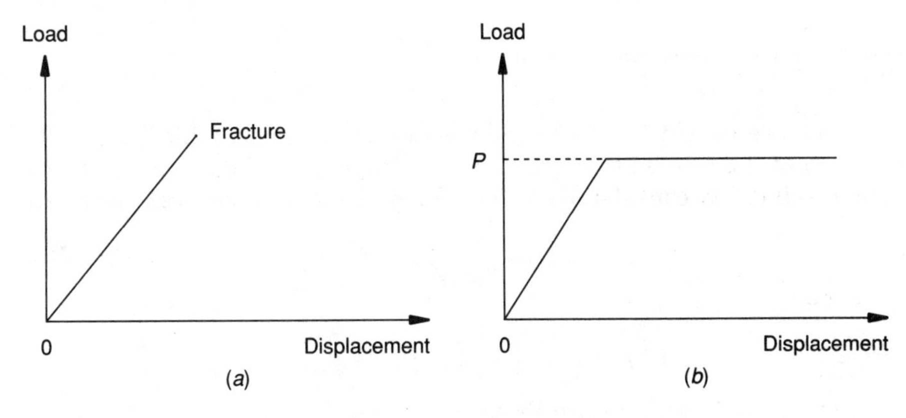

Figure 10.3

ductile, it will not carry any additional load after failure; but it will continue to carry the load that it was carrying at the time of failure (see Fig. 10.3*b*). This chapter deals with the reliability analysis of weakest-link and fail-safe systems. Many of the concepts presented in this chapter are based on the works of Moses and his associates [10.7, 10.8, 10.15] and Cornell [10.2]. For thoroughness, systems consisting of only one element are also considered as a special case of weakest-link systems.

10.2 Fundamental One Member–One Load Case

10.2.1 Probability of failure

We consider a special case of a weakest-link system known as the fundamental problem [10.1, 10.4]. It consists of a single member of strength S subjected to a load L as shown in Fig. 10.4*a*. The strength of the member and the load

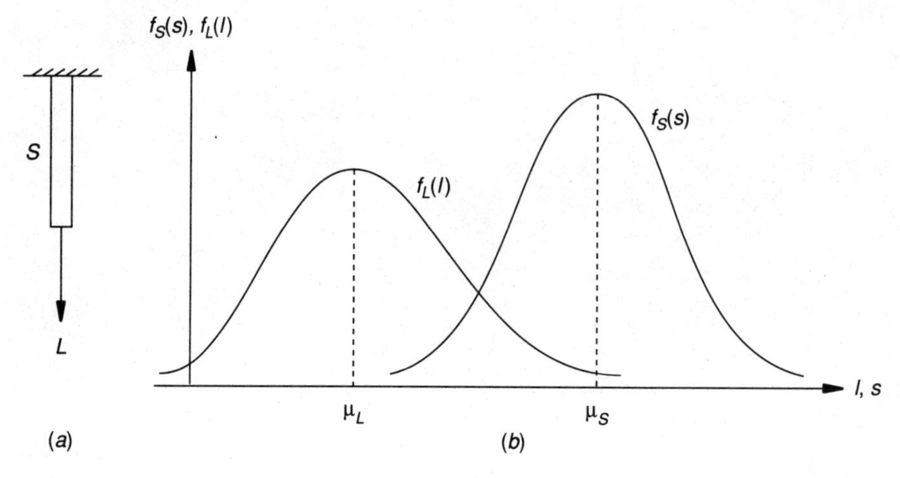

Figure 10.4 Fundamental (one member–one load) problem.

environment are considered to be independent random variables. Thus the computation of the reliability of the fundamental case will be identical to the component reliability considered earlier. The probability of failure is given by

$$P_f = P(S < L) = \int_{l=0}^{\infty} F_S(l) \cdot f_L(l) \cdot dl \tag{10.2}$$

or, equivalently,

$$P_f = 1 - P(L < S) = 1 - \int_{s=0}^{\infty} F_L(s) \cdot f_S(s) \cdot ds \tag{10.3}$$

If the strength and the load are discrete random variables, Eqs. (10.2) and (10.3) become

$$P_f = \sum_i F_S(l_i) \cdot f_L(l_i) \tag{10.4}$$

and

$$P_f = 1 - \sum_j F_L(s_j) \cdot f_S(s_j) \tag{10.5}$$

where l_i and s_j denote the ith value of the load and the jth value of the strength, respectively.

10.2.2 Reliability index

The probability of failure of a structural member can also be expressed as

$$P_f = P(M < 0) \tag{10.6}$$

where $M = S - L$ is called the safety margin. In the case of normally distributed S and L, the probability of failure of the member can be determined as

$$P_f = \Phi(-\beta) \tag{10.7}$$

where Φ denotes the value of the distribution function of the standard normal variate corresponding to

$$\beta = \frac{\mu_S - \mu_L}{\sqrt{\sigma_S^2 + \sigma_L^2}} \tag{10.8}$$

In the literature of structural reliability, β is known as the *reliability index*. The reliability index denotes the shortest distance from the origin to the failure surface given by the equation, $f(s,l) = s - l = 0$, as shown in Fig. 10.5*a*. The concept of the reliability index can be generalized to structural systems whose failure surface is given by the nonlinear equation, $f(z_1, z_2, \ldots, z_N) = 0$, where z_i is the normalized random variable given by $z_i = \frac{Z_i - \mu_{Z_i}}{\sigma_{Z_i}}$ with μ_{Z_i} and σ_{Z_i} denoting respectively, the mean value and standard deviation of Z_i; $i = 1, 2, \ldots, N$. In this case also, the reliability index, β, denotes the shortest distance from the origin to the failure surface as indicated in Fig. 10.5*b*. The reliability index, in the general case, is known as the Hasofer and Lind reliability index [10.4].

10.2.3 Bounds on the probability of failure

If the strength (S) and the load (L) are represented along two axes as shown in Fig. 10.6, the bounds on the probability of failure can be derived as follows. The probability of failure is the same as the probability of realizing the strength less than the load. Thus, for a factor of safety of one, we obtain

P_f = probability of realizing the point (S,L) in areas $(F_1 + F_2 + F_3)$ in Fig. 10.6

$$= P(S < L < L_{\max};\quad S < S_{\min}) + P(L > L_{\max};\quad S < S_{\min})$$

$$+ P(L > L_{\max};\quad S_{\min} < S < L) \tag{10.9}$$

that is,

$P_f >$ probability of realizing the point (S,L) in area F_2

$$> P(L > L_{\max};\quad S < S_{\min})$$

$$> p_1 p_2 \tag{10.10}$$

where

$$p_1 = P(S < S_{\min}) \tag{10.11}$$

$$p_2 = P(L > L_{\max}) \tag{10.12}$$

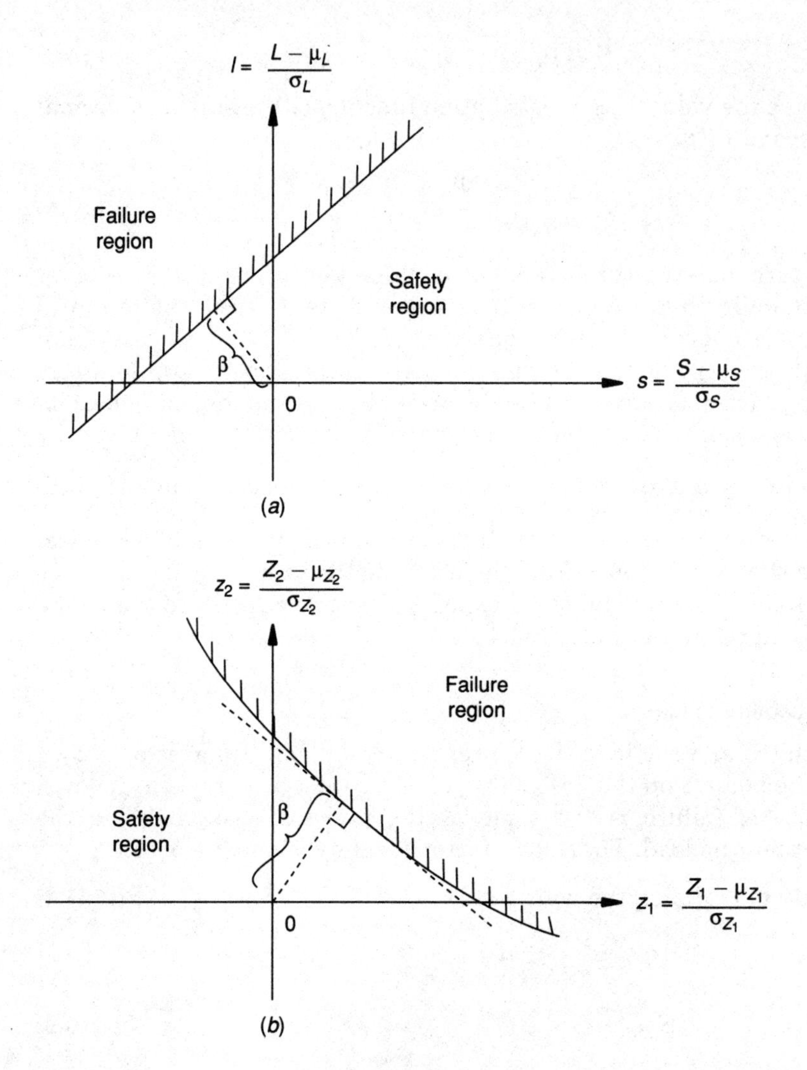

Figure 10.5

L_{max} is the maximum value of load and S_{min} is the minimum value of strength used in the deterministic design procedure.

Similarly the probability of survival or reliability can be expressed as

$$R_0 = \text{probability of realizing the point } (S,L) \text{ in areas } (S_1 + S_2 + S_3) \text{ in Fig. 10.6}$$

$$= P(L < S < S_{min}; L < L_{max}) + P(S > S_{min}; L < L_{max})$$

$$+ P(L > L_{max}; L < S < \infty) \tag{10.13}$$

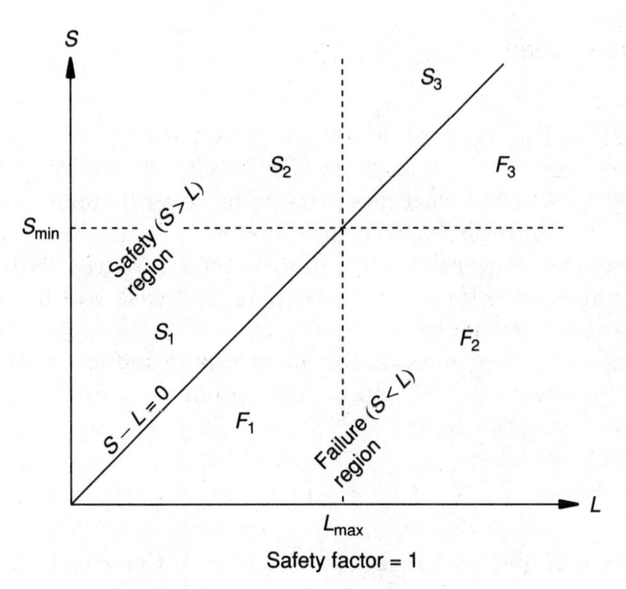

Figure 10.6 Bounds on the probability of failure.

that is,

$$R_0 > \text{probability of realizing the point } (S,L) \text{ in area } S_2$$

$$> P(S > S_{min}; L < L_{max})$$

$$> (1 - p_1)(1 - p_2) \tag{10.14}$$

Since $P_f = 1 - R_0$, the inequality (10.14) can be written as

$$P_f < 1 - (1 - p_1)(1 - p_2) \tag{10.15}$$

Combining the inequalities (10.10) and (10.15), the bounds on the probability of failure can be expressed as

$$(p_1 p_2) < P_f < (p_1 + p_2 - p_1 p_2) \tag{10.16}$$

Example 10.1 Find the bounds on the probability of failure of a system for which the strength is found to be less than the design strength once in 10 cases and the load is found to exceed the design load once in 500 cases.

solution Here $p_1 = 1 / 10$ and $p_2 = 1 / 500$, and hence Eq. (10.16) can be used to find the bounds on the probability of failure as

$$0.0002 < P_f < 0.1018$$

Note that this range is too wide for design purposes and we need to use the upper bound to ensure a conservative design.

10.3 Single Member–Several Loads Case

10.3.1 Probability of failure

Although the study of the fundamental problem is based on a single member, it is useful in investigating the sensitivity of reliability to the input statistical parameters. Most practical structures consist of several members and are often subjected to multiple-load conditions and will have numerous potential-failure modes. The reliability analysis of structures that are more complex than the fundamental one member-one load case will be considered in this and the next two sections.

Consider a single member subjected to m mutually independent loads L_j, $j = 1, 2, \ldots, m$ as shown in Fig. 10.7. The probability of failure of the member [10.8] can be expressed as

$$P_f = 1 - R_0 = 1 - \int_{s=0}^{\infty} \left\{ \prod_{j=1}^{m} F_{L_j}(s) \right\} f_S(s)\, ds \tag{10.17}$$

where $F_{L_j}(l)$ is the probability distribution function of the load L_j, $f_S(s)$ is the probability density function of the strength of the member, and $\prod_{j=1}^{m} F_{L_j}(s)$ is the probability of realizing each of the m loads less than s. If the loads and strength are discrete random variables, Eq. (10.17) becomes

$$P_f = 1 - \sum_{s_i = s_{\min}}^{s_{\max}} \left\{ \prod_{j=1}^{m} P(L_j < s_i) \right\} P(s = s_i)$$

$$= 1 - \sum_{s_i = s_{\min}}^{s_{\max}} \left\{ \prod_{j=1}^{m} F_j(s_i) \right\} p_S(s_i) \tag{10.18}$$

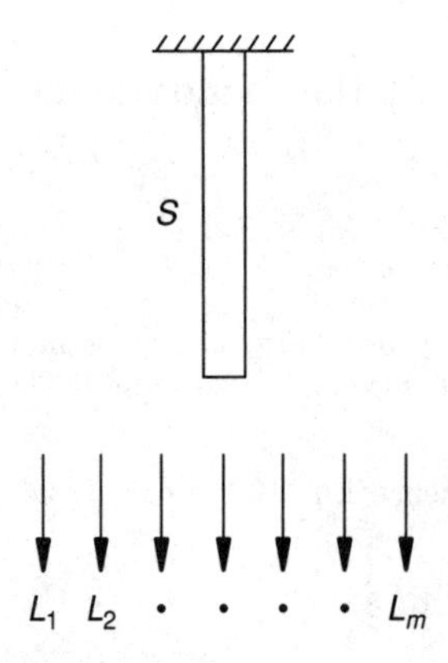

Figure 10.7 One member–m load case.

where $p_S(s)$ denotes the probability mass function of S, s_i is the ith discrete value of the strength, and $s_{\min}$ and $s_{\max}$ are the minimum and maximum values of the strength. If the loads $L_1, L_2, \ldots, L_m$ denote the independent repetitions of the same load, the probability distribution functions of L_j will be the same,

$$F_{L_j}(l) = F_L(l), \quad j = 1, 2, \ldots, m \tag{10.19}$$

and Eq. (10.17) reduces to

$$P_f = 1 - \int_0^\infty [F_L(s)]^m f_S(s)\, ds \tag{10.20}$$

10.3.2 Bounds on the probability of failure [10.2, 10.15]

Equation (10.20) can be rewritten as

$$R_0 = \int_{s=0}^\infty [F_L(s)]^m f_S(s)\, ds = \int_{s=0}^\infty [1 - H_L(s)]^m f_S(s)\, ds \tag{10.21}$$

where

$$H_L(s) = 1 - F_L(s) \tag{10.22}$$

is the complementary cumulative distribution function of L. For most practical structures, $H_L(s)$ will be very small compared to unity for all the values of s for which $F_S(s)$ is significantly greater than zero. In this case, Eq. (10.21) can be approximated as[1]

$$\begin{aligned} R_0 &\geq \int_{s=0}^\infty [1 - m\, H_L(s)]\, f_S(s)\, ds \\ &\geq 1 - m \int_0^\infty H_L(s)\, f_S(s)\, ds \\ &\geq 1 - m\, P_{f_1} \end{aligned} \tag{10.23}$$

where P_{f_1} is the probability of failure of the structure under any particular load and is given by

$$P_{f_1} = \int_0^\infty H_L(s)\, f_S(s)\, ds = \int_0^\infty [1 - F_L(s)]\, f_S(s)\, ds \tag{10.24}$$

To find an upper bound on R_0, we assume that the successive loads are statistically dependent on one another with perfect correlation. Then, if the structure survives the first load, it will survive all the subsequent loads.

---

[1] $(1 - x)^m \geq 1 - mx$ for $x \ll 1$.

Hence

$$R_0 = 1 - P_{f_1} \tag{10.25}$$

Except for the unlikely case of negative correlation (in which case, low loads imply high successive loads), for any degree of statistical dependence among the successive loads, we can write

$$R_0 \le 1 - P_{f_1} \tag{10.26}$$

Equations (10.23) and (10.26) lead to

$$1 - m\, P_{f_1} \le R \le 1 - P_{f_1} \tag{10.27}$$

It is to be noted that the more nearly the loads are independent of one another (i.e., the more chaotic the environment), the higher the probability of failure.

10.4 Several Members–One Load Case (Series System)

10.4.1 Probability of failure

Statically determinate trusses subjected to a single loading condition, for example, can be idealized as weakest-link structures with several members as shown in Fig. 10.8. If $F_{S_i}(s_i)$ denotes the probability distribution function of the strength of member i, the probability distribution function of the

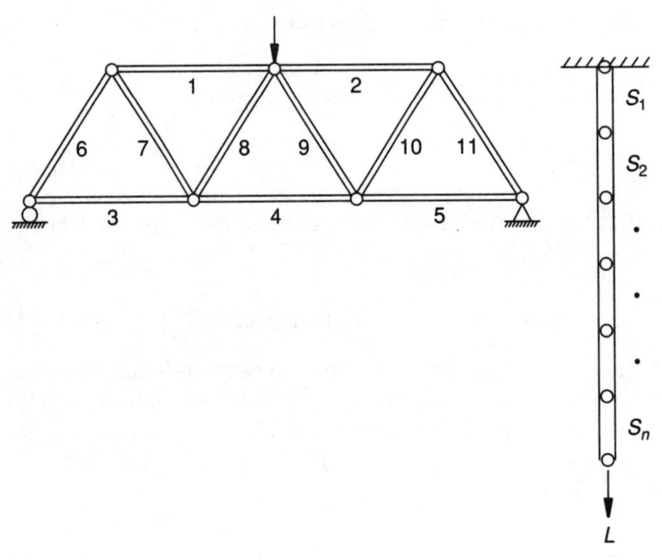

Figure 10.8 n member–one load case.

strength of the series system, $F_S(s)$, can be expressed as

$$F_S(s) = P(S \le s) = 1 - P(S > s) = 1 - P[S_1 > s_1 \cap S_2 > s_2 \cap \cdots \cap S_n > s_n]$$
$$= 1 - [1 - F_{S_1}(s_1)][1 - F_{S_2}(s_2)] \cdots [1 - F_{S_n}(s_n)]$$
$$= 1 - \prod_{i=1}^{n} [1 - F_{S_i}(s_i)] \tag{10.28}$$

where S denotes the strength of the series system and the strengths of the individual members (S_i) are assumed to be independent. If a load l is applied to the system, the force (or stress) induced in member i will be $a_i l$ where a_i denotes the force (or stress) induced in member i when a unit load is applied to the system. The value of a_i can be determined by using the methods of structural analysis. In this case, the probability of failure can be found as

$$P_f = 1 - R_0 = 1 - \int_{l=0}^{\infty} [P(S > l)] f_L(l)\, dl$$
$$= 1 - \int_{l=0}^{\infty} \left[\prod_{i=1}^{n} \{1 - F_{S_i}(a_i l)\} \right] f_L(l)\, dl \tag{10.29}$$

where the expression in the square brackets denotes the probability of all members surviving the load l. If the strengths and load are discrete random variables, Eq. (10.29) can be written as

$$P_f = 1 - \sum_{l_j = l_{\min}}^{l_{\max}} \left[\prod_{i=1}^{n} \{1 - F_{S_i}(a_i l_j)\} \right] f_L(l_j) \tag{10.30}$$

where l_j denotes the jth discrete value of the load, and $l_{\min}$ and $l_{\max}$ represent, respectively, the minimum and maximum values of the load, L.

10.4.2 Simple bounds on the probability of failure

Let a variable x_i be defined as follows:

$$x_i = \begin{cases} 1 & \text{if member } i \text{ has survived} \\ 0 & \text{if member } i \text{ has failed} \end{cases} \quad ; \quad i = 1, 2, \ldots, n \tag{10.31}$$

The probability of failure of the series system can be expressed as

$$P_f = 1 - R_0 = 1 - P[x_1 = 1 \cap x_2 = 1 \cap \cdots \cap x_n = 1] \tag{10.32}$$

Using the multiplicative rule for the probability of intersection of n events $A_1, A_2, \ldots, A_n$

$$P(A_1 \cap A_2 \cap \cdots \cap A_n) = P(A_1) P(A_2 |_{A_1}) P(A_3 |_{A_1 A_2}) \cdots P(A_n |_{A_1 A_2 \cdots A_{n-1}}) \tag{10.33}$$

Eq. (10.32) can be rewritten as

$$P_f = 1 - P(x_1=1)\frac{P(x_1=1\cap x_2=1)}{P(x_1=1)}\cdots\frac{P(x_1=1\cap x_2=1\cap\cdots\cap x_n=1)}{P(x_1=1\cap x_2=1\cap\cdots\cap x_{n-1}=1)} \tag{10.34}$$

If the safety margins of the members are positively correlated, it can be shown that the following relationship is valid

$$P(x_1=1\cap x_2=1\cap\cdots\cap x_{k+1}=1) \geq P(x_1=1\cap x_2=1\cap\cdots\cap x_k=1)P(x_{k+1}=1); \qquad 1\leq k\leq n-1 \tag{10.35}$$

Then Eq. (10.34) can be used to obtain an upper bound on the probability of failure of the system as

$$P_f \leq 1-\prod_{k=1}^{n} P(x_k=1) = 1-\prod_{k=1}^{n}[1-P(x_k=0)] \tag{10.36}$$

To obtain a lower bound on the probability of failure, we assume that the member failures are perfectly dependent on one another (correlated). Then the survival of any member implies the survival of all the members, and hence, that of the total system. In this case, the probability of failure of the series system is given by

$$P_f = P_{f_k} = P(x_k=0) \tag{10.37}$$

Equation (10.37) will be a lower bound to the true value of the probability of failure and hence the bounds on the probability of failure can be expressed as

$$\max_k P(x_k=0) \leq P_f \leq 1-\prod_{k=1}^{n}[1-P(x_k=0)]$$

or,

$$\max_k P_{f_k} \leq P_f \leq 1-\prod_{k=1}^{n}[1-P_{f_k}] \tag{10.38}$$

The right-hand side of the inequality (10.36) can be used to obtain another upper bound on the probability of failure of the system as follows:

$$\begin{aligned}1-\prod_{k=1}^{n}[1-P_{f_k}] &= 1-(1-P_{f_1})(1-P_{f_2})\cdots(1-P_{f_n})\\ &= 1-[1-(P_{f_1}+P_{f_2}+\cdots+P_{f_n})+(P_{f_1}P_{f_2}+P_{f_1}P_{f_3}+\cdots)\\ &\qquad -(P_{f_1}P_{f_2}P_{f_3}+\cdots)+\cdots]\\ &= \sum_{k=1}^{n}P_{f_k}-\sum_{j=1}^{n}\sum_{\substack{k=1\\ j\neq k}}^{n}P_{f_j}P_{f_k}+\cdots\end{aligned} \tag{10.39}$$

In most practical situations, P_{f_k} are very small compared to unity, and hence the terms involving their products will be negligible. Thus Eq. (10.39) can be written as

$$1 - \prod_{k=1}^{n} \left[1 - P_{f_k}\right] \approx \sum_{k=1}^{n} P_{f_k} \tag{10.40}$$

Since the sum of the terms dropped is nonnegative, we have

$$1 - \prod_{k=1}^{n} (1 - P_{f_k}) \leq \sum_{k=1}^{n} P_{f_k} \tag{10.41}$$

10.4.3 Ditlevsen bounds on the probability of failure

The bounds given by Eq. (10.38) will be too wide for many practical applications. Tighter bounds, known as *Ditlevsen bounds*, can be derived for series systems as follows [10.3]. Defining a variable y as

$$y = \begin{cases} 1 & \text{if the series system has survived} \\ 0 & \text{if the series system has failed} \end{cases} \tag{10.42}$$

we can express y in terms of $x_1, x_2, \ldots, x_n$ as

$$y = x_1 x_2 \cdots x_n \tag{10.43}$$

By replacing x_n as $1 - (1 - x_n)$, Eq. (10.43) can be rewritten as

$$y = x_1 x_2 \cdots x_{n-1} - x_1 x_2 \cdots x_{n-1}(1 - x_n) \tag{10.44}$$

By proceeding in a similar manner, Eq. (10.44) can be written as

$$y = x_1 - x_1(1 - x_2) - x_1 x_2(1 - x_3) - \cdots x_1 x_2 \cdots x_{n-1}(1 - x_n) \tag{10.45}$$

Noting that the reliability of the system R_0 is same as $E[y]$, Eq. (10.45) can be used to obtain

$$\begin{aligned} P_f &= 1 - R_0 = 1 - E[y] \\ &= E[1 - x_1] + E[x_1 (1 - x_2)] + \cdots + E[x_1 x_2 \cdots x_{n-1}(1 - x_n)] \end{aligned} \tag{10.46}$$

where all the terms on the right-hand side of Eq. (10.46) are nonnegative. Noting that the inequality

$$x_1 x_2 \cdots x_{k-1} \geq 1 - [(1 - x_1) + (1 - x_2) + \cdots + (1 - x_{k-1})] \tag{10.47}$$

is valid for any $k = 2, 3, \ldots, n$, we obtain

$$E[x_1 x_2 \cdots x_{k-1} (1 - x_k)] \geq E[1 - x_k] - \sum_{i=1}^{k-1} (1 - x_i)(1 - x_k) \tag{10.48}$$

Using Eqs. (10.48) and (10.46), the following inequality can be obtained:

$$P_f \geq P_{f_1} + \sum_{k=2}^{n} \max \left[\left\{ P_{f_k} - \sum_{i=1}^{k-1} (P_{f_i} \cap P_{f_k}) \right\}, \quad 0 \right] \tag{10.49}$$

Similarly, by noting that

$$x_1 x_2 \cdots x_{i-1} \leq x_k = 1 - (1 - x_k); \qquad \text{for } i \geq 2; \qquad k < i \tag{10.50}$$

we obtain an upper bound on the probability of failure of the system as

$$P_f \leq \sum_{i=1}^{n} P_{f_i} - \sum_{i=2}^{n} \max_{j<i} P[(1 - x_i) \cap (1 - x_j)] \tag{10.51}$$

Thus, the Ditlevsen bounds are given by

$$P_{f_1} + \sum_{k=2}^{n} \max \left[\left\{ P_{f_k} - \sum_{i=1}^{k-1} (P_{f_i} \cap P_{f_k}) \right\}; \quad 0 \right] \leq P_f \leq \sum_{i=1}^{n} P_{f_i} - \sum_{i=2}^{n} \max_{j<i} [P_{f_i} \cap P_{f_j}] \tag{10.52}$$

Example 10.2 The cross-sectional areas of the members of the two-bar truss shown in Fig. 10.9*a* are given by $A_1 = A_2 = A$. The probability mass functions of the strengths of the members and the load acting on the truss are shown in Figs. 10.9*b*, *c*, and *d*. Determine the following:

1. Probability of failure of the truss
2. Simple bounds on the probability of failure of the truss
3. Ditlevsen bounds on the probability of failure of the truss

solution

1. The force in member i is given by $F/2$ $(i = 1, 2)$ and the stress by $F_i / A_i = F/(2A)$ $(i = 1, 2)$. The probability mass function of the forces in the members is shown in Fig. 10.9*d*. The exact value of the probability of failure of the truss can be computed as:

$$P_f = P_f|_{F=2000}\, P(F = 2000) + P_f|_{F=3000}\, P(F = 3000) + P_f|_{F=4000}\, P(F = 4000) \tag{E1}$$

If S_i denotes the strength of member i $(i = 1, 2)$, Eq. (E1) gives

$$\begin{aligned} P_f &= \{1 - P(S_1 > 1000,\ S_2 > 1000)\}\, P(L_i = 1000) \\ &\quad + \{1 - P(S_1 > 1500,\ S_2 > 1500)\}\, P(L_i = 1500) \\ &\quad + \{1 - P(S_1 > 2000,\ S_2 > 2000)\}\, P(L_i = 2000) \\ &= \left\{1 - \left(\frac{5}{8}\right)\left(\frac{5}{8}\right)\right\}\frac{1}{8} + \left\{1 - \left(\frac{5}{8}\right)\left(\frac{5}{8}\right)\right\}\frac{1}{2} + \{1 - (0)(0)\}\frac{3}{8} \\ &= \frac{39}{64}\left(\frac{1}{8}\right) + \frac{39}{64}\left(\frac{1}{2}\right) + 1\left(\frac{3}{8}\right) = \frac{387}{512} = 0.7559 \end{aligned}$$

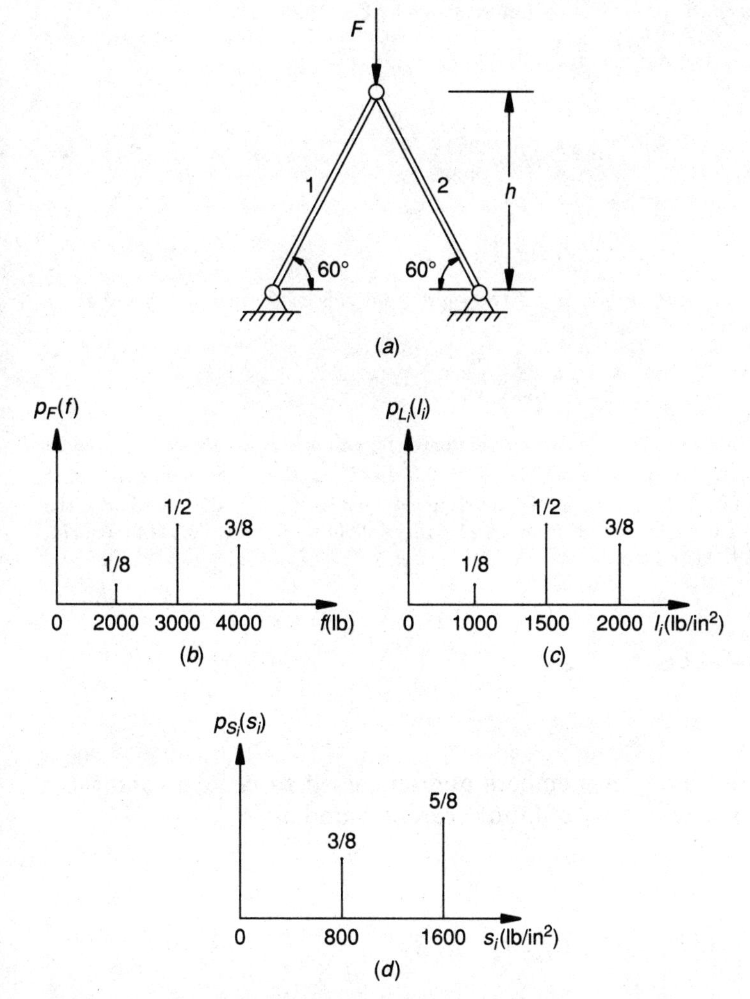

Figure 10.9 (*a*) A two-bar truss; (*b*) probability mass function of the load acting on the truss; (*c*) probability mass function of the force in member i ($i = 1, 2$); and (*d*) probability mass function of strength of member i ($i = 1, 2$).

2. The probability of failure of individual members P_{f_i} can be computed as

$$P_{f_1} = P_{f_2} = P(S_1 < L_1)$$

$$= \{1 - P(S_1 > 1000)\}\, P(L_1 = 1000) + \{1 - P(S_1 > 1500)\}\, P(L_1 = 1500)$$

$$+ \{1 - P(S_1 > 2000)\}\, P(L_1 = 2000)$$

$$= \left\{1 - \frac{5}{8}\right\} \frac{1}{8} + \left\{1 - \frac{5}{8}\right\} \frac{1}{2} + \{1 - 0\}\, \frac{3}{8} = \frac{39}{64} = 0.6094$$

The simple bounds on the probability of failure, given by Eq. (10.38), are

$$\max\left(\frac{39}{64}, \frac{39}{64}\right) \le P_f \le 1 - \left(1 - \frac{39}{64}\right)\left(1 - \frac{39}{64}\right)$$

that is,

$$\frac{39}{64} < P_f < \frac{3471}{4096}$$

that is,

$$0.6094 < P_f < 0.8474$$

while the exact value of P_f is 0.7559. If the right-hand side of the inequality (10.41) is used, the upper bound on P_f becomes

$$\sum_{k=1}^{2} P_{f_k} = \frac{39}{64} + \frac{39}{64} = 1.2188$$

3. To determine the Ditlevsen bounds, we approximate the value of $P_{f_1} \cap P_{f_2}$ as the product of P_{f_1} and P_{f_2}, which gives a value of $(39/64)^2 = 0.3714$. Thus, the lower bound on P_f can be determined from Eq. (10.49) as $0.6094 + \max[0.6094 - 0.3714, 0] = 0.8474$. Similarly, the upper bound on P_f can be computed from Eq. (10.51) as $0.6094 + 0.6094 - 0.3714 = 0.8474$. It can be seen that the two bounds coincide in this case.

10.5 Several Members–Several Loads Case (Series System)

10.5.1 Probability of failure

Consider a structure having n members and subjected to m load conditions (Fig. 10.10). Then the probability of failure can be found as

$$P_f = 1 - R_0$$

$$= 1 - \int_{l_1}\int_{l_2} \cdots \int_{l_m} \left\{\prod_{i=1}^{n} [1 - F_{S_i}(r_i)]\right\} f_{L_1}(l_1) \cdot f_{L_2}(l_2) \cdots f_{L_m}(l_m) \cdot dl_1\, dl_2 \cdots dl_m \tag{10.53}$$

where the quantity in the braces indicates the probability of all members surviving under the loads $l_1, l_2, \ldots, l_m$ and $r_i = a_i\, l_{\max}$ is the force (or stress) in member i due to the load $l_{\max}$ with

$$l_{\max} = \max_j (l_j) \tag{10.54}$$

The values of $r_i = a_i\, l_{\max}$, $i = 1, 2, \ldots, n$ can be found by the methods of linear structural analysis for any given value of $l_{\max}$. In the case of structures, such as the truss structure shown in Fig. 10.1, each element will have an independent strength (S_i) corresponding to its resistance against failure which could be yielding-, deflection-, or stability-failure mode. The load

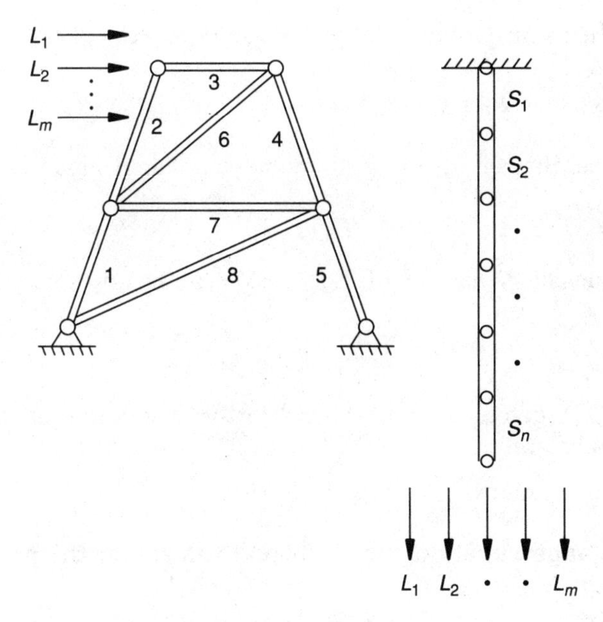

Figure 10.10 n member–m load case.

conditions $L_1, L_2, \ldots, L_m$ are similar to those used in the conventional deterministic design. $L_1, L_2, \ldots, L_m$ are considered to be statistically independent random variables with each load occurring only once. For example, L_1 may represent the extreme wind load plus average live load, L_2 may denote the average wind load plus extreme live load, etc. The structure must be designed to withstand all the loads L_j, $j = 1, 2, \ldots, m$. If the member strengths and loads are discrete random variables, Eq. (10.53) becomes

$$P_f = 1 - \sum_{l_{1j}=l_{1\min}}^{l_{1\max}} \sum_{l_{2j}=l_{2\min}}^{l_{2\max}} \cdots \sum_{l_{mj}=l_{m\min}}^{l_{m\max}} \left\{ \prod_{i=1}^{n} [1 - F_{S_i}(r_i)] \right\} \cdot f_{L_1}(l_{1j}) \cdot f_{L_2}(l_{2j}) \cdots f_{L_m}(l_{m_j}) \tag{10.55}$$

10.5.2 Bounds on the probability of failure [10.2, 10.15]

To derive an approximate expression for the probability of failure, Eq. (10.53) is rewritten as

$$P_f = 1 - \prod_{i=1}^{n} \left[\int_{l_1} \int_{l_2} \cdots \int_{l_m} \{ f_{L_1}(l_1) f_{L_2}(l_2) \cdots f_{L_m}(l_m)\, dl_1\, dl_2 \cdots dl_m \right.$$

$$\left. - F_{S_i}(r_i) f_{L_1}(l_1) f_{L_2}(l_2) \cdots f_{L_m}(l_m)\, dl_1\, dl_2 \cdots dl_m \} \right] \tag{10.56}$$

If the element failures are statistically independent, Eq. (10.56) can be approximated as

$$P_f \cong 1 - \prod_{i=1}^{n} \prod_{j=1}^{m} (1 - P_{f_{ij}}) \tag{10.57}$$

where $P_{f_{ij}}$ is the probability of failure of ith member under jth load given by

$$P_{f_{ij}} = P\,(S_i < a_{ij}\,L_j) = \int_0^\infty F_{S_i}(a_{ij}\,l_j) \cdot f_{L_j}(l_j) \cdot dl_j \tag{10.58}$$

If the individual values of $P_{f_{ij}}$ are small, Eq. (10.57) can be approximated as

$$P_f \cong \sum_{i=1}^{n} \sum_{j=1}^{m} P_{f_{ij}} \tag{10.59}$$

In practice, Eq. (10.59), gives an upper bound to the true value of P_f. Thus, the bounds on P_f can be stated as

$$\max_{i,j} (P_{f_{ij}}) \le P_f \le \sum_{i=1}^{n} \sum_{j=1}^{m} (P_{f_{ij}}) \tag{10.60}$$

Some other possible approximations (upper bound values) for the probability of failure are given by

$$P_f \cong \sum_{i=1}^{n} P_{f_i} \tag{10.61}$$

and

$$P_f \cong \sum_{j=1}^{m} P_{f_j} \tag{10.62}$$

where P_{f_i} is the probability of failure of member i due to any load, and P_{f_j} is the probability of failure of the system under the load L_j with

$$P_{f_i} = 1 - \int_{s=0}^{\infty} \left[\prod_{j=1}^{m} \{F_{L_j}(s)\} \right] f_{S_i}(s)\, ds \tag{10.63}$$

and

$$P_{f_j} = 1 - \int_{l=0}^{\infty} \left[\prod_{i=1}^{n} \{1 - F_{S_i}(l)\} \right] f_{L_j}(l)\, dl \tag{10.64}$$

All the bounds on P_f can be summarized as follows:

$$\max_{i,j} (P_{f_{ij}}) \le P_f \le \begin{Bmatrix} \displaystyle\sum_{i=1}^{n} \sum_{j=1}^{m} P_{f_{ij}} \\ \text{or} \\ \displaystyle\sum_{i=1}^{n} P_{f_i} \\ \text{or} \\ \displaystyle\sum_{j=1}^{m} P_{f_j} \end{Bmatrix} \tag{10.65}$$

10.5.3 Brittle materials

Since the total system fails as soon as the first element fails in a series system, no distinction need to be made between ductile and brittle members.

10.6 Reliability Analysis of Parallel Systems

A simplified model of a parallel system is shown in Fig. 10.11. A redundant structure with ductile members can be modeled as a parallel system since the structure can only fail when all its members reach their limit states. A redundant system might have several failure modes. In such a case, each of the failure modes can be modeled as a parallel system and all the modes, in turn, can be modeled as a series system to find the reliability of the complete system.

10.6.1 Probability of failure

The probability of failure of a parallel system consisting of n ductile elements (shown in Fig. 10.11) can be defined as

$$P_f = P(S < L) \tag{10.66}$$

where L is the load and S is the total strength of the system given by

$$S = S_1 + S_2 + \cdots + S_n \tag{10.67}$$

where S_i denotes the strength of member i ($i = 1, 2, \ldots, n$). The probability of failure of the system can be evaluated using the probability density functions of the strengths of individual members, $f_{S_i}(s_i)$, as follows.

By defining a set of new random variables R_j as

$$R_j = S_1 + S_2 + \cdots + S_j; \qquad j = 1, 2, \ldots, n \tag{10.68}$$

the probability distribution function of R_1 can be identified as

$$F_{R_1}(r) = F_{S_1}(r) = \int_{t=0}^{r} f_{S_1}(t)\, dt \tag{10.69}$$

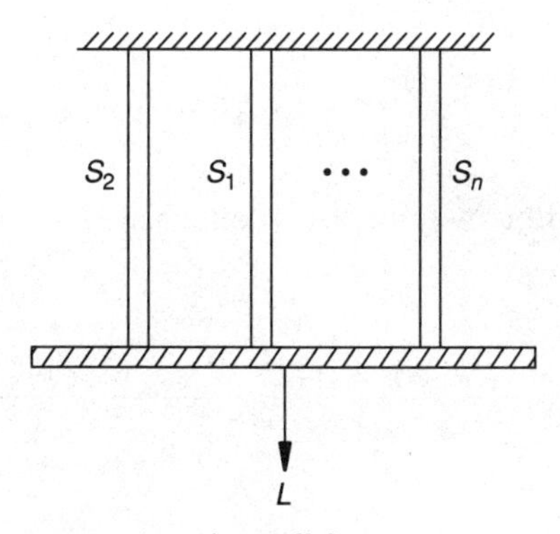

Figure 10.11 A parallel system.

The probability distribution function of R_2 is given by

$$F_{R_2}(r) = P(R_2 \le r) = P(S_1 + S_2 \le r) \tag{10.70}$$

Since $S_1 + S_2 \le r$ implies that $S_1 \le r - t$ when $S_2 = t$, Eq. (10.70) can be rewritten as

$$F_{R_2}(r) = \int_{R_2=t=0}^{\infty} F_{R_1}(r-t) f_{S_2}(t)\, dt \tag{10.71}$$

Similarly, the probability distribution function of R_3 is given by

$$F_{R_3}(r) = P(R_3 \le r) = \int_{R_3=t=0}^{\infty} F_{R_2}(r-t) f_{S_3}(t)\, dt \tag{10.72}$$

This procedure can be continued until the probability distribution function of $S = R_n$ is found as

$$F_S(r) = \int_{t=0}^{\infty} F_{R_{n-1}}(r-t)\, f_{S_n}(t)\, dt \tag{10.73}$$

Once $F_S(r)$ is known, the probability of failure of the system, Eq. (10.66), can be evaluated as

$$P_f = \int_{l=0}^{\infty} F_S(l) f_L(l)\, dl \tag{10.74}$$

and the reliability R_0 as

$$R_0 = 1 - P_f \tag{10.75}$$

Equation (10.74) shows that the probability of failure of the system can be computed by integrating the probability density functions of $S_1, S_2, \ldots,$ and S_n recursively. It can be seen that the recursive integration involves a large amount of computation. Hence, it is desirable to investigate alternate methods.

10.6.2 Simple bounds on the probability of failure

Simple bounds on the probability of failure of a parallel system can be derived by proceeding as in the case of a series system. Let a variable x_i be defined as follows:

$$x_i = \begin{cases} 1 & \text{if member } i \text{ has survived} \\ 0 & \text{if member } i \text{ has failed} \end{cases} ; \qquad i = 1, 2, \ldots, n \tag{10.76}$$

The probability of failure of the parallel system can be expressed as

$$P_f = P\left[x_1 = 0 \cap x_2 = 0 \cap \cdots \cap x_n = 0\right]$$

$$= P(x_1 = 0)\, \frac{P(x_1 = 0 \cap x_2 = 0)}{P(x_1 = 0)} \cdots \frac{P(x_1 = 0 \cap x_2 = 0 \cap \cdots \cap x_n = 0)}{P(x_1 = 0 \cap x_2 = 0 \cap \cdots \cap x_{n-1} = 0)} \tag{10.77}$$

By assuming that

$$P(x_1 = 0 \cap x_2 = 0) \geq P(x_1 = 0)\, P(x_2 = 0) \tag{10.78}$$

or, in general,

$$P(x_1 = 0 \cap x_2 = 0 \cap \cdots \cap x_i = 0) \geq P(x_1 = 0 \cap x_2 = 0 \cap \cdots \cap x_{i-1} = 0)\, P(x_i = 0); \qquad 2 \leq i \leq n \tag{10.79}$$

we can obtain from Eq. (10.77)

$$P_f \geq \prod_{i=1}^{n} P(x_i = 0) \tag{10.80}$$

The probability of failure of the member that has the lowest value will serve as a simple upper bound on P_f. Thus, the bounds on the probability of failure of the parallel system can be expressed as

$$\prod_{i=1}^{n} P(x_i = 0) \leq P_f \leq \min_{i=1,2,\ldots,n} P(x_i = 0)$$

or

$$\prod_{i=1}^{n} P_{f_i} \leq P_f \leq \min_{i=1,2,\ldots,n} P_{f_i} \tag{10.81}$$

Note that the exact probability of failure P_f will be equal to the lower bound value if there is no correlation between member failures ($\rho_{ij} = 0$ for all i and j with $i \neq j$) and will be equal to the upper bound if there is complete correlation between member failures ($\rho_{ij} = 1$ for all i and j).

10.6.3 Brittle materials

In a parallel system consisting of n brittle elements, if an element fails, it will not be able to carry any load. Thus the total load will have to be carried by the remaining elements. If the system has a low degree of indeterminacy (i.e., it is closer to being a series system), the redistribution of the loads causes the other members to fail as well. Thus the system can be modeled as a series system. On the other hand, if the system has a high degree of indeterminacy, the load carrying capacity (or strength) of the system can be expressed as

$$S = \max\{n\, S_1, (n-1)\, S_2, (n-2)\, S_3, \ldots, S_n\}; \qquad S_1 < S_2 < S_3 < \cdots < S_n \tag{10.82}$$

Example 10.3 The probability mass functions of the strengths of the members and the load acting on the parallel system of Fig. 10.12*a* are shown in Figs. 10.12*b* and *c*. Determine the following:

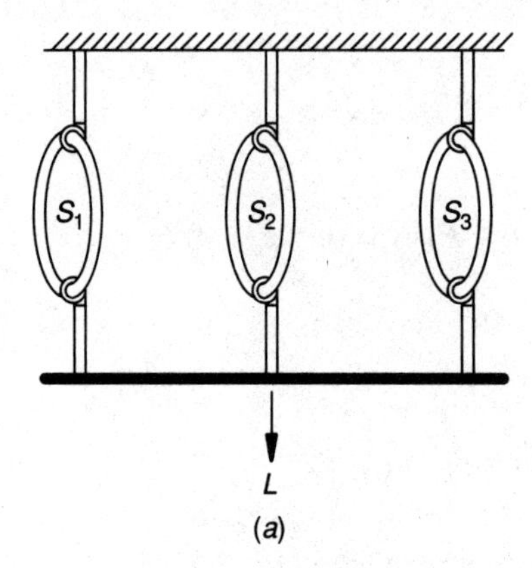

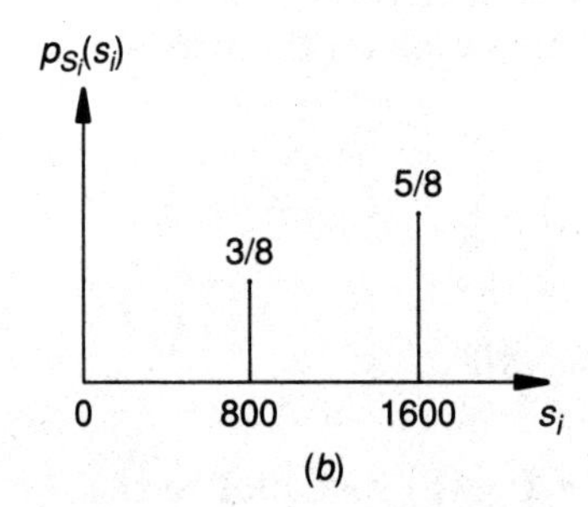

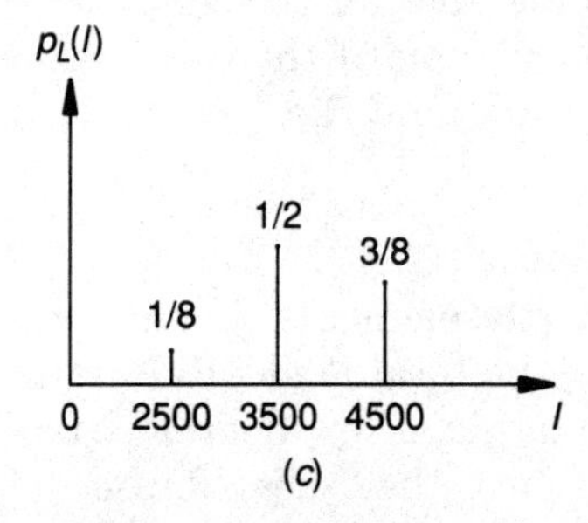

Figure 10.12 (*a*) A parallel system with three members; (*b*) probability mass function of strength of member i ($i = 1, 2, 3$); and (*c*) probability mass function of load.

1. Probability of failure of the system.
2. Bounds on the probability of failure of the system.

solution

1. Defining $R_1 = S_1$, $R_2 = S_1 + S_2$ and $R_3 = S_1 + S_2 + S_3$, the probability mass functions of R_i can be determined from the known probability mass functions of S_i as follows:

Since $R_1 = S_1$,

$$p_{R_1}(r) = p_{S_1}(r) = \begin{cases} \dfrac{3}{8} & \text{if } r = 800 \\ \dfrac{5}{8} & \text{if } r = 1600 \end{cases}$$

By definition, $R_2 = S_1 + S_2$, and hence

$$p_{R_2}(r) = p_{S_1+S_2}(r) = \begin{cases} \dfrac{9}{64} & \text{if } r = 1600 \\ \dfrac{30}{64} & \text{if } r = 2400 \\ \dfrac{25}{64} & \text{if } r = 3200 \end{cases}$$

Similarly, $R_3 = S_1 + S_2 + S_3$, and hence

$$p_R(r) = p_{R_3}(r) = p_{S_1+S_2+S_3}(r) = \begin{cases} \dfrac{27}{512} & \text{if } r = 2400 \\ \dfrac{135}{512} & \text{if } r = 3200 \\ \dfrac{225}{512} & \text{if } r = 4000 \\ \dfrac{125}{512} & \text{if } r = 4800 \end{cases}$$

The probability distribution functions of R_1, R_2, and R_3 are shown in Fig. 10.13. The probability of failure of the parallel system can be evaluated using Eq. (10.74) as

$$P_f = \int_{l=0}^{\infty} F_R(l)\, f_L(l)\, dl = \sum_{i=1}^{3} F_R(l_i)\, p_L(l_i)$$

$$= F_R(2500)p_L(2500) + F_R(3500)p_L(3500) + F_R(4500)p_L(4500)$$

$$= \frac{27}{512}\left(\frac{1}{8}\right) + \frac{162}{512}\left(\frac{1}{2}\right) + \frac{387}{512}\left(\frac{3}{8}\right)$$

$$= \frac{27}{4096} + \frac{162}{1024} + \frac{1161}{4096} = \frac{1836}{4096} = 0.4482$$

The reliability of the system is given by

$$R_0 = 1 - P_f = 0.5518$$

2. To determine the bounds on the probability of failure, we first compute the probability of failure of a single member. Since all the elements have the same strength and applied load, the probability of failure of member i can be computed as

$$P_{f_i} = \int_{l=0}^{\infty} F_{R_1}(l)\, f_{L_1}(l)\, dl$$

where the probability distribution function of R_1 is given in Fig. 10.13. Assuming that the load is shared equally by all the members, the probability mass function of the load acting

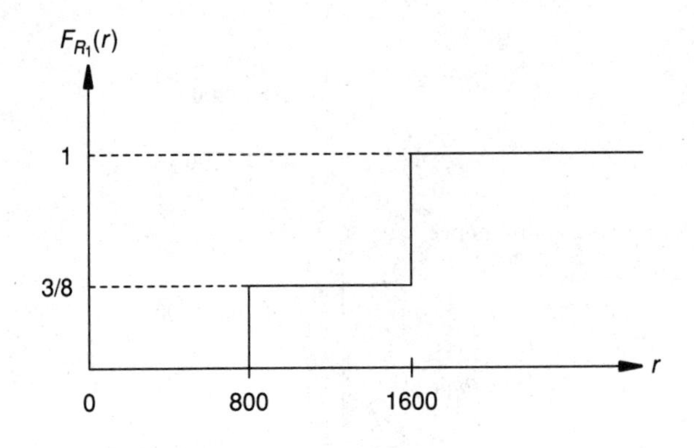

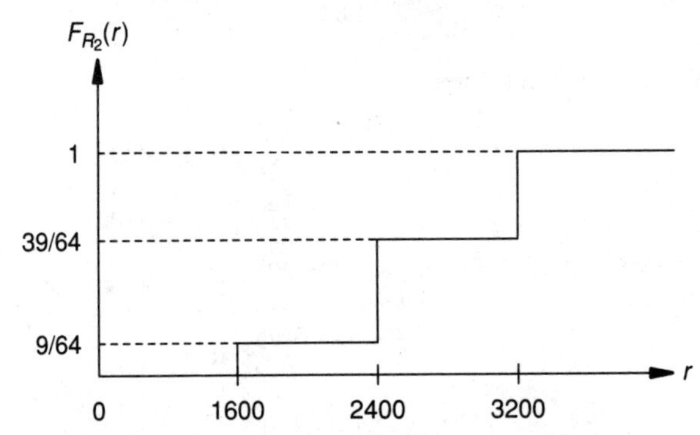

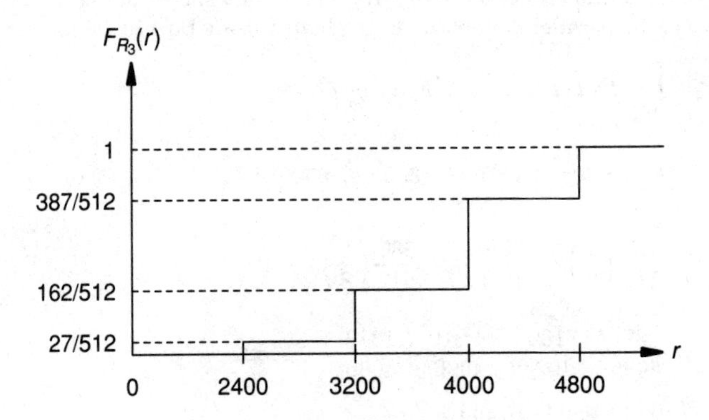

Figure 10.13 Probability distributions of R_1, R_2, and R_3.

on member 1, L_1, can be taken to be same as the one shown in Fig. 10.12*c* with the discrete values of L taken to be one-third of those shown in Fig. 10.12*c*. This gives

$$P_{f_i} = \sum_{i=1}^{3} F_{R_1}(l_i) p_{L_1}(l_i) = \left(\frac{3}{8}\right)\left(\frac{1}{8}\right) + \left(\frac{3}{8}\right)\left(\frac{1}{2}\right) + (1)\left(\frac{3}{8}\right) = \frac{39}{64}$$

The bounds on the probability of failure can be found from Eq. (10.81) as

$$\left(\frac{39}{64}\right)^3 \le P_f \le \frac{39}{64}$$

that is,

$$0.2263 \leq P_f \leq 0.6094$$

These bounds can be compared with the exact value of $P_f = 0.4482$.

10.7 Elastic-Plastic Analysis of Structures

Structures built up of elastic-plastic material, such as mild steel, can be modeled as plastic structures. Such structures will fail (collapse) due to the formation of a plastic-hinge mechanism. In general, there will be several possible mechanisms in which a structure can fail. For example, the possible failure mechanisms of the rigid frame shown in Fig. 10.14*a* are indicated in Figs. 10.14*b* to *k*. The probability of failure of the frame can be found from the probability of occurrence of any of the collapse mechanisms. For this, the "reserve strength" of the frame in the ith collapse mechanism, Z_i [10.8], can be defined as

$$Z_i = \sum_{j=1}^{n} a_{ij} M_j - \sum_{k=1}^{m} b_{ik} S_k \tag{10.83}$$

such that $Z_i < 0$ implies the occurrence of the mechanism i with a_{ij} = resistance coefficient determined by the position and condition of the jth point related to the ith failure mode or mechanism, M_j = resistance of the member at the jth point (the fully plastic-moment capacity at the plastic hinge j), b_{ik} = load coefficient determined by the position and magnitude of the kth load on the structure related to ith failure mode (mechanism), S_k = loads that are active in producing mechanism i, n = number of plastic hinges, and m = number of loads acting on the structure.

10.7.1 Probability of failure in ith collapse mode [10.7, 10.15]

By rewriting Eq. (10.83) more conveniently as

$$Z_i = \sum_{l=1}^{q} d_{il}\, y_l \tag{10.84}$$

where d_{il} is used to denote a_{ij} and $-b_{ik}$, and where y_l is used to represent M_j and S_k, the probability of failure of the structure in the ith collapse mode can be computed as

$$P_{f_i} = P(Z_i < 0) = \int_{-\infty}^{0} f_{Z_i}(z)\, dz \tag{10.85}$$

If the variables M_j and S_k follow normal distribution, Z_i will also follow normal distribution since Z_i is a linear function of M_j and S_k. However, for arbitrary distributions of M_j and S_k, the probability density functions of Z_i will be unknown. Hence, a recursive integration procedure is to be used to evaluate the probability of failure indicated in Eq. (10.85). For this, we define a

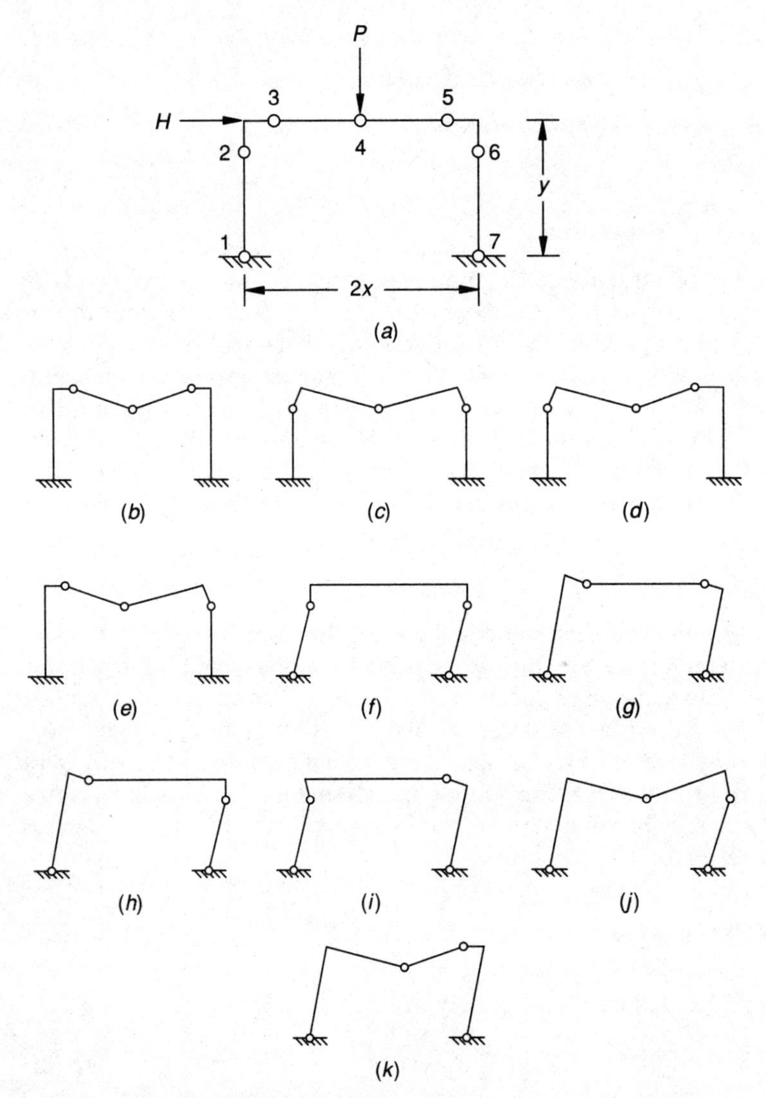

Figure 10.14 o denotes a plastic hinge.

series of functions α_i as follows:

$$\alpha_1(Y) = P(d_{i1}y_1 < Y) = \int_{-\infty}^{Y} P(d_{i1}\,y_1 = x)\,dx \tag{10.86}$$

$$\alpha_2(Y) = P(d_{i1}y_1 + d_{i2}y_2 < Y) = \int_{-\infty}^{Y} \alpha_1(Y - x)\,P(d_{i2}y_2 = x)\,dx \tag{10.87}$$

$$\vdots$$

$$\alpha_q(Y) = P(d_{i1}y_1 + d_{i2}y_2 + \cdots + d_{iq}y_q < Y)$$

$$= \int_{-\infty}^{Y} \alpha_{q-1}(Y - x)\, P(d_{iq}y_q = x)\, dx \tag{10.88}$$

Finally the desired probability of failure of the structure in the ith collapse mode is found as

$$P_{f_i} = P(Z_i < 0) = \alpha_q(Y = 0) \tag{10.89}$$

Note that the function $\alpha_1(Y)$ is determined by integrating the probability density function of x_1, which is assumed to be known. Once $\alpha_1(Y)$ is determined, $\alpha_2(Y)$ can be evaluated from the known density function of x_2, and a similar procedure holds true for the determination of subsequent $\alpha_i(Y)$.

10.7.2 Overall probability of failure of the structure [10.7, 10.15, 10.19]

If there are N possible failure modes or collapse mechanisms for the system, the probability of failure (or collapse) of the system can be determined as

$$P_f = P[Z_1 < 0 \cup Z_2 < 0 \cup \cdots \cup Z_N < 0] \tag{10.90}$$

Since some of the Z_i's may be correlated, the determination of the probability of failure of the system involves the evaluation of the N-dimensional joint probability density function of $Z_1, Z_2, \ldots, Z_N$. Equation (10.90) can be evaluated more conveniently as

$$P_f = P(Z_1 < 0) + P(Z_1 > 0,\, Z_2 < 0) + \cdots$$

$$+ P(Z_1 > 0,\, Z_2 > 0, \ldots, Z_{K-1} > 0,\, Z_K < 0) \tag{10.91}$$

The first term on the right-hand side of Eq. (10.91) denotes the probability of failure of the structure in the first mode, which can be computed by setting $i = 1$ in Eqs. (10.84) or (10.89). The second term on the right-hand side of Eq. (10.91) denotes the probability of failure of the structure in the second mode given that it has survived the first mode. It can be expressed as

$$P(Z_1 > 0,\, Z_2 < 0) = P(Z_1 > 0\,|_{Z_2 < 0})\, P(Z_2 < 0)$$

$$= \int_{-\infty}^{0} P(Z_1 > 0|_{Z_2 = t})\, P(Z_2 = t)\, dt$$

$$= \int_{t_2 = -\infty}^{0} \int_{t_1 = 0}^{\infty} P(Z_1 = t_1|_{Z_2 = t_2})\, P(Z_2 = t_2)\, dt_1\, dt_2 \tag{10.92}$$

The procedure can be extended to the case in which the structure has survived in $(K - 1)$ modes and failed in Kth mode to obtain

$$P(Z_1 > 0,\, Z_2 > 0, \ldots, Z_{K-1} > 0,\, Z_K < 0) = P(Y_1)\, P(Y_2)$$

$$\cdots\, P(Y_{K-1})\, P(Y_K) \tag{10.93}$$

where

$$P(Y_1) = P(Z_K < 0) \tag{10.94}$$

$$P(Y_2) = P(Z_{K-1} > 0 |_{Z_K < 0}) \tag{10.95}$$

$$\vdots$$

$$P(Y_K) = P(Z_1 > 0 |_{Z_2 > 0,\, Z_3 > 0, \ldots,\, Z_{K-1} > 0,\, Z_K < 0}) \tag{10.96}$$

With the values of the various terms on the right-hand side of Eq. (10.93) available, the overall probability of failure can be determined without any difficulty.

The procedure just described can be used for any distributions of the functions $Z_1, Z_2, \ldots, Z_q$. However, in general, the conditional and multivariate distribution functions of Z_i needed in Eqs. (10.94) to (10.96) cannot be found easily even if the individual probability distribution functions of Z_i are known, except in the case of normal distribution. To overcome these difficulties, the following simplifications have been suggested [10.7] to obtain reasonably good, but approximate, results: (1) Assume the conditional distributions of Z_i to be normal, and (2) Approximate the multiple integrals encountered in Eqs. (10.92) and (10.93) by a single integral. For example, Eq. (10.93) can be approximated by a single integral when the first $K-1$ terms are evaluated at the mean values of $Z_2, Z_3, \ldots, Z_K$ and the integration can be performed on the Kth term only as

$$P(Z_1 > 0, Z_2 > 0, \ldots, Z_{K-1} > 0, Z_K < 0)$$

$$\approx \int_{-\infty}^{0} P(E_2)\, P(E_3) \cdots P(E_K)\, P(Z_K = t_K)\, dt_K \tag{10.97}$$

where

$$P(E_2) = P(Z_{K-1} > 0 |_{Z_K = t_K}) \tag{10.98}$$

$$\vdots$$

$$P(E_{K-1}) = P(Z_2 > 0 |_{Z_3 = \bar{Z}_3, \ldots,\, Z_{K-1} = \bar{Z}_{K-1},\, Z_K = t_K}) \tag{10.99}$$

$$P(E_K) = P(Z_1 > 0 |_{Z_2 = \bar{Z}_2,\, Z_3 = \bar{Z}_3, \ldots,\, Z_{K-1} = \bar{Z}_{K-1},\, Z_K = t_K}) \tag{10.100}$$

and $\bar{Z}_i$ is the mean value of Z_i.

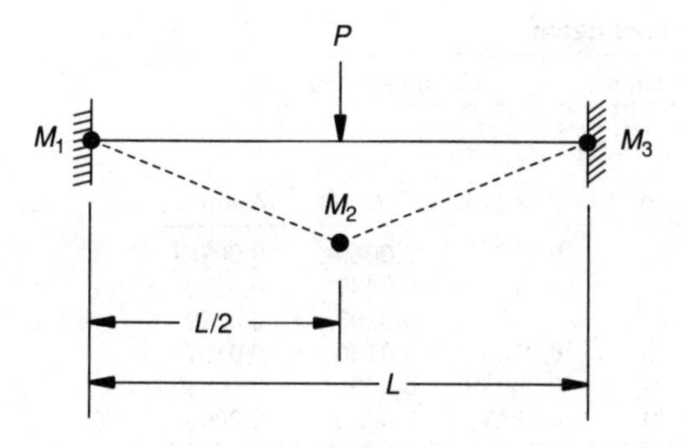

Figure 10.15 A fixed-fixed beam.

Example 10.4 Find the probability of failure of the beam shown in Fig. 10.15 whose failure mode involves hinges (or joints) at the critical sections 1, 2, and 3 [10.7].

solution This beam has a single-failure mode as indicated in Fig. 10.15 and fails if the strength of the joints in the failure mechanism is smaller than the applied moment; that is, if

$$M_1 + 2M_2 + M_3 < \frac{PL}{2} \tag{E1}$$

Thus the probability of the failure of the beam is given by

$$P_f = P(Z_1 < 0) \tag{E2}$$

where

$$Z_1 = M_1 + 2M_2 + M_3 - \frac{PL}{2} \tag{E3}$$

The mean value and standard deviation of Z_1, assuming all the variables to be independent, are given by

$$\overline{Z}_1 = \overline{M}_1 + 2\overline{M}_2 + \overline{M}_3 - \frac{\overline{P}L}{2} \tag{E4}$$

and

$$\sigma_{Z_1} = \left\{\sigma_{M_1}^2 + 4\,\sigma_{M_2}^2 + \sigma_{M_3}^2 + \sigma_P^2\,\frac{L^2}{4}\right\}^{1/2} \tag{E5}$$

The probability of failure, given by Eq. (E2), can be evaluated once the probability distributions of the moments M_1, M_2 and M_3 and the load P are known. Some typical results [10.7] are given in Table 10.1. The effects of the mean value and the coefficient of variation on the probability of failure can also be seen from these results. Figure 10.16 shows the variation of the probability of failure with the factor of safety of the beam when all the variables (M_i and P) follow normal distribution. The inconsistency of the factor of safety as a measure of reliability (discussed in Chapter 1) can be seen from this figure also. For example, compare the probability of failure corresponding to a factor of safety of 1.5. Curves 1, 2, and 3 correspond to different coefficients of variation of the moments and load, and indicate the probabilities of failure of 6×10^{-2}, 1×10^{-3} and 3×10^{-2}, respectively. The coefficients of variation of M_i and P, which are not even specified in the conventional design specifications, differ only by a factor of two; but the corresponding probabilities of failure differ by a factor of ≈ 325.

TABLE 10.1 Failure Probability of the Fixed-Fixed Beam

Mean values				Coefficient of variation		Failure probability		
Moments (kip·ft)			Load (kip)					
M_1	M_2	M_3	L	Moments	Load	Normal	Log Normal*	Weibull†
350	350	350	120	0.10	0.10	0.00875	0.00894	0.00781
350	350	350	120	0.20	0.20	0.0170	0.0237	0.0253
350	350	350	120	0.05	0.20	0.0206	0.0193	0.0212
350	350	350	120	0.20	0.05	0.0135	0.0130	0.0137
400	400	400	120	0.10	0.10	0.00049	0.00064	0.00040
400	400	400	120	0.20	0.20	0.0041	0.0098	0.0095

*Standard normal distribution tables used.

†Recursive function integration used.

SOURCE: J. Stevenson and F. Moses, "Reliability Analysis of Frame Structures," *Journal of the Structural Division, Proceedings of the ASCE*, vol. 96, no. ST 11, Nov. 1970, pp. 2409–2427.

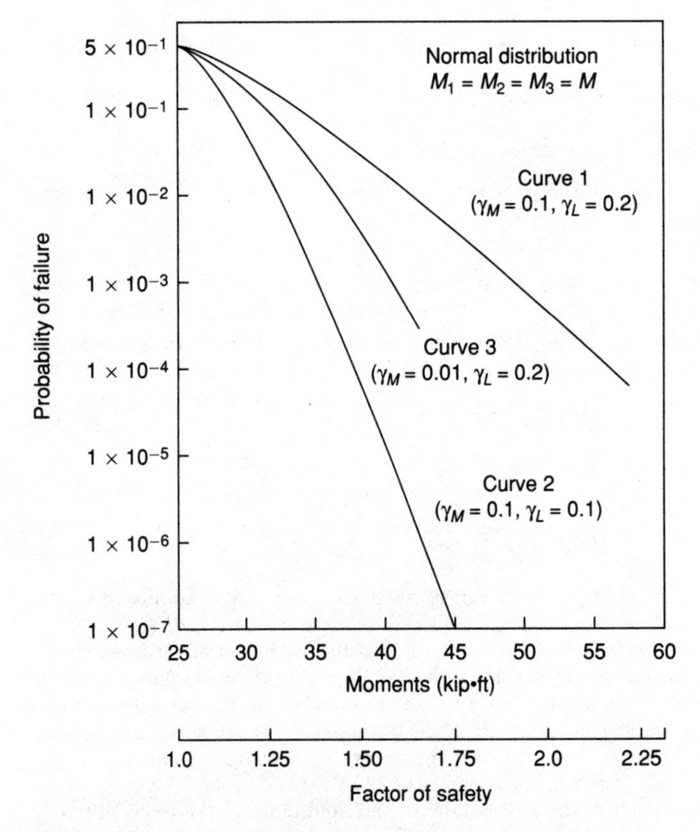

Figure 10.16 Reprinted with permission from J. Stevenson and F. Moses, "Reliability Analysis of Frame Structures," *Journal of the Structural Division, Proceedings of the ASCE*, vol. 96, no. ST 11, Nov. 1970, pp. 2409–2427.

Example 10.5 Compute the bounds on the probability of failure of the portal frame shown in Fig. 10.14*a*. Assume that the moment capacities of the plastic hinges, $M_1, M_2, \ldots, M_7$, as well as the loads H and P follow normal distribution with

$$\mu_{M_i} = 400 \text{ in·lb and } \sigma_{M_i} = 80 \text{ in·lb}; \qquad i = 1, 2, 6, 7$$

$$\mu_{M_i} = 500 \text{ in·lb and } \sigma_{M_i} = 100 \text{ in·lb}; \qquad i = 3, 4, 5$$

$$\mu_H = 50 \text{ lb}, \sigma_H = 20 \text{ lb}, \mu_P = 100 \text{ lb}, \sigma_P = 20 \text{ lb}$$

The values of x and y are 10 in and 15 in, respectively.

solution The safety margins in the various mechanisms shown in Figs. 10.14*b* to *k* can be found using the methods of plastic structural analysis [10.18] as

$$Z_1 = M_3 + 2M_4 + M_5 - xP$$

$$Z_2 = M_2 + 2M_4 + M_6 - xP$$

$$Z_3 = M_2 + 2M_4 + M_5 - xP$$

$$Z_4 = M_3 + 2M_4 + M_6 - xP$$

$$Z_5 = M_1 + M_2 + M_6 + M_7 - yH$$

$$Z_6 = M_1 + M_3 + M_5 + M_7 - yH$$

$$Z_7 = M_1 + M_3 + M_6 + M_7 - yH$$

$$Z_8 = M_1 + M_2 + M_5 + M_7 - yH$$

$$Z_9 = M_1 + 2M_4 + 2M_6 + M_7 - xP - yH$$

$$Z_{10} = M_1 + 2M_4 + 2M_5 + M_7 - xP - yH$$

Using the mean values and standard deviations of M_i ($i = 1, 2, \ldots, 7$), P and H, the mean and standard deviations of Z_1 can be computed as

$$\mu_{Z_1} = \mu_{M_3} + 2\,\mu_{M_4} + \mu_{M_5} - x\mu_P = 500 + 2(500) + 500 - 10(100) = 1000 \text{ in·lb}$$

$$\sigma_{Z_1}^2 = \sigma_{M_3}^2 + (2)^2\,\sigma_{M_4}^2 + \sigma_{M_5}^2 + (x)^2\,\sigma_P^2 = (100)^2 + 4(100)^2 + (100)^2 + 100(20)^2 = 64{,}000$$

$$\therefore\ \sigma_{Z_1} = 244.9490 \text{ in·lb}$$

The probability of failure in this failure mode or mechanism can be found using standard normal tables as

$$P_{f_1} = P(Z_1 \le 0) = \Phi\left(-\frac{\mu_{Z_1}}{\sigma_{Z_1}}\right) = \Phi(-4.0825) = 0.22 \times 10^{-4}$$

The probabilities of failure in the remaining failure modes can be determined in a similar manner. The results are shown in Table 10.2. Since the system fails with the occurrence of

TABLE 10.2 Computation of Probabilities of Failure in Different Failure Modes

Failure mode (mechanism) i	μ_{Z_i} (in-lb)	σ_{Z_i} (in-lb)	Value of $-\left(\frac{\mu_{Z_i}}{\sigma_{Z_i}}\right)$	P_{f_i}
1	1000	244.9490	−4.0825	0.22×10^{-4}
2	800	304.6309	−2.6261	0.43×10^{-2}
3	900	310.4836	−2.8987	0.19×10^{-2}
4	900	310.4836	−2.8987	0.19×10^{-2}
5	850	340.0000	−2.5000	0.62×10^{-2}
6	1050	350.4282	−2.9963	0.14×10^{-2}
7	950	345.2534	−2.7516	0.30×10^{-2}
8	950	345.2534	−2.7516	0.30×10^{-2}
9	850	456.5085	−1.8620	0.31×10^{-1}
10	1050	472.0171	−2.2245	0.13×10^{-1}

any of the failure modes, the failure modes can be modeled to be in series and hence Eq. (10.38) can be used to determine the bounds on the probability of failure of the complete system. This gives

$$\max_{i=1,2,\ldots,10} P_{f_i} \le P_f \le 1 - \prod_{i=1}^{10} (1 - P_{f_i})$$

or

$$0.0310 \le P_f \le 0.0642$$

10.8 Structures with Correlation of Element Strengths

In some multicomponent structures, correlation may be present between element strengths. Although the reliability analysis of systems with arbitrary correlation between element strengths is very difficult, Grigoriu and Turkstra [10.11] derived expressions for the reliability of n-member series and parallel systems whose element strengths follow normal distribution with the same correlation between any two member strengths. In addition, the load acting on the system is assumed to be deterministic with the same reliability index β_i for all elements

$$\beta_i = \frac{\mu_{S_i} - L_i}{\sigma_{S_i}}; \qquad i = 1, 2, \ldots, n \tag{10.101}$$

or

$$\mu_{S_i} = L_i + \beta_i \, \sigma_{S_i}; \qquad i = 1, 2, \ldots, n \tag{10.102}$$

where S_i and L_i (deterministic) are the strength and load of member i. Using the coefficient of variation of strength of element i, $\gamma_i = \mu_{S_i} / \sigma_{S_i}$, Eq. (10.102) can be expressed as

$$\mu_{S_i} = \frac{L_i}{1 - \dfrac{\beta_i}{\gamma_i}}; \qquad i = 1, 2, \ldots, n \tag{10.103}$$

10.8.1 Series systems

For a series system, the reliability of the system [10.12] can be expressed as

$$R_0 = \int_{-\infty}^{\infty} \left[\Phi \left(\frac{\beta_i + z\sqrt{\rho}}{\sqrt{1-\rho}} \right) \right]^n \phi_Z(z)\, dz \tag{10.104}$$

where ρ is the correlation coefficient between the strengths of any two elements, and $\phi_Z(z)$ and Φ denote the probability density and distribution functions of the standard normal variate z. It has been observed (see Problem 10.11) that the reliability of a series system increases with increasing correlation coefficient ρ.

10.8.2 Parallel systems

For a parallel system with ductile members, the overall strength of the system (S) is given by

$$S = \sum_{i=1}^{n} S_i \tag{10.105}$$

Assuming that all S_i are identical and normally distributed with mean values $\mu_i = \mu_0$, standard deviations $\sigma_{S_i} = \sigma_0$, and correlation coefficient ρ, we obtain

$$\mu_S = \sum_{i=1}^{n} \mu_{S_i} = n\mu_0 \tag{10.106}$$

and

$$\sigma_S^2 = \sum_{i=1}^{n} \sigma_{S_i}^2 + \rho \sum_{i=1}^{n} \sum_{j=1}^{n} \sigma_{S_i} \sigma_{S_j} = n\,\sigma_0^2 + n(n-1)\rho\sigma_0^2 \tag{10.107}$$

If L denotes the deterministic load acting on the structure and all the members have the same reliability index β_i we have (see Eq. (10.102))

$$\mu_{S_i} = L_i + \beta_i \sigma_{S_i} \tag{10.108}$$

Multiplication of Eq. (10.108) throughout by n gives

$$n\mu_{S_i} = nL_i + \beta_i n \sigma_{S_i}$$

or

$$L = nL_i = n\mu_0 - \beta_i n \sigma_0 \tag{10.109}$$

The reliability index for the parallel system β_0 can be expressed as

$$\beta_0 = \frac{\mu_S - L}{\sigma_S} = \frac{n\mu_0 - (n\mu_0 - \beta_i n\sigma_0)}{\sqrt{n\sigma_0^2 + n(n-1)\rho\sigma_0^2}}$$

$$= \beta_i \left[\frac{n}{1+\rho(n-1)}\right]^{1/2} \tag{10.110}$$

and the overall reliability R_0 as

$$R_0 = 1 - \Phi(-\beta_0) \tag{10.111}$$

It has been found (see Problem 10.12) that the reliability of a parallel system decreases with increasing correlation coefficient, ρ.

References and Bibliography

10.1. P. Thoft-Christensen and M. J. Baker, *Structural Reliability Theory and Its Applications,* Springer-Verlag, Berlin, 1982.

10.2. C. A. Cornell, "Bounds on the Reliability of Structural Systems," *J. of the Structural Div., ASCE*, vol. 93, 1967, pp. 171–200.

10.3. O. Ditlevsen, "Narrow Reliability Bounds for Structural Systems," *J. of Structural Mech.*, vol. 7, 1979, pp. 453–472.

10.4. P. Thoft-Christensen and Y. Murotsu, *Application of Structural Systems Reliability Theory,* Springer-Verlag, Berlin, 1986.

10.5. R. E. Melchers, *Structural Reliability: Analysis and Prediction,* Ellis Horwood Ltd., Chichester, England, 1987.

10.6. Y. Murotsu, H. Okada, K. Niwa, and S. Miwa, "Reliability Analysis of Truss Structures by Using Matrix Method," *ASME J. of Mechanical Design*, vol. 102, 1980, pp. 749–756.

10.7. J. D. Stevenson and F. Moses, "Reliability Analysis of Frame Structures," *J. of the Structural Div., Proc. of the ASCE*, vol. 96, 1970, pp. 2409–2427.

10.8. F. Moses and J. D. Stevenson, "Reliability-Based Structural Design," *J. of the Structural Div., Proc. of the ASCE*, vol. 96, 1970, pp. 221–244.

10.9. D. M. Frangopol, "Sensitivity Studies in Reliability Based Analysis of Redundant Structures," *Structural Safety*, vol. 3, 1985, pp. 13–22.

10.10. R. E. Melchers, "Reliability of Parallel Structural Systems," *J. of the Structural Div., Proc. of the ASCE*, vol. 109, 1983, pp. 2651–2665.

10.11. M. Grigoriu and C. Turkstra, "Safety of Structural Systems with Correlated Resistances," *Applied Mathematical Modelling*, vol. 3, 1979, pp. 130–136.

10.12. A. J. Stuart, "Equally Correlated Variates and the Multinormal Integral," *J. of Royal Statistical Soc.*, Series B, vol. 20, 1958, pp. 373–378.

10.13. B. S. Dhillon, "Bibliography of Literature on Safety Factors," *Microelectronics and Reliability*, vol. 29, 1989, pp. 267–280.

10.14. B. S. Dhillon and J. S. Belland, "Bibliography of Literature on Reliability in Civil Engineering," *Microelectronics and Reliability*, vol. 26, 1986, pp. 99–121.

10.15. F. Moses and D. E. Kinser, "Analysis of Structural Reliability," *J. of the Structural Div., Proc. of ASCE*, vol. 93, no. ST5, 1967, pp. 147–164.

10.16. A. C. King and C. B. Read, *Pathways to Probability*, Holt, Rinehart & Winston, Inc., New York, 1963.

10.17. C. C. Gillispie (Ed.-in-chief), *Dictionary of Scientific Biography*, vol. 13, Charles Scribner's Sons, New York, 1980.

10.18. R. Nakib and D. M. Frangopal, "RBSA and RBSA-OPT: Two Computer Programs for Structural System Reliability Analysis and Optimization," *Computers & Structures*, vol. 36, 1990, pp. 13–27.

10.19. F. B. Song, "A Numerical Integration Method for Computing Structural System Reliability," Computers & Structures, vol. 36, 1990, pp. 65–70.

Review Questions

10.1 What is the fundamental case in structural reliability?

10.2 What is a weakest-link system?

10.3 Define a fail-safe system.

10.4 State the bounds on the probability of failure of the fundamental case.

10.5 Give two examples of each of the statically determinate and indeterminate structures.

10.6 State the bounds on the probability of failure of a multimember–multiload weakest-link system.

10.7 What is a collapse mechanism in the elasto-plastic analysis of structures?

10.8 Suggest a method for finding the probability of failure of an elasto-plastic structure in a given collapse mode.

10.9 What is the influence of brittle and ductile materials in a series system?

10.10 How is the reliability of a parallel system found if the material is (1) brittle, and (2) ductile?

10.11 Define the reliability index and state its relationship to the reliability of a structure.

Problems

10.1 The frame structure shown in Fig. 10.14*a* can fail in any of the ten failure modes indicated in Figs. 10.14*b*–*k*. Develop (a) a fault tree, and (b) an event tree for the failure of the structure.

10.2 A single member–single load problem has a strength S and load L. The mean value and coefficient of variations of S and L are given by (μ_S, γ_S) and (μ_L, γ_L), respectively. If both S and L follow lognormal distribution, show that the standard normal variate corresponding to a specified reliability R_0 is given by

$$z_1 = \frac{\ln\left\{\left(\frac{\mu_S}{\mu_L}\right)\left(\frac{1+\gamma_S^2}{1+\gamma_L^2}\right)^{1/2}\right\}}{\left\{\ln\left[(1+\gamma_S^2)(1+\gamma_L^2)\right]\right\}^{1/2}}$$

where $R_0 = 1 - \Phi(-z_1)$, and Φ is the cumulative (normal) distribution function.

10.3 The elasto-plastic analysis of the frame shown in Fig. 10.14*a* involves consideration of the failure modes shown in Figs. 10.14*b*–*k*. The reserve strengths or safety margins in the various collapse modes are given by $Z_1, Z_2, \ldots,$ and Z_7. The plastic moment capacities of the hinges are normally distributed, each with a mean value of 200,000 ft·lb and a standard deviation of 20,000 ft·lb. The loads are also normally distributed, each with a mean value of 20,000 lb and a standard deviation of 4,000 lb. The values of x and y are 10 ft and 20 ft, respectively. (a) Find the probabilities of failure of the frame in each of the failure modes. (b) Assuming a series model for the failure modes, determine the upper and lower bounds on the exact probability of failure.

10.4 Find the bounds on the probability of failure of a structure corresponding to its (a) central factor of safety, and (b) its characteristic factor of safety.

10.5 A chain, consisting of three identical links, carries a load L as shown in Fig. 10.17*a*. The probability density functions of the strengths of the links and

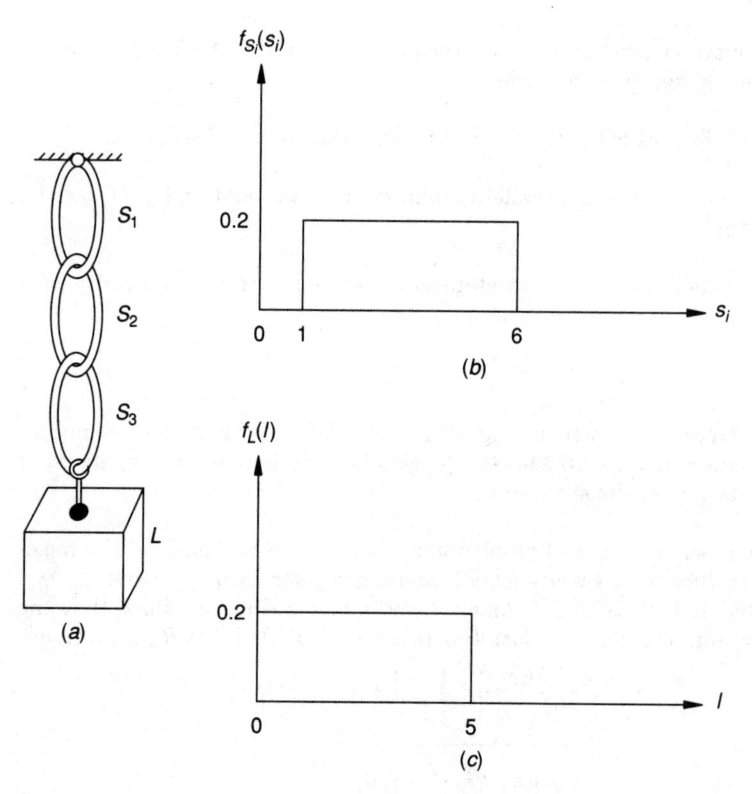

Figure 10.17

the load are shown in Figs. 10.17*b* and *c*, respectively. Determine (a) the probability of failure of a link, (b) the probability of failure of the chain, and (c) the bounds on the probability of failure of the chain.

10.6 A chain, consisting of three identical links, is subjected to two loads as shown in Fig. 10.18*a*. The probability mass functions of the strengths of the links (S_i) and the loads (L_j) are given, respectively, in Figs. 10.18*b* and *c*. Find the following: (a) the probability of failure of any link i under any load j; (b) the probability of failure of the chain under any load j; (c) the probability of failure of any link i under the two loads; (d) the bounds on the probability of failure of the chain; and (e) the exact probability of failure of the chain.

10.7 A parallel system, consisting of three identical members, is subjected to a probabilistic load as shown in Fig. 10.19*a*. Find the probability of failure of the system corresponding to the probability mass functions shown in Figs. 10.19*b* and *c*. Also, determine the bounds on the probability of failure of the system.

10.8 Find the probability of failure of the fixed-fixed beam shown in Fig. 10.20 based on the limit design procedure. The mean values and standard deviations

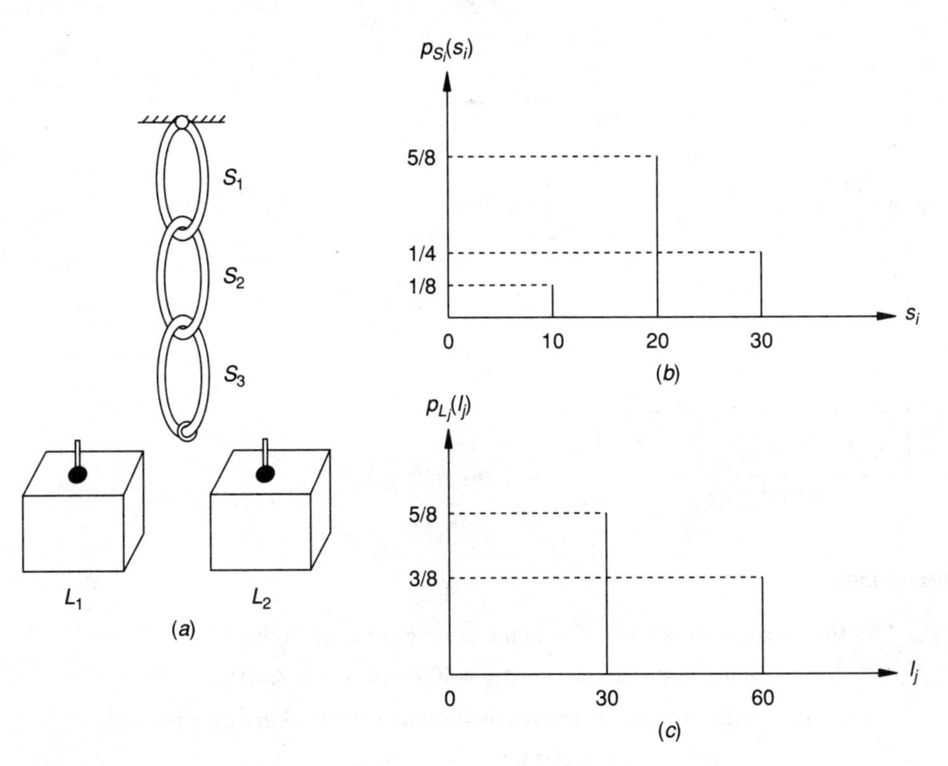

Figure 10.18

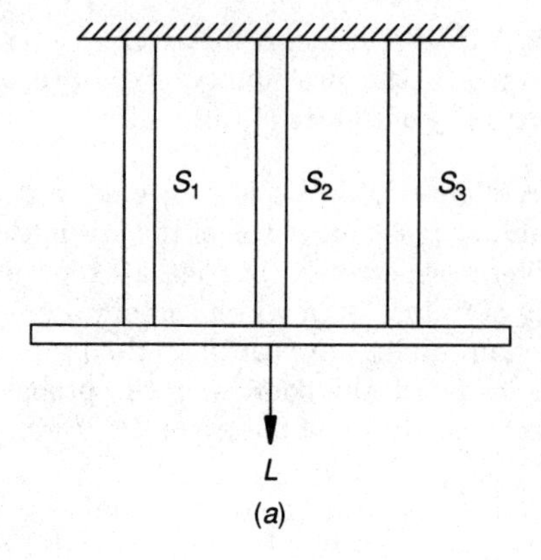

(a)

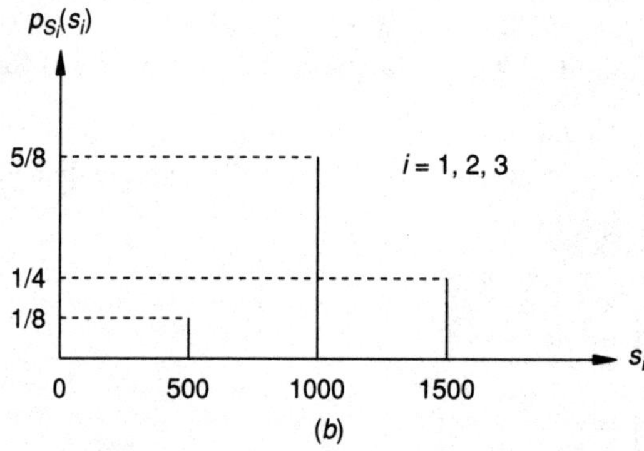

(b)

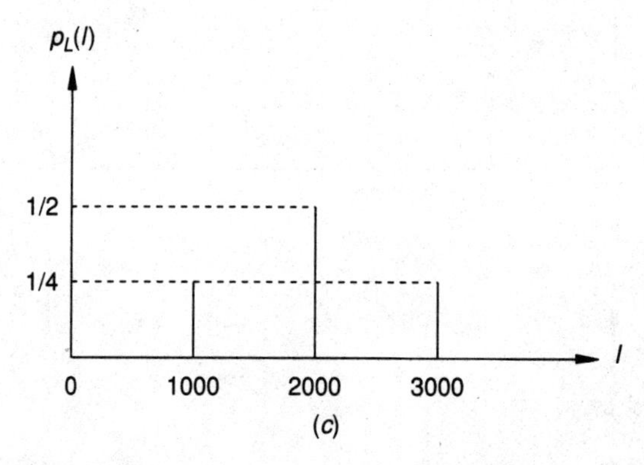

(c)

Figure 10.19

of the moment capacities of the plastic hinges are given by

$$\mu_{M_i} = 400 \text{ lb} \cdot \text{in}, \qquad \sigma_{M_i} = 40 \text{ lb} \cdot \text{in } (i = 1, 2, 3)$$

The mean value and the standard deviation of the load are given by

$$\mu_P = 100 \text{ lb}; \qquad \sigma_P = 10 \text{ lb}$$

Assume normal distribution for the moments and the load.

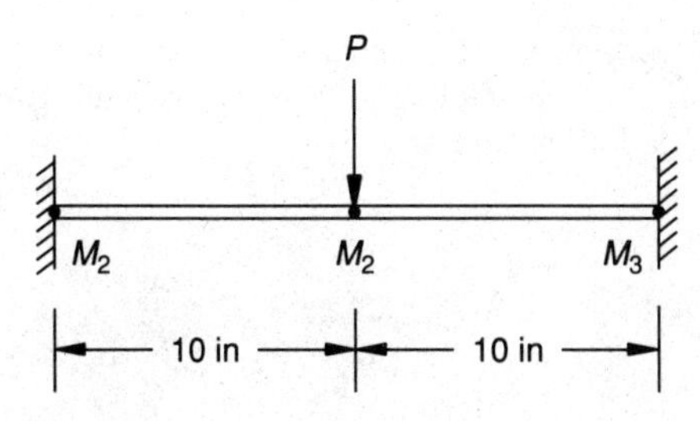

Figure 10.20

10.9 Solve Problem 10.8 assuming lognormal distribution for the total resisting moment ($M_1 + 2M_2 + M_3 = 4M$) and the applied moment ($PL/2 = 10\,P$).

10.10 Find the probability of failure of the rigid frame shown in Fig. 10.21 based on the limit-design procedure. The mean values and standard deviations of the moment capacities of the plastic hinges is given by

$$\mu_{M_i} = 200 \text{ lb} \cdot \text{in}; \qquad \sigma_{M_i} = 50 \text{ lb} \cdot \text{in } (i = 1, 2, \ldots, 7)$$

The mean values and standard deviations of the loads are given by

$$\mu_{P_i} = 30 \text{ lb}, \sigma_{P_i} = 5 \text{ lb}; \qquad i = 1, 2$$

Assume normal distribution for all the hinge moments and loads.

10.11 Plot the graph between the reliability of a series system, R_0, and the correlation coefficient between the member strengths, ρ [Eq. (10.104)] for the following data:

(a) $\beta_i = 2.0$, n = 1
(b) $\beta_i = 2.0$, n = 3
(c) $\beta_i = 2.0$, n = 6

Consider the values of ρ between 0 and 1.

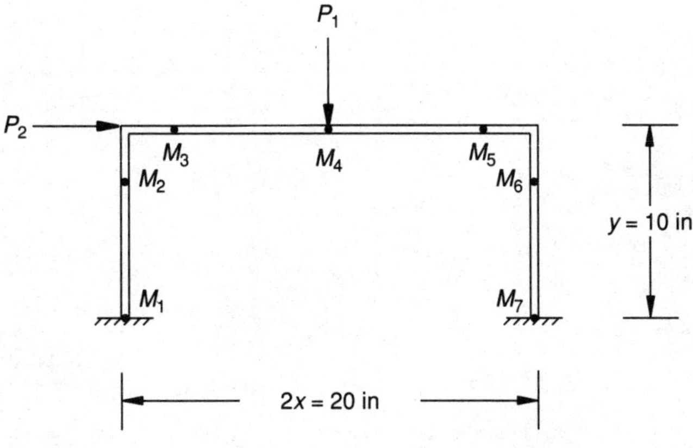

Figure 10.21

10.12 Plot the graph between the reliability of a parallel system with n identical ductile members, R_0, and the correlation coefficient between the member strengths, ρ (Eq. 10.111) for the following data:

(a) $\beta_i = 2.0$, n = 1

(b) $\beta_i = 2.0$, n = 3

(c) $\beta_i = 2.0$, n = 6.

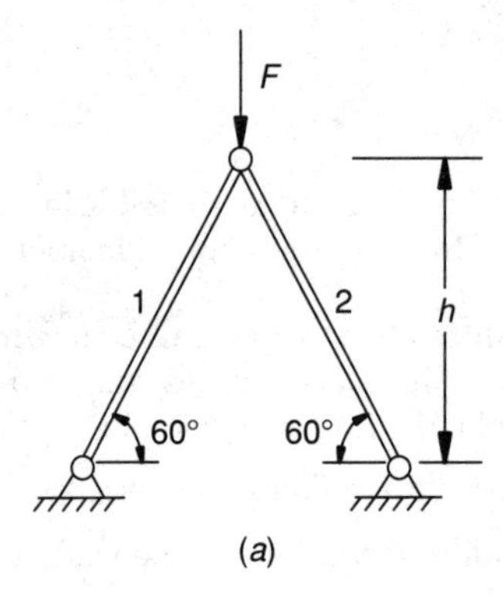

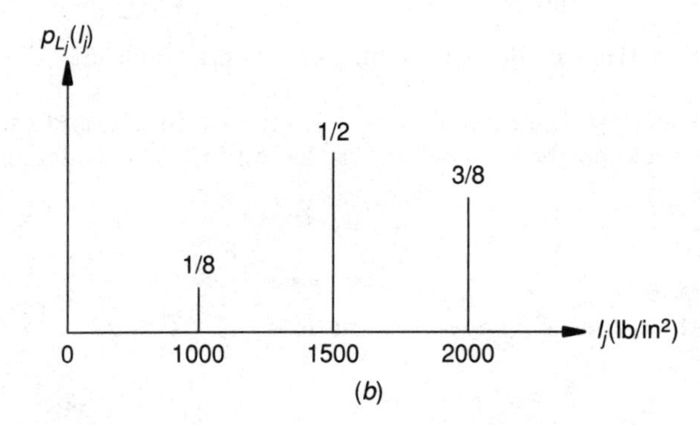

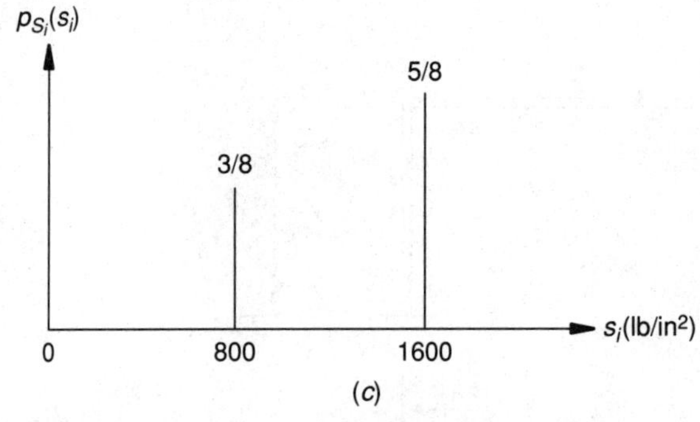

Figure 10.22 (*a*) A two-bar truss; (*b*) probability mass function of load L_j (j = 1, 2); and (*c*) probability mass function of strength of member i (i = 1, 2).

10.13 The two-bar truss shown in Fig. 10.22*a* is subjected to two independently applied probabilistic loads described in Fig. 10.22*b*. The probability mass functions of the strengths of the members are shown in Fig. 10.22*c*. Determine the following: (a) The probability of failure of the truss, and (b) The bounds on the probability of failure of the truss.

Chapter

11

Reliability-Based Optimum Design

Biographical Note

Joseph Louis Lagrange

Joseph Louis Lagrange was a French mathematician who was born on January 25, 1736. He was the youngest and the only one among his parents' eleven children to survive beyond infancy. He studied at the Royal Artillery School in Turin, Italy, and became a professor of mathematics there at the age of eighteen. He left for Berlin in 1766 and remained there until 1786. Later he went to Paris, became a professor at the newly established Ecole Normale, and taught there until his death on April 10, 1813. Lagrange was very brilliant and invented the calculus of variations. He solved the vibrating-string problem and explained the phenomena of beats, echos, and compound sounds. He contributed to the theory of numbers and differential equations and developed the principles of virtual velocities and least action. The method he presented for finding the extremum solution of a multivariable function in the presence of constraints is known as the Lagrange multiplier method and plays an important role in the literature of optimization [11.10, 11.11].

11.1 Introduction

Optimization plays an important role in system design. In the design of any component or system, we will be interested either in maximizing the reliability subject to a constraint on the cost or in minimizing the cost with a restriction on the reliability of the system. In structural and mechanical design problems, the strength-based reliability becomes important. In this chapter, the standard-optimization problem is stated first and then a variety of reliability-based design problems are formulated as optimization problems. The various techniques available for the solution of optimum-design problems are summarized. The computational details of some of the commonly used optimization techniques are given along with illustrative examples. The basic principles of graphical techniques, calculus-based methods and nonlinear programming-based strategies are explained. The illustrative examples deal with the reliability-based optimum design of machine and structural components and the optimal-reliability allocation in multicomponent systems.

11.2 Optimization Problem

Optimization is the process of obtaining the best result under given circumstances. The conventional design procedures aim at finding an adequate design that merely satisfies the functional requirements within the confines of the existing limitations. The optimum-design procedures seek to find the design variables in such a way that the design will be the best one (among the many adequate designs) that satisfies all of the limitations and restrictions placed on it.

A standard optimization problem can be stated in the following form:

$$\text{Find } \mathbf{X} = \begin{Bmatrix} x_1 \\ x_2 \\ \cdot \\ \cdot \\ \cdot \\ x_n \end{Bmatrix}$$

$$\text{which minimizes } f(\mathbf{X})$$

subject to

$$g_j(\mathbf{X}) \le 0; \qquad j = 1, 2, \ldots, m$$

and

$$l_k(\mathbf{X}) = 0; \qquad k = 1, 2, \ldots, p \tag{11.1}$$

where x_i = ith design variable, $\mathbf{X}$ = vector of design variables, f = objective or criterion function, g_j = inequality-constrained function, l_k = equality-constrained function, n = number of design variables, m = number of inequality constraints, and p = number of equality constraints. The problem stated in Eq. (11.1) is also known as a mathematical programming problem. Several methods are available for solving the problem stated in Eq. (11.1). However, no single method can solve all types of optimization problems efficiently. Depending on the type of optimization problem, only certain methods can be used for the efficient solution of Eq. (11.1). The classification of optimization problems is given in Table 11.1.

As indicated in Table 11.1, if one or more of the functions f, g_j, and l_k are nonlinear in terms of the design variables x_i, the problem stated in Eq. (11.1) is called a constrained-nonlinear programming problem. In most reliability-based optimum-design problems, many of the functions f, g_j, and l_k are nonlinear, and hence we shall study nonlinear optimization techniques in this chapter. We shall consider four techniques. The first one is a graphical method that can be used when the number of design variables (unknowns) is one or two. The second technique is a calculus-based method, called the Lagrange multiplier method, which can be used to solve problems containing only equality constraints. The third technique is termed the Sequential Unconstrained Minimization Technique (SUMT) or the penalty-function method in which the solution of a general optimization problem, as expressed in Eq. (11.1), is found by solving a series of unconstrained minimization problems. The fourth method is known as the dynamic-programming approach that can be used to solve problems whose physical structure is in series form. We also note that the maximization of $f(\mathbf{X})$ can be achieved by the minimization of $-f(\mathbf{X})$.

11.3 Formulation of Optimization Problems

11.3.1 Reliability allocation problems

Reliability allocation is the process of assigning reliabilities to individual components so as to attain a specified overall reliability for the system. The specified overall system reliability can be achieved in several ways. However, we will be interested in minimizing a criterion or objective function such as cost, weight, or volume while achieving the overall system reliability over a specified length of time, T. Thus the problem can be formulated in different ways as illustrated by the following examples.

Example 11.1 Formulate a reliability allocation problem using the minimization of the total cost as the objective function.

TABLE 11.1 Classification of Optimization Problems

	Type of problem	Characteristics of problem in Eq. (11.1): Nature of design variables (x_i)	Nature of the objective function $f(\mathbf{X})$	Nature of the constraint functions $g_j(\mathbf{X})$ and $l_j(\mathbf{X})$	Solution method(s) available	Name(s) of the person(s) associated with the problem
1.	Linear programming problem	Real	Linear function of x_i	Linear functions of x_i	Simplex method	Dantzig
2.	Integer linear programming problem	Integers	Linear function of x_i	Linear functions of x_i	Cutting-plate method Branch-bound method Zero-one programming method	Gomory, Land and Doig, Balas
3.	Quadratic programming problem	Real	Quadratic function of x_i	Linear functions of x_i	Modified simplex method	Wolfe
4.	Nonlinear programming problem	Real	General nonlinear function of x_i	General nonlinear functions of x_i	Penalty-function methods (SUMT), Feasible-directions method	Fiacco and McCormick, Zoutendikj
5.	Geometric programming problem	Real	Posynomial in x_i	Posynomials in x_i	Method based on arithmetic-geometric inequality	Duffin, Zener and Peterson
6.	Dynamic programming problem	Real or integers	Staged nonlinear function of x_i	Staged nonlinear functions of x_i	Multistage optimization based on the principle of optimality	Bellman
7.	Calculus of variations	Real functions, $x_i(t)$ where t is some parameter	Nonlinear function of x_i in the form of an integral	Nonlinear functions of x_i in the form of integrals	Calculus of variations	Bernoulli, Euler, Lagrange and Weirstrass

solution Assuming the cost of the ith component to be a function of its reliability as $c_i = c_i\ (R_i)$, the optimization problem can be stated as follows:

$$\text{Find } \mathbf{X} = \begin{Bmatrix} R_1 \\ R_2 \\ \cdot \\ \cdot \\ \cdot \\ R_n \end{Bmatrix}$$

$$\text{which minimizes } f(\mathbf{X}) = \sum_{i=1}^{n} c_i(R_i) \tag{E1a}$$

subject to

$$R_{\text{system}}\ (R_1, R_2, \ldots, R_n) \geq R_{\text{specified}} \tag{E1b}$$

$$0 \leq R_i \leq 1; \qquad i = 1, 2, \ldots, n \tag{E1c}$$

where R_i indicates the reliability of ith component and n denotes the number of components. As seen earlier, the expression for system reliability, R_{system}, depends on the type of system. For example, for a series system,

$$R_{\text{system}}\ (t) = \prod_{i=1}^{n} R_i(t) \tag{E2}$$

If the failure rates of the components (λ_i) are constant, Eq. (E2) becomes

$$R_{\text{system}}\ (t) = \prod_{i=1}^{n} e^{-\lambda_i t} = e^{-\lambda_{\text{system}}\ t} \equiv e^{-(\lambda_1 + \lambda_2 + \cdots + \lambda_n)\ t} \tag{E3}$$

In this case, the reliability constraint, Eq. (E1b), can be expressed in equivalent form as

$$\sum_{i=1}^{n} \lambda_i - \lambda_{\text{specified}} \leq 0 \tag{E4}$$

where $\lambda_{\text{specified}}$ is the failure rate corresponding to the specified reliability of the system.

Example 11.2 Formulate a reliability-allocation problem using the maximization of system reliability as the objective function.

solution In this case, the problem can be stated as follows:

$$\text{Find } \mathbf{X} = \begin{Bmatrix} R_1 \\ R_2 \\ \cdot \\ \cdot \\ \cdot \\ R_n \end{Bmatrix}$$

which maximizes

$$f(\mathbf{X}) = R_{\text{systems}}\ (R_1, R_2, \ldots, R_n)$$

subject to

$$c_{\text{system}} = \sum_{i=1}^{n} c_i(R_i) \leq c_{\text{specified}}$$

and

$$0 \le R_i \le 1; \qquad i = 1, 2, \ldots, n$$

where c_i is the cost of the ith component and $c_{\text{specified}}$ is the maximum specified cost of the system.

Example 11.3 Formulate the reliability-allocation problem of maximizing the system reliability for the parallel-series system shown in Fig. 11.1.

solution In this case, the optimization problem can be formulated as follows. Assuming that all the components of stage i are identical ($i = 1, 2, \ldots, N$), we have $R_1, R_2, \ldots, R_N$ as design variables. Thus, the problem can be formulated as

$$\text{Find } \mathbf{X} = \begin{Bmatrix} R_1 \\ R_2 \\ \cdot \\ \cdot \\ \cdot \\ R_N \end{Bmatrix}$$

which maximizes

$$R_{\text{system}}(\mathbf{X}) = \prod_{i=1}^{N} \left[1 - (1 - R_i)^{n_i} \right]$$

subject to

$$\sum_{i=1}^{N} c_i \, n_i \le c_o$$

where c_i is the cost of one component of stage i and c_o is the maximum permissible cost of the system.

11.3.2 Structural and mechanical design problems

The following examples illustrate the formulation of reliability-based structural and mechanical design problems as optimization problems.

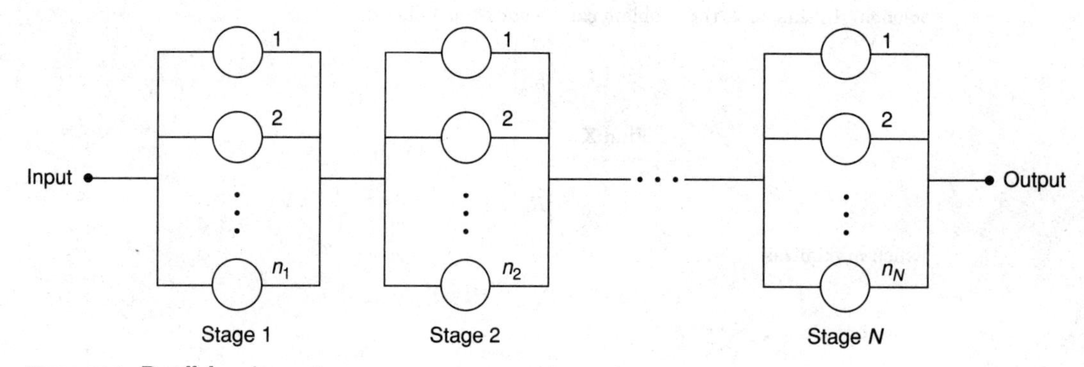

Figure 11.1 Parallel-series system.

Example 11.4 A two-bar truss, subjected to the loads F_v and F_h, is shown in Fig. 11.2. The members have a tubular section with mean diameter d and thickness t. Each member has two failure modes, namely, yielding and buckling. Formulate the problem of finding the mean values of d and t to minimize the weight of the truss subject to constraints on the reliability of the members in yielding and buckling. The data of the problem are given in the following list:

$F_v = N(50{,}000,\ 5000)$ lb

$F_h = N(30{,}000,\ 3000)$ lb

$t = N(\bar{t},\ 0.01\,\bar{t})$ in

$d = N(\bar{d},\ 0.01\,\bar{d})$ in

s_y = yield stress = $N(30{,}000,\ 3000)$ lb/in^2

E = Young's modulus = $N(3,\ 0.3) \times 10^7$ lb/in^2

$B = N(20,\ 1)$ in

$H = N(30,\ 1)$ in

$\rho = N(0.3,\ 0.03)$ lb/in ^{3}

Minimum acceptable reliability = 0.9973

Permissible ranges of $\bar{d}$ and $\bar{t}$ are 1.0 in to 5.0 in and 0.1 in to 1.0 in, respectively

solution Assuming that the stress induced in a member and the Euler buckling stress follow normal distribution, the minimum acceptable reliability in each failure mode can be seen to correspond to a value of $z_1 = -3.0$. Thus, the reliability constraints can be expressed as

$$z_1 = -\frac{\bar{s}_y - \bar{s}_c}{\sqrt{\sigma_{s_y}^2 + \sigma_{s_c}^2}} \tag{E1}$$

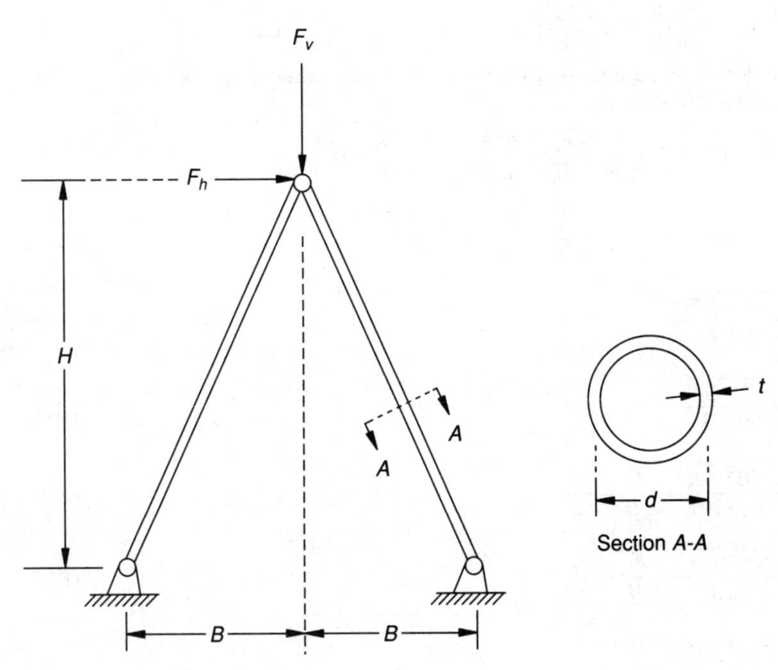

Figure 11.2 Two-bar truss

and

$$\bar{z}_2 = -\frac{\bar{s}_b - \bar{s}_c}{\sqrt{\sigma_{s_b}^2 + \sigma_{s_c}^2}} \tag{E2}$$

where s_c is the compressive stress induced in a member and s_b is the buckling stress of the member

$$s_c = \frac{\sqrt{(B^2 + H^2)}}{2\pi dt} \left(\frac{F_v}{H} + \frac{F_h}{B} \right) \tag{E3}$$

$$s_b = \frac{\pi^2 EI}{l^2 \pi dt} = \frac{\pi^2 E (d^2 + t^2)}{8 (B^2 + H^2)} \tag{E4}$$

where I = area moment of inertia and l = length of the bar. The mean values of s_c and s_b are given by

$$\bar{s}_c = \frac{\sqrt{(\bar{B}^2 + \bar{H}^2)}}{2\pi \bar{d}\bar{t}} \left(\frac{\bar{F}_v}{\bar{H}} + \frac{\bar{F}_h}{\bar{B}} \right) \tag{E5}$$

$$\bar{s}_b = \frac{\pi^2 \bar{E} (\bar{d}^2 + \bar{t}^2)}{8 (\bar{B}^2 + \bar{H}^2)} \tag{E6}$$

The standard deviations of s_c and s_b can be found using the partial derivative rule as

$$\sigma_{s_c}^2 = \left(\frac{\partial s_c}{\partial B} \right)^2 \sigma_B^2 + \left(\frac{\partial s_c}{\partial H} \right)^2 \sigma_H^2 + \left(\frac{\partial s_c}{\partial F_v} \right)^2 \sigma_{F_v}^2 + \left(\frac{\partial s_c}{\partial F_h} \right)^2 \sigma_{F_h}^2 + \left(\frac{\partial s_c}{\partial d} \right)^2 \sigma_d^2 + \left(\frac{\partial s_c}{\partial t} \right)^2 \sigma_t^2 \tag{E7}$$

$$\sigma_{s_b}^2 = \left(\frac{\partial s_b}{\partial E} \right)^2 \sigma_E^2 + \left(\frac{\partial s_b}{\partial d} \right)^2 \sigma_d^2 + \left(\frac{\partial s_b}{\partial t} \right)^2 \sigma_t^2 + \left(\frac{\partial s_b}{\partial B} \right)^2 \sigma_B^2 + \left(\frac{\partial s_b}{\partial H} \right)^2 \sigma_H^2 \tag{E8}$$

where the partial derivatives are to be evaluated at the mean values of the random variables. The partial derivatives are given by

$$\frac{\partial s_c}{\partial B} = (B^2 + H^2)^{-1/2} \frac{B}{2\pi dt} \left(\frac{F_v}{H} + \frac{F_h}{B} \right) - \frac{\sqrt{B^2 + H^2}}{2\pi dt} (F_h B^{-2}) \tag{E9}$$

$$\frac{\partial s_c}{\partial H} = (B^2 + H^2)^{-1/2} \frac{H}{2\pi dt} \left(\frac{F_v}{H} + \frac{F_h}{B} \right) - \frac{\sqrt{B^2 + H^2}}{2\pi dt} (F_v H^{-2}) \tag{E10}$$

$$\frac{\partial s_c}{\partial F_v} = \frac{\sqrt{B^2 + H^2}}{2\pi dt H} \tag{E11}$$

$$\frac{\partial s_c}{\partial F_h} = \frac{\sqrt{B^2 + H^2}}{2\pi dt B} \tag{E12}$$

$$\frac{\partial s_c}{\partial d} = -\frac{\sqrt{B^2 + H^2}}{2\pi d^2 t} \left(\frac{F_v}{H} + \frac{F_h}{B} \right) \tag{E13}$$

$$\frac{\partial s_c}{\partial t} = -\frac{\sqrt{B^2 + H^2}}{2\pi dt^2} \left(\frac{F_v}{H} + \frac{F_h}{B} \right) \tag{E14}$$

$$\frac{\partial s_b}{\partial E} = \frac{\pi^2 (d^2 + t^2)}{8 (B^2 + H^2)} \tag{E15}$$

$$\frac{\partial s_b}{\partial d} = \frac{\pi^2 E (2d)}{8(B^2 + H^2)} \tag{E16}$$

$$\frac{\partial s_b}{\partial t} = \frac{\pi^2 E (2t)}{8(B^2 + H^2)} \tag{E17}$$

$$\frac{\partial s_b}{\partial B} = -\frac{\pi^2 E (d^2 + t^2)}{64(B^2 + H^2)^2}(2B) \tag{E18}$$

$$\frac{\partial s_b}{\partial H} = -\frac{\pi^2 E (d^2 + t^2)}{64(B^2 + H^2)^2}(2H) \tag{E19}$$

The weight of the truss (W) is given by

$$W = 2\rho\pi dt\sqrt{B^2 + H^2} \tag{E20}$$

By treating the mean values of d and t as design variables and the mean value of the weight as the objective function, the optimization problem can be stated as follows:

$$\text{Find } \mathbf{X} = \begin{Bmatrix} x_1 \\ x_2 \end{Bmatrix} = \begin{Bmatrix} \overline{d} \\ \overline{t} \end{Bmatrix}$$

which minimizes

$$f(\mathbf{X}) = \overline{W} = 2\overline{\rho}\pi\overline{d}\overline{t}\sqrt{\overline{B}^2 + \overline{H}^2}$$

subject to

$$z_1 \le -\frac{\overline{s}_y - \overline{s}_c}{\sqrt{\sigma_{s_y}^2 + \sigma_{s_c}^2}} \tag{E21}$$

$$z_1 \le -\frac{\overline{s}_b - \overline{s}_c}{\sqrt{\sigma_{s_b}^2 + \sigma_{s_c}^2}}$$

$$0.1 \le \overline{t} \le 1.0$$

$$1.0 \le \overline{d} \le 5.0$$

Example 11.5 The single-shoe brake, shown in Fig. 11.3, is to be designed to have a braking capacity (torque) of $T_c \pm \Delta T_c$ = (350 ± 35) lb·in. The applied force F, the coefficient of friction between the brake drum and the brake shoe f, the dimensions a, b, c, and r are known to follow normal distributions as $F = N(50, 5)$ lb, $f = N(0.3, 0.03)$, $a = N(10, 1)$ in, $b = N(\overline{b}, \sigma_b)$ in, $c = N(20, 2)$ in and $r = N(\overline{r}, \sigma_r)$ in. Find the mean values of b and r to maximize the reliability without exceeding a mean pressure of 120 lb/in^2 on the brake pad. The brake-shoe has a contact area of $\overline{A} \pm \Delta A = 4 \pm 0.4$ in^2. Assume that the coefficient of variation of b and r is 0.1 and the tolerances given for the braking capacity and the brake-shoe area correspond to 3σ values. Formulate the optimization problem including the following constraints:

$$\frac{\overline{a}}{2} \le \overline{r} \le \overline{a} \le \overline{b} \le \overline{c}$$

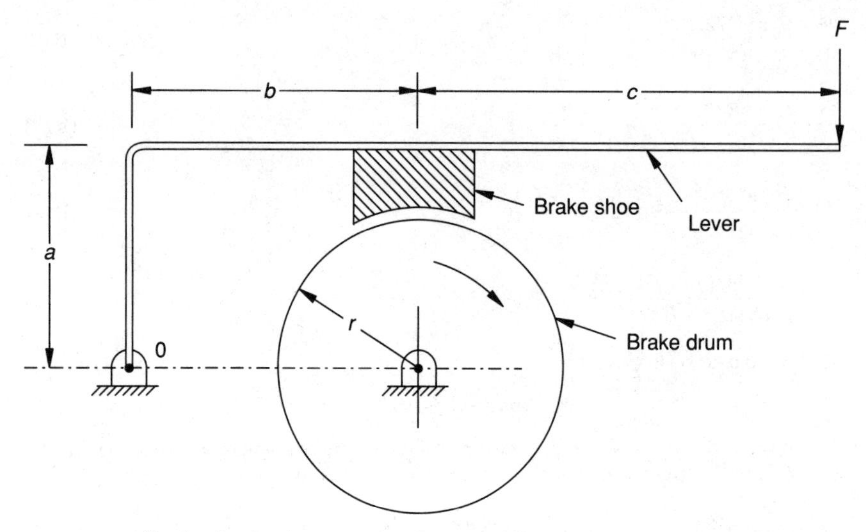

Figure 11.3 Single-shoe brake.

solution The free-body diagram of the lever and the brake shoe is shown in Fig. 11.4. Taking moments about the pivot point of the lever, O, we obtain

$$F(b+c)+rfN-bN=0 \quad \text{i.e.;} \qquad N=\frac{F(b+c)}{(b-fr)} \tag{E1}$$

where N and fN denote, respectively, the normal and tangential forces acting on the brake shoe. The torque acting on the brake drum (T_b) is given by

$$T_b=fNr=\frac{frF(b-c)}{(b-fr)} \tag{E2}$$

Assuming normal distribution for all the quantities, the reliability of the brake can be expressed as (by substituting T_b for load L and T_c for strength S in Eq. 7.23)

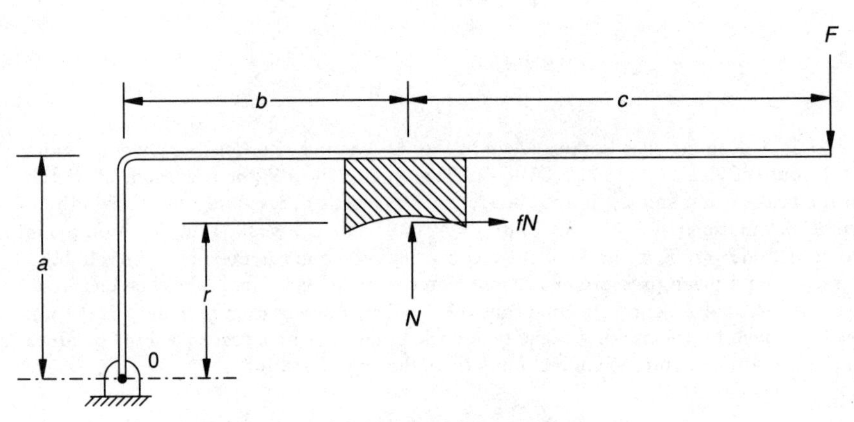

Figure 11.4 Free-body diagram of the lever and the brake shoe.

$$z_1 = \frac{\overline{T}_b - \overline{T}_c}{\sqrt{\sigma_{T_b}^2 + \sigma_{T_c}^2}} \tag{E3}$$

The mean values of T_c and T_b are given by $\overline{T}_c = 350$ lb·in and

$$\overline{T}_b = \frac{\overline{f}\,\overline{r}\,\overline{F}\,(\overline{b}+\overline{c})}{(\overline{b}-\overline{f}\,\overline{r})} = \frac{0.3\,\overline{r}\,(50)(\overline{b}+20)}{(\overline{b}-0.3\,\overline{r})} = \frac{15\,\overline{r}\,(\overline{b}+20)}{(\overline{b}-0.3\overline{r})} \tag{E4}$$

The standard deviation of T_c is $\sigma_{T_c} = 35/3 = 11.67$ lb · in and σ_{T_b} can be found using the partial derivative rule as

$$\sigma_{T_b}^2 = \left(\frac{\partial T_b}{\partial f}\right)^2 \sigma_f^2 + \left(\frac{\partial T_b}{\partial r}\right)^2 \sigma_r^2 + \left(\frac{\partial T_b}{\partial F}\right)^2 \sigma_F^2 + \left(\frac{\partial T_b}{\partial b}\right)^2 \sigma_b^2 + \left(\frac{\partial T_b}{\partial c}\right)^2 \sigma_c^2 \tag{E5}$$

where the partial derivatives, evaluated at the mean values of the random variables, give

$$\frac{\partial T_b}{\partial f} = \frac{rF\,(b+c)}{(b-fr)} - \frac{frF\,(b+c)}{(b-fr)^2}(-r) = \frac{50\,\overline{r}\,(\overline{b}+20)}{(\overline{b}-0.3\overline{r})} + \frac{15\,\overline{r}^2\,(\overline{b}+20)}{(\overline{b}-0.3\overline{r})^2} \tag{E6}$$

$$\frac{\partial T_b}{\partial r} = \frac{fF\,(B+c)}{(b-fr)} - \frac{frF\,(B+c)}{(b-fr)^2}(-f) = \frac{15\,(\overline{b}+20)}{(\overline{b}-0.3\overline{r})} + \frac{4.5\,\overline{r}\,(\overline{b}+20)}{(\overline{b}-0.3\overline{r})^2} \tag{E7}$$

$$\frac{\partial T_b}{\partial F} = \frac{fr\,(b+c)}{(b-fr)} = \frac{0.3\,\overline{r}\,(\overline{b}+20)}{(\overline{b}-0.3\overline{r})} \tag{E8}$$

$$\frac{\partial T_b}{\partial b} = \frac{frF}{(b-fr)} - \frac{frF\,(b+c)}{(b-fr)^2} = \frac{15\,\overline{r}}{(\overline{b}-0.3\overline{r})} - \frac{15\,\overline{r}\,(\overline{b}+20)}{(\overline{b}-0.3\overline{r})^2} \tag{E9}$$

$$\frac{\partial T_b}{\partial c} = \frac{frF}{(b-fr)} = \frac{15\,\overline{r}}{(\overline{b}-0.3\overline{r})} \tag{E10}$$

Substitution of Eqs. (E6) to (E10) into (E5), along with $\sigma_f = 0.03$, $\sigma_r = 0.1\overline{r}$, $\sigma_F = 5$, $\sigma_b = 0.1\overline{b}$, and $\sigma_c = 2.0$, gives

$$\sigma_{T_b}^2 = 2\left(\frac{1.5\,\overline{r}\,(\overline{b}+20)}{(\overline{b}-0.3\overline{r})} + \frac{0.45\,\overline{r}^2\,(\overline{b}+20)}{(\overline{b}-0.3\overline{r})^2}\right)^2 + \left(\frac{1.5\,\overline{r}\,(\overline{b}+20)}{(\overline{b}-0.3\overline{r})}\right)^2 + \left(\frac{1.5\,\overline{r}\,\overline{b}}{(\overline{b}-0.3\overline{r})} - \frac{1.5\,\overline{r}\,\overline{b}\,(\overline{b}+20)}{(\overline{b}-0.3\overline{r})}\right)^2 + \left(\frac{30\,\overline{r}}{\overline{b}-0.3\overline{r}}\right)^2 \tag{E11}$$

The pressure on the brake pad (p) is given by

$$p = \frac{N}{A} = \frac{F(b + c)}{(b - fr)A} \tag{E12}$$

where A is the area of the brake shoe. Hence, the mean value of pressure is given by

$$\bar{p} = \frac{\bar{F}(\bar{b} + \bar{c})}{(\bar{b} - \bar{f}\bar{r})\bar{A}} = \frac{50(\bar{b} + 20)}{(\bar{b} - 0.3\bar{r})} \tag{E13}$$

The optimization problem can be stated as follows:

Find $\mathbf{X}$ which minimizes

$$f(\mathbf{X}) = z_1 = \frac{\bar{T}_b - 350}{\sqrt{\sigma_{T_b}^2 + 136.1111}}$$

subject to

$$\begin{aligned} \frac{50(\bar{b} + 20)}{(\bar{b} - 0.3\bar{r})} - 120 &\le 0 \\ 5 - \bar{r} &\le 0 \\ \bar{r} - 10 &\le 0 \\ 10 - \bar{b} &\le 0 \\ \bar{b} - 20 &\le 0 \end{aligned} \tag{E14}$$

where $\bar{T}_b$ and σ_{T_b} are given by Eqs. (E4) and (E11), respectively.

11.4 Solution Techniques

11.4.1 Graphical-optimization method

Consider the optimization problem

Find $\mathbf{X}$ which minimizes $f(\mathbf{X})$

subject to the constraints

$$g_j(\mathbf{X}) \le 0, \quad j = 1, 2, \ldots, m \tag{11.2}$$

If each of the n design variables, x_i, is represented along a coordinate axis, the resulting n-dimensional cartesian space is known as the *design space*. Each point in this n-dimensional design space is called a design point and represents either a possible (feasible) or an impossible (infeasible) solution to the design problem. A *feasible design* satisfies all the constraints, while an *infeasible design* violates one or more of the constraints. The feasible and infeasible design points are separated by a surface, known as the *constraint surface*. To understand the concept of *constraint surface*, consider the inequality constraint $g_j(\mathbf{X}) \le 0$. The set of design points $\mathbf{X}$ that satisfy the equation

$g_j(\mathbf{X}) = 0$ forms a hypersurface in the design space and is called a constraint surface. The constraint surface divides the design space into two parts; one in which $g_j(\mathbf{X}) < 0$, and the other in which $g_j(\mathbf{X}) > 0$. Thus the points lying on the hypersurface will satisfy the constraint $g_j(\mathbf{X})$ critically, whereas the points lying in the region where $g_j(\mathbf{X}) > 0$ are infeasible or unacceptable, and the points lying in the region where $g_j(\mathbf{X}) < 0$ are feasible or acceptable. When the constraint surfaces corresponding to all the constraints $g_j(\mathbf{X}) \le 0$, $j = 1, 2, \ldots, m$ are considered, they collectively divide the design space into two regions; one called the feasible space in which every design point $\mathbf{X}$ corresponds to a value of $g_j(\mathbf{X})$ less than or equal to zero, and the other called the infeasible space in which every design point $\mathbf{X}$ gives a value of $g_j(\mathbf{X})$ greater than zero for some j.

When the objective function is plotted with $f(\mathbf{X}) = c =$ constant, the resulting graph forms a hypersurface in the design space. If several graphs are plotted with different values of the constant c, the resulting surfaces are known as the *objective-function surfaces.* Once the objective-function surfaces are known, the optimum solution of the problem can be identified as the design point in the feasible space that gives the minimum value of the objective function.

The constraint surfaces, objective-function surfaces, along with the optimum solution of a typical two-variable optimization problem, are shown in Fig. 11.5. It can be noted that as the number of design variables exceeds two or three, the constraints and objective function surfaces become complex even for visualization and the problem has to be solved purely as a mathematical problem. Example 11.6 illustrates the graphical optimization-procedure.

Example 11.6 Two components in series are shown in Fig. 11.6. The cost of the first component is given by $100R_1^2$ where R_1 denotes its reliability. Similarly, the cost of the second component is given by $200R_2^2$ where R_2 represents its reliability. Find the reliabilities of the components to minimize the total cost. The system reliability as well as the individual component reliabilities are required to be at least 0.7.

solution The design variables are the reliabilities of the components so that

$$\mathbf{X} = \begin{Bmatrix} x_1 \\ x_2 \end{Bmatrix} = \begin{Bmatrix} R_1 \\ R_2 \end{Bmatrix}$$

The objective is to minimize the total cost

$$f(\mathbf{X}) = 100\,R_1^2 + 200\,R_2^2$$

The reliability constraints can be stated as

$$R_1 R_2 \ge 0.7 \quad \text{i.e.;} \quad g_1(\mathbf{X}) = 0.7 - R_1 R_2 \le 0$$

$$0.7 \le R_1 \le 1.0 \quad \text{i.e.;} \quad g_2(\mathbf{X}) = 0.7 - R_1 \le 0 \quad \text{and} \quad g_3(\mathbf{X}) = R_1 - 1.0 \le 0$$

$$0.7 \le R_2 \le 1.0 \quad \text{i.e.;} \quad g_4(\mathbf{X}) = 0.7 - R_2 \le 0 \quad \text{and} \quad g_5(\mathbf{X}) = R_2 - 1.0 \le 0$$

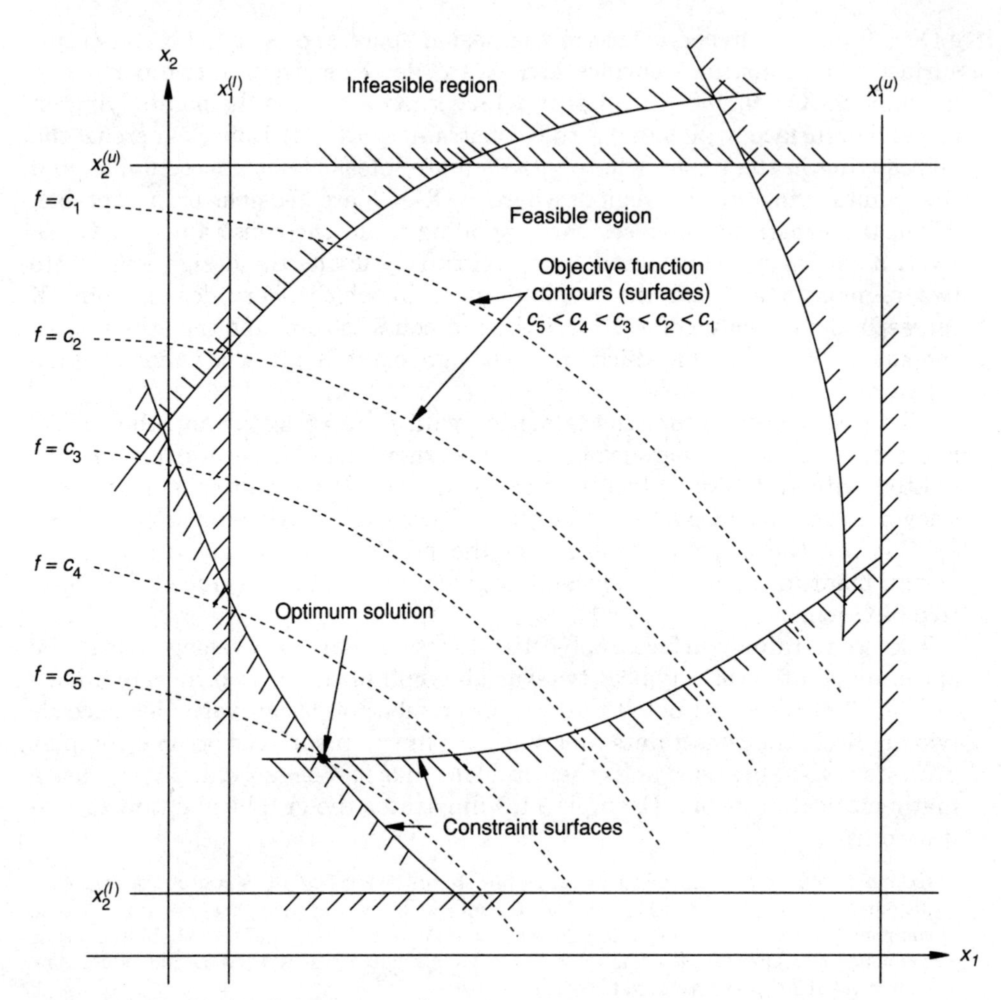

Figure 11.5 Graphical optimization procedure.

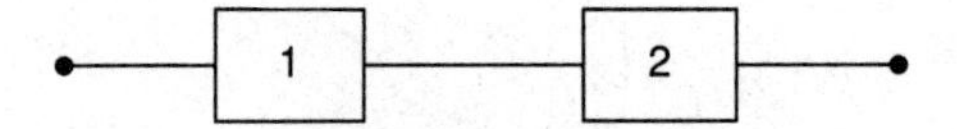

Figure 11.6 Two components in series.

The lower and upper bounds (i.e., the constraint boundaries of g_2 to g_5) are shown in Fig. 11.7. To plot the overall reliability constraint boundary, the equation $g_1 = R_1R_2 - 0.7 = 0$ is to be plotted. For this, a series of values are given to R_1 and the corresponding values of R_2 are obtained as $R_2 = 0.7 / R_1$

R_1	0.6	0.65	0.7	0.75	0.8	0.85	0.9	0.95	1.0
R_2	1.167	1.077	1.000	0.933	0.875	0.824	0.778	0.737	0.700

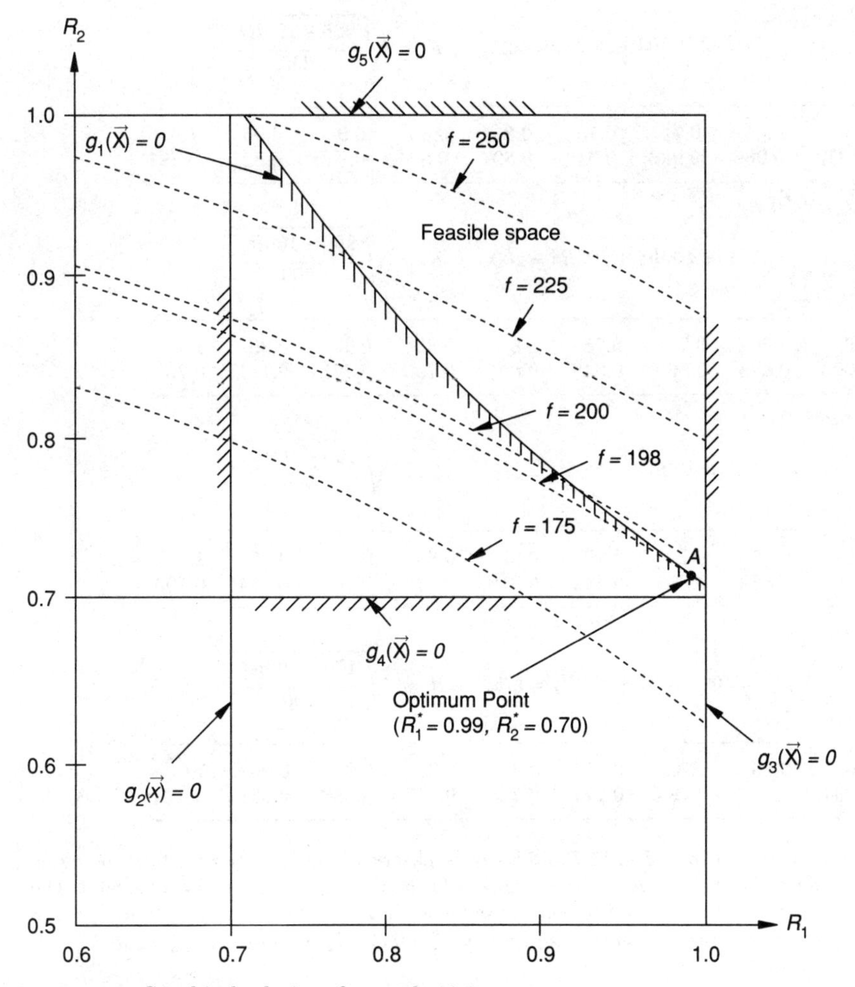

Figure 11.7 Graphical solution of example 11.6.

Next, the contours of the objective function,

$$f(\mathbf{X}) = 100R_1^2 + 200R_2^2 = c$$

are to be plotted for different values of c. The points on the contours are found as indicated.

$$\text{For } 100R_1^2 + 200R_2^2 = 250; \qquad R_2 = \sqrt{\frac{250 - 100R_1^2}{200}}$$

R_1	0.6	0.65	0.7	0.75	0.8	0.85	0.9	0.95	1.0
R_2	1.034	1.019	1.003	0.984	0.964	0.943	0.919	0.894	0.866

$$\text{For } 100R_1^2 + 200R_2^2 = 225; \qquad R_2 = \sqrt{\frac{225 - 100R_1^2}{200}}$$

R_1	0.6	0.65	0.7	0.75	0.8	0.85	0.9	0.95	1.0
R_2	0.972	0.956	0.938	0.919	0.897	0.874	0.849	0.821	0.791

$$\text{For } 100R_1^2 + 200R_2^2 = 200; \qquad R_2 = \sqrt{\frac{200 - 100R_1^2}{200}}$$

R_1	0.6	0.65	0.7	0.75	0.8	0.85	0.9	0.95	1.0
R_2	0.906	0.888	0.869	0.848	0.825	0.799	0.771	0.741	0.707

$$\text{For } 100R_1^2 + 200R_2^2 = 198; \qquad R_2 = \sqrt{\frac{198 - 100R_1^2}{200}}$$

R_1	0.6	0.65	0.7	0.75	0.8	0.85	0.9	0.95	1.0
R_2	0.900	0.882	0.863	0.842	0.819	0.793	0.765	0.734	0.700

$$\text{For } 100R_1^2 + 200R_2^2 = 175; \qquad R_2 = \sqrt{\frac{175 - 100R_1^2}{200}}$$

R_1	0.6	0.65	0.7	0.75	0.8	0.85	0.9	0.95	1.0
R_2	0.834	0.815	0.794	0.771	0.745	0.717	0.686	0.651	0.612

These contours are shown in Fig. 11.7 and it can be observed that the objective-function value cannot be reduced below a value of 198 without violating the overall reliability constraint. The optimum point can be seen to be the point of tangency between the curves $g_1 = 0$ and $f = 198$ (point A). Thus, the optimum solution is given by $R_1^* = 0.99$, $R_2^* = 0.70$ with $f_{\min} = 198$.

11.4.2 Lagrange multiplier method

The Lagrange multiplier method is an analytical method that can be used to find the minimum of a multivariable function in the presence of equality constraints. Thus, the problem to be solved can be stated as

$$\text{Find } \mathbf{X} = \begin{Bmatrix} x_1 \\ x_2 \\ \cdot \\ \cdot \\ \cdot \\ x_n \end{Bmatrix}$$

which minimizes

$$f(\mathbf{X})$$

subject to

$$l_j(\mathbf{X}) = 0; \qquad j = 1, 2, \ldots, p \tag{11.3}$$

The solution procedure involves construction of a new function, called the Lagrange function, $L(\mathbf{X}, \boldsymbol{\lambda})$, as

$$L(\mathbf{X}, \boldsymbol{\lambda}) = f(\mathbf{X}) + \sum_{j=1}^{p} \lambda_j \, l_j(\mathbf{X}) \tag{11.4}$$

where λ_j are unknown constants, called the Lagrange multipliers. Notice that there are as many Lagrange multipliers as there are constraints. It has been proved [11.1] that the solution of the problem stated in Eq. (11.3) can be obtained by finding the unconstrained minimum of the function $L(\mathbf{X}, \boldsymbol{\lambda})$. The necessary conditions for the minimum of $L(\mathbf{X}, \boldsymbol{\lambda})$ are

$$\frac{\partial L}{\partial x_i}(\mathbf{X}, \boldsymbol{\lambda}) = 0; \qquad i = 1, 2, \ldots, n$$

$$\frac{\partial L}{\partial \lambda_j}(\mathbf{X}, \boldsymbol{\lambda}) = 0; \qquad j = 1, 2, \ldots, p \tag{11.5}$$

The number of unknowns is $n + p$ (n design variables x_i and p Lagrange multipliers λ_j) and the number of available equations, Eq. (11.5), is also $n + p$. Thus the solution of the $(n + p)$ simultaneous nonlinear equations, Eq. (11.5), gives $\mathbf{X}^*$ and $\boldsymbol{\lambda}^*$. Although this method appears to be simple and straightforward, the solution of the $(n + p)$ simultaneous nonlinear equations, Eq. (11.5), may be tedious. Further, if the functions f and l_j are not available in explicit form in terms of x_i, it will be extremely difficult to solve Eq. (11.5). Hence, a suitable numerical method of optimization, such as the penalty function method discussed in the next section, is to be used for the solution of the optimization problem. The Lagrange multiplier method is illustrated with Examples 11.7 and 11.8.

Example 11.7 Two components are in parallel as shown in Fig. 11.8. It is desired to find the component reliabilities R_1 and R_2 to minimize the total cost of the components while achieving a system reliability of 0.7. Assume that the costs of the components 1 and 2 are given by $100R_1$ and $200R_2$, respectively.

solution The problem can be stated as follows:

$$\text{Find } \mathbf{X} = \begin{Bmatrix} x_1 \\ x_2 \end{Bmatrix} = \begin{Bmatrix} R_1 \\ R_2 \end{Bmatrix}$$

to minimize $f(\mathbf{X}) = 100R_1 + 200R_2$
subject to

$$l(\mathbf{X}) = R_1 + R_2 - R_1R_2 - 0.7 = 0$$

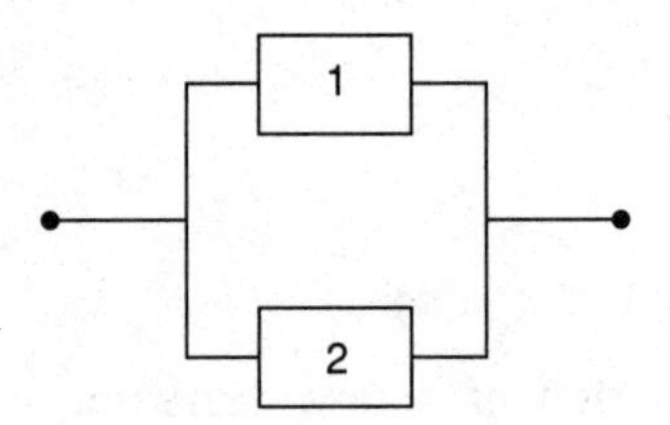

Figure 11.8 Two components in parallel.

The Lagrange function can be formulated as

$$L = 100R_1 + 200R_2 + \lambda(R_1 + R_2 - R_1R_2 - 0.7) \tag{E1}$$

where λ is the Lagrange multiplier. The necessary conditions of optimality give

$$\frac{\partial L}{\partial R_1} = 100 + \lambda\,(1 - R_2) = 0 \quad \text{i.e.;} \quad \lambda = -\frac{100}{1 - R_2} \tag{E2}$$

$$\frac{\partial L}{\partial R_2} = 200 + \lambda\,(1 - R_1) = 0 \quad \text{i.e.;} \quad \lambda = -\frac{200}{1 - R_1} \tag{E3}$$

$$\frac{\partial L}{\partial \lambda} = R_1 + R_2 - R_1R_2 - 0.7 = 0 \tag{E4}$$

Equating (E2) and (E3), we obtain

$$\lambda = \frac{1}{1 - R_2} = \frac{1}{1 - R_1} \quad \text{i.e.;} \quad R_1 = 2R_2 - 1 \tag{E5}$$

Substitution of Eq. (E5) into (E4) gives

$$(2R_2 - 1) + R_2 - R_2(2R_2 - 1) = 0.7 \quad \text{i.e.;} \quad 2R_2^2 - 4R_2 + 1.7 = 0 \tag{E6}$$

The solution of Eq. (E6) gives

$$R_2 = \frac{4 \pm \sqrt{16 - 13.6}}{4} = 0.6127; \quad 1.3873 \tag{E7}$$

Thus, the physically realizable optimum solution is given by $R_1^* = 0.2254$ and $R_2^* = 0.6127$ with $f_{\min} = 145.08$.

Example 11.8 Find the component failure rates λ_i, $i = 1, 2, \ldots, n$ that minimize the cost (objective function) $f = \sum_{i=1}^{n} f_i(\lambda_i)$ subject to the constraint $\sum_{i=1}^{n} \lambda_i = \lambda_0$. Assume the cost functions to be proportional to the component MTBF, that is, $f_i(\lambda_i) = a_i / \lambda_i$ where a_i is a constant.

solution The Lagrangian function can be constructed as

$$L(\lambda_1, \ldots, \lambda_n; \alpha) = \sum_{i=1}^{n} \frac{a_i}{\lambda_i} + \alpha \left(\sum_{i=1}^{n} \lambda_i - \lambda_0 \right) \tag{E1}$$

where α is the Lagrange multiplier. The necessary conditions for the extremum of L give

$$\frac{\partial L}{\partial \lambda_i} = -\frac{a_i}{\lambda_i^2} + \alpha = 0; \qquad i = 1, 2, \ldots, n$$

that is,

$$\lambda_i = \left(\frac{a_i}{\alpha}\right)^{1/2}; \qquad i = 1, 2, \ldots, n \tag{E2}$$

$$\frac{\partial L}{\partial \alpha} = \sum_{i=1}^{n} \lambda_i - \lambda_0 = 0$$

that is,

$$\sum_{i=1}^{n} \left(\frac{a_i}{\alpha}\right)^{1/2} = \lambda_0$$

that is,

$$\alpha^{1/2} = \frac{1}{\lambda_0} \sum_{i=1}^{n} \sqrt{a_i} \tag{E3}$$

Thus, the optimum values of the design variables are given by

$$\lambda_i = \frac{\lambda_0 \sqrt{a_i}}{\sum_{j=1}^{n} \sqrt{a_j}}; \qquad i = 1, 2, \ldots, n \tag{E4}$$

and the optimum value of the objective function by

$$f_{\min} = \sum_{i=1}^{n} \frac{a_i}{\lambda_i} = \frac{1}{\lambda_0} \left(\sum_{i=1}^{n} \sqrt{a_i}\right)^2$$

11.4.3 Penalty function method (SUMT)

In this method, a new function, $\phi_k(\mathbf{X}, r_k)$, is defined as

$$\phi_k(\mathbf{X}, r_k) = f(\mathbf{X}) + r_k \sum_{j=1}^{m} \frac{1}{g_j(\mathbf{X})} - \frac{1}{\sqrt{r_k}} \sum_{j=1}^{p} l_j^2(\mathbf{X}) \tag{11.6}$$

where r_k is a positive constant, known as the penalty parameter. It has been proved [11.1] that the optimum solution of the problem stated in Eq. (11.6) can be obtained by minimizing the function ϕ_k for a sequence of decreasing values of r_k. The value of r_k is initially chosen as a small positive value, say 1.0, and then decreased toward zero after each unconstrained minimization. The starting point for minimization is assumed to be feasible with respect to the inequality constraints. The iterative procedure involved can be summarized as follows:

1. Start with an initial feasible point $\mathbf{X}_1$ satisfying all the inequality constraints with strict inequality sign, that is, $g_j(\mathbf{X}_1) < 0$ for $j = 1, 2, \ldots, m$, and an initial value of $r_1 > 0$. Set $k = 1$.
2. Minimize $\phi_k(\mathbf{X}, r_k)$ by using an unconstrained minimization method (described below) and obtain the solution $\mathbf{X}_k^*$.

3. Test whether $\mathbf{X}_k^*$ is the optimum solution of the original problem of Eq. (11.6). If $\mathbf{X}_k^*$ is found to be optimum, terminate the process; otherwise, go to the next step.
4. Find the value of the next penalty parameter, r_{k+1}, as $r_{k+1} = c \cdot r_k$ where $c < 1$ (usually taken as 0.1).
5. Take the new starting point as $\mathbf{X}_1 = \mathbf{X}_k^*$, set the new value of $k = k + 1$, and go to step (2).

The convergence of the procedure as r_k tends to zero is illustrated graphically in Fig. 11.9 for the simple problem

$$\text{Find } \mathbf{X} = \{x_1\}$$

which minimizes

$$f(\mathbf{X}) = ax_1$$

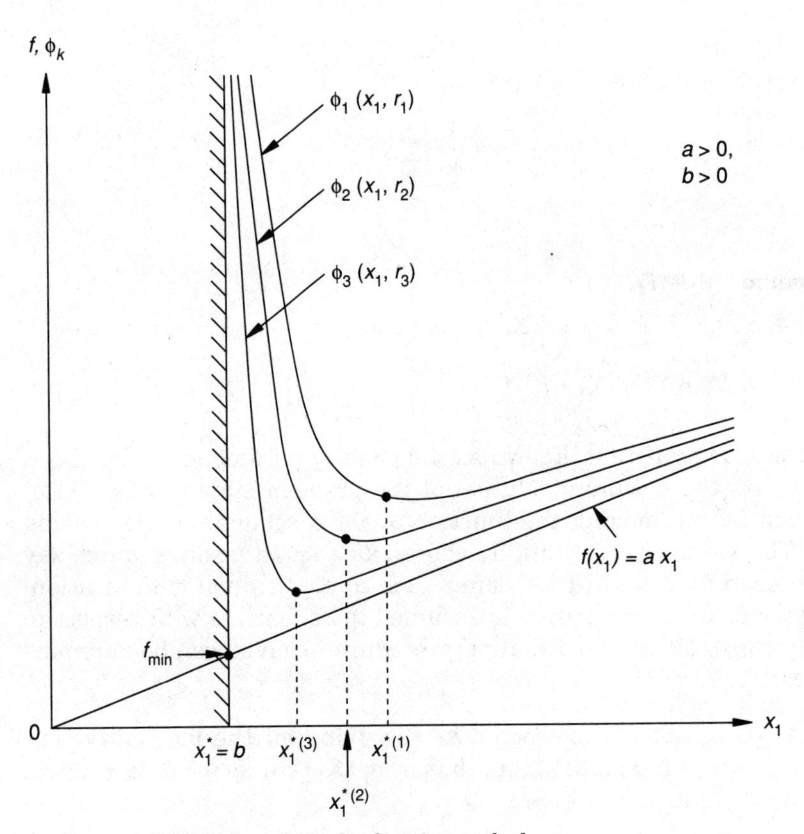

Figure 11.9 Convergence of penalty function method.

subject to

$$g_1(\mathbf{X}) = b - x_1 \le 0 \tag{11.7}$$

Unconstrained minimization. It can be seen that the penalty function method requires the solution of an unconstrained minimization problem for each value of the penalty parameter r_k. Although several unconstrained minimization methods have been developed in the literature [11.1], we shall consider one of the popular and efficient methods, known as the Davidon-Fletcher-Powell (DFP) method, in this section. For convenience of notation, we state the problem to be solved as follows:

$$\text{Find } \mathbf{X} = \begin{Bmatrix} x_1 \\ x_2 \\ \cdot \\ \cdot \\ \cdot \\ x_n \end{Bmatrix} \tag{11.8}$$

which minimizes $f(\mathbf{X})$.

The solution procedure can be summarized through the following steps:

1. Start with any initial vector $\mathbf{X}_1$ and an initial positive definite-symmetric matrix $[H_1]$. Usually $[H_1]$ is taken as the identity matrix, [I].
2. Find the first search direction $\mathbf{S}_1$ as

$$\mathbf{S}_1 = -[H_1]\nabla f_1 \tag{11.9}$$

   where $\nabla f_1 = \nabla f(\mathbf{X}_1)$ is the gradient vector evaluated at $\mathbf{X}_1$. Set the iteration number as $i = 1$.
3. Compute the new design vector

$$\mathbf{X}_{i+1} = \mathbf{X}_i + \alpha_i^* \mathbf{S}_i \tag{11.10}$$

   where α_i^* is the minimizing step length along the search direction $\mathbf{S}_i$, that is, α_i^* is chosen so that $f(\mathbf{X}_i + \alpha_i \mathbf{S}_i) = f(\alpha_i)$ attains a minimum at $\alpha_i = \alpha_i^*$. A method of finding α_i^* is described in the next section.
4. Test for convergence of the process. If the magnitude of the gradient of the function at $\mathbf{X}_{i+1}$ is zero, that is, if $||\nabla f(\mathbf{X}_{i+1})|| \approx 0$, set $\mathbf{X}^* = \mathbf{X}_{\text{opt}} \approx \mathbf{X}_{i+1}$ and stop the process. Otherwise, go to the next step.
5. Compute the new matrix $[H_{i+1}]$ as

$$[H_{i+1}] = [H_i] + [M_i] + [N_i] \tag{11.11}$$

where

$$[M_i] = \alpha_i^* \frac{\mathbf{S}_i \, \mathbf{S}_i^T}{\mathbf{S}_i^T \, \mathbf{Y}_i} \tag{11.12}$$

$$[N_i] = -\frac{([H_i]\,\mathbf{Y}_i)\,([H_i]\,\mathbf{Y}_i)^T}{\mathbf{Y}_i^T\,[H_i]\,\mathbf{Y}_i} \tag{11.13}$$

and

$$\mathbf{Y}_i = \nabla f(\mathbf{X}_{i+1}) - \nabla f(\mathbf{X}_i) \tag{11.14}$$

6. Compute the new search direction as

$$\mathbf{S}_{i+1} = -\,[H_{i+1}]\,\nabla f_{i+1} \tag{11.15}$$

7. Set the new iteration number as $i = i + 1$, and repeat step (3) onwards until the convergence in step (4) is satisfied.

One-dimensional minimization. The unconstrained-minimization method just presented requires the computation of α_i^* in finding an improved vector $\mathbf{X}_{i+1}$ (Eq. 11.10). Since only one variable α_i^* is to be found by minimizing the function $f(\mathbf{X}_i + \alpha_i\,\mathbf{S}_i) = f(\boldsymbol{\alpha}_i)$, the problem is called a one-dimensional minimization problem. One of the popular methods, known as the cubic-interpretation method, is considered in this section. This method makes use of the derivative of f with respect to α, which can be computed as (by neglecting the subscript i for simplicity)

$$f'\Big|_{\alpha} = \frac{df}{d\alpha}\Big|_{\alpha} = \frac{d}{d\alpha} f(\mathbf{X} + \alpha\,\mathbf{S}) = \mathbf{S}^T\,\nabla f(\mathbf{X} + \alpha\,\mathbf{S}) \tag{11.16}$$

In this method, the lower and upper bounds on the optimal step length, α^*, are determined first. Always $A = \alpha = 0$ can be taken as a lower bound on α^*. At $A = \alpha = 0$,

$$f'\Big|_{\alpha=0} = \mathbf{S}^T\,\nabla f(\mathbf{X}) < 0 \tag{11.17}$$

because $\mathbf{S}$ is assumed to be a descent direction. The upper bound on α^*, B, is taken as the first value among 1, 2, 4, 8, . . . , at which $f'\,|_B = \mathbf{S}^T\,\nabla f\,(\mathbf{X} + B\,\mathbf{S})$ is nonnegative. Since f' changes sign between $\alpha = A$ and $\alpha = B$, the minimum of f lies between A and B.

Since the exact functional relation between f and α is not known in most practical problems, we approximate $f(\alpha)$ by a cubic equation $h(\alpha) = a + b\alpha + c\alpha^2 + d\alpha^3$ (a, b, c, and d are unknown constants) between $\alpha = A$ and $\alpha = B$. This is the reason the method is called the cubic interpretation method. By using the relations $h(\alpha = A) = f(\alpha = A)$, $h'(\alpha = A) = f'(\alpha = A)$, $h(\alpha = B) = f(\alpha = B)$ and $h'(\alpha = B) = f'(\alpha = B)$, the constants a, b, c, and d can be determined. Then the minimum of $h(\alpha)$, obtained by solving the equation,

$$h'(\underset{\sim}{\alpha}^*) = b + c\,\underset{\sim}{\alpha}^* + d\,\underset{\sim}{\alpha}^{*^2} = 0, \tag{11.18}$$

is tested for convergence to see whether $\tilde{\alpha}^*$ can be used as an approximation for the minimum of $f(\alpha)$, that is, $\alpha^* \approx \tilde{\alpha}^*$. If the convergence test is satisfied, the procedure is terminated by taking $\alpha^* = \tilde{\alpha}^*$. Otherwise, the best two points out of $\alpha = A$, $\alpha = B$ and $\alpha = \tilde{\alpha}^*$ are used to fit another cubic equation. This procedure is continued until convergence is achieved.

Example 11.9 Determine the component reliabilities of the system shown in Fig. 11.10 to maximize the system reliability subject to an upper bound on the total cost of the system. The cost of component $i(c_i)$ is given by $c_i = k_i\,(R_i)^{\alpha_i}$ where R_i = reliability of component i, $k_1 = k_3 = 50$, $k_2 = 200$, $k_4 = 75$ and $\alpha_i = 0.5$ (i = 1, 2, 3, 4). The maximum permissible cost of the system is 550.

solution The system reliability (R_s) can be found as $R_s = 1 - P_{fs}$ where P_{fs} is the probability of failure of the system given by

$$P_{fs} = P_{fs}\Bigg|_{\substack{\text{component 2}\\ \text{operating}}} \cdot P\begin{pmatrix}\text{component 2}\\ \text{operating}\end{pmatrix} + P_{fs}\Bigg|_{\substack{\text{component 2}\\ \text{failed}}} \cdot P\begin{pmatrix}\text{component 2}\\ \text{failed}\end{pmatrix}$$

$$= \{(1 - R_3)(1 - R_4)\}^2 \cdot R_2 + \{1 - R_1\,[1 - (1 - R_3)(1 - R_4)]\}^2 \cdot (1 - R_2) \tag{E1}$$

The optimization problem can be stated as follows:

$$\text{Find } \mathbf{x} = \begin{Bmatrix} R_1 \\ R_2 \\ R_3 \\ R_4 \end{Bmatrix}$$

which maximizes

$$R_s = 1 - P_{fs} = 1 - R_2\,\{(1 - R_3)(1 - R_4)\}^2 - (1 - R_2)\,\{1 - R_1\,[1 - (1 - R_3)(1 - R_4)]\}^2 \tag{E2}$$

subject to

$$2c_1 + c_2 + 2c_3 + 2c_4 \le c_{\max}$$

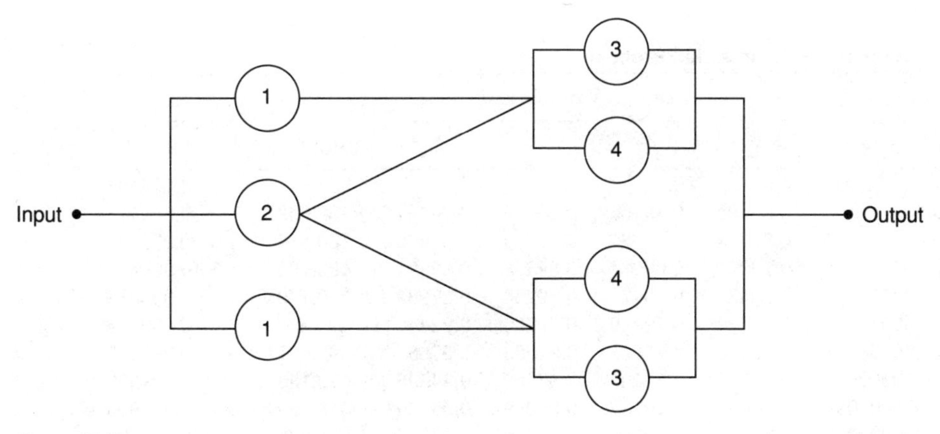

Figure 11.10

that is,

$$100\, R_1^{0.5} + 200\, R_2^{0.5} + 100\, R_3^{0.5} + 150\, R_4^{0.5} - 550 \le 0 \tag{E3}$$

and

$$0 \le R_i \le 1; \qquad i = 1, 2, 3, 4 \tag{E4}$$

This problem has been solved using the interior-penalty function method and the results are shown in Table 11.2. The optimum solution can be seen to be

$$\mathbf{x}^* = \begin{Bmatrix} R_1^* \\ R_2^* \\ R_3^* \\ R_4^* \end{Bmatrix} = \begin{Bmatrix} 0.99538 \\ 0.70189 \\ 0.99682 \\ 0.95339 \end{Bmatrix}$$

$$\mathbf{R}_s \Big|_{\max} = 0.999993$$

and total cost = 513.63.

Example 11.10 Optimum Design of a Machine Tool Gear Train The problem of the design of the nine-speed machine tool gear train shown in Fig. 9.3 is to be formulated as an optimization problem. The gear train consists of a cluster gear set on the input shaft as shown in Fig. 9.3. The intermediate shaft carries three fixed gears and a cluster gear set. The output shaft carries three fixed gears. Depending on the gear pair in contact, the intermeditate shaft will have three different speeds. For each intermediate shaft speed, three speeds can be realized at the output shaft. Thus nine speeds are possible for the output shaft. The speed diagram of the gear train is shown in Fig. 9.4. A gear tooth may fail due to excessive bending stress or surface-wear stress. The mass of the gear train is to be minimized by treating the face widths of the mating gear pairs as design variables. Formulate the design problem as an optimization problem.

solution The reliability analysis of this problem was considered in Chapter 9. The optimization problem is formulated first in a deterministic format and then in a probabilistic format.

TABLE 11.2 Solution of the Optimization Problem

Iteration k	Value of r_k	Design Variables R_1	R_2	R_3	R_4	Total Cost	Objective Function (R_s)
—	—	0.5	0.5	0.5	0.5	388.911	0.773438
1	0.05	0.65098	0.49455	0.58076	0.54162	407.929	0.8681115
2	0.005	0.82769	0.57822	0.73809	0.68962	453.536	0.971967
3	0.0005	0.90923	0.65037	0.84353	0.80631	483.181	0.994507
4	0.00005	0.95428	0.70478	0.89284	0.87983	500.779	0.998889
5	0.000005	0.97722	0.70210	0.91760	0.92442	506.450	0.999725
6	0.0000005	0.99303	0.70305	0.91986	0.92687	507.667	0.999927
7	0.00000005	0.99655	0.70103	0.97837	0.94596	512.136	0.999993
8	0.000000005	0.99550	0.70188	0.99669	0.95339	513.628	0.999993
9	0.0000000005	0.99538	0.70189	0.99682	0.95339	513.630	0.999993

Deterministic formulation. The face widths of the gear pairs are treated as design variables and the optimum-design problem is stated as a nonlinear programming problem as follows:

$$\text{Find } X = \begin{Bmatrix} x_1 \\ x_2 \\ \cdot \\ \cdot \\ \cdot \\ x_6 \end{Bmatrix} \tag{E1}$$

which minimizes the mass of the gear train

$$f(X) = \sum_{i=1}^{12} \rho \frac{\pi D_i^2}{4} b_i \tag{E2}$$

subject to

$$S_{b_{ij}} - S_b \leq 0; \qquad i = 1, 2, \ldots, k, \qquad j = 1, 2, \ldots, l \tag{E3}$$

$$S_{w_{ij}} - S_w \leq 0; \qquad i = 1, 2, \ldots, k, \qquad j = 1, 2, \ldots, l \tag{E4}$$

$$x_i^{(l)} \leq x_i; \qquad i = 1, 2, \ldots, 6 \tag{E5}$$

where $x_i = b_i$ = face width of ith gear pair (mm), ρ = density of the material (kg/mm^3), D_i = pitch circle diameter of ith gear (mm), $x_i^{(l)}$ = lower bound on the face width of ith gear (mm), k = number of gears in engagement while delivering any particular output speed, l = total number of output shaft speeds (9), $S_{b_{ij}}(S_{w_{ij}})$ = bending (surface-wear) stress induced in ith gear at jth output shaft speed (N/mm^2), and $S_b(S_w)$ = permissible bending (surface wear) stress for material (N/mm^2).

The stresses induced in a gear, $S_{b_{ij}}$ and $S_{w_{ij}}$, can be expressed as indicated in Eqs. (9.24) to (9.27).

Probabilistic formulation. To formulate the optimum design problem in the probabilistic format, the power transmitted, the input shaft speed, the allowable and induced stresses, the center distances of the gear pairs, and the face widths of the gears are treated as normally distributed random variables. The mean values of the face widths are considered as the design variables. A linear combination of the mean value and standard deviation of the mass of the gear train is considered as the objective for minimization. Two failure modes, namely the bending and surface wear modes, are considered in the design process. The reliability of the gear train is required to be greater than or equal to a specified quantity in bending and surface wear failure modes. The gear train is modeled as a weakest-kinematic chain in both the failure modes as indicated in Chapter 9. The optimization problem can be stated as follows:

$$\text{Find } X = \begin{Bmatrix} x_1 \\ x_2 \\ \cdot \\ \cdot \\ \cdot \\ x_6 \end{Bmatrix} \tag{E6}$$

which minimizes

$$f(X) = C_1 \bar{f}(X) + C_2 \sigma_f(X) \tag{E7}$$

subject to

$$R_{bj} \geq R_s; \qquad j = 1, 2, \ldots, l \tag{E8}$$

$$R_{wj} \geq R_s; \qquad j = 1, 2, \ldots, l \tag{E9}$$

$$x_i \geq x_i(l); \qquad i = 1, 2, \ldots, 6 \tag{E10}$$

where $\bar{f}$ and σ_f are the mean and standard deviations of the mass of the gear box, C_1 and C_2 are constants denoting the relative importances of the mean value and standard deviations of the mass during minimization, R_{bj} (R_{wj}) is the reliability of the gear train in jth speed in bending (surface wear) failure mode and R_s is the specified reliability. Since the weakest-link hypothesis is used, the reliabilities R_{bj} and R_{wj} can be expressed as

$$R_{bj} = \prod_{i=1}^{k} R_{b_{ij}} \tag{E11}$$

$$R_{wj} = \prod_{i=1}^{k} R_{w_{ij}} \tag{E12}$$

where $R_{b_{ij}}(R_{w_{ij}})$ indicates the reliability of ith gear in jth speed in bending (surface wear) failure mode.

If the strength (S) and the load acting (s) on a typical link of a weakest-link chain are assumed to be independent normally distributed random variables, the reliability of the link (R_0) can be determined as follows. A new random variable u is defined as

$$u = S - s \tag{E13}$$

which implies that u is also normally distributed. If z denotes the standard normal variate

$$z = \frac{u - \bar{u}}{\sigma_u} \tag{E14}$$

the normality relation becomes

$$\int_{-\infty}^{\infty} \frac{1}{\sqrt{2\pi}} \exp\left(\frac{-t^2}{2}\right) dt = 1 \tag{E15}$$

and the reliability of the link (R_0) can be expressed as

$$R_0 = \frac{1}{\sqrt{2\pi}} \int_{-(\bar{u}/\sigma_u)}^{\infty} \exp\left(\frac{-z^2}{2}\right) \cdot dz = \int_{z_1}^{\infty} \frac{1}{\sqrt{2\pi}} \cdot \exp\left(\frac{-z^2}{2}\right) \cdot dz \tag{E16}$$

where the lower limit of integration can be expressed in terms of the expected values ($\bar{S}$ and $\bar{s}$) and the standard deviations (σ_S and σ_s) of the random variables S and s as

$$z_1 = -\frac{\bar{u}}{\sigma_u} = -\frac{(\bar{S} - \bar{s})}{\sqrt{\sigma_S^2 + \sigma_s^2}} \tag{E17}$$

If the reliability of the gear pair is known, the normal probability tables provide the value of the lower limit of integration, $z_1 = z_{10}$. By assuming equal reliabilities for all the gears participating in the jth output speed, z_{10} can be determined from the relation

$$(R_0)^{1/k} = \frac{1}{\sqrt{2\pi}} \int_{z_{10}}^{\infty} \exp\left(\frac{-z^2}{2}\right) dz \tag{E18}$$

Thus, the constraints (E8) and (E9) can be rewritten as

$$(z_{1b})_{ij} \le z_{10} \tag{E19}$$

$$(z_{1w})_{ij} \le z_{10} \tag{E20}$$

where

$$(z_{1b})_{ij} = -\frac{(\overline{S}_b - \overline{s}_{b_{ij}})}{\sqrt{\sigma^2_{S_b} + \sigma^2_{s_{b_{ij}}}}} \tag{E21}$$

and

$$(z_{1w})_{ij} = -\frac{(\overline{S}_w - \overline{s}_{w_{ij}})}{\sqrt{\sigma^2_{S_w} + \sigma^2_{s_{w_{ij}}}}} \tag{E22}$$

The mean and standard deviations of f, $S_{b_{ij}}$, and $S_{w_{ij}}$ can be computed using the procedure of section 5.5.5.

The optimization problems have been solved using the penalty function method. The following data are assumed for the deterministic design calculations: p = power = 7.457 kW, m = module = 4 mm, $\rho = 7.75 \times 10^{-6}$ kg/mm^3, K_c = stress concentration factor = 1.5, K_d = dynamic load factor = 1.1, $x_i^{(l)}$ = 5 mm for $i = 1$ to 6, $E = 2.0601 \times 10^6$ N/mm^2, input shaft speed = 268 rpm, gear tooth profile = involute type, pressure angle = 20°, material of the gears = case-hardened steel, and strength of the material = 245.25 N/mm^2 in bending and 1716.75 N/mm^2 in the surface wear. The values of l and k are 9 and 4, respectively. The factor of safety (FS) is applied to the material strength and the permissible stresses are takes as $S_b = (245.25 / FS)$, and $S_w = (1716.75 / FS)$. The results obtained with deterministic optimization procedure with $FS = 1$ are given in Table 11.3. The minimum mass of the gear train can be seen to be 85.901 kg.

TABLE 11.3 Results of Deterministic Optimization (with $FS = 1$)

Quantity	Initial design	Optimum design
Design variables		
x_1 (mm)	60	17.339
x_2 (mm)	40	12.527
x_3 (mm)	40	9.311
x_4 (mm)	150	48.563
x_5 (mm)	60	13.044
x_6 (mm)	60	18.655
Objective function		
$f(X)$, kg	293.0505	85.9010

The data of the deterministic problem are taken as the mean values for the probabilistic problem. The coefficients of variation of all the random variables are assumed to be the same for simplicity. Further, the values of C_1 and C_2 are chosen such that the contributions from the mean value and the standard deviation of the mass are the same at the starting design point, that is, $C_1 f(X_s) = C_2 \sigma_f(X_s)$, where X_s denotes the starting design vector. The reliability-based optimum-design results obtained with three sets of values of C_1 and C_2 and two different values of the coefficient of variation of the random variables are given in Table 11.4. It can be observed that all combinations of C_1 and C_2 yield essentially the same optimum-design vector. The mean value of the mass ($\bar{f}$ with $C_1 = 1$ and $C_2 = 0$) increased by more than 200 percent when the coefficient of variation of all the random variables is increased form 0.01 to 0.10. This indicates that the optimum value of the objective functions increases at a rapid rate as the uncertainty in the design parameters increases.

TABLE 11.4 Results of reliability-based optimization (with $R_0 = 0.9999$)
Initial design: $\bar{x}_1 = \bar{x}_5 = \bar{x}_6 = 60$ mm, $\bar{x}_2 = \bar{x}_3 = 40$ mm, $\bar{x}_4 = 130$ mm, $\bar{f} = 274.9867$ kg

Quantity	With coefficient of variation of all random variables = 0.01		
Design variables	$C_1 = 1, C_2 = 7.524$	$C_1 = 1, C_2 = 0$	$C_1 = 0, C_2 = 7.514$
x_1 (mm)	19.127	19.121	19.147
x_2 (mm)	10.266	10.269	12.575
x_3 (mm)	13.820	13.820	13.975
x_4 (mm)	53.550	53.550	53.553
x_5 (mm)	14.382	14.383	14.383
x_6 (mm)	19.640	19.642	19.657
Objective function			
$f(X)$, kg	104.530	94.097	10.453
Quantity	With coefficient of variation of all random variables = 0.10		
Design variables	$C_1 = 1, C_2 = 7.524$	$C_1 = 1, C_2 = 0$	$C_1 = 0, C_2 = 7.514$
x_1 (mm)	50.415	50.416	50.416
x_2 (mm)	24.635	24.632	24.805
x_3 (mm)	32.896	32.905	32.915
x_4 (mm)	127.014	127.017	127.014
x_5 (mm)	34.123	34.115	34.116
x_6 (mm)	42.711	42.717	42.712
Objective function			
$f(X)$, kg	470.722	223.171	247.560

SOURCE: S. S. Rao and G. Das, "Reliability Based Optimum Design of Gear Trains," *ASME Journal of Mechanisms, Transmissions, and Automation in Design*, vol. 106, 1984, with permission from ASME.

11.4.4 Dynamic programming

The dynamic programming technique, developed by Richard Bellman in the early 1950s, solves a multistage decision problem as a sequence of single-stage decision problems [11.1, 11.3, 11.6]. In most cases, the solution of a set of N single-stage optimization problems is much easier compared to the solution of a N-stage optimization problem. An additional advantage of dynamic programming is that the method can handle discrete variables, nonconvex, continuous and nondifferentiable functions without any difficulty. The main disadvantage of dynamic programming is that it suffers from the so called "curse of dimensionality," which necessitates a large storage for large problems. In spite of this disadvantage, the method is suitable for solving a wide range of decision-making problems, including those arising in the field of reliability. An example of a multistage decision problem is shown in Fig. 11.11 where, for convenience, the stages $n, n-1, \ldots, i, \ldots, 2, 1$, are labeled in the decreasing order. The input state of stage i is assumed to be s_{i+1} and the output state as s_i. The design parameter of stage i is denoted as x_i and its contribution to the objective function as $f_i = f_i(\mathbf{X})$.

The problem is to find $\mathbf{X} = (x_1, x_2, \ldots, x_n)^T$ which minimizes the objective function

$$F(\mathbf{X}) = f_1(\mathbf{X}) + f_2(\mathbf{X}) + \cdots + f_n(\mathbf{X})$$

subject to

$$s_i = s_i(s_{i+1}, x_i); \qquad i = 1, 2, \ldots, n \tag{11.19}$$

and other constraints specified on x_i and $s_i (i = 1, 2, \ldots, n)$.

The dynamic programming makes use of the concept of suboptimization and the principle of optimality in solving the multistage decision problem. The basic procedure can be stated as follows. The last component (the component labelled 1 in Fig. 11.11) is used first for suboptimization. If the input state of this component, s_2, is specified, x_1 must be chosen to minimize $f_1(x_1, s_2)$. Denoting the minimum value of f_1 as $F_1^*(s_2)$

$$F_1^*(s_2) = \min_{x_1} \left[f_1(x_1, s_2) \right] \tag{11.20}$$

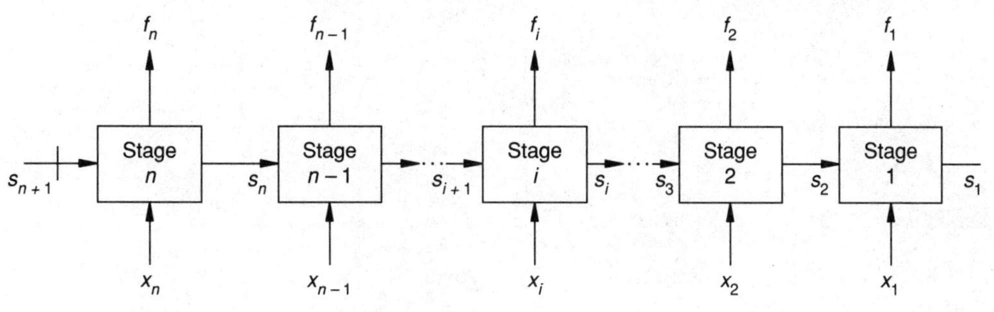

Figure 11.11 Multistage decision problem.

This equation gives the minimum value of f_1 as a function of the input state s_2 in a parametric form. Next, we consider the last two components (components labeled 2 and 1 in Fig. 11.11) together for suboptimization. Here we choose x_1 and x_2 such that

$$F_2(s_3) = f_2(x_2, s_3) + f_1(x_1, s_2) \tag{11.21}$$

is minimized for any specified value of s_3 to obtain

$$F_2^*(s_3) = \min_{x_1, x_2} \left[f_2(x_2, s_3) + f_1(x_1, s_2) \right] \tag{11.22}$$

Since s_2 can be computed once s_3 and x_2 are specified, Eq. (11.22) can be rewritten, with the help of Eq. (11.20), as

$$F_2^*(s_3) = \min_{x_2} \left[f_2(x_2, s_3) + F_1^*(s_2) \right] \tag{11.23}$$

Thus F_2^* denotes what is known as the *optimal policy* for the two-stage subproblem. This process of suboptimization can be continued by considering one more component each time until all the n components are considered in the nth suboptimization problem. In general, the ith suboptimization problem can be stated as

$$F_i^*(s_{i+1}) = \min_{x_i} \left[f_i(x_i, s_{i+1}) + F_{i-1}^*(s_i) \right] \tag{11.24}$$

In the last (nth) suboptimization problem, the input state, s_{n+1}, is usually specified and hence F_n^* need to be found just for one input, namely, s_{n+1}. Also note that only one design variable is involved at any given stage; hence the suboptimization is expected to be very simple. The multistage optimization procedure of dynamic programming is illustrated with the following example.

solution

Example 11.11 A series-parallel system is composed of three types of components in three stages as shown in Fig. 11.12. The type, number, reliability, and cost of components used in

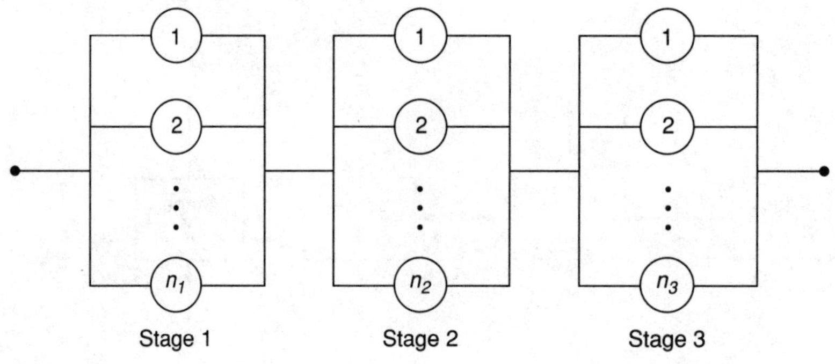

Figure 11.12 Three-stage problem.

TABLE 11.5

Component	Stage i=1	Stage i=2	Stage i=3
Type of component	1	2	3
Number of components	n_1	n_2	n_3
Reliability of each component (R_i)	0.9	0.7	0.8
Cost of each component (c_i)	4	2	3

the various stages are given Table 11.5. If the minimum required reliability of the system is 0.99, and the permissible total cost is 40, determine the number of components to be used in stages 1, 2, and 3 for maximizing the system reliability. The reliability of the system can be expressed as

$$R_0 = \prod_{i=1}^{3} \left\{1 - (1 - R_i)^{n_i}\right\} \tag{E1}$$

Thus the optimization problem involves finding

$$\mathbf{X} = \begin{Bmatrix} n_1 \\ n_2 \\ n_3 \end{Bmatrix}$$

to maximize $R_0(\mathbf{X})$ given by Eq. (E1) subject to the constraints

$$R_0 \geq 0.99$$

and

$$\sum_{i=1}^{3} c_i n_i \leq 40$$

Since the three stages are in series, the minimum required overall system reliability (0.99) dictates that each stage should have a reliability of at least 0.99. Since the reliability of stage i (with n_i components in parallel) is given by

$$1 - (1 - R_i)^{n_i} \tag{E2}$$

the reliability of each stage with different number of components can be obtained as indicated in Table 11.6.

TABLE 11.6

Number of components in the stage (n_i)	Stage 1	Stage 2	Stage 3
1	0.9	0.7	0.8
2	0.99	0.91	0.96
3	0.999	0.973	0.992
4	0.9999	0.9919	0.9984
5	0.99999	0.99757	0.99968
6	—	0.999271	0.999936
7	—	0.9997813	0.9999872
8	—	0.9999344	0.9999975
9	—	0.9999804	—
10	—	0.9999942	—
11	—	0.9999983	—

The results shown in Table 11.6 indicate that the minimum number of components required to achieve the specified overall reliability will be 2, 4, and 3 for stages 1, 2, and 3, respectively. The cost of these components will be $2 \times 4 + 4 \times 2 + 3 \times 3 = 25$. In order to apply the dynamic programming procedure, we first consider the suboptimization of stage 1. Since the minimum values of n_3, n_2, and n_1 have to be 3, 4, and 2, respectively, with an associated cost of 25, we investigate all possible values of cost between 26 and 40 (in increments of 1) and find the additional number of type 1 components that can be added in stage 1 (with the additional money available) to improve the overall reliability of the system. The results are shown in Table 11.7. For example, a total cost of 26 implies that only $26 - 25 = 1$ is available to procure additional type-1 components. Since each component of type-1 costs 4 units of money, the value of n_1 can not be increased from 2. This situation persists until the total cost increases to 29 units at which time n_1 can be increased from 2 to 3 by spending $29 - 25 = 4$ units of money to procure an additional component of type-1.

Next, we consider the suboptimization of stages 1 and 2 together by fixing the value of n_3 at 3. By varying the total cost from 26 to 40 in increments of 1, we find the additional number of components of type 1 and 2 (beyond $n_1 = 2$ and $n_2 = 4$) that can be procured with the money available after meeting the cost of 25 units for obtaining the minimum number of components $n_1 = 2$, $n_2 = 4$, and $n_3 = 3$. The results are summarized in Table 11.8. For example, when the total money available (cost) is 31 units, there will be $31 - 25 = 6$ units available for procuring additional components beyond $n_1 = 2$ and $n_2 = 4$. Since each component of type-1 and type-2 costs 4 and 2 units, respectively, we have $n_1 = 3$ and $n_2 = 5$, or $n_1 = 2$ and $n_2 = 7$.

Finally, the optimization of stages 1, 2, and 3 is considered. Since the total permissible cost is 40 units, the additional money available, namely, $40 - 25 = 15$ units, is used to procure additional components beyond $n_1 = 2$, $n_2 = 4$, and $n_3 = 3$. For example, if n_3 is increased from 3 to 6 (by spending $3 \times 3 = 9$ units of money), there remains $15 - 9 = 6$ units of money to procure additional components of type-1 and type-2. This means that we can choose either $n_2 = 5$, $n_1 = 3$ or $n_2 = 7$, $n_1 = 2$. All these possibilities, along with their corresponding system reliabili-

TABLE 11.7 Result of Suboptimization of Stage 1

Total money (cost)	R_{03} Reliability of stage 3 ($n_3 = 3$ fixed)	R_{02} Reliability of stage 2 ($n_2 = 4$ fixed)	Value of n_1 ($n_1 = 2$ minimum)	R_{01} Reliability of stage 1	R_0 Overall system reliability
26	0.992	0.9919	2	0.99	0.9741251
27	0.992	0.9919	2	0.99	0.9741251
28	0.992	0.9919	2	0.99	0.9741251
29	0.992	0.9919	3	0.999	0.9829808
30	0.992	0.9919	3	0.999	0.9829808
31	0.992	0.9919	3	0.999	0.9829808
32	0.992	0.9919	3	0.999	0.9829808
33	0.992	0.9919	4	0.9999	0.9838664
34	0.992	0.9919	4	0.9999	0.9838664
35	0.992	0.9919	4	0.9999	0.9838664
36	0.992	0.9919	4	0.9999	0.9838664
37	0.992	0.9919	5	0.99999	0.9839550
38	0.992	0.9919	5	0.99999	0.9839550
39	0.992	0.9919	5	0.99999	0.9839550
40	0.992	0.9919	5	0.99999	0.9839550

TABLE 11.8 Result of Suboptimization of Stages 1 and 2

Total money	R_{03} Reliability of stage 3 ($n_3 = 3$ fixed)	Value of n_2 ($n_2 = 4$ minimum)	R_{02} Reliability of stage 2	Value of n_1 ($n_1 = 2$ minimum)	R_{01} Reliability of stage 1	R_0 = Overall system reliability
25	0.992	4	0.9919	2	0.99	0.9741251
26	0.992	4	0.9919	2	0.99	0.9741251
27	0.992	5	0.99757	2	0.99	0.9796935
28	0.992	5	0.99757	2	0.99	0.9796935
29	0.992	6	0.999271	2	0.99	0.9813640
29	0.992	4	0.9919	3	0.999	0.9829808
30	0.992	6	0.999271	2	0.99	0.9813640
30	0.992	4	0.9919	3	0.999	0.9829808
31	0.992	7	0.9997813	2	0.99	0.9818651
31	0.992	5	0.99757	3	0.999	0.9885998
32	0.992	7	0.9997813	2	0.99	0.9818651
32	0.992	5	0.99757	3	0.999	0.9885998
33	0.992	8	0.9999344	2	0.99	0.9820155
33	0.992	6	0.999271	3	0.999	0.9902855
33	0.992	4	0.9919	4	0.9999	0.9838664
34	0.992	8	0.9999344	2	0.99	0.9820155
34	0.992	6	0.999271	3	0.999	0.9902855
34	0.992	4	0.9919	4	0.9999	0.9838664
35	0.992	9	0.9999804	2	0.99	0.9820606
35	0.992	7	0.9997813	3	0.999	0.9907912
35	0.992	5	0.99757	4	0.9999	0.9894904
36	0.992	9	0.9999804	2	0.99	0.9820606
36	0.992	7	0.9997813	3	0.999	0.9907912
36	0.992	5	0.99757	4	0.9999	0.9894904
37	0.992	10	0.9999942	2	0.99	0.9820742
37	0.992	8	0.9999344	3	0.999	0.9909429
37	0.992	6	0.999271	4	0.9999	0.9911776
37	0.992	4	0.9919	5	0.99999	0.9839549
38	0.992	10	0.9999942	2	0.99	0.9820742
38	0.992	8	0.9999344	3	0.999	0.9909429
38	0.992	6	0.999271	4	0.9999	0.9911776
38	0.992	4	0.9919	5	0.99999	0.9839549
39	0.992	11	0.9999983	2	0.99	0.9820783
39	0.992	9	0.9999804	3	0.999	0.9909885
39	0.992	7	0.9997813	4	0.9999	0.9916838
39	0.992	5	0.99757	5	0.99999	0.9895795
40	0.992	11	0.9999983	2	0.99	0.9820783
40	0.992	9	0.9999804	3	0.999	0.9909885
40	0.992	7	0.9997813	4	0.9999	0.9916838
40	0.992	5	0.99757	5	0.99999	0.9895795

ties, are summarized in Table 11.9. From this table, the optimum solution of the problem can be identified as $n_1 = 3$, $R_{01} = 0.999$, $n_2 = 6$, $R_{02} = 0.999271$, $n_3 = 5$, $R_{03} = 0.99968$ and $R_0 = 0.99795228$.

TABLE 11.9 Result of Optimization of Stages 1, 2, and 3

Value of n_3 ($n_3 = 3$ minimum)	R_{03} = Reliability of stage 3	Value of n_2 ($n_2 = 4$ minimum)	R_{02} = Reliability of stage 2	Value of n_1 ($n_1 = 2$ minimum)	R_{01} = Reliability of stage 1	R_0 = Overall system reliability
8	0.9999975	4	0.9919	2	0.99	0.98197854
7	0.9999872	5	0.99757	2	0.99	0.98758167
6	0.999936	7	0.9997813	2	0.99	0.98972011
6	0.999936	5	0.99757	3	0.999	0.99650866
5	0.99968	8	0.9999344	2	0.99	0.98961824
5	0.99968	6	0.999271	3	0.999	0.99795228
5	0.99968	4	0.9919	4	0.9999	0.99148339
4	0.9984	10	0.9999942	2	0.99	0.98841023
4	0.9984	8	0.9999344	3	0.999	0.99733615
4	0.9984	6	0.999271	4	0.9999	0.99757236
4	0.9984	4	0.9919	5	0.99999	0.99030304
3	0.992	11	0.9999983	2	0.99	0.98207825
3	0.992	9	0.9999804	3	0.999	0.99098861
3	0.992	7	0.9997813	4	0.9999	0.99168384
3	0.992	5	0.99757	5	0.99999	0.98957950

References and Bibliography

11.1. S. S. Rao, *Optimization: Theory and Applications*, 2d ed., John Wiley, New York, 1984.

11.2. S. S. Rao and G. Das, "Reliability Based Optimum Design of Gear Trains," *ASME J. of Mechanisms, Transmissions, and Automation in Design*, vol. 106, 1984, pp. 17–22.

11.3. F. A. Tillman, C. L. Hwang and W. Kuo, *Optimization of Systems Reliability*, Marcel Dekker Inc., New York, 1980.

11.4. K. B. Misra and M. D. Ljubojevic, "Optimal Reliability Design of System: A New Look," *IEEE Trans. on Reliability*, vol. R-22, 1973, pp. 255–258.

11.5. R. M. Burton and G. T. Howard, "Optimal Design for System Reliability and Maintainability," *IEEE Trans. on Reliability*, vol. R-20, 1971, pp. 56–60.

11.6. R. E. Bellman and S. E. Dreyfus, *Applied Dynamic Programming*, Princeton University Press, Princeton, New Jersey, 1962.

11.7. R. C. Juvinall and K. M. Marshek, *Fundamentals of Machine Component Design*, 2d ed., John Wiley, New York, 1991.

11.8. S. S. Rao, "Minimum Cost Design of Concrete Beams with a Reliability-Based Constraint," *Building Sci.*, vol. 8, 1973, pp. 33–38.

11.9. G. Winter, L. C. Urquhart, C. E. O'Rourke, and A. H. Nilson, *Design of Concrete Structures*, McGraw-Hill, New York, 1964.

11.10. A. C. King and C. B. Read, *Pathways to Probability*, Holt, Rinehart & Winston, Inc., New York, 1963.

11.11. C. C. Gillispie (ed.-in-chief), *Dictionary of Scientific Biography*, vol. 7, Charles Scribner's Sons, New York, 1980.

Review Questions

11.1 Define a mathematical programming problem.

11.2 What is the difference between a linear programming problem and a nonlinear programming problem?

11.3 What is an integer programming problem?

11.4 What is the Lagrange multiplier method?

11.5 What is a SUMT method?

11.6 What is the role of the penalty parameter in the penalty function method?

11.7 How is the minimum of a function $f(x)$ found in the cubic interpolation method?

11.8 What is the role of Davidon-Fletcher-Powell method in the solution of a constrained optimization problem?

11.9 What is a multistage decision process?

11.10 State a typical reliability allocation problem as an optimization problem.

Problems

11.1 Consider a system consisting of N stages in series with n_j components in parallel in stage j; $j = 1, 2, \ldots, N$. Each component of stage j costs c_j, weighs w_j, and has a reliability of R_j.

1. Formulate the problem of maximizing the overall reliability subject to constraints on the cost and weight of the system.
2. Solve the problem formulated in (1) for the following data:

Stage j	R_j	c_j	w_j
1	0.9	5	3
2	0.7	6	1
3	0.8	8	2

Maximum cost permitted: 100. Maximum weight permitted: 50.

11.2 A system consists of N stages in parallel with n_j components in series in stage j; $j = 1, 2, \ldots, N$. Each component of stage j costs c_j, weighs w_j, and has a reliability of R_j.

1. Formulate the problem of maximizing the reliability of the system subject to constraints on the cost and weight of the system.
2. Solve the problem formulated in (1) for the following data:

Stage j	R_j	c_j	w_j
1	0.9	5	3
2	0.7	6	1
3	0.8	8	2

Maximum cost permitted: 100. Maximum weight permitted: 50.

11.3 The shear stress (τ) induced in a helical spring under a compressive load F [11.7] is given by

$$\tau = \left(\frac{2C+1}{2C}\right)\frac{8FD}{\pi d^3}$$

where $C = D/d$ = spring index, d = wire diameter and D = mean coil diameter. The stiffness of the spring k is given by

$$k = \frac{d^4 G}{8D^3 N}$$

where G = shear modulus and N = number of turns (coils). A spring is to be designed for a machine suspension system to achieve a stiffness of 10 ± 0.2 lb/in. The spring material has a permissible (yield) strength in shear of 90 ± 10 k(lb/in^2), a specific weight of 0.28 ± 0.028 lb/in^3, and a shear modulus of $11.5 \times 10^6 \pm 0.5 \times 10^6$ lb/in^2. Formulate the problem of finding $\bar{d}$, $\bar{D}$, and $\bar{N}$ to minimize the mean weight of the spring to achieve a reliability greater than or equal to 0.99 in strength as well as stiffness of the spring. The axial load on the spring is given by 400 ± 100 lb. Assume that all the variables follow normal distribution and the tolerances, where specified, correspond to 3σ-band. Also, assume that the coefficient of variation of d, D, and N is 0.1.

11.4 Solve the reliability-based spring optimization problem described in Problem 11.3 for $\bar{C} = \bar{D} / \bar{d} = 10$ using a graphical procedure.

11.5 A hollow circular shaft is to be designed for minimum weight to achieve a reliability of 0.999 when subjected to a torque, $T = N(1{,}000{,}000,\ 10{,}000)$ lb·in. The permissible shear stress, τ_0, of the material is given by $\tau_0 = N(50{,}000,\ 5000)$ lb/in^2, while the maximum induced stress, τ, is given by

$$\tau = \frac{T r_0}{J}$$

where r_0 is the outer radius of the shaft, and J is the polar moment of inertia of the cross section. The manufacturing tolerances on the inner and outer radii of the shaft are to be specified as ± 0.045 in. The length of the shaft is given by 50 ± 1 in and the specific weight of the material by 0.3 ± 0.03 lb/in^3. Formulate the design problem and solve it using a graphical procedure. Assume normal distribution for all the random variables and 3σ values for the specified tolerances.

11.6 A flywheel is to be designed for a forging press to store an energy of $10{,}000 \pm 1000$ J. The minimum and maximum operating speeds of the flywheel are 600 ± 60 and 800 ± 80 rpm, respectively. The kinetic energy (E) of the flywheel is given by $E = 1/2 I \omega^2$ where I is the mass moment of inertia and ω is the rotational speed of the flywheel. By idealizing the flywheel as a rotating annular disk with inner radius r_i, outer radius r_0 and thickness t, the mass moment of inertia can be expressed as

$$I = \frac{\pi (r_0^4 - r_i^4) t \rho}{2}$$

where ρ is the density of the material. The tangential and radial stresses developed in the flywheel at a radius r, $r_i \le r \le r_0$, is given by

$$s_t = \rho \omega^2 \left(\frac{3 + \nu}{8} \right) \left(r_i^2 + r_0^2 + \frac{r_i^2 r_0^2}{r^2} - \frac{1 + 3\nu}{3 + \nu} r^2 \right)$$

$$s_r = \rho \omega^2 \left(\frac{3 + \nu}{8} \right) \left(r_i^2 + r_0^2 - \frac{r_i^2 r_0^2}{r^2} - r^2 \right)$$

where ν is the Poisson ratio. Formulate the problem of determining $\bar{r}_0$, $\bar{r}_i$ and $\bar{t}$ of the flywheel to minimize the mass while achieving a reliability of 0.99 in both strength and energy storage. The following data are applicable to the design problem: $\rho = N(7700,\ 770)$ kg/m^3, $s_y = N(30{,}000,\ 3000)$ lb/in^2, and $\nu = N(0.3, 0.03)$. Assume that all the variables follow normal distribution and the tolerances, where specified, represent the 3σ values. Use a coefficient of variation of 0.1 for r_i, r_0 and t.

11.7 Formulate the flywheel design problem described in Problem 11.6 with $\bar{r}_i$ and $\bar{r}_0$ as design variables assuming the value of $\bar{t}$ as $\bar{t} = 0.5\bar{r}_i$. Solve the resulting problem using a graphical procedure.

11.8 An electronic navigation system consists of n components in series, each following an exponential distribution as:

$$f_{T_i}(t_i) = \lambda_i\, e^{-\lambda_i t_i}; \qquad i = 1, 2, \ldots, n; \qquad \lambda_i > 0; \qquad t > 0$$

where T_i denotes the time to failure and λ_i represents the failure rate of component i. The cost of the ith component is given by (a_i / λ_i^2) where a_i is a constant. Find the values of $\lambda_1, \lambda_2, \ldots, \lambda_n$ to maximize the reliability of the system without exceeding a total cost of b using the Lagrange multiplier method.

11.9 The preload on each of the eight equally spaced 1 / 4-in. UNF bolts in the gasketed cylinder head shown in Fig. 11.13 is given by 800 ± 80 lb. The cylinder, cylinder head and the bolts are all made of steel with an Young's modulus of $(30\pm3)\times10^6$ lb/in^2. The proof strength of bolts is known to be 50 ± 5 k(lb/in^2). The gasket has a thickness of t_2 with a Young's modulus of 250 ± 25 lb/in^2. The stress induced in the bolt (s_b) is given by

$$s_b = \frac{CF}{A_t} + \frac{F_i}{A_t}$$

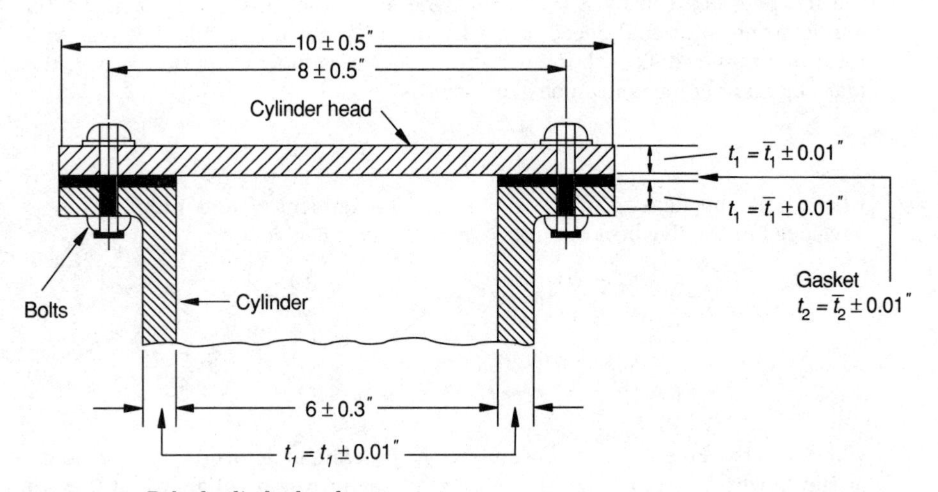

Figure 11.13 Bolted cylinder head.

where

$$C = \frac{k_b}{k_b + k_g}$$

is the joint constant, k_b is the stiffness of the bolt, k_g is the stiffness of the gasket, F = tensile load acting on one bolted joint, F_i = preload on the joint and A_t is the tensile area of the bolt which is equal to 0.0364 in^2 for the 1/4 in UNF bolt. The stiffness of the bolt and gasket are given by

$$k_b = \frac{A_b E_b}{l_b}; \qquad k_g = \frac{A_g E_g}{l_g}$$

where A_b (A_g) = area of cross section, E_b (E_g) = Young's modulus and l_b (l_g) = length of bolt (gasket) in the axial direction of the cylinder. The internal pressure of the cylinder is given by 500 ± 50 lb/in^2. Formulate the problem of finding $\bar{t}_1$ and $\bar{t}_2$ to minimize the mass of the cylinder head while achieving a reliability of 0.999 for the bolt strength. The bounds on $\bar{t}_i$ are specified as

$$0.5 \text{ in} \le \bar{t}_i \le 1.5 \text{ in}; \qquad i = 1, 2$$

11.10 Find the solution of Problem 11.9 using a graphical procedure.

11.11 The reliability of a system, consisting of n stages in series with x_j redundant units in each stage j ($j = 1, 2, \ldots, n$), is given by

$$R_0 = \prod_{j=1}^{n} R_j = \prod_{j=1}^{n} \{1 - (1 - r_j)^{x_j}\} \tag{E1}$$

where r_j is the reliability of a unit in stage j. The cost of each unit in stage j, c_j, is given by

$$c_j = a_j e^{b_j/(1-r_j)} \tag{E2}$$

and the total cost of the system is restricted as

$$\sum_{j=1}^{n} c_j x_j \le C \tag{E3}$$

where C is a constant. By eliminating x_j using Eq. (E1), Eq. (E3) takes the form

$$\sum_{j=1}^{n} a_j \frac{\ln(1 - R_j)}{(1 - r_j)} e^{b_j/(1-r_j)} \le C \tag{E4}$$

Using the Lagrange multiplier method, find the values of R_j and r_j which maximizes R_0 subject to the cost constraint of Eq. (E4) for a 4-stage system with the following data [11.4]:

Stage j	1	2	3	4
a_j	1.0	3.5	2.0	5.0
b_j	0.30	0.55	0.40	0.65

Assume the value of C as 300 and discuss the practicality of the solution found.

11.12 Minimize $f(x) = 2x^2 - 12x + 5$ subject to $x \le 2$ using the interior penalty function method. Plot the contours of the ϕ-function for different values of r_k and identify the optimum point in the graph.

11.13 Find the solution of the following optimization problem using a graphical procedure:

$$\text{Maximize } f(x_1, x_2) = 3x_1 + x_2$$

subject to

$$3x_1 + 2x_2 \ge 6$$

$$x_1 + x_2 \le 8$$

$$2x_1 - 4x_2 \le 2$$

$$-4x_1 + 3x_2 \le 12$$

$$x_1 \ge 0; \quad x_2 \ge 0$$

11.14 A parallel system, consisting of two components with constant failure rates of λ_1 and λ_2, is to be optimized for minimum cost. The costs of the components are given by $c_1 = 100/\lambda_1$ and $c_2 = 200/\lambda_2$. Find the values of λ_1 and λ_2 to minimize the total cost while achieving a reliability of 0.99.

11.15 Solve the problem formulated in Example 11.4 using a graphical procedure.

11.16 Solve the problem formulated in Example 11.5 using a graphical procedure.

11.17 Formulate the problem of minimum cost design of a rectangular under-reinforced concrete beam which can carry a bending moment M with a reliability of 0.99. The costs of concrete (c_c), steel (c_s) and form work (c_f) are given by 0.01 dollar/in^3, 0.20 dollar/in^3, and 0.1 dollar/in^2 of surface area. The bending moment acting on the beam is normally distributed with a mean value $\overline{M} = 10^6$ lb·in and a standard deviation of $\sigma_M = 10^5$ lb·in. The strengths of concrete (s_c) and steel (s_s) are normally distributed:

$$s_c = N(1500,\ 300)\ \text{lb/in}^2; \qquad s_s = N(30{,}000,\ 3000)\ \text{lb/in}^2$$

Treat the width (w) and depth (d) of the beam and the reinforcing steel area (A_s) as deterministic design variables (see Fig. 11.14). The resisting moment (M_r) of a reinforced beam section [11.8, 11.9] is given by

$$M_r = A_s s_s \left(d - 0.59 \frac{A_s s_s}{s_c w} \right)$$

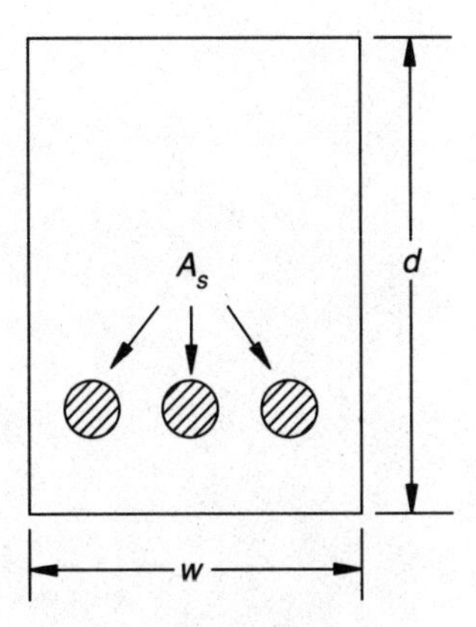

Figure 11.14 Reinforced concrete beam section.

and the balanced-steel area, which can be used as an upper bound on A_s to ensure under-reinforcement of the beam [11.9], is given by

$$A_s^{(b)} = 0.542 \frac{s_c w d}{s_s} \left(\frac{6117}{6117 + s_s} \right)$$

11.18 Solve the concrete beam design problem formulated in Problem 11.17 using a graphical method assuming $w = 0.5d$.

Chapter

12

Maintainability and Availability

Biographical Note

Andrei Andreevich Markov

Andrei Andreevich Markov (or Markoff) was a Russian mathematician who was born on June 14, 1856, and died on May 20, 1922. He graduated from St. Petersburg University in 1878 with a gold medal for his thesis on the solution of differential equations. He was a student of Chebyshev, and taught at St. Petersburg University, becoming a professor there in 1893. He developed the chain theory that has become popular in weather prediction, social sciences, biological sciences, molecular and statistical physics, quantum theory, and genetics. A chain is associated with a process, which changes from stage to stage. The chain theory provides the probability of being in the next stage once the probability of being in the current stage is known. Markov chains are applied to random walks, which are a succession of walks along a particular path. Entire books are written on Markov chains and separate courses are offered in colleges and universities on the subject. Markov extended many of the works of Chebyshev and made many contributions to statistics. He explained sampling theory within the framework of probability theory [12.17, 12.18].

12.1 Introduction

The previous chapters were concerned with the problem of assessing the reliability of a component or a system without considering the aspects of maintenance and repair. This chapter deals with maintainable and repairable systems that can be restored to service either at regular intervals of time or after a failure. Both the aspects of maintainability (preventive, as well as corrective maintainability) and availability are discussed. In general, the maintainability of a system provides a measure of the repairability of a system when it fails while the availability of a system provides a measure of the readiness of a system for use at any instant of time.

The availability analysis of a system requires a knowledge of the following aspects: (1) the system configuration describing how the components are functionally connected, (2) the failure process of the components, (3) the method of operation and the definition of failure of the system, and (4) the repair or maintenance policy. A variety of failure and repair time distributions can be used in the availability analysis. Lie, Hwang, and Tillman [12.1] gave a detailed classification of references on availability with regard to both failure and repair time distributions. The most frequently used distributions for failure process (or failure times) include exponential, Weibull, and normal distributions. For the repair process (or repair times), the most prominent ones are exponential, lognormal, gamma, Poisson, and uniform distributions. For example, if the components of a system that fail frequently have a relatively short repair times compared to those components that fail infrequently, the repair times can be assumed to follow exponential distribution. On the other hand, if every component of the system has the same failure rate and the same repair time, the repair times can be assumed to follow uniform distribution.

In this chapter, we assume, mostly, that the failure times as well as repair times of individual components follow exponential distribution. This permits us to use Markov models for the analysis.

12.2 Maintainability

12.2.1 Concept

The purpose of maintenance is to restore a deteriorating or failed system to the normal operating state. Maintainability $M(t)$ is defined as the probability of repairing a failed component or system in a specified period of time. Two types of maintenance are possible: preventive maintenance and corrective maintenance. In *preventive maintenance*, the system is periodically inspected, some components are replaced, lubrication checked, and adjustments are made before the system fails. The preventive maintenance is intended to eliminate costly repairs involved during the corrective maintenance stage when the system fails. On the other hand, *corrective maintenance* is used after the system fails. While the aim of preventive maintenance is to increase the reliability and prolong the life of the system by overcoming the effects of aging, fatigue, and wear, the objective of corrective maintenance is to bring the system from the failed state to the operating state as soon as possible to increase its availability. The preventive maintenance, by its nature, can be scheduled and controlled to minimize the cost, while the corrective maintenance cannot be controlled. The type of maintenance used depends on the type of system, cost involved, and the risk associated with the failure. For example, the type of maintenance used for an automobile engine is totally different from the one used for an aircraft engine.

12.2.2 Preventive maintenance

Reliability of the system. Let the reliability of the system with no maintenance be given by $R(t)$ and let preventive maintenance be used at times t_0, $2\,t_0$, $3\,t_0$, ...[1] We assume that each time a maintenance action is taken, the system is restored to "as good as new" condition as indicated in Fig. 12.1 [12.4]. The reliability of the maintained system, $R_m(t)$, during $0 \le t \le t_0$ can be stated as

$$R_m(t) = R(t); \qquad 0 \le t \le t_0 \tag{12.1}$$

since no maintenance action is taken until $t = t_0$. At any time during the next interval of time, $t_0 \le t \le 2\,t_0$, the reliability of the maintained system is given by the product of the probability of the original system operating at time $t = t_0$, and the probability of the system, restored to new condition at t_0, operating at $t = t - t_0$

$$R_m(t) = R(t_0)\,R(t - t_0); \qquad t_0 \le t \le 2\,t_0 \tag{12.2}$$

Next, the reliability of the maintained system during $2\,t_0 \le t \le 3\,t_0$ can be obtained as the product of the probability of the system operating at $t = 2\,t_0$ and the probability of the system, restored to new condition at $2\,t_0$, operating at time $t = t - 2\,t_0$

$$R_m(t) = R(t_0)R(t - t_0)\big|_{t=2t_0} R(t - 2\,t_0)$$

$$= R^2(t_0)R(t - 2\,t_0); \qquad 2\,t_0 \le t \le 3\,t_0 \tag{12.3}$$

By proceeding in a similar manner, the reliability of the maintained system during $it_0 \le t \le (i+1)t_0$ can be expressed as

$$R_m(t) = R^i(t_0)R(t - it_0); \qquad it_0 \le t \le (i+1)\,t_0; \qquad i = 0, 1, 2, \ldots \tag{12.4}$$

---

[1]Note that t denotes the actual operating time of the system and excludes the time periods during which the system is shut off for maintenance action.

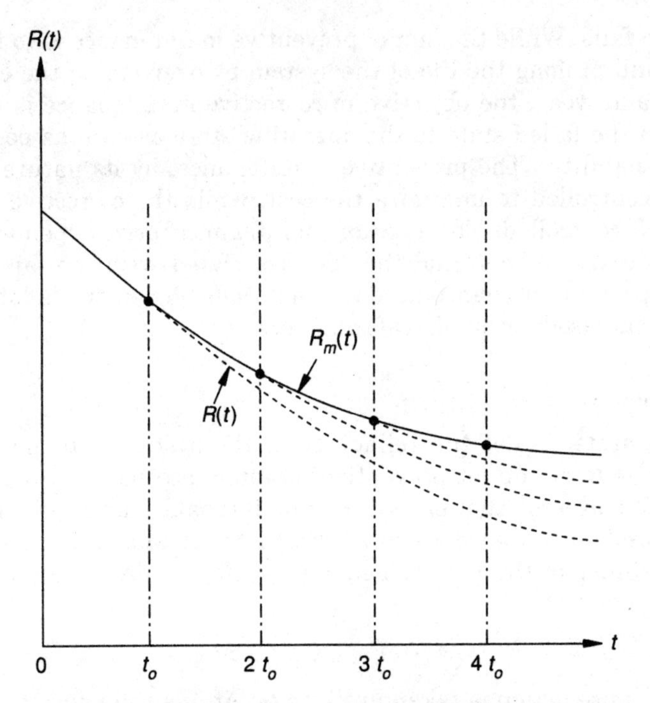

Figure 12.1 Reliability of a system with and without preventive maintenance.

Mean time to failure. The mean time to failure (see Eq. 6.26) of a maintained system (MTTF_m) is given by

$$\mathrm{MTTF}_m = \int_0^{\infty} R_m(t)\,dt = \int_0^{t_0} R_m(t)\,dt + \int_{t_0}^{2t_0} R_m(t)\,dt + \cdots + \int_{it_0}^{(i+1)t_0} R_m(t)\,dt + \cdots$$

$$= \sum_{i=0}^{\infty} \int_{it_0}^{(i+1)t_0} R_m(t)\,dt \tag{12.5}$$

By substituting Eq. (12.4) into Eq. (12.5) and noting that $R(t_0)$ is independent of t, we obtain

$$\mathrm{MTTF}_m = \sum_{i=1}^{\infty} R^i(t_0) \int_{t=it_0}^{(i+1)t_0} R(t - it_0)\,dt \tag{12.6}$$

Defining $\tau = t - it_0$, Eq. (12.6) can be rewritten as

$$\mathrm{MTTF}_m = \sum_{i=1}^{\infty} R^i(t_0) \int_{\tau=0}^{t_0} R(\tau)\,d\tau \tag{12.7}$$

In view of the equivalence [12.9]

$$\sum_{i=1}^{\infty} R^i(t_0) = \frac{1}{1 - R(t_0)} \tag{12.8}$$

Eq. (12.7) gives

$$\text{MTTF}_m = \frac{1}{1 - R(t_0)} \int_0^{t_0} R(\tau)\, d\tau \tag{12.9}$$

Example 12.1 Find the reliability of a system without and with preventive maintenance performed at regular intervals of time t_0 when the failure time distribution of the original system is given by

i. $f_T(t) = \lambda e^{-\lambda t}$ (exponential)

ii. $f_T(t) = \dfrac{1}{\sqrt{2\pi}\,\sigma} \exp\left\{-\dfrac{1}{2}\left(\dfrac{t-\mu}{\sigma}\right)^2\right\}$ (normal)

iii. $f_T(t) = \dfrac{\beta}{\eta}\left(\dfrac{t}{\eta}\right)^{\beta-1} \exp\left\{-\left(\dfrac{t}{\eta}\right)^{\beta}\right\}$ (Weibull)

solution

Without repair. The reliability of the system without repair (see Section 6.6) can be expressed as follows

i. $R(t) = e^{-\lambda t}$

ii. $R(t) = 1 - \Phi\left(\dfrac{t-\mu}{\sigma}\right)$

iii. $R(t) = \exp\left\{-\left(\dfrac{t}{\eta}\right)^{\beta}\right\}$

With repair. When preventive maintenance is performed at time intervals t_0, the reliability of the system is given by Eq. (12.4)

$$R_m(t) = R^i(t_0)R(t - it_0); \qquad it_0 \le t \le (i+1)\,t_0; \qquad i = 0, 1, 2, \ldots \tag{E1}$$

i. Here $R(t) = e^{-\lambda t}$ and Eq. (E1) gives

$$R_m(t) = (e^{-\lambda t_0})^i e^{-\lambda(t - it_0)} = e^{-i\lambda t_0} e^{-\lambda t} e^{i\lambda t_0} = e^{-\lambda t} \text{ for any } t$$

$$\therefore\ R_m(t) = R(t) = e^{-\lambda t}; \qquad t \ge 0$$

ii. In this case, $R(t) = 1 - \Phi(\{t - \mu\}/\sigma)$, and hence

$$R_m(t) = R^i(t_0)\,R(t - it_0)$$

$$= \left[1 - \Phi\left(\frac{t_0 - \mu}{\sigma}\right)\right]^i \left[1 - \Phi\left(\frac{t - it_0 - \mu}{\sigma}\right)\right]; \qquad it_0 \le t \le (i+1)\,t_0; \qquad i = 0, 1, 2, \ldots$$

iii. Since the reliability function of the system without maintenance (see Example 6.4) is given by

$$R(t) = \exp\left\{-\left(\frac{t}{\eta}\right)^{\beta}\right\}$$

the reliability of the maintained system becomes

$$R_m(t) = R^i(t_0)\,R(t - it_0) = \left(\exp\left\{-\left(\frac{t_0}{\eta}\right)^\beta\right\}\right)^i \exp\left\{-\left(\frac{t - it_0}{\eta}\right)^\beta\right\}$$

$$= e^{-i\left(\frac{t_0}{\eta}\right)^\beta} e^{-\left(\frac{t - it_0}{\eta}\right)^\beta}$$

$$= e^{-\left\{i\left(\frac{t_0}{\eta}\right)^\beta + \left(\frac{t - it_0}{\eta}\right)^\beta\right\}}; \qquad it_0 \le t \le (i+1)t_0; \qquad i = 0, 1, 2, \ldots$$

Example 12.2 Derive the condition under which preventive maintenance is beneficial to the system for the following cases:

1. when the failure time is governed by exponential distribution
2. when the failure time is governed by Weibull distribution

solution The ratio of reliabilities of the system with and without preventive maintenance is given by

$$\frac{R_m(t)}{R(t)} = \frac{R^i(t_0)\,R(t - it_0)}{R(t)}; \qquad it_0 \le t \le (i+1)\,t_0; \qquad i = 0, 1, 2, \ldots \tag{E1}$$

For specificness, we set $t = it_0$ so that Eq. (E1) becomes

$$\frac{R_m(it_0)}{R(it_0)} = \frac{R^i(t_0)\,R(0)}{R(it_0)} = \frac{(R(t_0))^i}{R(it_0)}; \qquad i = 0, 1, 2, \ldots \tag{E2}$$

Case 1: For exponential failure time distribution, the reliability function is given by

$$R(t) = e^{-\lambda t} \tag{E3}$$

and Eq. (E2) reduces to

$$\frac{R_m(it_0)}{R(it_0)} = \frac{(e^{-\lambda t_0})^i}{e^{-i\lambda t_0}} = 1 \tag{E4}$$

This shows that preventive maintenance is not beneficial in the case of exponential failure time distribution, which has a constant failure rate.

Case 2: For Weibull failure-time distribution, the reliability function (see Example 6.4) is given by

$$R(t) = \exp\left\{-\left(\frac{t}{\eta}\right)^\beta\right\} \tag{E5}$$

and Eq. (E2) leads to

$$\frac{R_m(it_0)}{R(it_0)} = \frac{\left[\exp\left\{-\left(\frac{t_0}{\eta}\right)^\beta\right\}\right]^i}{\exp\left\{-\left(\frac{it_0}{\eta}\right)^\beta\right\}} = \exp\left\{-i\left(\frac{t_0}{\eta}\right)^\beta + i^\beta\left(\frac{t_0}{\eta}\right)^\beta\right\} \tag{E6}$$

This indicates that $R_m(it_0) > R(it_0)$ if $-i + i^\beta > 0$, or $i^{\beta-1} > 1$. This inequality will be satisfied only when $\beta > 1$ (i.e., when the failure rate increases). This indicates that preventive maintenance is useful over the wearout or aging period when β increases with time.

Example 12.3 A system is scheduled for preventive maintenance at times t_0, $2t_0$, $3t_0$, ... Find the mean and standard deviation of the number of times preventive maintenance is performed before the occurrence of a system failure.

solution The probability of the system undergoing exactly i preventive maintenances, P_i, can be assumed to be same as the probability of the maintained system failing at time $t = (i + 1)t_0$. This gives, using Eq. (12.4):

$$P_i = [R(t_0)]^i[1 - R(t_0)]; \qquad i = 0, 1, 2, \ldots$$

The expected value of i can be found as

$$E[i] = \sum_{i=1}^{\infty} iP_i = \sum_{i=1}^{\infty} i[R(t_0)]^i\,[1 - R(t_0)] = R(t_0)\,[1 - R(t_0)] \sum_{i=1}^{\infty} i[R(t_0)]^{i-1}$$

$$= \frac{R(t_0)\,[1 - R(t_0)]}{[1 - R(t_0)]^2} = \frac{R(t_0)}{1 - R(t_0)}$$

The standard deviation of i can be evaluated as

$$\sigma_i^2 = E[(i - E[i])^2] = E[i^2] - (E[i])^2$$

$$= \sum_{i=1}^{\infty} i^2\,P_i - (E[i])^2 = \frac{R(t_0)}{[1 - R(t_0)]^2}$$

12.2.3 Imperfect maintenance

It has been assumed in the previous section that maintenance is performed perfectly so that the system can be considered to be "as-good-as-new" after the maintenance action. However, there is a possibility of performing imperfect maintenance due to human error. In such cases, the system will fail immediately after the preventive maintenance action. If the probability of performing imperfect maintenance is denoted as p, the reliability of the system is to be multiplied by $(1 - p)$ each time a preventive maintenance action is taken. Thus the reliability of a maintained system can be expressed, by modifying Eq. (12.4), as

$$R_m(t) = (1 - p)^i R^i(t_0) R(t - it_0); \qquad it_0 \le t \le (i + 1)t_0; \qquad i = 0, 1, 2, \ldots \qquad (12.10)$$

In order to find whether the replacement of wearing parts during preventive maintenance is worthwhile compared to the reduced reliability caused by faulty maintenance, we consider the ratio $R_m(t)/R(t)$ at $t = it_0$

$$\frac{R_m(it_0)}{R(it_0)} = \frac{(1 - p)^i\,R^i(t_0)\,R(0)}{R(it_0)} = \frac{(1 - p)^i\,R^i(t_0)}{R(it_0)}; \qquad i = 0, 1, 2, \ldots \qquad (12.11)$$

since $R(0) = 1$.

Example 12.4 Determine the condition under which faulty preventive maintenance is useful when the failure time of the original (unmaintained) system follows: (1) Exponential distribution and (2) Weibull distribution.

solution

1. When the failure time of the original (unmaintained) system follows exponential distribution,

$$R(t) = e^{-\lambda t} \tag{E1}$$

and Eq. (12.11) gives

$$\frac{R_m(it_0)}{R(it_0)} = (1-p)^i \frac{e^{-i\lambda t_0}}{e^{-i\lambda t_0}} = (1-p)^i \tag{E2}$$

Since the ratio given by Eq. (E2) is less than one for all practical values of p, the faulty preventive maintenance is not desirable.

2. When the failure time of the original system follows Weibull distribution, we have

$$R(t) = \exp\left\{-\left(\frac{t}{\eta}\right)^\beta\right\} \tag{E3}$$

and Eq. (12.11) gives

$$\frac{R_m(it_0)}{R(it_0)} = \frac{(1-p)^i \left(\exp\left\{-\left(\frac{t_0}{\eta}\right)^\beta\right\}\right)^i}{\exp\left\{-\left(\frac{it_0}{\eta}\right)^\beta\right\}}$$

$$= (1-p)^i \exp\left\{\left(\frac{it_0}{\eta}\right)^\beta - i\left(\frac{t_0}{\eta}\right)^\beta\right\}; \qquad i = 0, 1, 2, \ldots \tag{E4}$$

For small values of p, we can use the approximation

$$(1-p)^i \approx e^{-ip} \tag{E5}$$

and Eq. (E4) can be expressed as

$$\frac{R_m(it_0)}{R(it_0)} \approx \exp\left\{-ip - i\left(\frac{t_0}{\eta}\right)^\beta + \left(\frac{it_0}{\eta}\right)^\beta\right\}; \qquad i = 0, 1, 2, \ldots \tag{E6}$$

It was observed in Example 12.2 that ideal preventive maintenance will be beneficial only when $\beta > 1$ (i.e., when failure rate increases, as in the case of wearout period). Equation (E6) shows that faulty preventive maintenance will be useful, for $\beta > 1$, only when

$$-ip - i\left(\frac{t_0}{\eta}\right)^\beta + \left(\frac{it_0}{\eta}\right)^\beta > 0$$

that is,

$$p < \left(i^{\beta-1} - 1\right)\left(\frac{t_0}{\eta}\right)^\beta \tag{E7}$$

12.2.4 Repair-time distributions

Some of the distributions used for the repair times include exponential, normal, lognormal and Weibull distributions. A typical distribution of repair times is shown in Fig. 12.2. The maintainability of the system, in different cases, can be computed as follows.

When time to repair follows exponential distribution. The probability density function of repair time (T) is given by

$$f_T(t) = \mu e^{-\mu t} \tag{12.12}$$

where $\mu = 1/\text{MTTR}$ is the repair rate and MTTR = mean time to repair. If $t_1, t_2, \ldots, t_n$ denote the observed repair times, MTTR can be computed as

$$\text{MTTR} = \frac{1}{n} \sum_{i=1}^{n} t_i \tag{12.13}$$

In this case, the maintainability function, $M(t)$, can be determined as

$$M(t) = P[T \le t] = \int_0^t f_T(\tau) \cdot d\tau = 1 - e^{-\mu t} \tag{12.14}$$

When time to repair follows lognormal distribution. Let the observed repair times be $t_1, t_2, \ldots, t_n$. Then the sample mean ($\overline{X}$) and variance s^2 of $X = \ln t$ are given by

$$\overline{X} = \frac{1}{n} \sum_{i=1}^{n} \ln t_i \tag{12.15}$$

$$s^2 = \frac{1}{n-1} \sum_{i=1}^{n} (\ln t_i - \overline{X})^2 \tag{12.16}$$

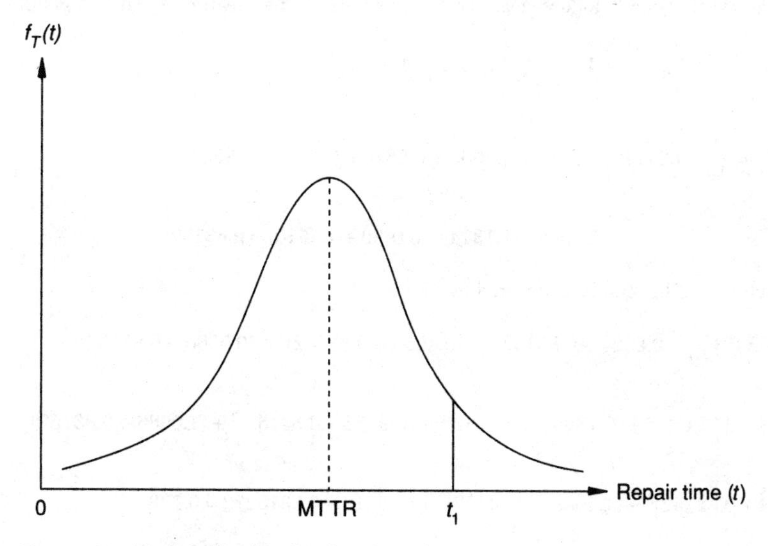

Figure 12.2 Probability density function of repair time.

If T_0 denotes the permissible repair time, the maintainability will be equal to the probability that the repair time is less than T_0. This can be found from the standard normal tables by finding the value of the standard normal variate z_1 corresponding to T_0 as

$$z_1 = \frac{\ln T_0 - \overline{X}}{s} \tag{12.17}$$

Example 12.5 The total number of failures of a numerically controlled milling machine in a year is 84. The total number of maintenance hours spent in correcting all these failures is 578. Compute the probability of repairing a failure in (1) 1 hour, (2) 5 hours, and (3) 20 hours assuming an exponential repair time.

solution The data gives $\mu = 84/578 = 0.1453$ maintenance actions per hour and MTTR $= 1/\mu = 578/84 = 6.8810$ hours (average time for one repair). Thus the maintainability or probability of repair can be determined as follows:

1. $M(1) = 1 - e^{-\mu t} = 1 - e^{-0.1453\,(1)} = 1 - 0.8648 = 0.1352$
2. $M(5) = 1 - e^{-0.1453\,(5)} = 1 - 0.4836 = 0.5164$
3. $M(20) = 1 - e^{-0.1453\,(20)} = 1 - 0.0547 = 0.9453$

These results indicate that a failure has only 13.52 percent chance of being repaired in 1 hour but a 94.53 percent chance of being repaired in 20 hours.

Example 12.6 The repair times t_i for a mainframe computer system are observed to be 1.3, 1.5, 1.7, 1.8, 2.2, 2.6, 3.0, 3.1, 3.3, and 3.9 hours. Assuming lognormal distribution for the repair times, determine the following:

1. Maintainability of the system for an allowed downtime of 5 hours.
2. Downtime required to achieve a maintainability of 0.99.

solution Since the repair time T follows lognormal distribution, the mean of $X = \ln t$ is given by

$$\overline{X} = \frac{1}{10}\sum_{i=1}^{10} \ln t_i = \frac{1}{10}(\ln 1.3 + \ln 1.5 + \cdots + \ln 3.9)$$

$$= \frac{1}{10}[0.2624 + 0.4055 + 0.5306 + 0.5878 + 0.7885 + 0.9555$$

$$+ 1.0986 + 1.1314 + 1.1939 + 1.3610] = 0.83152$$

The standard deviation of $\ln t$ can be computed as

$$s^2 = \frac{1}{9}\sum_{i=1}^{10} (\ln t_i - \overline{X})^2 = \frac{1}{9}[(0.2624 - 0.83152)^2 + (0.4055 - 0.83152)^2 + (0.5306 - 0.83152)^2$$

$$+ (0.5878 - 0.83152)^2 + (0.7885 - 0.83152)^2 + (0.9555 - 0.83152)^2 + (1.0986 - 0.83152)^2$$

$$+ (1.1314 - 0.83152)^2 + (1.1939 - 0.83152)^2 + (1.3610 - 0.83152)^2] = 0.1384$$

This gives $s = 0.3720$.

1. The maintainability of the system corresponding to a downtime of 5 hours is given by

$$M(5) = \Phi\left(\frac{\ln 5 - \bar{X}}{s}\right) = \Phi\left(\frac{1.6094 - 0.83152}{0.3720}\right) = \Phi(2.0911) = 0.9817$$

2. To achieve a maintainability of 0.99, the required downtime t can be determined from the relation

$$M(t) = 0.99 = \Phi\left(\frac{\ln t - \bar{X}}{s}\right) = \Phi\left(\frac{\ln t - 0.83152}{0.3720}\right)$$

From standard normal tables, we obtain

$$\frac{\ln t - 0.83152}{0.3720} = 2.327$$

that is,

$$\ln t = 1.69176 \quad \text{or} \quad t = 5.4584 \text{ hours}$$

Example 12.7 Solve Example 12.6 assuming that the repair times follow Weibull distribution.

solution The 10 sample repair times (t_i) can be used to find the parameters of Weibull distribution as (using the procedure outlines in Section 15.7.7) $\eta = 2.75$ and $\beta = 2.93$ so that the maintainability $M(t)$ is given by

$$M(t) = 1 - \exp\left\{-\left(\frac{t}{\eta}\right)^{\beta}\right\}$$

1. The maintainability corresponding to a downtime of 5 hours is given by

$$M(5) = 1 - \exp\left\{-\left(\frac{5}{\eta}\right)^{\beta}\right\} = 1 - \exp\left\{-\left(\frac{5}{2.75}\right)^{2.93}\right\}$$

$$= 1 - \exp(-5.764177) = 0.996862$$

2. The downtime required, t_0, to achieve a maintainability of 0.99 is given by

$$t_0 = \eta[-\ln(1 - 0.99)]^{\frac{1}{\beta}} = 2.75\,[-\ln 0.01]^{\frac{1}{2.93}} = 4.631231 \text{ hours}$$

12.2.5 Unrepaired failures

The number of unrepaired failures, that is, the number of failures that are each not likely to be repaired in a specified time T_0 will be of interest to the management. If λ denotes the failure rate, the expected number of failures during time T_0 will be λT_0. If the repair time follows exponential distribution with a repair rate of μ, the maintainability function is given by

$$M(t) = 1 - e^{-\mu t} \tag{12.18}$$

Thus the probability of not being able to repair one system failure in time t_r is given by $e^{-\mu t_r}$. Hence the number of unrepaired failures in time T_0, N_{ut}, will be equal to

$$N_{ut} = \lambda T_0 e^{-\mu t_r} \tag{12.19}$$

Example 12.8 The air conditioning unit used in a computer installation had an average of 0.1 failure per month. In a system life of 12 months, how many failures are likely not to get repaired if a repair time of 2 hours is allowed for each repair. Assume the average number of repairs per hour as 0.5.

solution The expected number of failures during the life, T_0, is given by $\lambda T_0 = 0.1(12) = 1.2$ failures. The number of failures which are not likely to get repaired is given by Eq. (12.19)

$$N_{ut} = \lambda T_0 e^{-\mu t_r} = 1.2\, e^{-0.5(2)} = 1.2(0.3679) = 0.4415$$

Thus, 0.4415 failures every 12 months, on the average, will take longer than 2 hours to repair.

12.2.6 Optimal replacement strategy

As seen in the previous sections, the preventive maintenance improves the reliability of a system during the wearout period. In some systems, a component is automatically replaced after an operational time of t_0. This strategy is known as age replacement. Changing the oil and oil filter in an automobile after every t_0 months is an example of age replacement. In some cases, a set or batch of components is replaced regularly at times t_0, $2t_0$, $3t_0$, . . . , and also whenever a failure occurs. This strategy, known as batch replacement, is used whenever a large number of inexpensive components are involved such as a set of light bulbs in a factory. In all the cases, the cost of improving the reliability through preventive maintenance should be compared with the cost of restoring a failed system through repair. Let a component be replaced at regular times t_0, $2t_0$, $3t_0$, . . . ,. Let the cost of replacing a component through preventive maintenance be c_m and the cost of replacing a failed component through corrective maintenance be c_r. In most cases, $c_r > c_m$, since the failure of a system involves not only the cost of replacing the failed component but also the costs associated with several additional factors. For example, if the transmission of an automobile breaks down during a long-distance trip, several additional costs, such as the costs of towing and extra lodging, are involved besides the cost of replacing the transmission.

Let the component be replaced n_m times through preventive maintenance (unfailed components replaced) and n_r times through corrective maintenance (failed component replaced) during a sufficiently long operating time t. Then the total number of times the component is replaced n is given by

$$n = n_m + n_r \tag{12.20}$$

and the total cost of replacement c by

$$c = c_m\, n_m + c_r\, n_r \tag{12.21}$$

Since the total operating time t is very large compared to the mean time between replacements (MTBR), we have

$$n = \frac{t}{\text{MTBR}} \tag{12.22}$$

where MTBR, similar to Eq. (6.26), is given by

$$\text{MTBR} = \int_0^{\infty} \tilde{R}(\tau)\, d\tau \tag{12.23}$$

and $\tilde{R}(\tau)$ is the reliability of the component which is automatically replaced after a time t_0

$$\tilde{R}(\tau) = \begin{cases} R(\tau) & \text{for} \quad \tau \le t_0 \\ 0 & \text{for} \quad \tau > t_0 \end{cases} \tag{12.24}$$

with $R(\tau)$ denoting the reliability of the original (unmaintained) component. Equations (12.24) and (12.23) yield

$$\text{MTBR} = \int_0^{t_0} \tilde{R}(\tau)\, d\tau + \int_{t_0}^{\infty} \tilde{R}(\tau)\, d\tau = \int_0^{t_0} R(\tau)\, d\tau \tag{12.25}$$

Since the fraction of components that survive until the next preventive maintenance is given by $R(t_0)$, the number of surviving components, n_m, that are replaced through preventive maintenance, are given by

$$n_m = R(t_0) n \tag{12.26}$$

and similarly, the number of failed components n_r, which are replaced through corrective maintenance, are given by

$$n_r = (1 - R(t_0)) n \tag{12.27}$$

By substituting Eqs. (12.22), (12.25), (12.26), and (12.27) into Eq. (12.21), we obtain

$$c = \frac{c_m R(t_0) t}{\int_0^{t_0} R(\tau)\, d\tau} + \frac{c_r (1 - R(t_0)) t}{\int_0^{t_0} R(\tau)\, d\tau} \tag{12.28}$$

To minimize the total cost of maintenance by selecting a proper value of the preventive maintenance interval t_0 we set

$$\frac{dc}{dt_0} = 0 \tag{12.29}$$

where c is given by Eq. (12.28) and solve for t_0. Equation (12.29) will be a nonlinear equation in most practical cases (see Problems 12.8 and 12.9).

12.2.7 Spare parts requirement

When the failure times of components of complex systems follow exponential distribution, the number of failures in a specified time follows Poisson distribution [12.2]. Thus the probability of having n failures in time t is given by

$$P(n) = \frac{(\lambda t)^n}{n!} e^{-\lambda t} \tag{12.30}$$

where $\lambda = 1/\text{MTBF}$ and MTBF = mean time between failures. The probability of having less than or equal to r failures in time t is given by

$$P(n \leq r) = \sum_{i=0}^{r} P(i) \tag{12.31}$$

The value given by Eq. (12.31) is also the probability of having adequate replacement parts available if r spare parts are stocked at the beginning of the period. Equation (12.31) can also be used to find the number of spare parts needed as follows.

If a desired cumulative probability is given for a particular item, the number of spare parts needed can be determined by summing the values of the individual probabilities of failure until this sum is equal to or greater than the desired cumulative probability. A simpler, but approximate, formula for determining the number of spare parts in a mission time T_0 is given by

$$\text{Number of spare parts} = \lambda T_0 + z_{1-\alpha}\sqrt{\lambda T_0} \tag{12.32}$$

where λT_0 and $\sqrt{\lambda T_0}$ denote the mean and standard deviation of the number of failures in time T_0, according to Poisson distribution, and $z_{1-\alpha}$ is the value of standard normal variate corresponding to the confidence level α.

Example 12.9 An electric surveillance system consists of three types of components. There are 20 components of type A, 130 components of type B, and 100 components of type C in the system. The failure rates (number of failures per hour) of types A, B, and C components are given by 0.002, 0.0001, and 0.0005, respectively. Find the number of spare parts required for each type of component to cover a mission period of 30 days with 99 percent probability.

solution The mission time is equal to $T_0 = 30 \times 24 = 720$ hours and the standard normal variate corresponding to 99 percent probability is given by $z_{1-\alpha} = 2.33$. Thus, the number of spare parts required can be determined as follows:

Type A. $n_A = \lambda_A T_0 + 2.33\sqrt{\lambda_A T_0} = (0.002)(720) + 2.33\sqrt{(0.002)(720)} = 4.2360$ per unit

Type B. $n_B = \lambda_B T_0 + 2.33\sqrt{\lambda_B T_0} = (0.0001)(720) + 2.33\sqrt{(0.0001)(720)} = 0.6972$ per unit

Type C. $n_C = \lambda_C T_0 + 2.33\sqrt{\lambda_C T_0} = (0.0005)(720) + 2.33\sqrt{(0.0005)(720)} = 1.758$ per unit

Thus the total number of spares of type A, B, and C required is given by 4.2360 (20) = 84.72, 0.6972(130) = 90.636 and 1.758(100) = 175.8, respectively.

12.3 Availability

The term availability is used to indicate the probability of a system or equipment being in operating condition at any time t, given that it was in operating condition at $t = 0$. In order to be in operating condition at time t, the system must not have failed or, if it had failed during the period t, it must have been repaired. Thus, availability includes both the aspects of reliability and maintainability.

12.3.1 Definitions [12.1, 12.3]

Availability can be defined in several ways as indicated by the following figure.

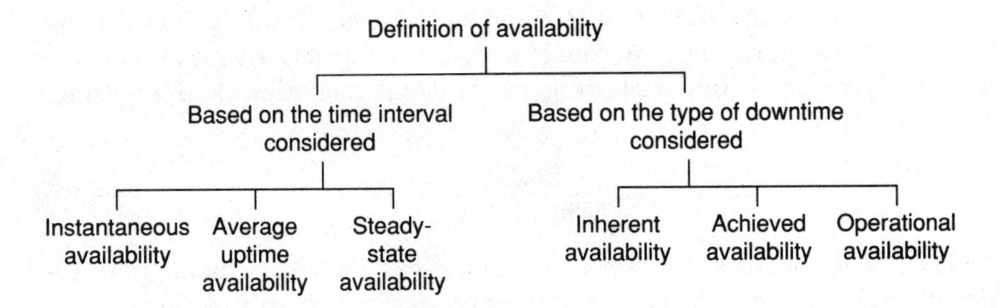

1. *Instantaneous Availability, $A(t)$*: It is the probability that the system is operational at any arbitrary time t. It is given by the expected up-time of the system

$$A(t) = E[z(t)] \tag{12.33}$$

where $z(t)$ is an indicator variable defined as

$$z(t) = \begin{cases} 0; & \text{if the system is in operating state at time } t \\ 1; & \text{if the system is in failed state at time } t \end{cases} \tag{12.34}$$

In view of Eq. (12.34), Eq. (12.33) can be rewritten as

$$A(t) = (1)\,P[z(t) = 0] + (0)\,P[z(t) = 1] = P[z(t) = 0] \tag{12.35}$$

2. *Average uptime availability, $A(T)$*: It is defined as the proportion of time during which the system is available for use in a specified interval $(0,T)$:

$$A(T) = \frac{1}{T}\int_0^T A(t)\,dt \tag{12.36}$$

3. *Steady state availability,* $A(\infty)$: It is defined as the probability that the system is operational when the time interval considered is very large:

$$A(\infty) = \lim_{T \to \infty} A(T) = \lim_{T \to \infty} \frac{1}{T} \int_0^T A(t)\, dt \tag{12.37}$$

4. *Inherent availability,* A_i: It is defined as the proportion of time during which the system is operational, by considering only corrective maintenance downtime and excluding ready time, preventive maintenance downtime, logistics (supply) time and waiting downtime:

$$A_i = \frac{\text{MTBF}}{\text{MTBF} + \text{MTTR}} \tag{12.38}$$

where MTBF = mean time between failures and MTTR = mean time to repair.

5. *Achieved availability,* A_a: It is defined as the proportion of time during which the system is operational by considering both corrective and preventive maintenance downtimes and excluding ready time, logistics (supply) time, and waiting downtime

$$A_a = \frac{\text{MTBM}}{\text{MTBM} + M} \tag{12.39}$$

where MTBM = mean time between maintenances and M = mean maintenance downtime due to breakdown and preventive maintenance actions.

6. *Operational availability,* A_o: It is defined as the proportion of time during which the system is operational, by considering ready time, logistics (supply) time, and waiting time along with corrective and preventive maintenance downtimes

$$A_o = \frac{\text{MTBF} + \text{ready time}}{\text{MTBF} + \text{ready time} + \text{MDT}} \tag{12.40}$$

where ready time = (operational cycle − MTBF − MDT), and MDT = mean downtime = M + delay time due to supply and administrative factors.

12.3.2 Availability analysis

Several different approaches have been used in the literature for the availability analysis of systems [12.1]. If prior information is available about the system parameters, a Bayesian approach can be used to obtain additional information about the system. The Bayesian approach in the formulation of availability models was used by several authors [12.10–12.12]. Brender [12.10] carried a statistical assessment of system availability by considering alternating sequence of independent exponentially distributed operational and repair intervals, with the failure time and repair time parameters described by distinct gamma distributions. The Monte Carlo simulation approach was suggested for the availability analysis of extremely complex systems and/or systems requiring costly experimentation [12.13].

The Markovian approach has been frequently used for the availability analysis using exponential distributions for failure times and repair times [12.14, 12.15]. Due to the mathematical complexities involved, relatively less work has been done using failure time and repair time distributions other than the exponential type. The Markovian approach is described in the following sections.

12.3.3 Development of the model

The following assumptions are made in developing a model for the availability analysis of a system:

1. At any given time the system is either in the operating state or in the failed state.
2. The state of the system changes as time progresses.
3. The transition of the system from one state to the other takes place instantaneously.
4. The failure and repair rates are constant.

The transition of the system from one state to the other is illustrated in Fig. 12.3. The probability of failure of the system and the probability of being returned to the operating state play important roles in the availability analysis. One of the best known state-space analysis techniques, namely, the Markov analysis, is presented in this section for the availability analysis of a system.

For a system consisting of one component, an indicator variable $z(t)$ can be defined as indicated in Eq. (12.34). If the system has a constant failure rate λ

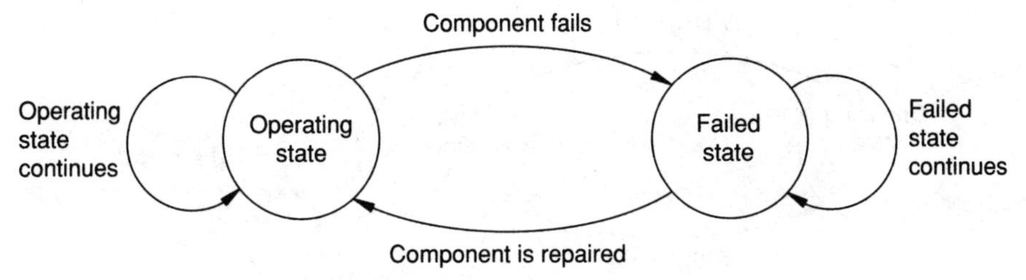

Figure 12.3 Transition diagram for the states of the system.

and a constant repair rate μ, the probabilities of the four possible events can be found as follows:

$$P\left[z(t+dt)=0\,|_{z(t)=1}\right] = \text{probability of the system being in operating state at time } t+dt \text{ given that it was in failed state at time } t$$
$$= \mu\, dt$$

$$P\left[z(t+dt)=0\,|_{z(t)=0}\right] = \text{probability of the system being in operating state at time } t+dt \text{ given that it was in operating state at time } t$$
$$= 1-\lambda\, dt$$

$$P\left[z(t+dt)=1\,|_{z(t)=0}\right] = \text{probability of the system being in failed state at time } t+dt \text{ given that it was in operating state at time } t$$
$$= \lambda\, dt$$

$$P\left[z(t+dt)=1\,|_{z(t)=1}\right] = \text{probability of the system being in failed state at time } t+dt \text{ given that it was in failed state at time } t$$
$$= 1-\mu\, dt$$

All the four transition probabilities are shown in the Markov diagram of Fig. 12.4. The transition probabilities are also shown in the form of a transition (or stochastic) matrix $[P]$ as

$$[P] = \begin{array}{c} \\ 0 \\ 1 \end{array}\begin{array}{c} \begin{array}{cc} 0 & 1 \end{array} \\ \begin{bmatrix} (1-\lambda) & \lambda \\ \mu & (1-\mu) \end{bmatrix} \end{array} \tag{12.41}$$

where the ijth entry denotes the probability that the system will be in state i at time $t+dt$ given that it was in state j at time t.

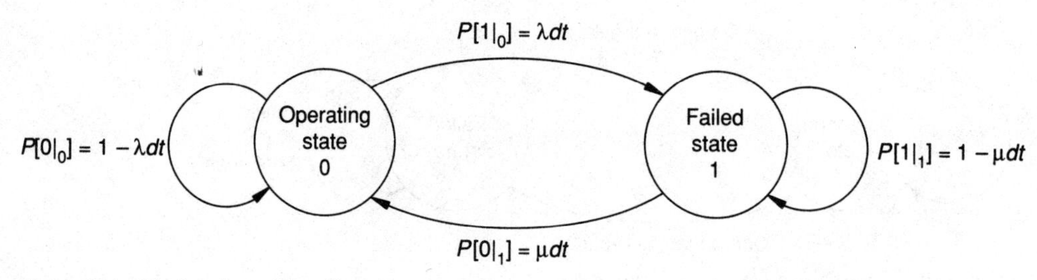

Figure 12.4 Markov transition diagram.

12.3.4 Systems with a single component

For a system consisting of one component, the probability of its being in operating state (state 0) at time $t + dt$, $P_0(t + dt)$, can be determined by adding the probability that it was in state 0 at time t and did not fail in $(t, t + dt)$ to the probability that it was in state 1 at time t and was brought to state 0 in $(t, t + dt)$. Thus,

$$P_0(t + dt) = P_0(t)(1 - \lambda\, dt) + P_1(t)\mu\, dt \tag{12.42}$$

Similarly, the probability of the system being in state 1 at time $t + dt$, $P_1(t + dt)$, is given by the sum of the probability that it was in state 0 at time t and failed in $(t, t + dt)$ and the probability that it was in state 1 at time t and the repair was not completed in $(t, t + dt)$. Thus,

$$P_1(t + dt) = P_0(t)\lambda\, dt + P_1(t)(1 - \mu\, dt) \tag{12.43}$$

Since the derivative, $dP_i(t)/dt$, can be defined as

$$\frac{dP_i(t)}{dt} = \lim_{dt \to 0} \frac{P_i(t + dt) - P_i(t)}{dt} \tag{12.44}$$

Eqs. (12.42) and (12.43) can be rewritten as

$$\frac{dP_0(t)}{dt} = -\lambda P_0(t) + \mu P_1(t) \tag{12.45}$$

$$\frac{dP_1(t)}{dt} = +\lambda P_0(t) - \mu P_1(t) \tag{12.46}$$

If the system was in operating state at time $t = 0$, the initial conditions are given by

$$P_0(0) = 1,\ P_1(0) = 0 \tag{12.47}$$

The solution of the simultaneous differential equations, Eqs. (12.45) and (12.46), with the initial conditions of Eq. (12.47) can be obtained as [12.3]

$$P_0(t) = A(t) = \frac{\mu}{\lambda + \mu} + \frac{\lambda}{\lambda + \mu} e^{-(\lambda + \mu)t} \tag{12.48}$$

$$P_1(t) = 1 - A(t) = \frac{\lambda}{\lambda + \mu} - \frac{\lambda}{\lambda + \mu} e^{-(\lambda + \mu)t} \tag{12.49}$$

On the other hand, when the system was in failed state at time $t = 0$, the initial conditions are

$$P_0(0) = 0,\ P_1(0) = 1 \tag{12.50}$$

The solution of Eqs. (12.45) and (12.46), with the initial conditions of Eq. (12.50), can be found as [12.3]

$$P_0(t) = A(t) = \frac{\mu}{\lambda + \mu} - \frac{\lambda}{\lambda + \mu} e^{-(\lambda + \mu)t} \tag{12.51}$$

$$P_1(t) = 1 - A(t) = \frac{\lambda}{\lambda + \mu} + \frac{\lambda}{\lambda + \mu} e^{-(\lambda + \mu)t} \tag{12.52}$$

It can be seen that as t becomes large, the solutions of Eqs. (12.48) and (12.51) converge to

$$A(\infty) = \frac{\mu}{\lambda + \mu} \tag{12.53}$$

Similarly, as $t \to \infty$, the solutions of Eqs. (12.49) and (12.52) converge to

$$1 - A(\infty) = \frac{\lambda}{\lambda + \mu} \tag{12.54}$$

Equations (12.53) and (12.54) indicate that the state of the system, after a long time, becomes independent of its starting (or initial) state.

The steady-state solution can also be found by setting derivatives of P_i equal to zero since the derivative of the steady-state value, which is a constant, is zero. By setting $dP_0/dt = 0$ and $dP_1/dt = 0$ in Eqs. (12.48) and (12.49), we obtain

$$-\lambda P_0 + \mu P_1 = 0 \tag{12.55}$$

$$\lambda P_0 - \mu P_1 = 0 \tag{12.56}$$

Since the sum of the probabilities of mutually exclusive events is one, we have

$$P_0 + P_1 = 0 \tag{12.57}$$

Equations (12.55) to (12.57) yield the steady state solutions as

$$P_0 = A(\infty) = \frac{\mu}{\lambda = \mu} \tag{12.58}$$

$$P_1 = 1 - A(\infty) = \frac{\lambda}{\lambda + \mu} \tag{12.59}$$

In some cases, we will be interested in finding the avarage uptime of the system over a specific period of time T. This can be found by integrating $A(t)$ over the time interval T and dividing by the total time:

$$A(T) = \frac{1}{T} \int_0^T A(t)\, dt \tag{12.60}$$

If the system was in operating state at $t = 0$, $A(t)$ is given by Eq. (12.48) and Eq. (12.60) yields

$$A(T) = \frac{\mu}{\lambda + \mu} + \frac{\lambda}{(\lambda + \mu)^2 T} - \frac{\lambda}{(\lambda + \mu)^2 T} e^{-(\lambda + \mu) T} \tag{12.61}$$

Similarly, if the system was in failed state at $t = 0$, $A(t)$ is given by Eq. (12.51) and Eq. (12.60) yields

$$A(T) = \frac{\mu}{\lambda + \mu} - \frac{\lambda}{(\lambda + \mu)^2 T} + \frac{\lambda}{(\lambda + \mu)^2 T} e^{-(\lambda + \mu) T} \tag{12.62}$$

The steady state or long term availability of the system can be determined by letting $T \to \infty$ in Eqs. (12.61) and (12.62). This gives

$$A(\infty) = \frac{\mu}{\lambda + \mu} \tag{12.63}$$

Equation (12.63) can also be expressed as

$$A(\infty) = \frac{\text{MTTF}}{\text{MTTF} + \text{MTTR}} \tag{12.64}$$

where MTTF $= 1/\lambda =$ mean time to failure and MTTR $= 1/\mu =$ mean time to repair.

Example 12.10 A system has a mean time between failures of 150 hours and a mean time to repair of 20 hours. What is the steady-state availability of the system?

solution The steady-state availability of the system can be found from Eq. (12.64) as

$$A(\infty) = \frac{150}{150 + 20} = 0.8824$$

12.3.5 Series systems

Consider a series system with two components A and B as shown in Fig. 12.5. Let the failure rate be λ and the repair rate be μ for each component. The system will be in any of the following three possible states at any time t:

1. Both components are in operating condition (state 0)
2. One component is in operating condition and the other under repair (state 1)
3. Both components are under repair (state 2).

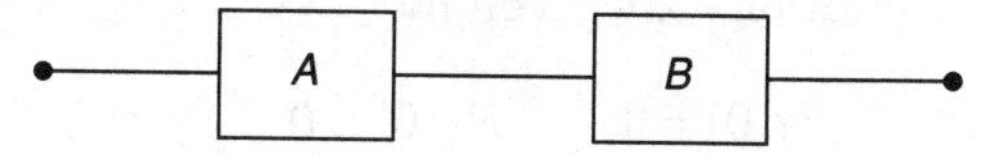

Figure 12.5 A series system.

Since both components must be in operating condition for the system to be in operating condition, the availability of the system is given by the probability of the system being in state 0 at time t, that is, $A(t) = P_0(t)$. The system will be down in states 1 and 2.

One repair person. The availability analysis of a multicomponent system depends on the number of repair persons available for servicing the failed components. First we consider the case with one repair person. The resulting transition probabilities are summarized in Table 12.1.

From this table, the transition matrix [P] can be identified as [12.3]

$$[P] = \begin{array}{c} \\ 0 \\ 1 \\ 2 \end{array} \begin{array}{c} \begin{array}{ccc} 0 & 1 & 2 \end{array} \\ \begin{bmatrix} (1-2\lambda) & 2\lambda & 0 \\ \mu & (1-\lambda-\mu) & \lambda \\ 0 & \mu & (1-\mu) \end{bmatrix} \end{array} \tag{12.65}$$

Thus the probability of the system being in state 0 or 1 or 2 at time $t + dt$ can be written as:

$$P_0(t+dt) = P_0(t)\,(1 - 2\lambda\, dt) + P_1(t)\,\mu\, dt \tag{12.66}$$

$$P_1(t+dt) = P_0(t)\, 2\lambda\, dt + P_1(t)\,(1 - \lambda\, dt - \mu\, dt) + P_2(t)\,\mu\, dt \tag{12.67}$$

$$P_2(t+dt) = P_1(t)\,\lambda\, dt + P_2(t)\,(1 - \mu\, dt) \tag{12.68}$$

These equations can be rewritten, using the definition given in Eq. (12.44), as

$$\frac{dP_0(t)}{dt} = -2\lambda P_0(t) + \mu P_1(t) \tag{12.69}$$

$$\frac{dP_1(t)}{dt} = 2\lambda P_0(t) - (\lambda + \mu) P_1(t) + \mu P_2(t) \tag{12.70}$$

$$\frac{dP_2(t)}{dt} = \lambda P_1(t) - \mu P_2(t) \tag{12.71}$$

These equations can be solved for any specified initial conditions to find P_0, P_1 and P_2 as functions of time. For example, if the system was in operating state at time $t = 0$, the initial conditions are given by

$$P_0(0) = 1; \qquad P_1(0) = 0; \qquad P_2(0) = 0 \tag{12.72}$$

TABLE 12.1 Transition Probabilities of a Two-Component Series System with One Repair Person

Serial number	State of the system at time t	State of the system at time $t + dt$	Transition of components	Transition probability
1	0 (operating)	0 (operating)	Both components remain in operating condition	$(1 - \lambda\, dt)^2 \approx 1 - 2\lambda\, dt$
2	0 (operating)	1 (one component failed)	Either of the components fails and the other remains in operating condition	$2\lambda\, dt(1 - \lambda\, dt) \approx 2\lambda\, dt$
3	0 (operating)	2 (two components failed)	Both components fail	$(\lambda\, dt)^2 \approx 0$
4	1 (one component failed)	0 (operating)	One remains in operating condition and repair is complete for failed one.	$\mu\, dt$
5	1 (one component failed)	1 (one component failed)	One remains in operating condition and repair is not complete for failed one.	$(1 - \mu\, dt)(1 - \lambda\, dt) \approx 1 - (\lambda + \mu)\, dt$
6	1 (one component failed)	2 (two components failed)	One component fails and repair of the other is not complete.	$\lambda\, dt\,(1 - \mu\, dt) \approx \lambda\, dt$
7	2 (two components failed)	0 (operating)	Repair of both components completed	$(\mu\, dt)^2 \approx 0$
8	2 (two components failed)	1 (one component failed)	Only one component repaired	$\mu\, dt$
9	2 (two components failed)	2 (two components failed)	Repair of two components not completed (Only one component is being repaired)	$1 - \mu\, dt$

In most cases, the steady state solution of Eqs. (12.69) to (12.71) will be of interest. For this, we set $dP_0/dt = 0$, $dP_1/dt = 0$ and $dP_2/dt = 0$ in Eqs. (12.69) to (12.71) and use the condition $P_0 + P_1 + P_2 = 1$:

$$-2\lambda P_0 + \mu P_1 = 0 \tag{12.73}$$

$$2\lambda P_0 - (\lambda + \mu) P_1 + \mu P_2 = 0 \tag{12.74}$$

$$\lambda P_1 - \mu P_2 = 0 \tag{12.75}$$

$$P_0 + P_1 + P_2 = 1 \tag{12.76}$$

The solution of Eqs. (12.73) to (12.76) gives

$$P_0 = \frac{\mu^2}{\mu^2 + 2\lambda\mu + 2\lambda^2} \tag{12.77}$$

$$P_1 = \frac{2\,\lambda\mu}{\mu^2 + 2\lambda\mu + 2\lambda^2} \tag{12.78}$$

$$P_2 = \frac{2\lambda^2}{\mu^2 + 2\lambda\mu + 2\lambda^2} \tag{12.79}$$

Thus the steady state availability of the system is given by

$$A(\infty) = P_0 = \frac{\mu^2}{\mu^2 + 2\lambda\mu + 2\lambda^2} \tag{12.80}$$

Two repair persons. In this case, each repair person can work on a different component. If the system was in state 2 at time t, it can return to state 1 at time $t + dt$ if repair of one of the components is complete, while that of the other is incomplete. Since the repair of either of the two components can be completed by the time $t + dt$, the probability is $2\mu\, dt(1 - \mu\, dt) \approx 2\mu\, dt$. On the other hand, if the system was in state 2 at time t, it continues to remain in state 2 if the repair of both the components is incomplete. The probability of this event is equal to $(1 - \mu\, dt)^2 \approx 1 - 2\mu\, dt$. Thus the transition matrix, $[P]$, of Eq. (12.65) will be modified as

$$[P] = \begin{array}{c} \\ 0 \\ 1 \\ 2 \end{array} \begin{array}{c} \begin{array}{ccc} 0 & 1 & 2 \end{array} \\ \begin{bmatrix} (1-2\,\lambda) & 2\,\lambda & 0 \\ \mu & (1-\lambda-\mu) & \lambda \\ 0 & 2\,\mu & (1-2\,\mu) \end{bmatrix} \end{array} \tag{12.81}$$

This matrix can be used to derive the steady state equations as

$$-2\lambda P_0 + \mu P_1 = 0 \tag{12.82}$$

$$2\lambda P_0 - (\lambda + \mu)P_1 + 2\mu P_2 = 0 \tag{12.83}$$

$$\lambda P_1 - 2\mu P_2 = 0 \tag{12.84}$$

$$P_0 + P_1 + P_2 = 1 \tag{12.85}$$

The solution of Eqs. (12.82) to (12.85) is given by

$$P_0 = \frac{\mu^2}{\mu^2 + 2\lambda\mu + \lambda^2} \tag{12.86}$$

$$P_1 = \frac{2\lambda\mu}{\mu^2 + 2\lambda\mu + \lambda^2} \tag{12.87}$$

$$P_2 = \frac{\lambda^2}{\mu^2 + 2\lambda\mu + \lambda^2} \tag{12.88}$$

Thus the steady state availability of the system will be

$$A(\infty) = P_0 = \frac{\mu^2}{\mu^2 + 2\lambda\mu + \lambda^2} \tag{12.89}$$

12.3.6 Parallel systems

Consider a parallel system consisting of two components A and B as shown in Fig. 12.6. For this system, there are three possible states: state 0 (both components operating), state 1 (one component operating and the other under repair), and state 2 (both components under repair).

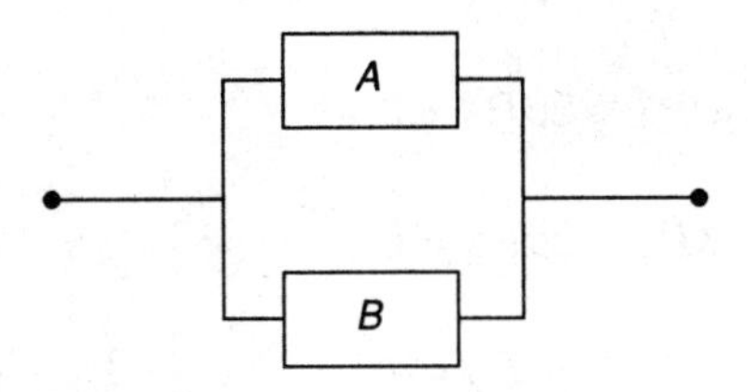

Figure 12.6 A parallel system.

One repair person. If there is only a single repair person to service the two components, the transition matrix can be derived as:

$$[P] = \begin{matrix} & \begin{matrix} 0 & 1 & 2 \end{matrix} \\ \begin{matrix} 0 \\ 1 \\ 2 \end{matrix} & \begin{bmatrix} (1-2\lambda) & 2\lambda & 0 \\ \mu & (1-\lambda-\mu) & \lambda \\ 0 & \mu & (1-\mu) \end{bmatrix} \end{matrix} \tag{12.90}$$

This gives the steady state equations as

$$-2\lambda P_0 + \mu P_1 = 0 \tag{12.91}$$

$$2\lambda P_0 - (\lambda + \mu) P_1 + \mu P_2 = 0 \tag{12.92}$$

$$\lambda P_1 - \mu P_2 = 0 \tag{12.93}$$

$$P_0 + P_1 + P_2 = 1 \tag{12.94}$$

The solution of these equations is given by Eqs. (12.77) to (12.79). Since states 0 and 1 constitute the operation of the system, the steady state availability of the system can be found as

$$A(\infty) = P_0 + P_1 = \frac{\mu^2 + 2\lambda\mu}{\mu^2 + 2\lambda\mu + 2\lambda^2} \tag{12.95}$$

Two repair persons. If there are two repair persons, one repair person can be assigned to each component and the transition matrix can be derived as

$$[P] = \begin{matrix} & \begin{matrix} 0 & 1 & 2 \end{matrix} \\ \begin{matrix} 0 \\ 1 \\ 2 \end{matrix} & \begin{bmatrix} (1-2\lambda) & 2\lambda & 0 \\ \mu & (1-\lambda-\mu) & \lambda \\ 0 & 2\mu & (1-2\mu) \end{bmatrix} \end{matrix} \tag{12.96}$$

From this matrix, the steady state equations can be obtained as

$$-2\lambda P_0 + \mu P_1 = 0 \tag{12.97}$$

$$2\lambda P_0 - (\lambda + \mu) P_1 + 2\mu P_2 = 0 \tag{12.98}$$

$$\lambda P_1 - 2\mu P_2 = 0 \tag{12.99}$$

$$P_0 + P_1 + P_2 = 1 \tag{12.100}$$

The solution of these equations is given by Eqs. (12.86) to (12.88). The steady state availability of the system can be determined as

$$A(\infty) = P_0 + P_1 = \frac{\mu^2 + 2\lambda\mu}{\mu^2 + 2\lambda\mu + \lambda^2} \tag{12.101}$$

12.4 Optimization Approaches

Several optimization methods have been used for solving maintainability and availability problems. The allocation of availability parameters (repair times and failure rates) to various components of a system for minimum total cost was formulated and solved using the Lagrange multiplier method in Ref. [12.5]. In Ref. [12.6], the optimal values of MTBF, MTTR, and the number of redundant units to achieve a specified availability for minimum cost were found using a dynamic programming approach. The steady-state availability of a repairable system with cold standbys and nonzero replacement time is maximized under constraints of total cost and total weight in Ref. [12.7]. The provision of spare parts significantly influences the operating costs, maintainability and availability of a system. The optimal determination of spare parts can be carried for maximum availability or reliability, and minimum spare part and operating costs or weight of spare parts. Different models have been developed for the optimal spare part determination and some of the methods are being routinely used by airlines, rental car agencies, etc. [12.8].

References and Bibliography

12.1. C. H. Lie, C. L. Hwang, and F. A. Tillman, "Availability of Maintained Systems: A State-of-the-art Survey," *AIIE Trans.*, vol. 9, 1977, pp. 247–259.

12.2. M. O. Locks, "Reliability, Maintainability, and Availability Assessment," Hayden Book Co., Rochelle Park, New Jersey, 1973.

12.3. G. H. Sandler, *System Reliability Engineering*, Prentice-Hall, Englewood Cliffs, New Jersey, 1963.

12.4. E. E. Lewis, *Introduction to Reliability Engineering*, John Wiley, New York, 1987.

12.5. R. J. McNichols and G. H. Messer, Jr., "A Cost-Based Availability Allocation Algorithm," *IEEE Transactions on Reliability*, vol. R-20, 1971, pp. 178–182.

12.6. B. K. Lambert, A. G. Walvekar and J. P. Hirmas, "Optimal Redundancy and Availability Allocation in Multistage Systems," *IEEE Trans. on Reliability*, vol. R-20, 1971, pp. 182–185.

12.7. M. Sasaki, S. Kaburaki, and S. Yanagi, "System Availability and Optimum Spare Units," *IEEE Trans. on Reliability*, vol. R-26, 1977, pp. 182–188.

12.8. E. G. Frankel, *Systems Reliability and Risk Analysis*, Martinus Nijhoff Publishers, The Hague, The Netherlands, 1984.

12.9. S. M. Selby, *Standard Mathematical Tables*, The Chemical Rubber Co., Cleveland, 17th ed., 1969.

12.10. D. M. Brender, "The Prediction and Measurement of System Availability: A Bayesian Treatment," *IEEE Trans. on Reliability*, vol. R-17, 1968, pp. 127–138.

12.11. D. P. Gover, Jr. and M. Mazumdar, "Some Bayes Estimates of Long-Run Availability in a Two-State System," *IEEE Trans. on Reliability*, vol. R-18, 1969, pp. 184–189.

12.12. W. E. Thompson and M. D. Springer, "A Bayes Analysis of Availability for a System Consisting of Several Independent Subsystems," *IEEE Trans. on Reliability*, vol. R-21, 1972, pp. 212–214.

12.13. W. E. Faragher and H. S. Watson, "Availability Analysis—A Realistic Methodology," 10th Nat. Symp. on Reliability and Quality Control, Washington, D.C., January 1964, pp. 365–378.

12.14. R. E. Barlow and F. Proschan, *Mathematical Theory of Reliability*, John Wiley, New York, 1965.

12.15. L. T. Hunt, "Reliability Prediction Techniques for Complex Systems," *IEEE Trans. on Reliability*, vol. R-15, 1966, pp. 58–69.

12.16. M. A. Moss, *Designing for Minimal Maintenance Expense*, Marcel Dekker, New York, 1985.

12.17. A. C. King and C. B. Read, *Pathways to Probability*, Holt, Rinehart & Winston, Inc., New York, 1963.

12.18. C. C. Gillespie (ed.-in-chief), *Dictionary of Scientific Biography* , vol. 9, Charles Scribner's Sons, New York, 1980.

Review Questions

12.1 Define the terms maintainability and availability.

12.2 What is the difference between preventive maintenance and corrective maintenance?

12.3 What is MTTR?

12.4 How is the reliability of a maintained system defined?

12.5 Define instantaneous availability of a system?

12.6 What is average uptime availability?

12.7 How is steady state availability defined?

12.8 Explain the difference between inherent, achieved, and operational availabilities.

12.9 What is a transition matrix?

12.10 What is Markov analysis?

12.11 What is the effect of imperfect maintenance on the reliability of a system?

12.12 State the factors that influence the optimal replacement strategy.

12.13 How is the number of spare parts determined for a single-component system?

Problems

12.1 The repair time data for a boiler in a thermal power plant are given by 4.2, 4.5, 4.9, 5.0, 5.2, 5.6, 5.8, 6.1, 6.3, 6.4, 6.7, 6.8, 7.0, 7.3 and 7.5 hours. Find

(a) the maintainability of the boiler for an allowed downtime of 6 hours.

(b) the downtime required to achieve a maintainability of 0.95.

(c) the required MTTR when the allowable downtime is 6 hours for a maintainability of 0.95. Assume that the repair times follow exponential distribution.

12.2 Solve Problem 12.1 assuming that the repair times follow normal distribution.

12.3 Solve Problem 12.1 assuming that the repair times follow lognormal distribution.

12.4 Solve Problem 12.1 assuming that the repair times follow Weibull distribution.

12.5 The times required to restore power outages by an electric power company are 0.5, 0.7, 0.8, 1.0, 1.2, 1.3, 1.6, 1.7, 1.9, 2.1, 2.2, 2.6, 2.9, 3.4, and 4.2 hours. Determine the following:

(a) The maintainability of the power supply if the outage is permitted for 5 hours.

(b) The duration of the outage corresponding to a maintainability of 0.99. Solve the problem assuming (i) normal distribution, and (ii) lognormal distribution for the outage restoring times.

12.6 Find the reliability of a system for which preventive maintenance is performed after every 500 hours of operation when the failure time T in hours, of the original system is given by the following distribution:

(a) $f_T(t) = 0.001\, e^{-0.001\, t}$

(b) $f_T(t) = \dfrac{1}{\sqrt{2\,\pi}\,(50)} \exp\left\{-\dfrac{1}{2}\left(\dfrac{t-1000}{50}\right)^2\right\}$

12.7 A system has a mean time between failures of 200 hours, mean time to repair of 10 hours, mean ready time of 5 hours and a mean downtime of 15 hours. Find the inherent, achieved, and operational availabilities of the system.

12.8 Find the time interval, t_0, needed for replacing the components in preventive maintenance if the reliability of the original (unmaintained) components is given by

$$R(t) = \exp\left\{-\left(\frac{t}{100}\right)^2\right\}; \qquad t \text{ in hrs}$$

when c_m and c_r are given by 100 and 1000 dollars, respectively.

12.9 Find the time interval, t_0, needed for replacing the components in preventive maintenance if the reliability of the original (unmaintained) components is given by

$$R(t) = \exp\left\{-\left(\frac{t}{50}\right)^4\right\}; \qquad t \text{ in hrs}$$

where c_m and c_r are given by 100 and 5000 dollars, respectively.

12.10 A system consists of three types of components with 10 units of type-A, 20 units of type-B, and 30 units of type-C. The mean times before failure of type-A, -B and -C components are, respectively, 2000, 1500 and 3000 hours. Find the number of spare parts required over a mission time of 5000 hours for a confidence of (a) 90 percent (b) 95 percent, and (c) 99 percent.

12.11 Find the mean time to failure of a system for which preventive maintenance is performed at times $t_0,\ 2t_0,\ 3t_0, \ldots,$ when the failure time follows (a) exponential distribution, $f_T(t) = \lambda e^{-\lambda t}$, (b) Weibull distribution,

$$f_T(t) = \frac{m}{\theta}\left(\frac{t}{\theta}\right)^{m-1} \exp\left\{-\left(\frac{t}{\theta}\right)^m\right\}$$

12.12 Find the condition under which the ideal preventive maintenance would be beneficial for a system for which the failure time distribution is given by

$$f_T(t) = k^2 t e^{-kt}$$

12.13 The number of failures in a thermal power station in a year is 65. The total time spent in restoring the system is 220 hours. Find the probability of repairing a failure in (a) 1 hour, (b) 2 hours, and (c) 10 hours.

12.14 A welding robot in an assembly shop is found to fail once in every 100 hours of operation and a repair takes, on the average, 1 hour. Determine the number of failures that are likely not to get repaired over a period of one year if a repair time of 1 hour is allowed for each repair. Assume that the robot is operated 8 hours per day for 350 days in a year.

12.15 A system has a mean time between failures of 200 hours and a mean time to repair of 30 hours. Find the steady state availability of the system.

12.16 A series system consists of two identical components each with a failure rate of 0.005 per hour and a repair rate of 0.1 per hour. Find the steady state availability of the system if the number of repair persons is (a) one, and (b) two.

12.17 A parallel system consists of two identical components each with a failure rate of 0.005 per hour and a repair rate of 0.1 per hour. Find the steady state availability of the system if the number of repair persons is (a) one, and (b) two.

12.18 A turbine is known to sustain two types of failures: bearing failures and blade failures. The bearing failure times follow exponential distribution with a failure rate of 0.0005 per hour and the blade failure times follow Weibull distribution with

$$f_T(t) = \frac{1}{100}\left(\frac{t}{200}\right) \exp\left\{-\left(\frac{t}{200}\right)^2\right\}; \qquad t \text{ in hours}$$

(a) Find the reliability of the unmaintained system after 10,000 hours of operation of the turbine.

(b) If the reliability of the turbine is to be increased by 20 percent at the end of 10,000 hour operating period by replacing the turbine blades at times $t_0, 2t_0, 3t_0, \ldots,$ find the value of t_0.

12.19 The failure rate of a system is given by $h(t) = 0.001\sqrt{t}$; t in hours. If the design life of the system is 10,000 hours, determine the following:

(a) Reliability of the unmaintained system at the end of the design life.

(b) Reliability of the system assuming perfect maintenance after every 1000 hours of operation.

(c) Reliability of the system assuming imperfect preventive maintenance after every 1000 hours of operation with a 1 percent probability of failure of the system immediately after the maintenance.

12.20 The MTTF and MTTR of a spot welding robot are 60 hours and 3 hours, respectively. Find the availability of the robot. If the MTTR is reduced to 1 hour, find the smallest value of MTTF which gives at least the same value of availability.

12.21 The times at which a multicomponent system failed and the times at which the system is restored through repair over a mission period of 1000 hours are shown in the following table.

Time at which failure occurred (hours)	Time at which system restored (hours)
74.2	77.5
102.1	104.2
193.4	194.1
245.5	247.8
381.0	381.9
415.3	418.2
464.6	467.1
499.8	502.9
580.2	583.4
633.1	635.1
678.9	680.3
711.7	714.2
788.5	791.6
891.0	893.3
982.4	985.8

Find the following: (a) Average availability of the system, (b) Average failure and repair rates of the system, and (c) Availability of the system using the failure and repair rates found in part (b).

12.22 Plot the availability functions given by Eqs. (12.48), (12.51), and (12.63) for a system having a failure rate of 0.05 per hour and a repair rate of 0.5 per hour.

Chapter

13

Failure Modes, Event-Tree, and Fault-Tree Analyses

Biographical Note

John von Neumann

John von Neumann was born in Budapest, Hungary on December 28, 1903. He received his B.A. in 1925 in Chemical Engineering from the Zurich Institute and his Ph.D. in 1926 in Mathematics from the University of Budapest. He emigrated to the United States in 1930 and taught at Princeton University for the rest of his career. Von Neumann is best known for his work in game theory, quantum mechanics, and computing. He worked on the government's atomic bomb project at Los Alamos Scientific Laboratory during World War II. At that time, the scientists at Los Alamos wanted to find the distance the neutrons travel through various materials in connection with shielding and other important aspects. Although the physicists had most of the data, such as the average distance a neutron of a given speed would travel in a given material before it collides with an atomic nucleus and the probability of a neutron bouncing off instead of being absorbed by the nucleus, they found it to be an extremely complicated problem beyond the reach of theoretical calculations; the experimental determination by trial and error would have been expensive, time-consuming, and hazardous. To that crisis, von Neumann, along with Stanislas Ulam, suggested a remarkably simple, workable answer based on the use of a roulette wheel. When the method was revived for the secret work at Los Alamos, von Neumann gave it the code name "Monte Carlo." The method was found to be so successful on neutron diffusion problems that it later was applied to numerous other fields. Von Neumann was appointed as a member of the Atomic Energy Commission in 1955 and died in Washington, D.C. on February 8, 1957 [13.19–13.21].

13.1 Introduction

The failure-mode, event-tree and the fault-tree analyses fall under the category of "design and safety review techniques." Although the considerations of reliability are included in the design stage, it is desirable to conduct a review of the complete system from safety point of view. The failure of a complex system may not necessarily always be due to the failure of its components, but sometimes due to the faults of operators and maintenance crews. Thus the consideration of human errors is also important in the safety analysis. Some system failures may simply imply inconvenience and loss of productivity, while some others imply loss of human life. For example, the breakdown of a household refrigerator may not cause a serious loss, whereas the breakdown of an artificial respirator in a hospital emergency room might cause a loss of human life.

Several design and safety review methods have been developed to understand as to how accidents occur, how their probabilities can be estimated and how to reduce the probability of their occurrences. These methods aid in the systematic analysis of a system for safety, in identifying potential weak spots that might lead to a safety hazard, and in highlighting areas that need special attention in the design, manufacture, operation, and maintenance of the system. In most cases, the design review might reveal that nearly all aspects of the design are satisfactory. However, in some cases, it might reveal certain deficiencies of design from the point of view of system safety. It is desirable to discover any possible design deficiencies at preliminary design stages, even at the expense of considerable effort, rather than having to modify the design at a later stage or live with the consequences of a defective design. Thus, a well-planned and well-conducted design review will prove to be extremely cost effective in the design and development of complex engineering systems. Among the various methods available for the design and safety review of complex systems, the failure modes and effects analysis, the event-tree analysis, and the fault-tree analysis are particularly popular in industry.

13.2 System Safety Analysis

Both qualitative and quantitative methods are available for system safety analysis. The qualitative methods help in understanding the logical structure of the various failure modes of a system and their interrelationship. This, in turn, might lead to clues as to how the likelihood of certain accidents can be reduced or eliminated. The quantitative methods, on the other hand, make use of the available component failure data, estimates of repair times and human errors, and predict the probabilities of occurrence of certain accidents. In practice, the type of system safety analysis to be used depends on the complexity of the system, the availability of failure data, and the degree to which human factors are involved. The various types of system safety analysis procedures used for the identification and quantification of safety and risk of a system are shown in the following figure [13.9]. As shown in the figure, all the available methods can be broadly classified into inductive and deductive techniques.

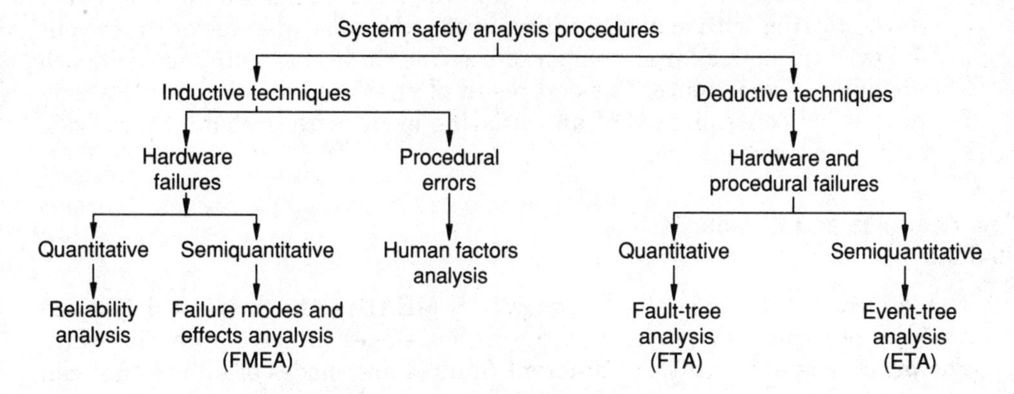

Basic types of system-safety analysis.

In the inductive procedures, the analysis begins at the component level, identifies the failure modes of each component, and establishes the effect of each component failure on the overall system. Thus, the inductive analysis works its way up in the system through higher assemblies until the complete system is analyzed. The failure modes and reliability analyses belong to the category of inductive hardware-failure analysis techniques. The failure-modes analysis is used to examine each component of the system and find how various modes of failure affect the system. This technique is mostly subjective due to the qualitative nature of the analysis. An extension of this procedure takes human factors into account and results in a semiquantitative method. The reliability analysis deals with the methods of finding the probability of survival of the system over a specified period under prescribed operating conditions.

In the deductive techniques, the analysis starts with an enumeration of the potential hazards and works down through the system to identify the system hardware failures or human errors which could have caused the potential hazards. The fault-tree and event-tree analyses belong to deductive failure-analysis techniques. In the fault-tree analysis, a top event, such as an explosion, fire or release of toxic material which signifies the total failure of the system, is identified first and then all the possible causes (precursor events or basic causes) are identified. This method of analysis is probably the most powerful and accurate technique for the quantification of risk [13.1]. The event-tree analysis is a logical procedure, based on a form of binary logic, in which an event is deemed to have either happened or not, or a component is either deemed either to have worked properly or not. A major limitation of the method is that it does not consider partially degraded components. This method uses a logic that is essentially the reverse of that used in fault-tree analysis. Given the occurrence of a particular type of failure (called the top event), a fault tree identifies the various combinations and sequences of other failures that lead to the top event. On the other hand, in the event-tree analysis, starting with some initiating event, the consequences of that event are followed through a large number of possible chains assigning to each path a probability of occurrence. The end result of an event-tree analysis is a long list of possible consequences of an initiating event with probabilities associated with each consequence.

13.3 Failure Modes and Effects Analysis (FMEA)

The *failure modes and effects analysis* (FMEA) is the most widely used analysis procedure in practice at the initial stages of system development. The purpose is to identify the different failures and modes of failure that can occur at the component, subsystem, and system levels and to evaluate the consequences of these failures. It involves an analysis of the system to determine the effect of component or subsystem failure (1) on the overall performance of the system and (2) on the ability to meet performance requirements or objectives. The FMEA is usually performed during the conceptual and initial design phases of the system in order to assure that all possible failure modes have been considered and that proper provisions have been made to eliminate all the potential failures. Sometimes, a criticality analysis is also made along with the FMEA. In this case, an index related to the criticality or severity of the failure is also included. The analysis is then called a *failure modes, effects, and criticality analysis* (FMECA). A comprehensive FMECA for a complex system is a sizable task. However, this procedure is a very effective reliability design-review technique and is mandatory for many government funded projects. US MIL-STD-1629 [13.18] describes in detail how FMECA must be conducted, documented, and presented for military contracts. It is also cited in several other reliability program standards in USA and other countries.

The FMECA aids in producing block-diagram reliability analysis and diagnostic charts for repair purposes. In fact, the work involved in generating diagnostic procedures and user maintenance handbooks can be eased considerably if FMECA details are known to the designer. The level of detail to which FMECA is performed should be based on the purpose for which it is performed. For example, the feasibility study for a large system may consider only the functional subsystems with much of the detailed design yet to be performed. Then the initial FMECA will consider only the failures associated with the subsystems or blocks. As the detailed design progresses, the FMECA will be updated, which aids in the identification of detailed features of the system. Most FMECAs deal with single failure modes. If several critical failure modes are identified, their simultaneous effect will be studied separately.

The procedure involved in FMECA can be summarized as follows:

1. Identify all potential failure modes of the system.
2. Relate the causes, effects, and hazards of each mode of failure.
3. Prioritize the modes relative to their probabilities of occurrence, failure criticality (or severity), and detection capability.
4. Provide suitable follow-up or corrective actions for each type of failure mode.

The FMECA procedure can be implemented systematically using a standardized FMECA form. Although the details of the form should reflect the specific needs and characteristics of the particular system being analyzed, a typical FMECA form is shown in Fig. 13.1. In general, the following information is included in a FMECA form:

1. System. Name of the system or component under consideration.
2. Analyst. Name of the person conducting the analysis.
3. Function. Broad description of the function of the system.
4. Modes of failure. The modes in which the system can fail.
5. Cause of failure. The possible reason for the failure of the system.
6. Frequency of occurrence. The frequency of occurrence of this type of failure on a scale of 1 to 100 with 1 denoting a very rare occurrence and 100 denoting a very frequent occurrence.
7. Effects of failure. The local and overall consequences of failure, including possible loss or damage to humans and property.
8. Failure detection method. The method(s) of detecting/isolating the failure.

<table>
<tr><td colspan="4">System:
Function/Mission:</td><td colspan="9">Identification Number:
Date:
Analyst:</td></tr>
<tr><td colspan="4">Modes of failure</td><td colspan="5">Failure effects</td><td>Criticality</td><td colspan="3">Corrective action</td></tr>
<tr><td>Serial number</td><td>Failure mode</td><td>Cause</td><td>Frequency of occurence</td><td>Local effects</td><td>Overall effects</td><td>Failure detection method</td><td>Degree of severity</td><td>Probability of detection</td><td>Priority of risk</td><td>Design modification</td><td>Design verification</td><td>Remarks</td></tr>
<tr><td></td><td></td><td></td><td></td><td></td><td></td><td></td><td></td><td></td><td></td><td></td><td></td><td></td></tr>
</table>

Figure 13.1 Typical FMECA form.

9. Degree of severity. The expected degree of severity of the failure on a scale of 1 to 100 with 1 indicating almost no loss and 100 denoting a major loss to humans and property.

10. Probability of detection. The chance of detecting the failure before any damage is caused, on a scale of 1 to 100 with 1 denoting sure detection and 100 indicating impossible detection.

11. Priority of risk. The relative importance of the failure for corrective action on a scale of 1 to 100 with 1 denoting least priority and 100 indicating highest priority. This number can be obtained by multiplying the numbers in steps 6, 9 and 10 and dividing the result by 10,000.

12. Design modification. The action taken or the design modification made to reduce the priority of risk in step 11.

13. Design verification. The procedure to be used to verify the action taken or design modification made in step 12.

13.4 Event-Tree Analysis

An event tree is a graphical representation of all possible events in a system. It is called a "tree" because the graphical representation gradually expands in the form of a tree with more and more branches created as the number of events increases. It is based on binary logic in which an event is assumed to have either happened or not, or a component is assumed to have either worked properly or not. A major limitation of this analysis is that components that are partially degraded cannot be considered as such. This method proceeds essentially in reverse order compared to that used in the fault-tree analysis [13.2]. Given that a particular top event (fault) occurred, a fault tree identifies the various combinations and sequences of other failures that might have led to the top event. The event-tree analysis was originally devised to assess the safety of nuclear reactors.

In an event-tree analysis, we begin with some initiating event and follow the consequences of that event through a large number of possible paths (chains) until a final outcome is encountered. By assigning to each path a probability of occurrence, the end result is a long list of possible consequences of the initiating event in which each consequence has a certain probability associated with it. Example 13.1 is given to illustrate the procedure.

Example 13.1 A room consists of two light bulbs operated by a single switch as shown in Fig. 13.2. Considering the initial event as the "room without electric light," develop the event tree of the system.

solution In this case, we can proceed to trace a series of events by following through Fig. 13.2. Each time a component operates successfully, the corresponding path in the event tree is constructed upwards as shown in Fig. 13.3. Failures of components are indicated using descending paths. We start with two branches at the left of the event tree—one ("up") corresponding to the presence of light in the room and the other ("down") corresponding to the absence of

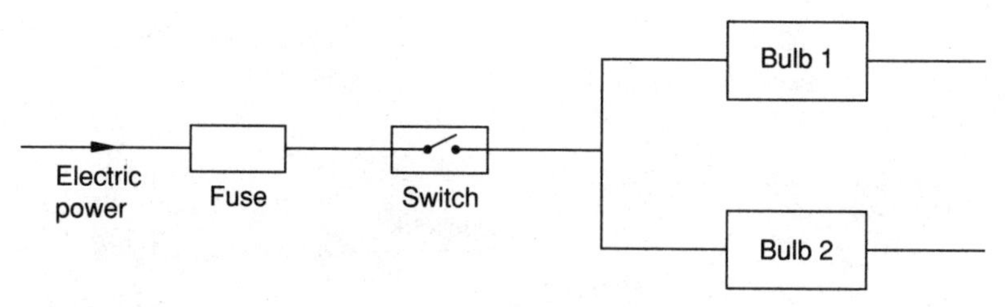

Figure 13.2

light in the room. For the branch corresponding to "no light in the room," we notice that the fuse must function properly for the electric power to reach the switch. Thus, we get two branches: "up" for fuse functioning and "down" for fuse nonoperating (blown off). The next set of branches are produced by the switch being in operating (closed) or failed (open) condition. Finally we produce branches for bulbs 1 and 2 being in good or burnt condition. After completing the event tree as shown in Fig. 13.3, we can label the final outcome with letters denoting which failures produced each branch. The second branch, with final event labeled "A," results from only the initial event, "room without electric light" (event A), with all components operating in the desired condition. The third branch represents event A followed by the failed condition of bulb 2 (event E) and thus is labeled "AE." From the event tree, we can also determine the physical outcome of each output event (percent of light in room corresponding to the event). For example, the second branch results in the operation of both the bulbs (100 percent light), the third and fourth branches represent the operation of one bulb only (50 percent light), and the remaining branches correspond to no light at all in the room (0 percent light). Figure 13.3, thus, shows how an event tree resulting from an initially defined event results in several different consequences.

Once the event tree is constructed, we can compute the probability of each of the final events. If the components are assumed to be statistically independent, the probability of each outcome is simply the product of probabilities of each event in the sequence. Note that in Fig. 13.3, an outcome such as "AE" implicitly includes the events $\overline{B}$, $\overline{C}$ and $\overline{D}$ where, for example, $\overline{B}$ denotes the complement of event B (B indicates the event "fuse in good condition," while $\overline{B}$ indicates the the event "fuse blown off"). Thus, the probability of the event "AE" (that is, the occurrence of "no electric light in the room" followed by the "failure of bulb 2") can be determined as

$$P_{AE} = P_A \, P_{\overline{B}} \, P_{\overline{C}} \, P_{\overline{D}} \, P_E \tag{13.1}$$

where P_X denotes the probability of occurrence of the event X. It can be seen that the probability of realizing "50% light in the room" can be found by adding P_{AE} and P_{AD}.

In many cases, the construction of the event tree can be simplified further [13.11]. These situations occur when a specific event leads to a final outcome regardless of any subsequent events. For example, if the fuse is in nonoperating condition in Example 13.1, the room will not have any light, irrespective of the conditions of the switch and the bulbs. Thus, it is unnecessary to proceed any further in the chain of events, and hence the branch corresponding to "failed fuse" can be immediately terminated. A similar reduction occurs if the switch fails, eliminating the need to check the conditions of bulbs 1 and 2. The resulting event tree, shown in Fig. 13.4, now has only six branches for the "no electric light in the room" situation, rather than the 16 such outcomes indicated in Fig. 13.3.

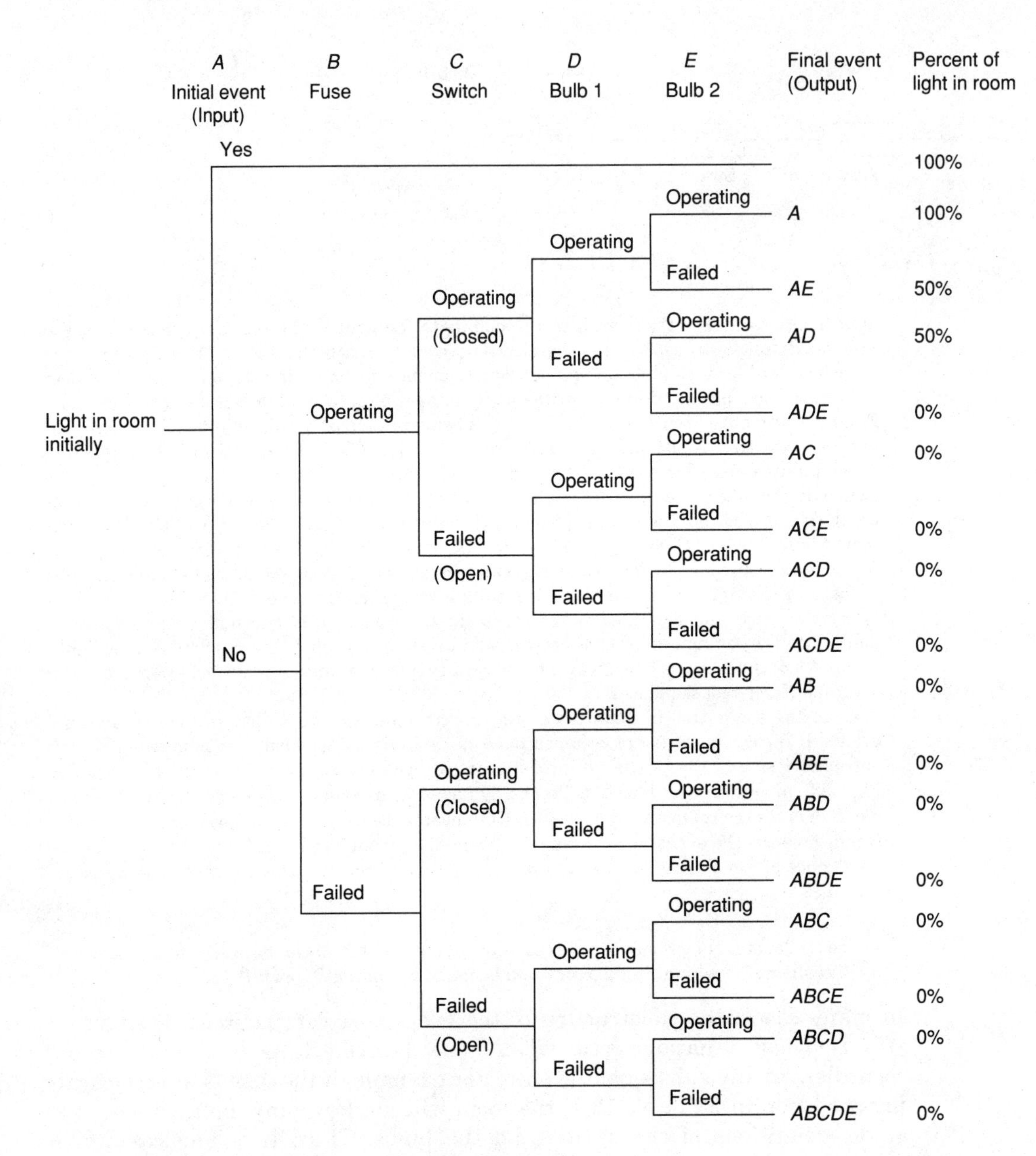

Figure 13.3 Event tree for the system depicted in Fig. 13.2.

The event tree thus provides a pictorial definition of the different possible sequences of events when a major failure occurs. Because of the tree-like appearance, each sequence has separate end points and it is practical to analyze the importance of the sequences by supplementing the diagram with

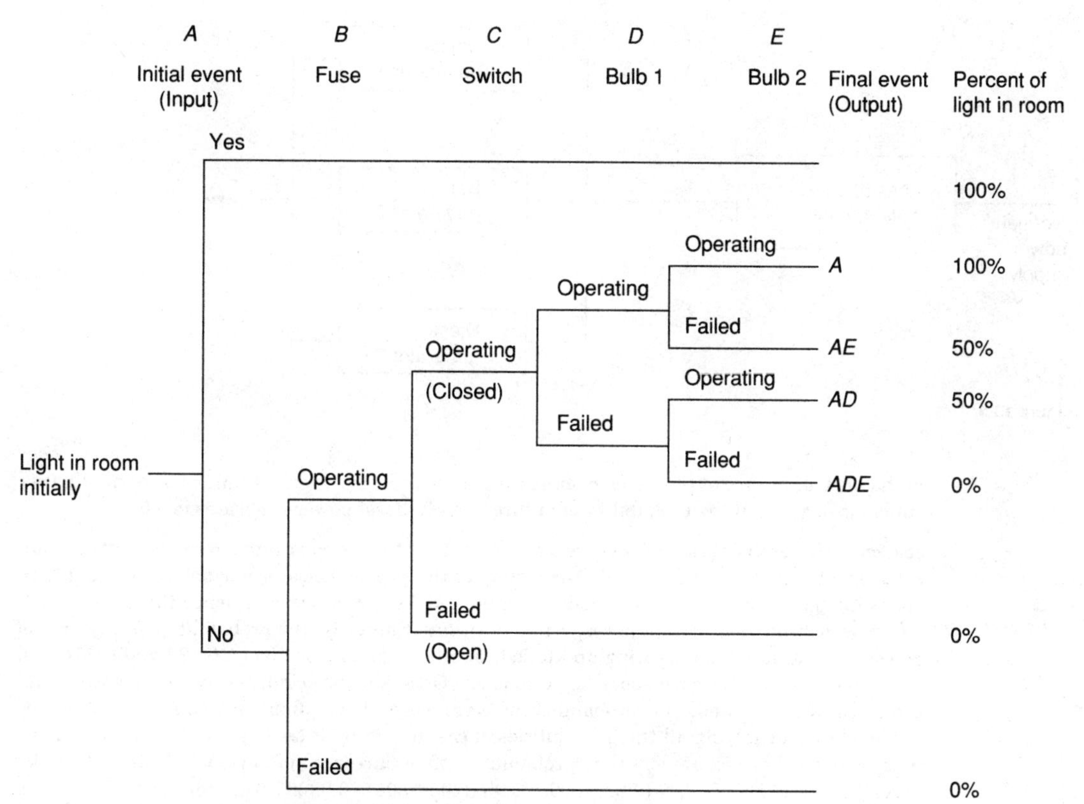

Figure 13.4 Reduced-event tree for the system depicted in Fig. 13.2.

a simple table. In practice, the probability of occurrence of the various paths of the event tree are not quantifiable during the synthesis of a system, but the approach provides a quick means of comparing different system configurations. A more detailed examination of the subsystems would be necessary to identify subsystem failure modes and effects. Such a detailed examination could be achieved by extending the event tree methodology to the fault tree analysis.

Example 13.2 Consider a diesel power generating system in a hospital (see Fig. 13.5). When normal power supply fails, a "loss of power detector unit" switches on and starts three diesel power generators. Each diesel power generator satisfies one-third of the needs of the hospital. Develop the event tree of the system.

solution When the initiating event occurs, that is, when the normal power supply fails, each of the "loss of power detector unit" and the "diesel power generator" may function or fail. The consideration of operation and failure of each component leads to the event tree shown in Fig. 13.6.

Example 13.3 Determine the probability of occurrence of each outcome of the system described in Example 13.2 if the probability of failure of the normal power supply is 0.01, the

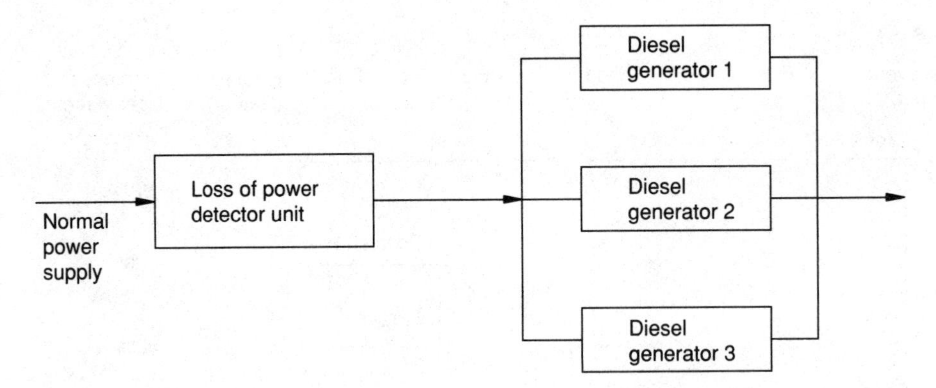

Figure 13.5

probability of failure of the detector unit switching on in the event of failure of normal power supply is 0.001, and the probability of failure of each diesel power generator is 0.05.

solution The probabilities of occurrence of each branch (for each unit or component) is indicated in parentheses in Fig. 13.6. For example, the probabilities of normal power supply in operating (good) and failed conditions are 0.99 and 0.01, respectively; hence these values are shown in parentheses along the respective branches. Similarly, the probabilities of the "loss of power detector unit" in operating and failed conditions are, respectively, 0.999 and 0.001, and are shown along the corresponding branches. Once the probabilities of occurrence of all branches are indicated, the probability of occurrence of any final outcome (event) can be obtained by multiplying all the probabilities of the individual branches that lead to the particular event. For example, the probability of occurrence of event #5 is given by $P(\bar{A}) \cdot P(B) \cdot P(C) \cdot P(\bar{D}) P(\bar{E}) = (0.01)(0.999)(0.95)(0.05)(0.05) = 0.00002373$. Similar calculations yield results as shown in Table 13.1. The fractional power supply available for each outcome is also indicated in Fig. 13.6 and Table 13.1.

13.5 Fault-Tree Analysis (FTA)

The fault-tree techniques have received a widespread attention in the reliability and safety analysis of complex systems. The techniques were extensively used in aerospace and nuclear industries [13.3, 13.11]. The importance of fault-tree analysis in the nuclear reactor design was discussed at length in the Reactor Safety Study [13.4]. The main reason for the widespread use of fault-tree analysis in nuclear and aerospace industries is due to the concern for human safety. The fault-tree analysis procedure was first introduced at Bell Telephone Laboratories in connection with the safety analysis of the Minuteman missile launch control system in 1962. Subsequently, the method was developed at the Boeing Company in the mid-1960's [13.5]. Since that time, fault-trees have been widely used to investigate the reliability and safety of complex and large systems and for diagnostic applications. A comprehensive bibliography on the fault-tree analysis and its applications was presented in Ref. [13.6].

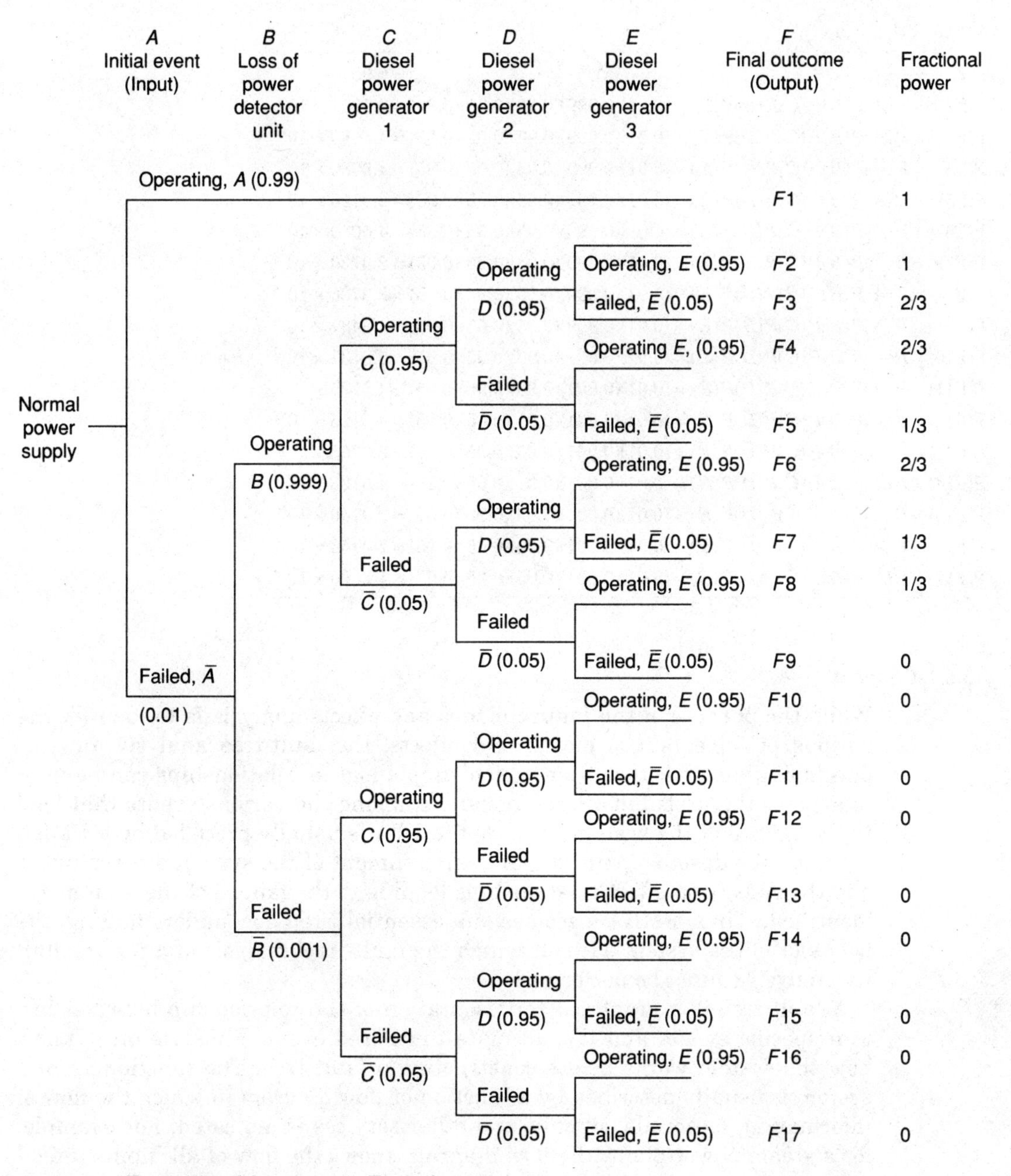

Figure 13.6 Event tree of the system in Example 13.2.

TABLE 13.1

Event	Expression	Value
P(F1)	$= P(A)$	$= 0.99$
P(F2)	$= P(\bar{A})P(B)P(C)P(D)P(E) = (0.01)(0.999)(0.95)(0.95)(0.95)$	$= 8.5652 \times 10^{-3}$
P(F3)	$= P(\bar{A})P(B)P(C)P(D)P(\bar{E}) = (0.01)(0.999)(0.95)(0.95)(0.05)$	$= 4.5080 \times 10^{-4}$
P(F4)	$= P(\bar{A})P(B)P(C)P(\bar{D})P(E) = (0.01)(0.999)(0.95)(0.05)(0.95)$	$= 4.5080 \times 10^{-4}$
P(F5)	$= P(\bar{A})P(B)P(C)P(\bar{D})P(\bar{E}) = (0.01)(0.999)(0.95)(0.05)(0.05)$	$= 2.3726 \times 10^{-5}$
P(F6)	$= P(\bar{A})P(B)P(\bar{C})P(D)P(E) = (0.01)(0.999)(0.05)(0.95)(0.95)$	$= 4.5080 \times 10^{-4}$
P(F7)	$= P(\bar{A})P(B)P(\bar{C})P(D)P(\bar{E}) = (0.01)(0.999)(0.05)(0.95)(0.05)$	$= 2.3726 \times 10^{-5}$
P(F8)	$= P(\bar{A})P(B)P(\bar{C})P(\bar{D})P(E) = (0.01)(0.999)(0.05)(0.05)(0.95)$	$= 2.3726 \times 10^{-5}$
P(F9)	$= P(\bar{A})P(B)P(\bar{C})P(\bar{D})P(\bar{E}) = (0.01)(0.999)(0.05)(0.05)(0.05)$	$= 1.2487 \times 10^{-6}$
P(F10)	$= P(\bar{A})P(\bar{B})P(C)P(D)P(E) = (0.01)(0.001)(0.95)(0.95)(0.95)$	$= 8.5737 \times 10^{-6}$
P(F11)	$= P(\bar{A})P(\bar{B})P(C)P(D)P(\bar{E}) = (0.01)(0.001)(0.95)(0.95)(0.05)$	$= 4.5125 \times 10^{-7}$
P(F12)	$= P(\bar{A})P(\bar{B})P(C)P(\bar{D})P(E) = (0.01)(0.001)(0.95)(0.05)(0.95)$	$= 4.5125 \times 10^{-7}$
P(F13)	$= P(\bar{A})P(\bar{B})P(C)P(\bar{D})P(\bar{E}) = (0.01)(0.001)(0.95)(0.05)(0.05)$	$= 2.3750 \times 10^{-8}$
P(F14)	$= P(\bar{A})P(\bar{B})P(\bar{C})P(D)P(E) = (0.01)(0.001)(0.05)(0.95)(0.95)$	$= 4.5125 \times 10^{-7}$
P(F15)	$= P(\bar{A})P(\bar{B})P(\bar{C})P(D)P(\bar{E}) = (0.01)(0.001)(0.05)(0.95)(0.05)$	$= 2.3750 \times 10^{-8}$
P(F16)	$= P(\bar{A})P(\bar{B})P(\bar{C})P(\bar{D})P(E) = (0.01)(0.001)(0.05)(0.05)(0.95)$	$= 2.3750 \times 10^{-8}$
P(F17)	$= P(\bar{A})P(\bar{B})P(\bar{C})P(\bar{D})P(\bar{E}) = (0.01)(0.001)(0.05)(0.05)(0.05)$	$= 1.2500 \times 10^{-9}$

13.5.1 Concept

While the purpose of the failure modes and effects analysis is to identify the various possible failure modes and effects, the fault-tree analysis aims at developing the structure from which simple logical relationships can be used to express the probabilistic relationships among the various events that lead to the failure of the system. In fact, the FTA is usually preceded by a FMEA in which the design, operation, and environment of the system are evaluated and the cause and effect relationships leading to the failure of the system are identified. Thus FMEA becomes an essential step for understanding the behavior of the system without which the fault-tree analysis and the reliability analysis cannot be performed.

A fault tree is a graphical representation of the relationship between certain specific events and the ultimate undesired event. Thus we must know how the system works before constructing a fault tree. The functioning of a system is usually described by a function or flow diagram in which the flow of information, materials, signals, and other services is indicated. For example, for a steam power plant, the flow diagram shows the flow of all inputs affecting the power plant such as fuel (from receipt to the emission of flue gases from the smokestack), water (from the intake to the final discharge), and so on. The flow diagram is then used to identify the various functional sequences from input to output. Next, a logic diagram is prepared by translating the functional relationships into logical relationships among the various components of the system. Once the logical relationships among the components of the system are established, the cause and effect relationships in the operation of the system can be studied, and the fault tree of the system can be developed.

13.5.2 Procedure

The fault-tree analysis starts with the identification of the top events, usually the set of most serious system failure events that could occur. Next, the events which contribute directly to the top events are identified and connected to the top events by logical links. This process is continued until the lowest level or most basic events are reached. The fault tree is constructed using the symbols shown in Fig. 13.7. The two basic symbols involved are AND and OR gates. For an AND gate, the output will be present only if all of the inputs are present. For an OR gate, the output will be present if one or more of the input events are present. Among the other symbols used in the fault tree analysis, the rectangle is simply used to label an intermediate event in a tree. For instance, if the output from a specific gate is present, a rectangle might be used to describe what the event would physically represent. A rectangle is also used to denote the top event from which the construction of the entire fault tree starts. The triangles are used to indicate the transfer of parts of the fault tree to other locations. A circle is used to specify a "primary input event," usually the failure of a specific component, whenever the quantitative behavior of that component is known sufficiently well so that no further analysis of it is required.

The following points aid in the construction of fault trees [13.11]:

1. The first step in the fault-tree analysis is to define a suitable TOP event that constitutes a serious system failure.
2. Usually several different, but equivalent, fault-trees can be constructed for a given system. Also, different TOP events lead to different fault-trees.
3. For any specified TOP event, each possible event is examined to see whether it can, either alone or in conjunction with some other event(s), cause the TOP event.
4. The primary events that lead to the TOP event and the secondary events that cause each of the primary events are determined. The procedure is continued until all the basic failures are identified.
5. The set of events that are all required to produce an event of interest are connected to AND gates.
6. The set of events that can individually produce an event of interest are connected to OR gates.

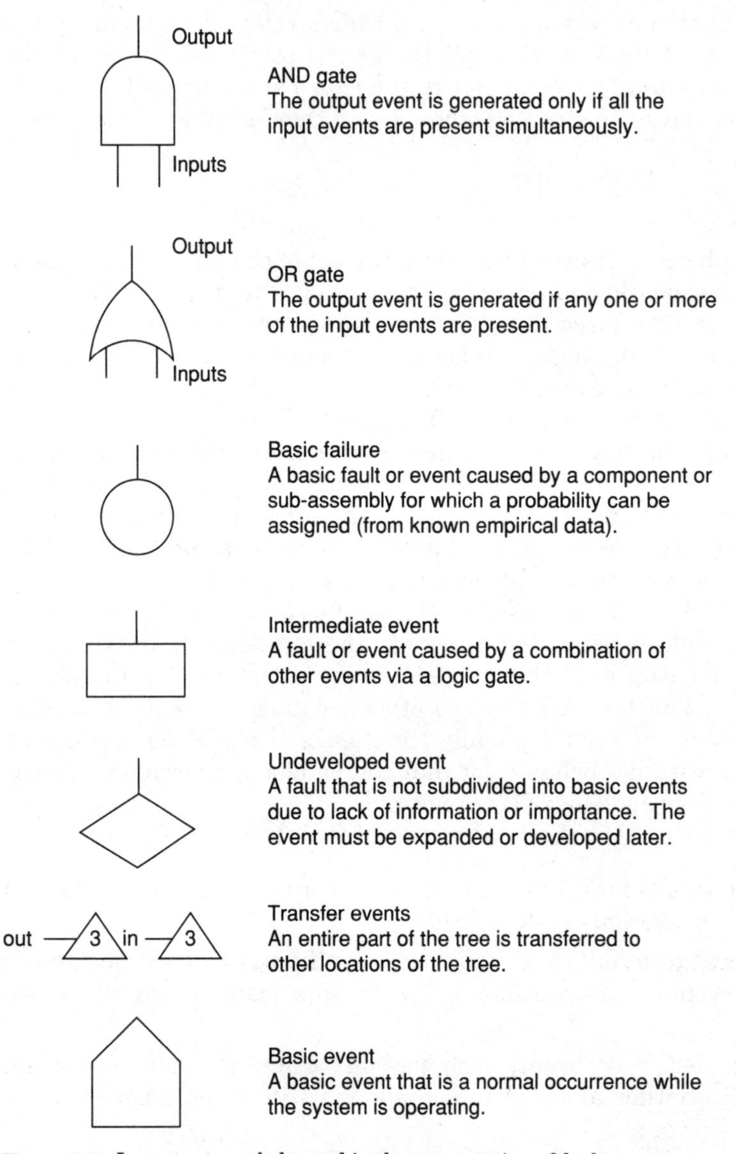

Figure 13.7 Important symbols used in the construction of fault trees.

Example 13.4 Construct a fault tree for the series-parallel system shown in Fig. 13.8.

solution The top (or the most undesirable) event is the failure of the system. When the system fails, either one or both of the parallel units $P1$ and $P2$ must have failed. Thus the units $P1$ and $P2$ are connected to the top event through an OR gate. The unit $P1$ ($P2$) will fail only when both the components A and B (C and D) fail. Hence the components A and B (C and D) are connected to an AND gate as shown in Fig. 13.9. Note that a bar over a letter is used to

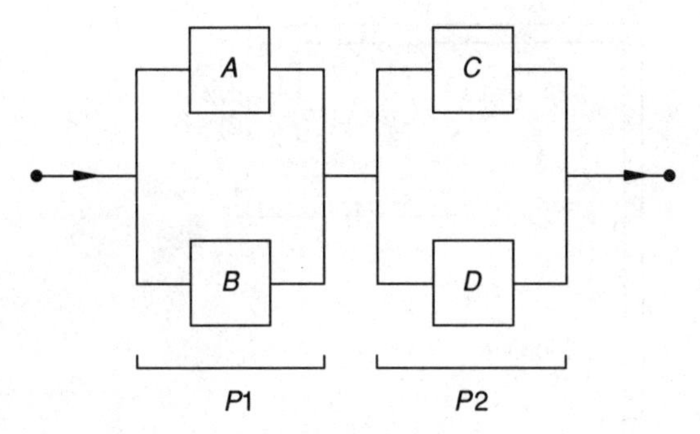

Figure 13.8 A series-parallel system.

indicate the failure of the corresponding component. The complete fault tree is also shown in Fig. 13.9.

Example 13.5 Consider the system shown in Fig. 13.10 which consists of three valves, a pump, a pipeline, and a tank to collect water pumped from the pond. Construct a fault tree corresponding to the top event: "no flow of water into the tank" [13.11].

solution For simplicity, we will ignore the possibility of loss of electric power (to run the pump), lack of water in the pond or failure of the pipeline. To construct the fault-tree, we first note that there are two primary events, namely, "both valves B and C closed (failed)," and "no flow of water to valves B and C," either one of which could cause the TOP event. Thus, an OR gate is introduced with the two stated primary events as inputs and the TOP event as the

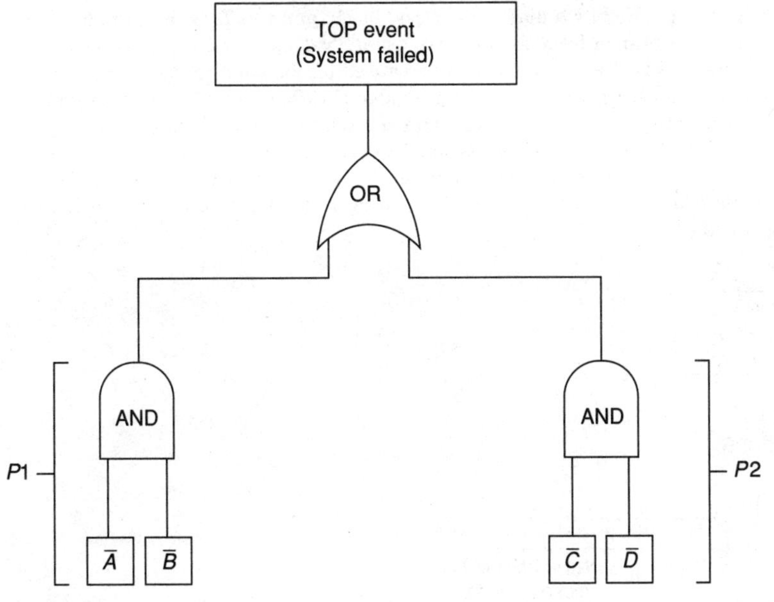

Figure 13.9

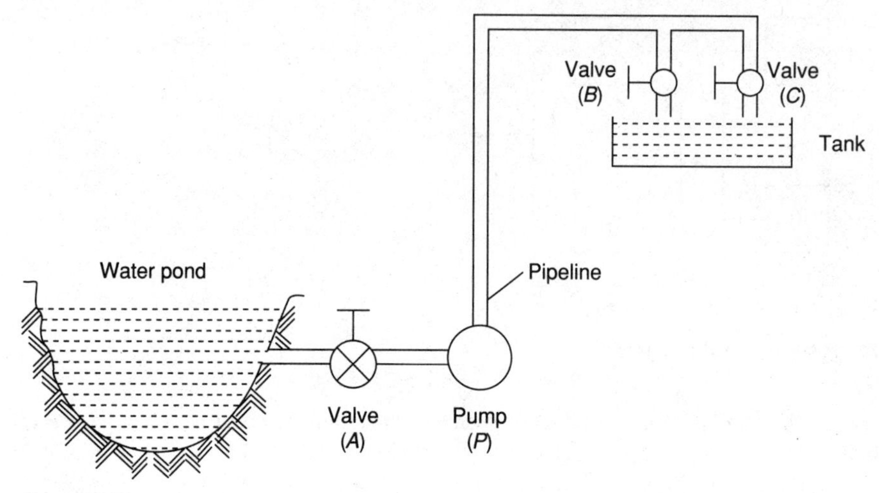

Figure 13.10

output (Fig. 13.11). Since the primary event, "no flow of water to valves B and C," in Fig. 13.11 can be caused by the occurrence of any of the secondary events "valve A closed (failed)" and "pump P failed," an OR gate is used to connect these primary and secondary events as shown in Fig. 13.12. In addition, the event "pump P failed" can be caused by either "pump not started (human failure)" or "pump broken" and hence an OR gate is used to connect them as indicated in Fig. 13.12. By assuming the events "valve A closed" and "pump P broken" as basic failures with known failure rates, these events are denoted by circles. The event "pump not started" is assumed to be an undeveloped event (basic events causing this event are unknown), hence is represented by a diamond-shaped block, and this branch of the fault-tree is terminated. Next, the branch from the primary event, "valves B and C closed (failed)," of Fig. 13.11 is developed. This primary event can be caused by the simultaneous occurrence of the secondary events, "valve B closed (failed)" and "valve C closed (failed)," and thus an AND gate is to be introduced to connect the primary and secondary events. Further, if the events "valve B closed" and "valve C closed" are assumed to be basic failures with known failure rates, this branch of the fault-tree can be terminated by using circles to represent the basic failures. The complete fault-tree, with each branch ending with basic or undeveloped failures, is shown in Fig. 13.13.

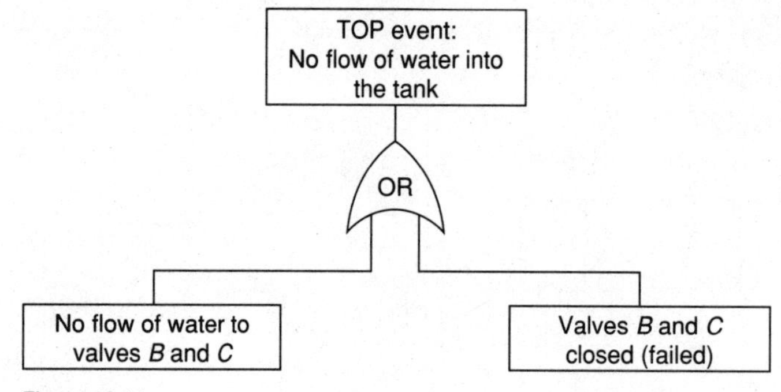

Figure 13.11

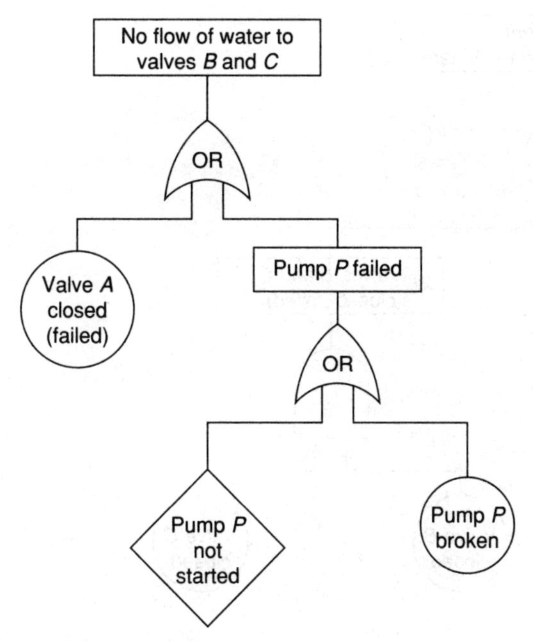

Figure 13.12

Example 13.6 Consider a room containing two light bulbs operated by a single switch (see Fig. 13.2). By assuming the "room without electric light" as the undesirable (top) event, construct a fault tree of the system.

solution The top event can occur due to the occurrence of any of the three input events, namely, "bulbs burnt out," "switch failed to close," and "no electricity." This solution can be represented by a three-input OR gate. The occurrence of the events "bulbs burnt out" and "no electricity" can be investigated further. The event "bulbs burnt out" occurs when both the bulbs are burnt out and so the situation can be represented by a two-input AND gate. The event "no electricity" occurs when electric power is cut off or the fuse is blown, and the situation can be represented by a two-input OR gate. The resulting fault tree is shown in Fig. 13.14.

Example 13.7 The schematic diagram of an electromagnetic brake for elevators [13.17] is shown in Fig. 13.15. The brake is situated between the motor and the gearbox where it has the greatest mechanical advantage. When the operating switch is in the "on" position, the solenoid will be energized and will pull the brake shoe towards the drum thereby stopping the drum and the elevator cage. When the operating switch is in the "off" position, the solenoid will be de-energized and release the brake shoe, thereby permitting the motion of the drum and the elevator cage. Construct a fault-tree diagram corresponding to the top event, "failure of the brake to stop the motion of the elevator cage."

solution The major components of the system are shown in Fig. 13.15. The following failures can be associated with the various components:

Operating switch: Not applied (human error), fails to make contact, wiring damaged
Electromagnetic solenoid: Coil failure, solenoid sticks open

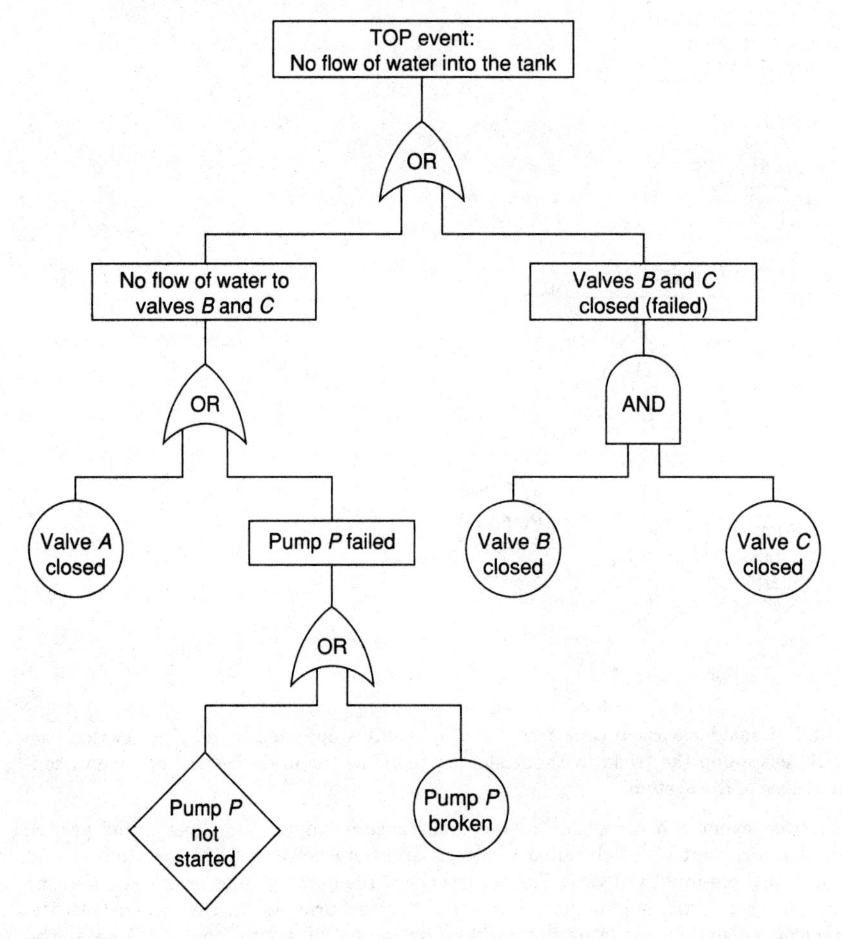

Figure 13.13

Brake shoe: Excessive pad wear
Brake shoe pivot bearings: Seizure, lock nut failure

The failure of the brake to stop the motion of the elevator cage can cause a potential safety problem and hence is considered as the top event. By considering the various modes of failure of the components, the fault tree can be developed as shown in Fig. 13.16.

The fault-tree structure, which relates the top event to the basic events, can be used to perform a quantitative analysis of the failure of the system. This is done by first reducing the fault-tree structure to a logically equivalent form and then using the "minimal cut-set" theory discussed below.

13.6 Minimal Cut-Sets

As defined in Section 6.11.3, a cut-set is a group of primary events whose occurrence causes a system to fail, that is, the occurrence of the TOP event. A minimal cut-set is a set of minimum number of primary events that produces

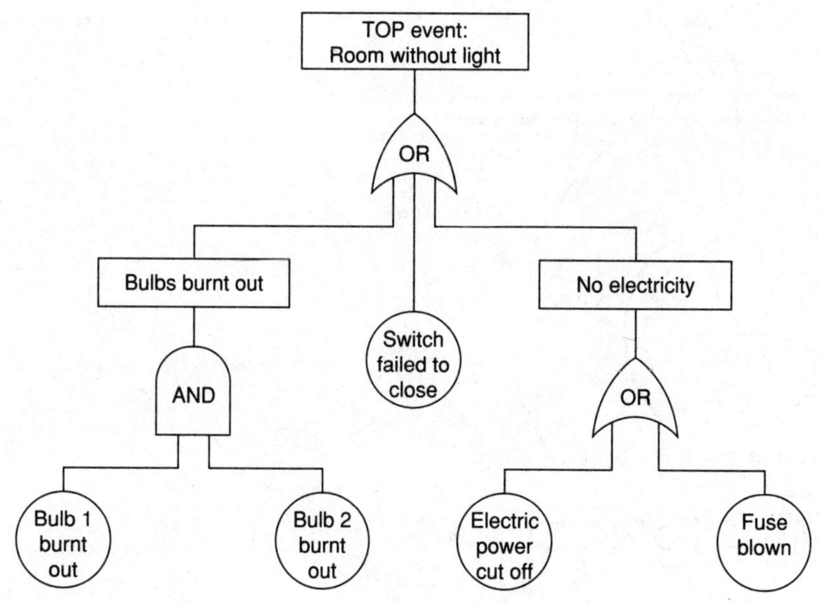

Figure 13.14 Construction of a fault tree.

the TOP event if and only if all the events of the set occur. This also implies that there are no extraneous events of the set, whose failures do not contribute to the TOP event. Enumerating the minimal cut-sets is an essential step in evaluating the system reliability. Interpretation of the minimal cut-sets gives a number of qualitative results such as the weak points of the system, false redundancies or the effect of a given component on the overall system reliability. Different methods can be used for determining the minimal cut-sets [13.7]. In one of the methods, known as the inductive method, the minimal cuts are found directly by combining significant failures of the system components. In another procedure, known as the fault-tree approach, the minimal cut-set is derived directly from the fault tree. A third alternative is to proceed automatically from the reliability block diagram (or successful paths).

In the fault-tree approach, the minimal cut-sets are determined by converting the fault-tree to a boolean expression [13.11]. For this, a boolean variable is assigned to each basic event. The output event of an AND gate is assigned a boolean variable equal to the minimum of the boolean variables for the input events. The output event of an OR gate is assigned a boolean variable equal to the maximum of the input event variables. Finally, a boolean expression is obtained for the undesirable final event as a function of the boolean variables associated with each basic event. It will be convenient to define a "boolean variable" for a typical event A, denoted as X_A, such that $X_A = 1$ if the event A has occurred (true) and $X_A = 0$ if the event A has not occurred (false).

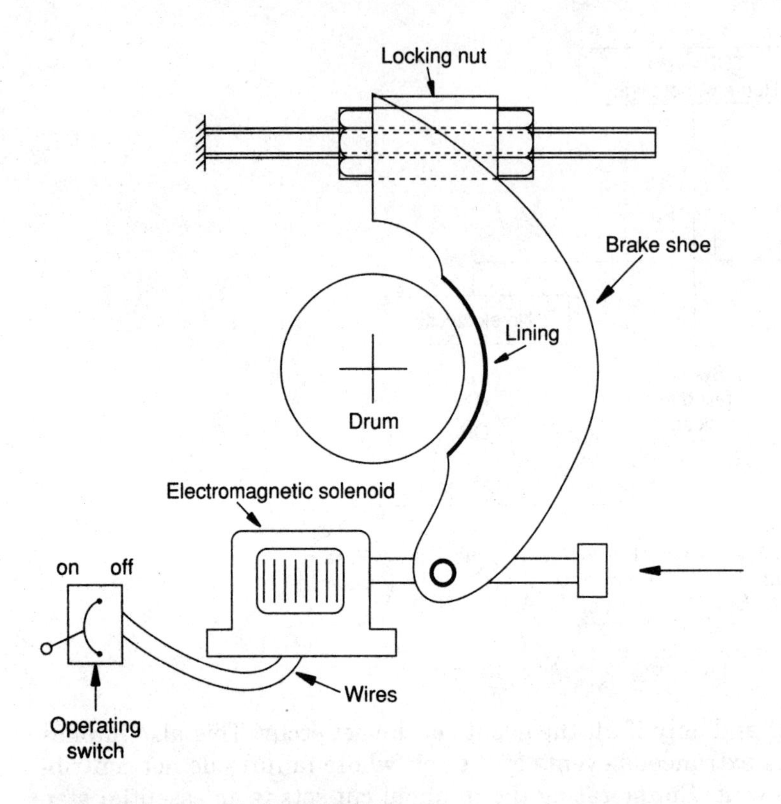

Figure 13.15 Schematic diagram of an electromagnetic brake for elevators (with permission from: H. M. N. Raafat, "Reliability in Design and Failure to Safety," *Engineering*, vol. 219, Nov. 1979, ©The Design Council, London, pp. 1425–1429).

Consider an AND gate with n input events and one output event A. If the n events 1, 2, . . . , n have the boolean variables $X_1, X_2, \ldots, X_n$, the output event A is defined as the product of these n events such that the event A will occur if and only if all the n input events occur. That is, $X_A = 1$ if and only if all $X_i = 1$, and hence we can write

$$X_A = \prod_{i=1}^{n} X_i \tag{13.2}$$

Since X_A is expressed as the product of zeros and ones, X_A will be either zero or one. Alternately, X_A can be defined as the minimum value of the X_is. Next, consider an OR gate with n input events and one output event B. Here the output event B is defined as the union of the n input events 1, 2, . . . , n such that the event B will occur if one or more of the input events occur. In other words, B will not occur if and only if all the input events do not occur. Thus we can write

$$X_B = 1 - \prod_{i=1}^{n} (1 - X_i) \tag{13.3}$$

It can be seen that X_B will be equal to one if any of the X_is are equal to one. Thus X_B can also be expressed as the maximum value of the X_is.

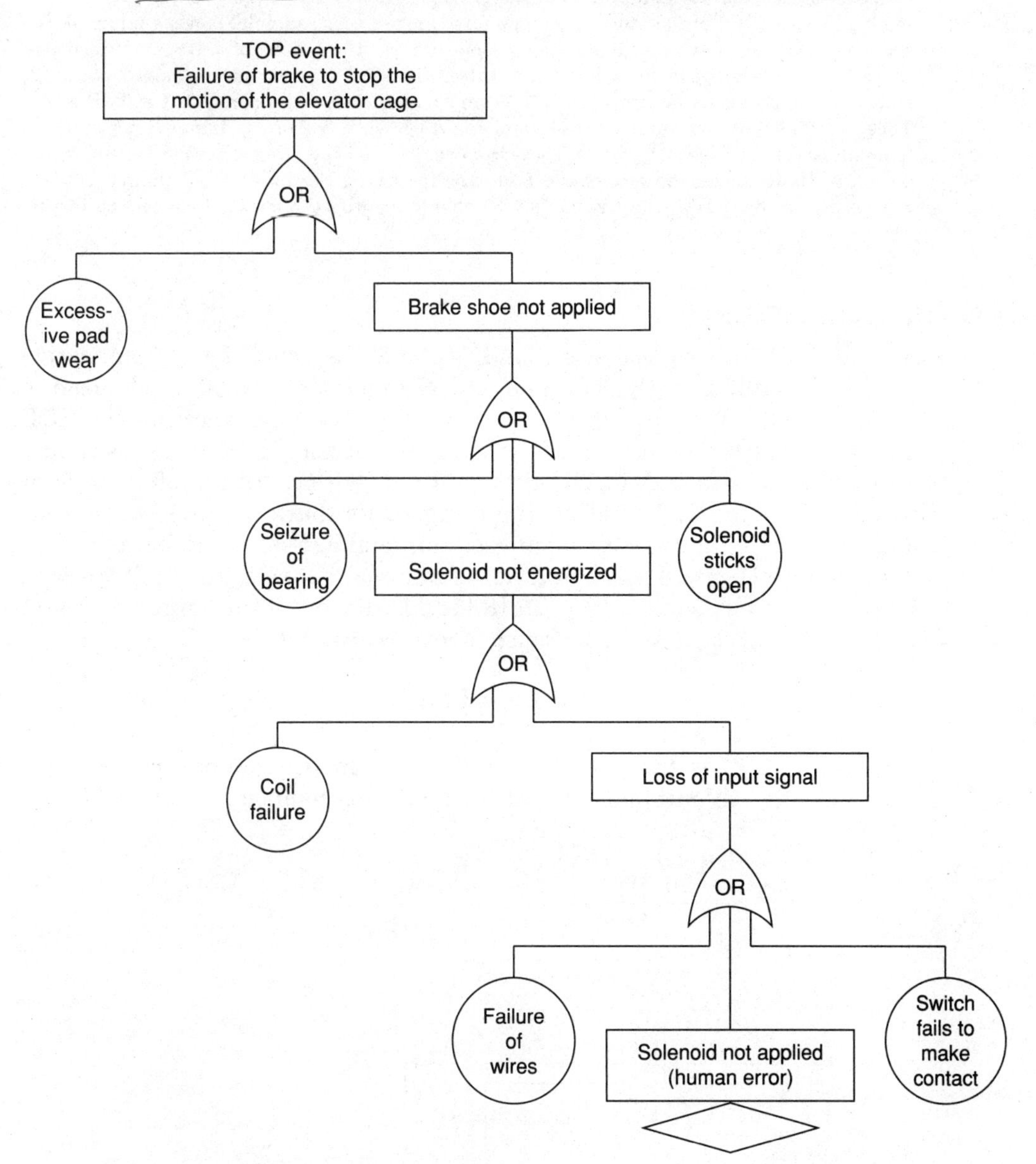

Figure 13.16 Fault tree of the brake system of Fig. 13.15. (With permission from: H. M. N. Raafat, "Reliability in Design and Failure to Safety," *Engineering*, vol. 219, Nov. 1979, ©The Design Council, London, pp. 1425–1429).

Example 13.8 Find the minimal cut-sets for the fault tree shown in Fig. 13.13.

solution It can be seen that if all indicator variables of a cut-set are true ($X_i = 1$), then the TOP event is true. It can be seen that for the fault-tree shown in Fig. 13.13, valve *A* by itself forms a cut-set and valves *B* and *C* together form another cut-set. Hence, valves *A*, *B*, and *C* also form a cut-set. Several other combinations are also possible. Since no extraneous members of the set, whose failures do not contribute to the TOP event, can be present in a minimal cut-set, "failure of valve *A* plus pump *P* broken," for example, is not a minimal cut-set. The reason is that even if the pump is not broken, the failure of valve *A* by itself will cause the TOP event. On the other hand, valves *B* and *C* together form a minimal cut-set, since the simultaneous failure of both the components is needed to produce the TOP event. Thus, we find that the minimal cut-sets of Fig. 13.13 are: (valve *A*), (pump *P* not started), (pump *P* broken), (valve *B*, valve *C*). These can even be identified from the way the tree is constructed. However, as the size of the fault-tree increases, the number of minimal cut-sets increases at a rapid rate, and we may not be able to identify the minimal cut-sets by inspection.

13.6.1 Probability of the TOP event

Once the minimal cut-sets are found, we can use them for a quantitative analysis of a fault tree. By definition, any Minimal Cut Set (MCS) can lead to the TOP event. Thus, we can reconstruct the tree by connecting the TOP event to a OR gate, with each of the minimal cut-sets represented as inputs to the OR gate (see Fig. 13.17). By definition, the occurrence of a minimal cut-set implies the failure of all the components (basic failures) of the set. Thus, the individual components of each minimal cut-set are to be connected through an AND gate. Hence, in terms of minimal cut-sets, any fault tree can be represented as shown in Fig. 13.18 [13.11]. For a typical minimal cut-set (labeled as MCS_i) consisting of *s* components, we can write

$$X_{\text{MCS}_i} = \prod_{j=1}^{s} X_j^{(i)} \tag{13.4}$$

In Eq. (13.4), $X_j^{(i)}$ is the boolean variable for the *j*th component (or basic failure) of the *i*th minimal cut-set and all the components of MCS_i are

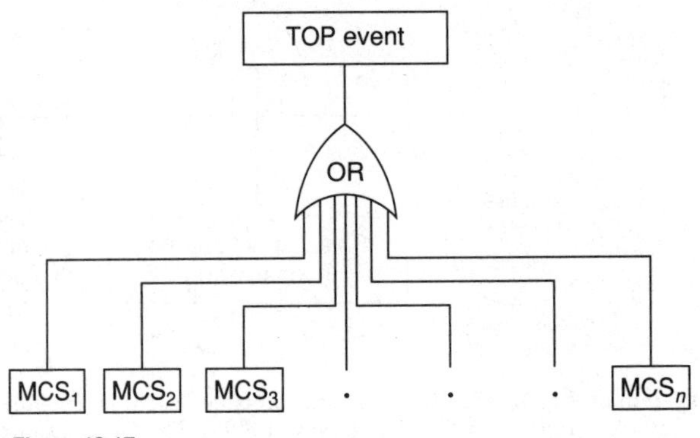

Figure 13.17

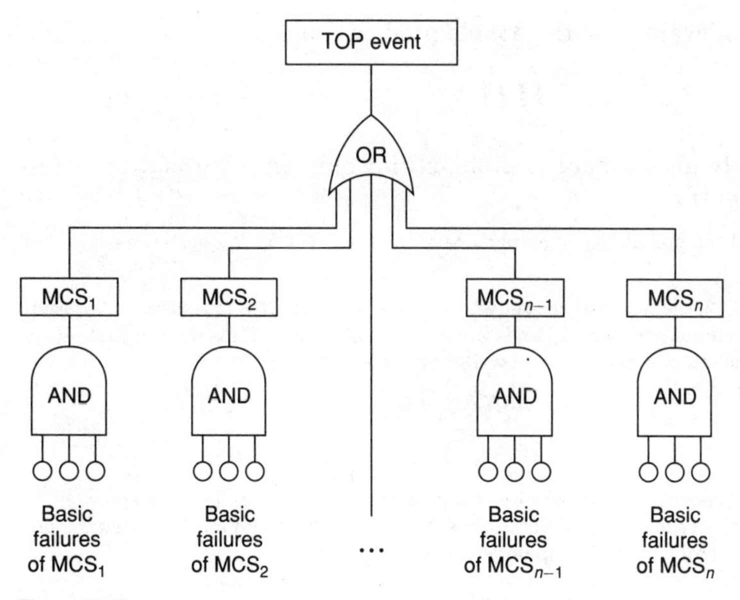

Figure 13.18

assumed to be independent. The probability of the occurrence of MCS_i is given by

$$P_{\mathrm{MCS}_i} = \prod_{j=1}^{s} P_j^{(i)} \tag{13.5}$$

where $P_j^{(i)}$ is the probability of failure of the jth component in ith minimal cut-set (also equal to the expected value of $X_j^{(i)}$). Thus, the boolean variable corresponding to the TOP event can be expressed, using OR for minimal cut-sets, as

$$X_{\mathrm{TOP}} = 1 - \prod_{i=1}^{n} (1 - X_{\mathrm{MCS}_i}) \tag{13.6}$$

Equations (13.6) and (13.4) yield

$$X_{\mathrm{TOP}} = 1 - \prod_{i=1}^{n} \left(1 - \prod_{j=1}^{s} X_j^{(i)} \right) \tag{13.7}$$

where the component j of $\mathrm{MCS}_i(X_j^{(i)})$ is assumed to be the same as the component l of $\mathrm{MCS}_k(X_l^{(k)})$. It is to be noted that even if the individual components are independent, the minimal cut-sets may not be independent always. Particularly, when several cut-sets share the same components, there will be interdependence between the minimal cut-sets. This dependence can be accounted by using the following boolean reduction relationship

$$X_j^{(i)} X_l^{(k)} = X_j^{(i)} \tag{13.8}$$

This implies that in evaluating the double products such as

$$\prod_{i=1}^{n}\prod_{j=1}^{s} X_j^{(i)}$$

we need to multiply all component probabilities only once without considering any duplicate entry.

Example 13.9 Find the probability of occurrence of the top event for the system represented by Fig. 13.13.

solution For Fig. 13.13, the minimal cut-set was found in Example 13.8 to be: (valve A closed), (pump P broken), (pump P not started), and (valves B and C closed). Thus we find that all the cut-sets are independent; and hence we can write, without reductions,

$$X_{\text{TOP}} = 1 - (1 - X_A)(1 - X_{P1})(1 - X_{P2})(1 - X_B X_C)$$

and

$$P_{\text{TOP}} = 1 - (1 - P_A)(1 - P_{P1})(1 - P_{P2})(1 - P_B P_C)$$

where P_A, P_B, and P_C denote the probability of failure of valves A, B, and C, respectively, P_{P1} represents the probability of occurrence of the event "pump broken" and P_{P2} indicates the probability of occurrence of the event "pump not started."

The fault-tree analysis and the method of finding the system reliability in mechanical design is illustrated through Example 13.10.

Example 13.10 The gear train shown in Fig. 13.19 consists of a cluster gear set mounted on the input shaft [13.8]. The intermediate shaft carries three fixed gears and a cluster gear set. The output shaft carries three fixed gears. Each shaft is mounted on two bearings. Depending on the gear pair in contact, the intermediate shaft will have three different speeds. For each intermediate shaft speed, three speeds can be realized at the output shaft. Thus the system has three shafts, two clusters, six gears, and six bearings and delivers nine output speeds for any particular input speed.

a. Develop a fault tree of the system corresponding to the top event: "no output speed from the gear train."

b. Find the failure rate of the top event corresponding to the failure rates of the individual components given in Table 13.2 [the notation used in Table 13.2 is introduced in the solution of part (a)].

c. Find the reliability of the gear train when the component reliabilities are given as follows:

R_1 = reliability of each bearing = 0.9999

R_2 = reliability of each gear in bending failure mode = 0.9998

R_3 = reliability of each gear in surface wear failure mode = 0.9998

R_4 = reliability of each cluster gear set = 0.9999

R_5 = reliability of each shaft = 0.99999

R_{PE1} = reliability of fuse = 0.999

R_{PE2} = reliability of electric switch = 0.9999

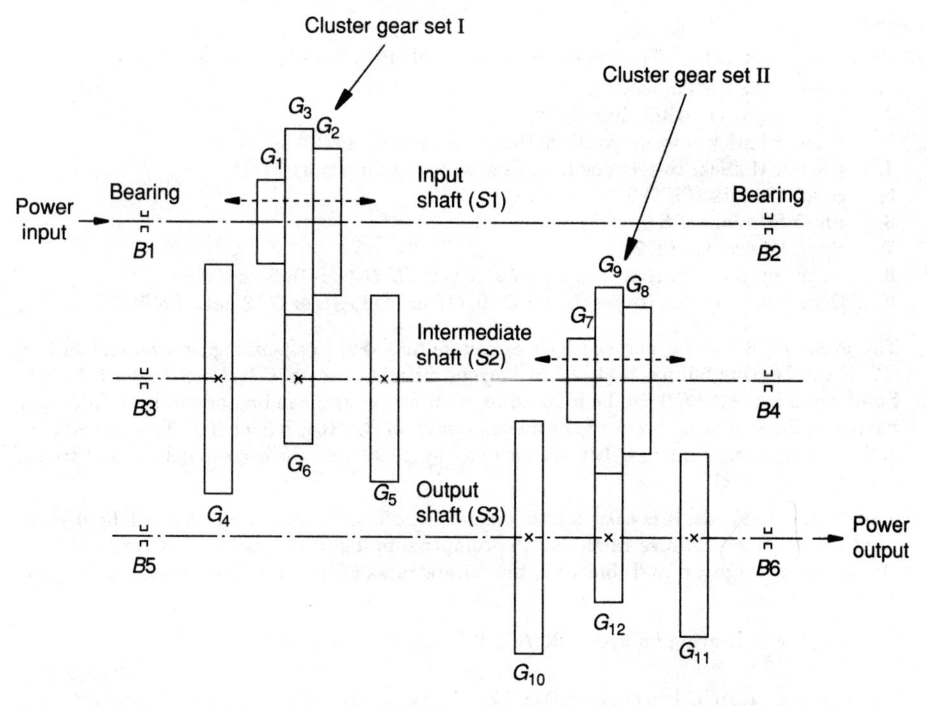

Figure 13.19 Layout of the gear train (Gi = ith gear, Bi = ith bearing, Si = ith shaft). (Reprinted with permission from S. S. Rao, "A Probabilistic Approach to the Design of Gear Trains," *Intern. J. of Machine Tool Design and Research*, vol. 14, copyright 1974, Pergamon Press PLC, pp. 267–278.)

TABLE 13.2 Failure Rates of the Components

Component (event)	Failure rate in hr^{-1} (λ)
PE 1	10^{-3}
PE 2	2×10^{-4}
PE 3	10^{-5}
PE 4	10^{-5}
PE 5	2×10^{-6}
PE 6	2×10^{-6}
PE 7	2×10^{-6}
PE 10	10^{-4}
PE 11	10^{-5}
PE 12 (*B* 1 to *B* 6)	10^{-5}

solution

(a) Fault-tree construction: The basic events can be identified [13.9, 13.10] as follows:

1. Fuse of the system blows (*PE* 1)
2. Electric power switch fails (*PE* 2)
3. Cluster I sticks to zero position, that is, no output (*PE* 3)
4. Cluster II sticks to zero position, that is, no output (*PE* 4)
5. Shaft I breaks (*PE* 5)
6. Shaft II breaks (*PE* 6)
7. Shaft III breaks (*PE* 7)
8. Gear group I fails, that is, *G* 1 or *G* 4, *G* 2 or *G* 5, *G* 3 or *G* 6 fails (*PE* 8)
9. Gear group II fails, that is, *G* 7 or *G* 10, *G* 8 or *G* 11, *G* 9 or *G* 12 fails (*PE* 9)

The event *PE* 8 can be induced by wearout failure (*PE* 10); bending or overload failure (*PE* 11), or bearing failure, that is, breakage or misalignment of *B* 1, *B* 2, *B* 3, or *B* 4 (*PE* 12). Similarly the event *PE* 9 can be induced by wearout failure; bending or overload failure; or bearing failure; that is, breakage or misalignment of *B* 3, *B* 4, *B* 5, or *B* 6. The connectivity of the components of the gear box is shown in Fig. 13.20, and the corresponding fault tree is shown in Fig. 13.21.

(b) Failure-rate analysis: It is assumed that component failure times are exponentially distributed with constant failure rates and the components (basic events) are not repaired. For the failure rates given in Table 13.2, the failure rates of intermediate events can be computed as follows:

$$\lambda_a = \text{bearing failure} = \text{OR}\,(B\,1, B\,2, B\,3, B\,4) = 4 \times 10^{-5}$$

$$\lambda_b = \text{failure rate of } G\,1 \text{ to } G\,12 = \text{OR}\,(\lambda_a, \lambda_{10}, \lambda_{11}) = 4 \times 10^{-5} + 10^{-4} + 10^{-5} = 1.5 \times 10^{-4}$$

$$\lambda\,(\text{any two pair of gears}) = \lambda_{\text{pair}} = \text{OR}\,(\lambda_{b\,1}, \lambda_{b\,2}) = 3 \times 10^{-4}$$

$$\lambda_8 = \lambda_9\ \text{AND}\,(\lambda_{\text{pair}}, \lambda_{\text{pair}}, \lambda_{\text{pair}}) = \frac{3 \cdot \lambda_{\text{pair}} \cdot (Z - 1)}{Z^3 - 1} = 2.848 \times 10^{-4}$$

where

$$Z = \{1 - \exp\,(-\lambda_{\text{pair}} \cdot t)\}^{-1} = 1.0524$$

$$\lambda_c = \lambda_1 + \lambda_2 = 1.2 \times 10^{-3}$$

$$\lambda_d = \lambda_5 + \lambda_6 + \lambda_7 = 6 \times 10^{-6}$$

$$\lambda_e = \lambda_3 + \lambda_4 = 2 \times 10^{-5}$$

$$\lambda_{\text{TOP}} = \text{OR}\,(\lambda_c, \lambda_d, \lambda_e, \lambda_8, \lambda_9) = \sum \lambda_i = 1.7956 \times 10^{-3}\ \text{hr}^{-1}$$

Thus, the total system failure rate is 1.7956×10^{-3}/hr. This implies that on the average there is one failure per 588.6 hr. It is to be noted that the major failure has been contributed by the "fuse of the system blows."

(c) Reliability analysis: The AND gate in a fault tree represents a weakest link or series system and the OR gate denotes a fail safe or parallel system. If the reliabilities of the components of the gear train are known, then the total system (gear train) reliability can be determined. From the fault-tree analysis and the expressions for the reliabilities of

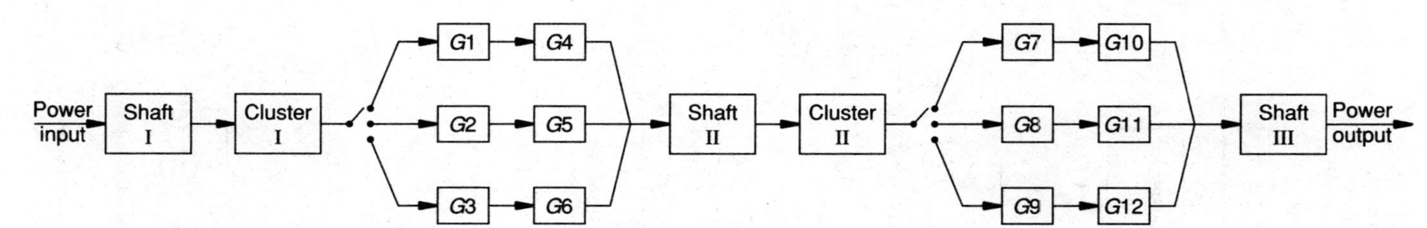

Figure 13.20 Connectivity of components of the gear train. (With permission from T. S. Pan and S. S. Rao, "Fault Tree Approach for the Reliability Analysis of Gear Trains," *Journal of Mechanisms, Transmissions, and Automation in Design*, ASME, vol. 110, 1988.)

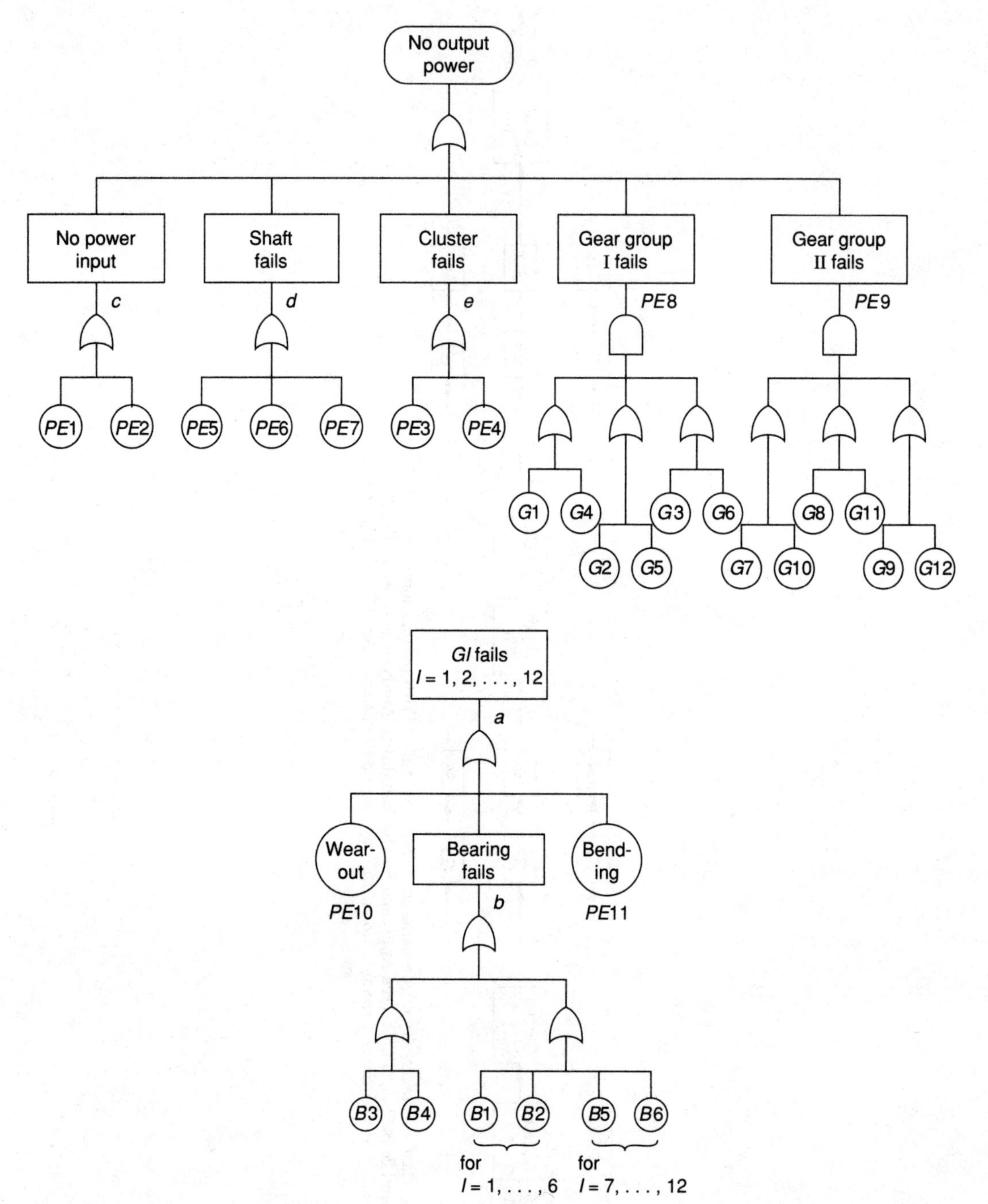

Figure 13.21 Fault tree of the gear train. (With permission from T. S. Pan and S. S. Rao, "Fault Tree Approach for the Reliability Analysis of Gear Trains," *Journal of Mechanisms, Transmissions, and Automation in Design*, ASME, vol. 110, 1988.)

weakest link and fail safe systems, the overall reliability of the gear train can be obtained as

$$R_{\text{gear train}} = R_{PE8}\, R_{PE9}\, R_e\, R_d\, R_c = \{1 - (1 - R_b^2)^3\}^2\, R_4^2\, R_5^3\, R_{PE1}\, R_{PE2}$$

$$= \{1 - (1 - R_2^2\, R_3^2\, R_1^8)^3\}^2\, R_4^2\, R_5^3\, R_{PE1}\, R_{PE2}$$

Thus, the failure probability of the gear train is given by $P_{f\,\text{system}} = 1 - R_{\text{system}}$. The reliabilities of the intermediate events can be computed as

$$R_a = R_1^4,\ R_b = R_a\, R_2\, R_3 = R_1^4\, R_2\, R_3, \qquad R_{PE9} = R_{PE8} = 1 - (1 - R_b^2)^3,$$

$$R_e = R_4^2, \qquad R_d = R_5^3 \qquad \text{and} \qquad R_c = R_{PE1} \cdot R_{PE2}$$

For the given component reliabilities,

$$R_1 = R_4 = 0.9999, \quad R_2 - R_3 = 0.9998, \quad R_5 = 0.99999, \quad R_{PE1} - 0.999, \quad R_{PE2} = 0.9999$$

the reliability of the gear train can be found as

$$R_{\text{system}} = R_{\text{gear train}} = 0.99867$$

and hence the probability of failure is given by

$$P_{f\,\text{system}} = 0.00133$$

References and Bibliography

13.1. H. M. N. Raafat, "The Quantification of Risk in System Design," *J. of Eng. for Industry*, vol. 105, 1983, pp. 223–233.

13.2. S. Levine and W. E. Vesely, "Important Event-Tree and Fault-Tree Considerations in the Reactor Safety Study," *IEEE Trans. on Reliability,* R-25, 1976, pp. 132–139.

13.3. H. R. Roberts, W. E. Vesely, D. F. Haast and F. F. Goldberg, "Fault Tree Handbook," U.S. Nuclear Regulatory Commission, NURE G-0492, 1981.

13.4. "Reactor Safety Study: An Assessment of Accident Risks in U.S. Commercial Nuclear Power Plants," U.S. Atomic Energy Commission, DRAFT WASH-1400, Washington, D.C., Aug. 1974.

13.5. R. E. Barlow and H. E. Lambert, "Introduction to Fault Tree Analysis," in *Reliability and Fault Tree Analysis*, SIAM, Philadelphia, 1975, pp. 7–37.

13.6. B. S. Dhillon and C. Singh, "Bibliography of Literature on Fault Trees," *Microelectronics and Reliability*, vol. 17, 1978, pp. 501–503.

13.7. J. B. Fussell, and W. E. Vesely, "A New Methodology for Obtaining Cut Sets for Fault Trees," *Trans. of American Nuclear Society*, vol. 15, 1972, pp. 262–263.

13.8. S. S. Rao, "A Probabilistic Approach to the Design of Gear Trains," *Int'l J. of Machine Tool Design and Research*, vol. 14, 1974, pp. 267–278.

13.9. T. S. Pan and S. S. Rao, "Fault Tree Approach for the Reliability Analysis of Gear Trains: Part I—Basic Theory, Part II—Example Applications," in *Advances in Design Automation*—1987 (Rao, S. S., ed.), publication no. DE-Vol-10-1, ASME, New York, 1987.

13.10. T. S. Pan and S. S. Rao, "Fault Tree Approach for the Reliability Analysis of Gear Trains," *ASME J. of Mechanisms, Transmissions and Automation in Design*, vol. 110, 1988, pp. 348–353.

13.11. S. L. Salem, G. E. Apostolakis and D. Okrent, "A Computer-Oriented Approach to Fault-Tree Construction," report no. UCLA–ENG-7635 and NSF/RA-760320, UCLA, Los Angeles, CA, April 1976.

13.12. T. W. Yellman, "Event-Sequence Analysis," *Proceedings of the Annual Reliability and Maintainability Symposium*, Washington, D.C., January 28–30, 1975, IEEE, New York, 1975, pp. 286–291.

13.13. R. J. Schroder, "Fault Trees for Reliability Analysis," *Proc. of the Annual Symposium on Reliability*, Los Angeles, CA, February 3–5, IEEE, New York, 1970, pp. 198–205.

13.14. W. F. Stoecker, *Design of Thermal Systems,* 3rd ed., McGraw-Hill, New York, 1989.

13.15. J. E. Shigley and C. R. Mischke, "Mechanical Engineering Design," 5th ed., McGraw-Hill, New York, 1989.
13.16. P. L. Crown, "Design Effective Failure Mode and Effect Analysis," *Proc. of the Annual Symp. on Reliability*, Chicago, IL, January 21–23, 1969, IEEE, New York 1969, pp. 514–521.
13.17. H. M. N. Raafat, "Reliability in Design and Failure to Safety," *Engineering*, vol. 219, 1979, pp. 1425–1429.
13.18. MIL-STD-1629, "Procedures for Performing a Failure Mode, Effects and Criticality Analysis," version 1629A, Nov. 24, 1980.
13.19. D. Tarwater (ed.), *The Bicentennial Tribute to American Mathematics 1776–1976*, The Mathematical Assn. of America, Washington D.C., 1977.
13.20. D. Abbott (ed.), *The Biographical Dictionary of Scientists: Mathematicians*, Peter Bedrick Books, New York, 1986.
13.21. J. von Neumann, *The Computer and the Brain*, Yale University Press, New Haven, 1958.

Review Questions

13.1 What is the purpose of FMEA?

13.2 What is the purpose of a design and safety review technique?

13.3 What is the difference between the inductive and deductive methods of system safety analysis?

13.4 What is a fault tree?

13.5 What is the difference between an event tree and a fault tree?

13.6 What is meant by "TOP event"?

13.7 Explain the significance of AND and OR gates in a fault tree.

13.8 What meaning is associated with a triangle and a rectangle in a fault tree?

13.9 How is the minimal cut-set theory used in the fault-tree analysis?

13.10 How was the fault tree analysis developed originally?

Problems

13.1 Construct a fault tree for the flashlight shown in Fig. 13.22. Assume the top event as "no light from the flashlight" and the basic events as (a) switch failure (switch sticks or loses contact), (b) bulb burns, (c) battery dies or discharges, (d) retaining spring fails (breaks or falls off), and (e) end cap fails (falls off or spring breaks) [13.16].

13.2 Consider the system shown in Fig. 13.23 which consists of a battery, two switches and two bulbs [13.12]. The operator is required to switch on S_1 to illuminate L_1. If L_1 fails to illuminate, the operator is expected to switch off S_1 and switch on S_2 to illuminate L_2. Construct the event tree for the system by considering the following possibilities: (a) The operator may initially forget to switch on S_1; (b) if L_1 does not illuminate, the operator may fail to switch off S_1 or fail to switch on S_2; (c) when the operator switches on, the switches S_1 and S_2 may fail to close; (d) when power is received, the bulbs L_1 and L_2

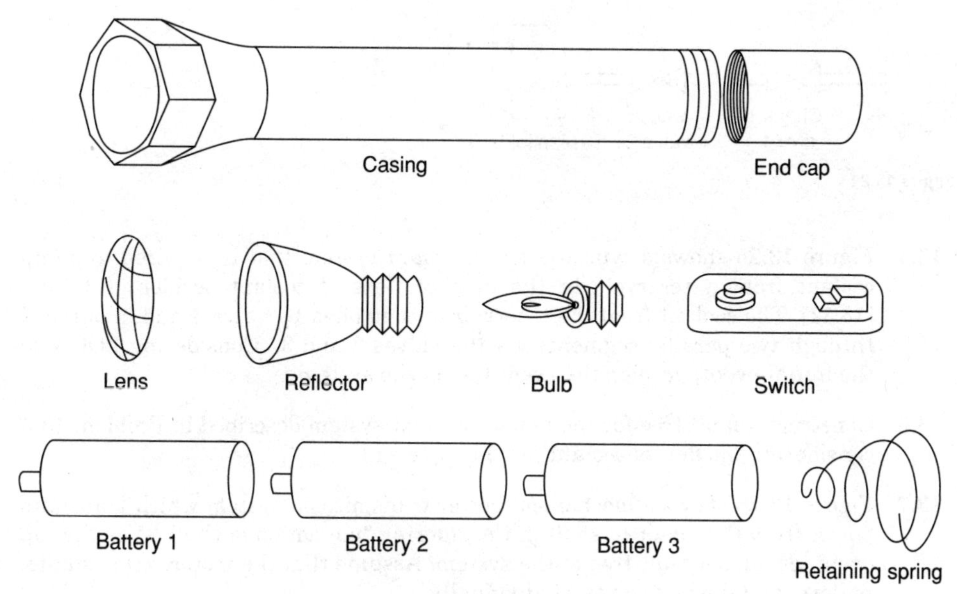

Figure 13.22 A flashlight. (Reprinted with permission from P. L. Crown, "Design Effective Failure Mode and Effect Analysis," *Proceedings of the 1969 Annual Symposium of Reliability*, Chicago, Jan. 1969, ©1969 IEEE.)

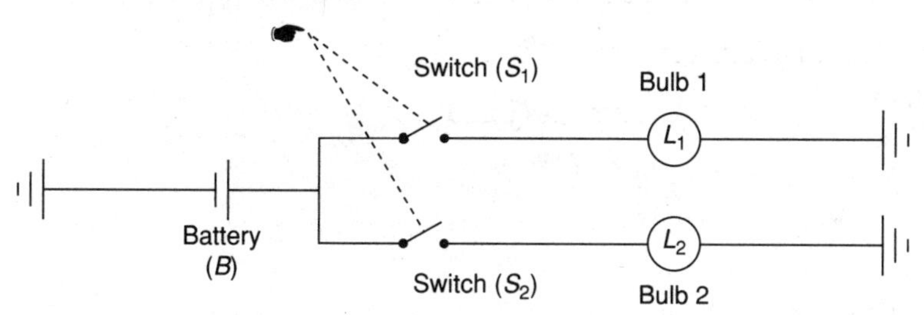

Figure 13.23 Reprinted with permission from T. W. Yellman; "Event-sequence Analysis," *Proc. of the Annual Reliability and Maintainability Symposium*, Washington, D.C., Jan. 1975, ©1975 IEEEE.

may fail to illuminate (burnt); (e) if both bulbs are made to illuminate accidentally, the battery may fail (become exhausted).

13.3 Construct a fault tree for the system described in Problem 13.2 assuming the TOP event is "no illumination."

13.4 Figure 13.24 shows an air-driven emergency hydraulic pump that is connected to a manifold through two check valves [13.12]. The pump is normally in a standby condition and is used to test it before each mission. The check valves are intended to prevent the pump, when de-energized, from being back-driven by pressure in the manifold. Back-driving of the pump causes a hazardous situation. Construct an event tree of the system.

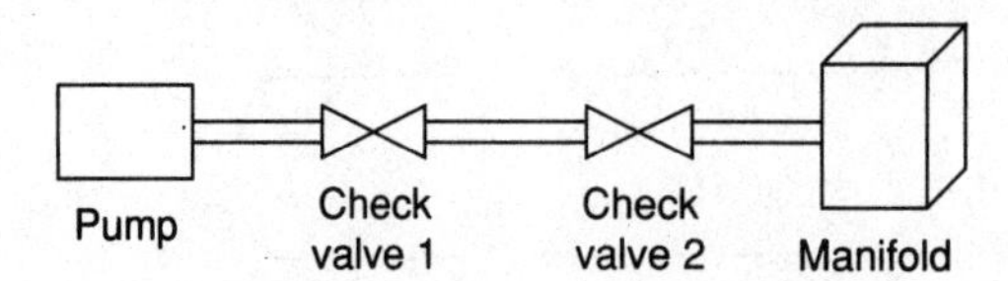

Figure 13.24

13.5 Figure 13.25 shows a typical reactor coolant system that is required to pump coolant from a reservoir in the case of "loss of coolant accident (LOCA)" [13.11]. The coolant from the reservoir is controlled by valve 1 and is pumped through two parallel segments via the valves 2 and 3. Considering LOCA as the initial event, develop the event tree of the system.

13.6 Construct a fault tree for the reactor coolant system described in Problem 13.5 considering "no flow of coolant" as the top event.

13.7 Figure 13.26 shows a mechanical power transmission system which transmits power from the motor to shaft 2. Considering "no motion of shaft 2" as the top event, develop a fault tree of the system. Assume that the motor, keys, shafts, pulleys, and the belt can fail individually.

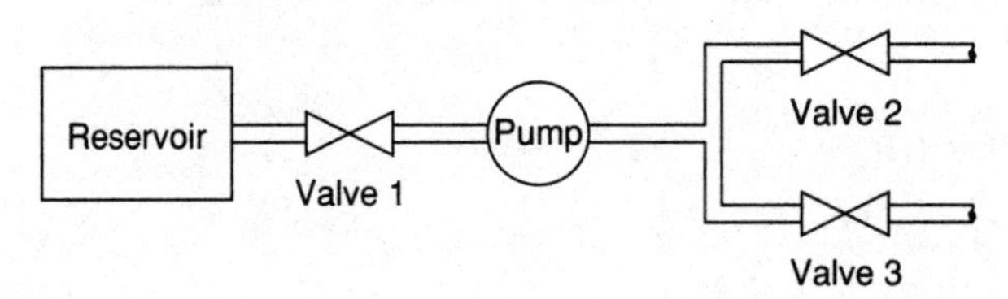

Figure 13.25

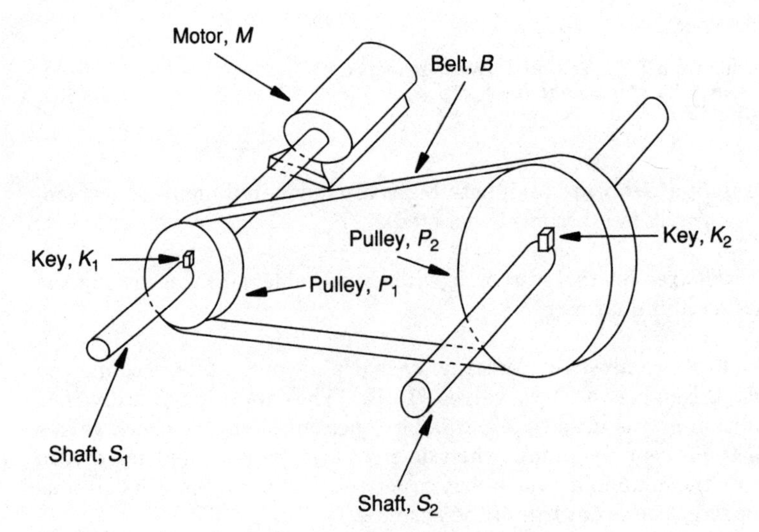

Figure 13.26 A mechanical power transmission system.

13.8 Develop an event tree for the mechanical transmission system described in Problem 13.7.

13.9 A schematic diagram of a thermal power plant, consisting of a heating system, feedwater pump, boiler, turbine, and generator, is shown in Fig. 13.27. Assuming the top event as "no electric power output," construct a fault tree of the system.

13.10 Construct a fault tree for the series-parallel network shown in Fig. 13.28.

13.11 Develop an event tree for the series-parallel network shown in Fig. 13.28.

13.12 An electric overhead traveling crane is shown in Fig. 13.29. The signal from the operator is transmitted to the electric motor, located in the trolley, which lifts the load through the rope and hook. Assuming that "the load is not lifted" when the operator gives the signal, construct a fault tree of the system.

13.13 Develop an event tree for the overhead traveling crane system described in Problem 13.12.

13.14 The various stages of the cement manufacturing process are shown schematically in Fig. 13.30. Develop an event tree of the system.

13.15 Figure 13.31 shows the arrangement of a solar air heater [13.14]. Assuming the component failures as (a) solar energy may not be available, (b) air blower failure, (c) insulation failure, (d) absorbing sheet failure, and (e) transparent sheet failure, develop the event tree of the system.

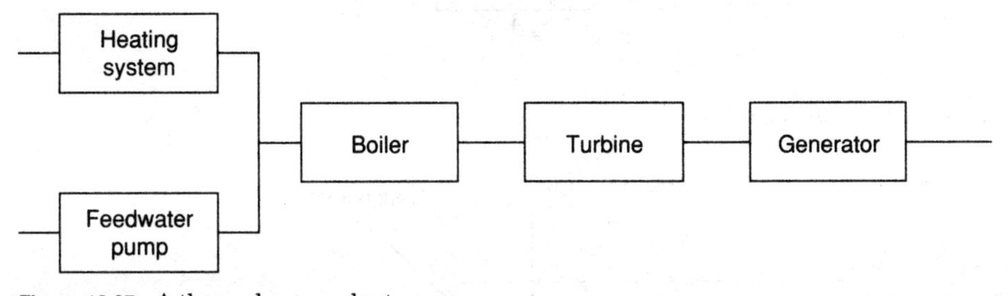

Figure 13.27 A thermal power plant.

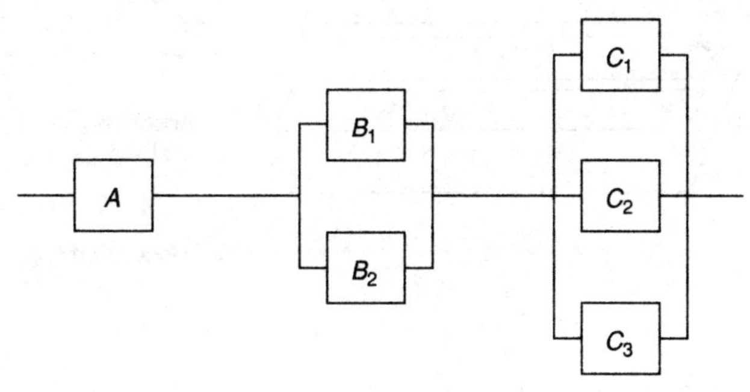

Figure 13.28

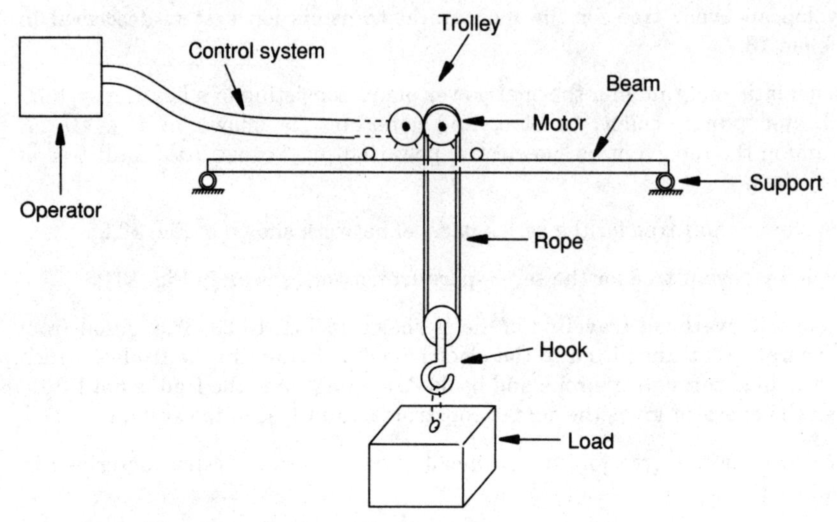

Figure 13.29 Electric overhead traveling crane.

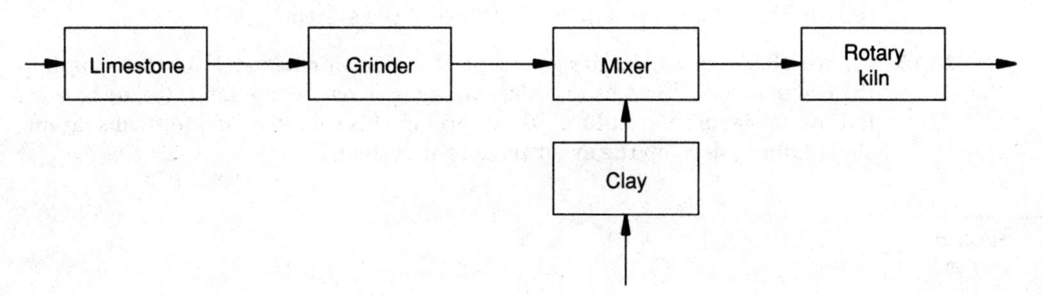

Figure 13.30 Cement manufacturing process.

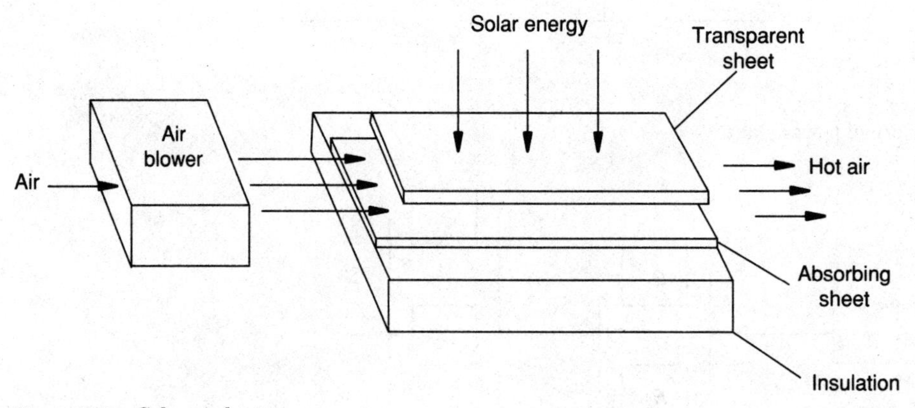

Figure 13.31 Solar air heater.

13.16 If the top event is "no hot air output" for the solar air heater considered in Problem 13.15, construct a fault tree of the system.

13.17 A mechanical press consists of four bearings, a motor, three gears, two screws, and a flat plate as shown in Fig. 13.32. If the TOP event is "object is not pressed," construct a fault tree of the system.

13.18 Construct an event tree for the mechanical press considered in Problem 13.17.

13.19 A compound gear train consists of eight bearings, four shafts, and eight gears as shown in Fig. 13.33. Assuming that a gear can fail either due to excessive bending or surface-wear stress, a shaft can fail either due to excessive stress or

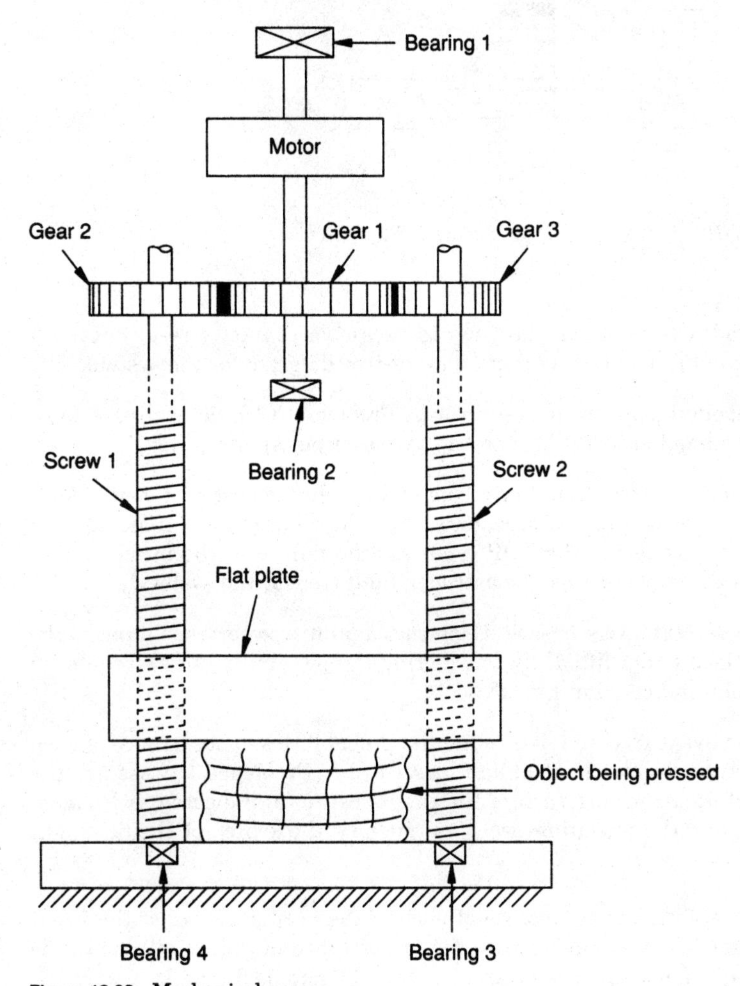

Figure 13.32 Mechanical press.

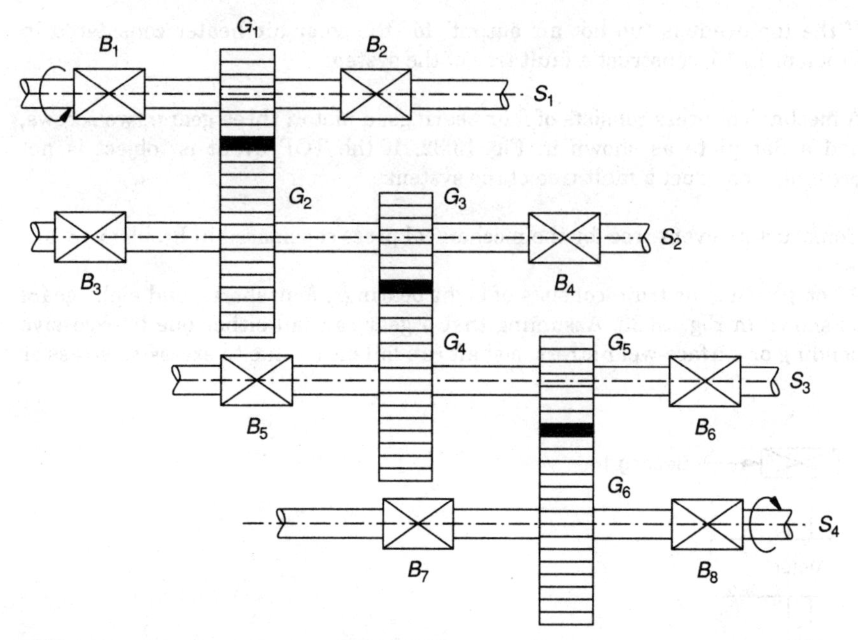

Figure 13.33 A compound gear train (B_i = ith bearing, G_i = ith gear, and S_i = ith shaft).

deflection, and a bearing can fail due to excessive contact stress, misalignment, or lack of lubricant, construct an event-tree diagram for the system.

13.20 For the compound gear train described in Problem 13.19, determine a fault tree corresponding to the TOP event, "no power output at shaft S_4."

13.21 In order to improve the reliability of the elevator brake system considered in Example 13.7, redundant components are introduced as indicated in Fig. 13.34. By considering the TOP event as "the failure of the brake to stop the motion of the elevator cage," construct a fault tree for the system.

13.22 Figure 13.35 depicts a scissor jack. Construct a fault tree corresponding to the TOP event: "load is not lifted" by considering the possibility of failure of each component/joint indicated in Fig. 13.35.

13.23 Construct the event tree and determine the probabilities of occurrence of each of the final outcomes for the flashlight described in Problem 13.1. Assume the probability of occurrence of each of the events (a) to (e) indicated in Problem 13.1 as 0.01, and the initiating event as "pushing the switch of the flashlight on."

13.24 Find the probabilities of occurrence of each of the output events of the event tree of Problem 13.19 assuming that the probabilities of failure of each of the bearings, shafts, and gears are given by 0.97, 0.98, and 0.99, respectively.

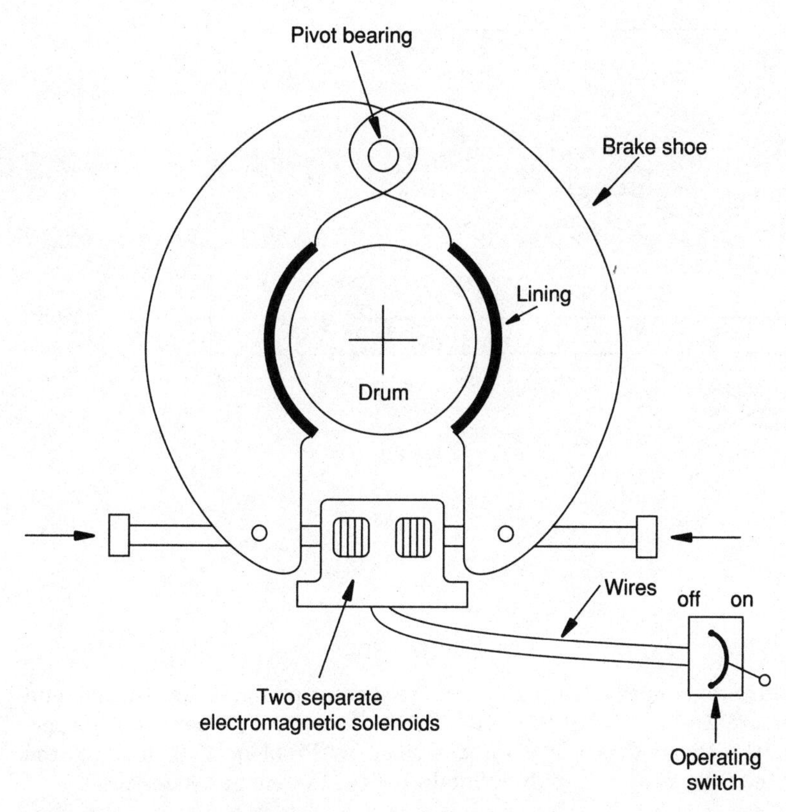

Figure 13.34 Electromagnetic brake for elevators (with permission from: H. M. N. Raafat, "Reliability in Design and Failure to Safety," *Engineering*, vol. 219, Nov. 1979, ©The Design Council, London, pp. 1425–1429).

13.25 Determine the probabilities of occurrence of each of the outcomes of the event tree considered in Problem 13.15 assuming the probabilities of occurrence of each of the events (a) to (e) as 0.05.

13.26 If the probability of occurrence of each of the events stated in (a) to (e) in Problem 13.2 is 0.99, determine the probability of occurrence of the top event: "no illumination."

13.27 If the probabilities of failure of the valves 1, 2, and 3 are 0.99, 0.98, and 0.97, respectively, and the probability of failure of the pump is 0.96, find the reliability of the system shown in Fig. 13.25.

13.28 Find the probability of failure of the system (i.e., the occurrence of the event, "no motion of shaft 2") described in Problem 13.7 assuming that the probability of failure of each of the components is 0.99.

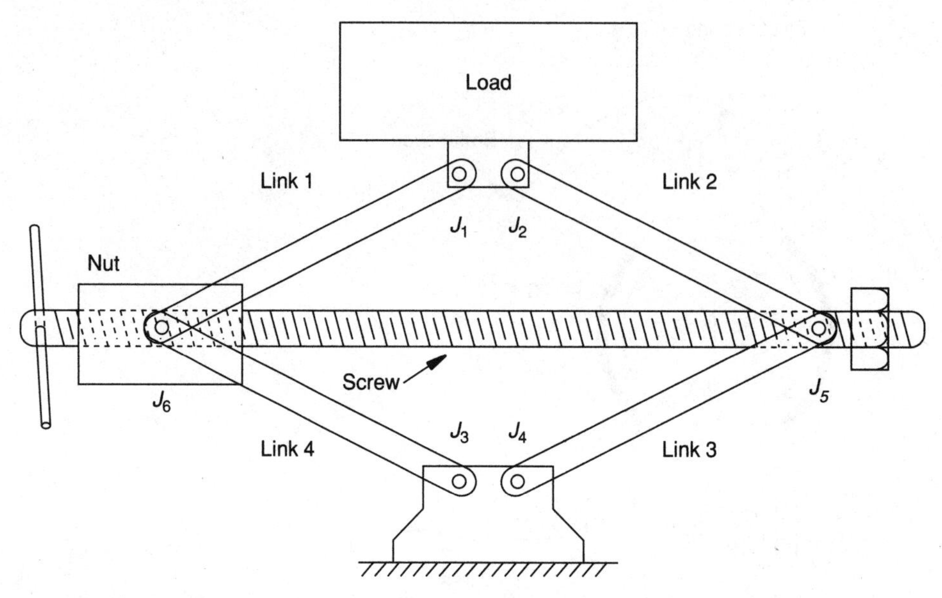

Figure 13.35 A scissor jack.

13.29 If the failure rates of the heating system, feedwater pump, boiler, turbine and generator are 2×10^{-6}, 5×10^{-7}, 4×10^{-5}, 8×10^{-6} and 1×10^{-7} per hour, respectively, in the thermal power plant described in Problem 13.9, find (a) the failure rate of the system, and (b) reliability of the system at 1000 hours.

13.30 Find the reliability of the mechanical press described in Problem 13.17 if the reliabilities of the bearing, motor, gear and screw are 0.99, 0.95, 0.98 and 0.92, respectively.

Chapter

14

Monte Carlo Simulation

Geographical Note

Monte Carlo

Monte Carlo is a town in Monaco with a population of 10,000 (1977 estimate) and is located about one mile north of Monaco Ville, the capital of the country. The name of the town was given by Prince Charles III. Although the citizens of Monaco are not admitted to the casino, the town is very famous for its luxurious casino which was founded in 1856 by the prince himself. Besides being famous as a gambling place, Monte Carlo is one of the most fashionable seaside resorts on the French Riviera. Two world famous automobile races are held annually in this town—the International Monaco Grand Prix and the Monte Carlo Rally. The name Monte Carlo method dates back to about 1944 when von Neumann and Ulam introduced it as a code name for their secret work on neutron diffusion problems, encountered in their work on the atomic bomb, at the Los Alamos Scientific Laboratory. Probably the name was chosen because roulette is one of the simplest tools available for generating random numbers and the roulette is traditionally associated with the casino town, Monte Carlo [14.14,14.15].

14.1 Introduction

Monte Carlo simulation is a powerful engineering tool that can be used for the statistical analysis of uncertainty in engineering problems. It is particularly useful for complex problems in which several random variables are related through nonlinear equations. The Monte Carlo analysis can be considered as an experiment performed on a computer rather than performed in an engineering laboratory. If a system parameter is known to follow certain probability distribution, the performance of the system is studied by considering several possible values of the parameter, each following the specified probability distribution. In the context of structural reliability, for example, the Monte Carlo method involves constructing trial structures on a computer according to generated random numbers and determining the percentage of structures that fail [14.1]. Of course, a large number of trial structures is needed if high confidence is required at small failure probability levels.

Since physical experimentation on real systems is a time-consuming and costly affair, this experimentation on trial systems generated with random numbers has proved to be very convenient. The Monte Carlo approach, thus, requires considerable calculations and is useful only for complex interrelated systems or for the verification of other analyses. Also, the solutions derived from Monte Carlo simulations can not provide generalizations or extrapolations about the system behavior. The term "Monte Carlo" is associated with the city of Monte Carlo in Monaco, which is well-known for gambling, especially the roulette wheel. Since the simulation process involves the generation and use of random or chance variables, it has been called the *Monte Carlo simulation*.

14.2 Generation of Random Numbers

The essential feature common to all Monte Carlo computations is that we need to substitute for each random variable a corresponding set of numbers having the statistical properties of that random variable. The numbers that we substitute are called random numbers, on the grounds that they could well have been produced by chance by a suitable random process. Special devices may be used for generating random variables. For example, a fair coin can be tossed for a discrete random variable having two equally likely possible values. Uniformly distributed random numbers within a specified range may be generated by spinning a frictionless disc or wheel with the range subdivided equally on the circumference of the disc. Before digital computers became popular, published tables of random numbers [14.2] were used for hand computations. These tables were generated by physical processes which are, as far as one can tell, random in the strict sense, and have been successfully subjected to a number of statistical tests. For Monte Carlo simulation using digital computers, it will be necessary to generate automatically the requisite random numbers with specified probability distributions. This is accomplished by first generating a uniformly distributed random number between 0 and 1 and then using a suitable transformation to obtain the corresponding random number following the specified probability distribution. For the standard uniform random variable, U, the density and distribution functions are given by (see Figs. 14.1a and b):

$$f_U(u) = \begin{cases} o; & u < o \\ 1; & o \leq u \leq 1 \\ o; & u > 1 \end{cases} \tag{14.1}$$

and

$$F_U(u) = \begin{cases} o; & u < o \\ u; & o \leq u \leq 1 \\ 1; & u > 1 \end{cases} \tag{14.2}$$

Let X be a random variable with distribution function $F_X(x)$. Then the value of x corresponding to a given value of the cumulative probability, say, u, is given by

$$x = F_X^{-1}(u) \tag{14.3}$$

where F_X^{-1} denotes the inverse function of F_X. The significance of Eq. (14.3) can be seen in Fig. 14.1c.

14.2.1 Generation of random numbers following standard uniform distribution

It is most convenient to calculate a sequence of random numbers, one at a time, by a completely specified rule that is devised such that no reasonable statistical test will detect any significant departure from randomness. Such numbers are called *pseudo-random numbers* [14.3]. The main advantage of a specified rule is that the random sequence can be reproduced for purposes of computational verification. The pseudo random numbers u_i, following standard uniform distribution, can be computed from a sequence of positive integers x_i using the relation

$$u_i = \frac{x_i}{m} \tag{14.4}$$

where m is a suitable positive integer. A pseudo-random sequence that can be generated by a recurrence relation is given by

$$x_{i+1} = ax_i \pmod{m} \tag{14.5}$$

where x_{i+1} is the remainder when ax_i is divided by m, a is any integer between 0 and $m - 1$, and m is a large integer whose magnitude is deter-

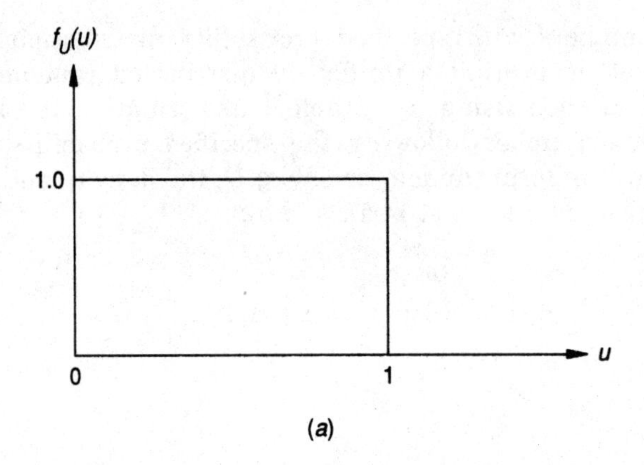

(a)

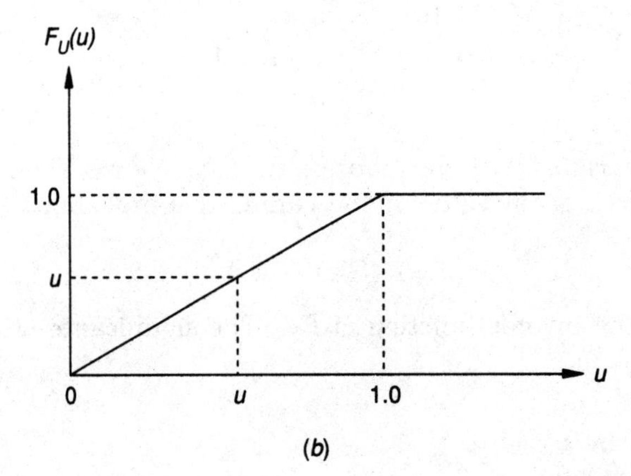

(b)

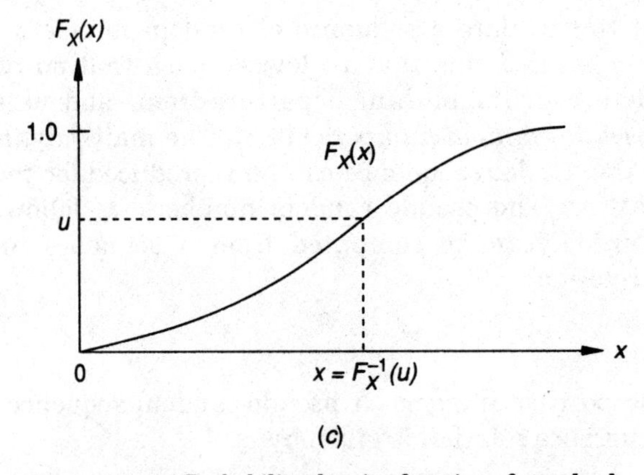

(c)

Figure 14.1 (*a*) Probability density function of standard uniform variate U; (*b*) probability distribution function of standard uniform variate U; and (*c*) probability distribution function of a general random variable X.

mined by the limitations of the computer used. Equation (14.5) can be expressed in a more general form as

$$x_{i+1} = (ax_i + b)\,(\text{mod } m) \tag{14.6}$$

where a and b are integers between 0 and $m - 1$. Thus, if n_i denotes the integer part of the ratio

$$\left(\frac{ax_i + b}{m}\right) \text{ i.e., } \quad n_i = \text{integer}\left(\frac{ax_i + b}{m}\right) \tag{14.7}$$

then the remainder x_{i+1} is given by

$$x_{i+1} = ax_i + b - m\,n_i \tag{14.8}$$

The sequence of numbers generated by Eqs. (14.6) and (14.4) are repetitive after, at most, m numbers. This means that the numbers are periodic. For example, the recurrence relation

$$x_{i+1} = (7x_i + 1) \quad (\text{mod } 15) \tag{14.9}$$

with $x_0 = 1$ generates the sequence 1, 8, 12, 10, 11, 3, 7, 5, 6, 13, 2, 0 ; 1, 8, ... The period of this pseudo-random sequence can be seen to be twelve. Obviously the period must be selected to be equal to or larger than the number of random numbers needed in a given problem. The uniformly distributed random numbers corresponding to the above pseudo random sequence can be found by dividing the numbers by $m = 15$ (i.e., by using Eq. 14.4) as 0.066667, 0.533333, 0.800000, 0.666667, 0.733333, 0.20000, 0.4666667, 0.333333, 0.400000, 0.866667, 0.133333, 0.000000; 0.066667, 0.533333, ... It can be shown that for large values of m, the numbers generated using the deterministic formula, Eq. (14.6), are uniformly distributed and statistically independent [14.4]. Representative values are $a = 2^7 + 1$, $b = 1$, and $m = 2^{35}$ for binary computers [14.11].

14.2.2 Random variables with nonuniform distribution

As stated earlier, if u denotes a particular value of a uniform random variate, the corresponding random number, x, following the distribution function $F_X(x)$ can be found as

$$x = F_X^{-1}(u) \tag{14.10}$$

The method of determining random numbers using Eq. (14.10) is known as the inverse-transformation method. If the function F can be inverted analytically (as in the cases of exponential and Weibull distributions), the application of Eq. (14.10) is simple and straightforward. On the other hand, if the probability distribution function cannot be inverted analytically (as in the cases of normal and lognormal distributions), other methods such as the composition method and the method of functions of random variables have to be used.

Example 14.1 If u_i , $i = 1, 2, \ldots, N$ denote a set of standard uniformly distributed random numbers, find the corresponding numbers x_i of a random variable X, which is uniformly distributed in the range a to b.

solution The probability density and distribution functions of X are given by

$$f_X(x) = \begin{cases} \dfrac{1}{b-a}; & a \le x \le b \\ 0; & \text{elsewhere} \end{cases} \tag{E1}$$

and

$$F_X(x) = \int_a^x f_X(x')\,dx' = \begin{cases} \dfrac{x-a}{b-a}; & a \le x \le b \\ 0; & \text{elsewhere} \end{cases} \tag{E2}$$

Since u_i is uniformly distributed in (0,1), we set

$$u_i = F_X(x_i) = \frac{x_i - a}{b-a} \tag{E3}$$

which can be inverted to obtain

$$x_i = a + (b-a)\,u_i; \qquad i = 1, 2, \ldots, N \tag{E4}$$

Example 14.2 If u_i, $i = 1, 2, \ldots, N$ represent a set of standard uniformly distributed random numbers, find the corresponding numbers x_i of a random variable X that follows exponential distribution.

solution The probability density and distribution functions of X are given by

$$f_X(x) = \begin{cases} \dfrac{1}{\alpha} e^{-(x/\alpha)}; & o \le x < \infty,\ \alpha > 0 \\ 0; & \text{elsewhere} \end{cases} \tag{E1}$$

$$F_X(x) = \int_o^x f_X(x')\,dx' = \frac{1}{\beta}\int_o^x e^{-(x'/\alpha)}\,dx' = 1 - e^{-(x/\alpha)} \tag{E2}$$

Equating the standard uniform number u_i to $F_X(x_i)$,

$$u_i = F_X(x_i) = 1 - e^{-(x_i/\alpha)} \tag{E3}$$

Equation (E3) can be inverted by first rewriting it as

$$1 - u_i = e^{-(x_i/\alpha)} \tag{E4}$$

and then taking logarithms on both sides of Eq. (E4) to obtain

$$x_i = -\alpha \ln (1 - u_i); \qquad i = 1, 2, \ldots, N \tag{E5}$$

Since $(1 - u_i)$ also can be considered as a standard uniform random number, Eq. (E5) can be rewritten as

$$x_i = -\alpha \ln u_i \;; \qquad i = 1, 2, \ldots, N \tag{E6}$$

Example 14.3 If u_i, $i = 1, 2, \ldots, N$ indicate a set of standard uniformly distributed random numbers, find the corresponding numbers x_i of a random variable X that follows Weibull distribution.

solution The probability density and distribution functions of X are given by

$$f_X(x) = \begin{cases} \dfrac{\alpha}{\beta^\alpha}\, x^{\alpha-1}\, e^{-(x/\beta)^\alpha}; & 0 \le x < \infty \\ 0; & \text{elsewhere} \end{cases} \tag{E1}$$

and

$$F_X(x) = \begin{cases} 1 - e^{-(x/\beta)^\alpha}; & 0 \le x < \infty \\ 0; & \text{elsewhere} \end{cases} \tag{E2}$$

Equating u_i to $F_X(x_i)$,

$$u_i = 1 - e^{-(x_i/\beta)^\alpha} \tag{E3}$$

we can solve for x_i to obtain

$$x_i = \beta\,[-\ln(1 - u_i)]^{1/\alpha}; \qquad i = 1, 2, \ldots, N \tag{E4}$$

Since $(1 - u_i)$ can also be considered as a standard uniform number, Eq. (E4) can be rewritten as

$$x_i = \beta(-\ln u_i)^{1/\alpha}; \qquad i = 1, 2, \ldots, N \tag{E5}$$

Example 14.4 If u_i denotes a standard uniformly distributed random number, determine the corresponding number x_i of a random variable X that follows normal distribution.

solution If X follows normal distribution with mean μ and standard deviation σ, the corresponding standard normal variate Z is given by

$$Z = \frac{x - \mu}{\sigma} \tag{E1}$$

with

$$f_Z(z) = \frac{1}{\sqrt{2\pi}}\, e^{-z^2/2} \tag{E2}$$

The standard normal tables list the values z versus $F_Z(z)$. Corresponding to the given uniform number $u_i = F_Z(z_i)$, we can find the value of z_i from standard normal tables. Once z_i is known, the random number x_i can be determined from Eq. (E1) as

$$x_i = \mu + \sigma\, z_i \tag{E3}$$

Instead of using standard normal tables, z_i corresponding to $u_i = F_Z(z_i)$ can be determined by making use of the approximate formula given by Eq. (8.28). Another approximate method, presented by Tocher [14.5], is based on the approximation

$$e^{z^2/2} \approx \frac{2\,e^{-kz}}{(1 + e^{-kz})^2}\;; \qquad z > 0\;; \qquad k = \sqrt{\frac{8}{\pi}} \tag{E4}$$

The integral of $e^{z^2/2}$ gives

$$F_Z(z) = \int \frac{1}{\sqrt{2\pi}} e^{z^2/2}\, dz = \frac{1}{\sqrt{2\pi}} \int \frac{2\, e^{-kz}}{(1+e^{-kz})^2}\, dz$$

$$= \frac{2}{1+e^{-kz}} - 1 \tag{E5}$$

Equating $F_Z(z)$, given by Eq. (E5), to u_i and inverting the relation yields

$$z = \frac{1}{k} \log\left(\frac{1+u_i}{1-u_i}\right) \tag{E6}$$

The desired random number z_i can be obtained by assigning randomly a sign to z in Eq. (E6).

Example 14.5 The cumulative probability distribution function of a random variable X is known at a finite number of points from experimental results as shown in Fig. 14.2. Generate a random number x_k corresponding to the uniformly distributed number u_k.

solution Since the distribution function of X, $F_X(x)$, is not known in the form of an expression, we use a linear interpolation technique to generate the random number x_i. Let the given uniformly distributed number $u_k \equiv F_k$ fall in between the known discrete distribution function values F_i and F_{i+1}, which correspond to the values x_i and x_{i+1}, respectively, of the random variable X. Then the linear interpolation of F_k between F_i and F_{i+1} gives

$$\frac{F_k - F_i}{x_k - x_i} = \frac{F_{i+1} - F_i}{x_{i+1} - x_i}; \quad x_i \le x_k \le x_{i+1} \tag{E1}$$

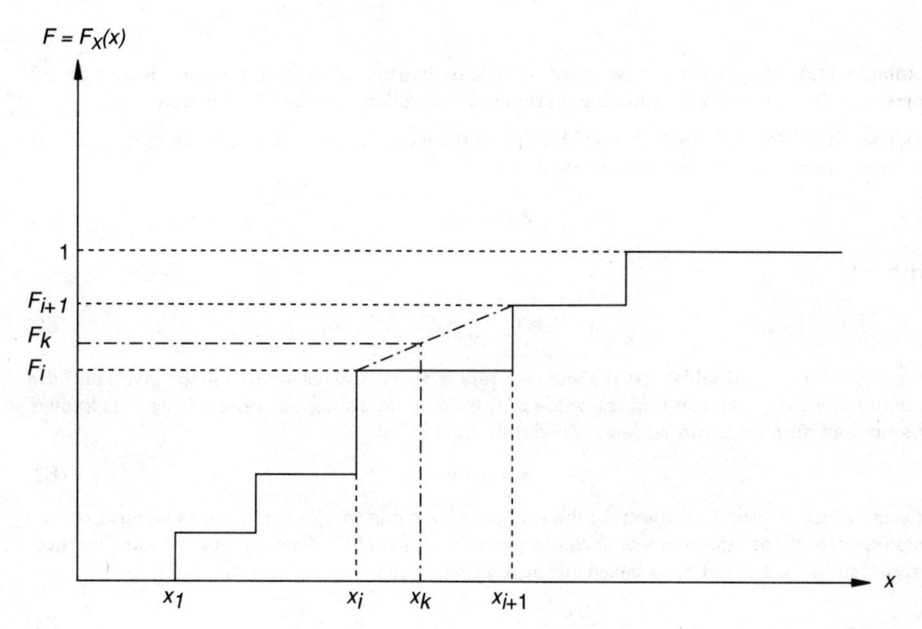

Figure 14.2

By equating F_k to u_k, Eq. (E1) gives

$$x_k = x_i + (u_k - F_i)\left(\frac{x_{i+1} - x_i}{F_{i+1} - F_i}\right)$$

14.2.3 Generation of discrete random variables

Let X be a discrete random variable that can take $n + 1$ distinct values with the probability mass function $p_X(x_i) = P(X = x_i)$, $i = 0, 1, \ldots, n$. The cumulative distribution function of X is given by $F_X(x_i) = P(X \le x_i) = \sum_{j=0}^{i} p_X(x_j)$. The standard uniformly distributed random numbers can be used to generate random numbers following the discrete distribution $F_X(x)$. Let u be a standard uniformly distributed number. The corresponding discrete random number x_i can be determined by identifying x_{i+1} and x_i such that

$$p_X(x_o) + p_X(x_1) + \cdots + p_X(x_{i-1}) < u \le p_X(x_o) + p_X(x_1) + \cdots + p_X(x_i)$$

that is

$$F_X(x_{i-1}) < u \le F_X(x_i) \tag{14.11}$$

It can be seen that the generation of discrete random numbers by this method requires the determination of the cumulative distribution function, in general, for all the possible values of the random variable. Each time a number u is generated, we need to search for the corresponding x_i which satisfies Eq. (14.11). This method is basically the inverse-transformation procedure requiring, in most cases, a numerical search to determine the inverse of the distribution function.

Example 14.6 Generate random integers following uniform distribution between the integers a and b.

solution If u_i denotes a standard (continuous) uniformly distributed number in (0,1), we can set

$$u_i = F_X(x_i) = \frac{x_i - a}{b - a} \tag{E1}$$

and get

$$x_i = a + (b - a)\, u_i \tag{E2}$$

Since x_i is restricted to take only integer values, we can use the relation

$$x_i = a + [u_i(b - a) + 1] \tag{E3}$$

where $[\cdot]$ is the largest integer value of $[u_i(b - a) + 1]$. For example, if $u_i = 0.2892, a = 1$, and $b = 15$, we obtain

$$x_i = 1 + [0.2892(15 - 1) + 1] = 1 + [4.0488 + 1]$$

$$= 1 + [5.0488] = 1 + 5 = 6$$

Example 14.7 Generate random numbers following binomial distribution with $n = 5$ and $p = 0.4$ corresponding to the standard uniformly distributed numbers 0.75300 and 0.22015.

solution The probability mass function of X is given by

$$p_X(x_i) = \binom{5}{i} (0.4)^i (0.6)^{5-i}\,; \qquad i = 0, 1, 2, 3, 4, 5 \tag{E1}$$

and the cumulative distribution function by

$$F_X(x_i) = \sum_{j=0}^{i} \binom{5}{j} (0.4)^j (0.6)^{5-j}\,; \qquad i = 0, 1, 2, 3, 4, 5 \tag{E2}$$

The numerical values given by Eqs. (E1) and (E2) are shown in Table 14.1. The probability mass and distribution functions are shown graphically in Fig. 14.3. By equating the first standard uniform random number to the distribution function as $u_1 = 0.75300 = F_X(x_1)$, we find that u_1 is bounded by $F_X(2)$ and $F_X(3)$. Hence, the value of x_1 is given by 3. Similarly, corresponding to the second standard uniform random number, we have $u_2 = 0.22015 = F_X(x_2)$. Since u_2 is bounded by $F_X(0)$ and $F_X(1)$, we choose the value of x_2 as 1.

TABLE 14.1 Probability mass and distribution functions of X

i	$p_X(i)$	$F_X(i)$
0	0.07776	0.07776
1	0.2592	0.33696
2	0.3456	0.68256
3	0.2304	0.91296
4	0.0768	0.98976
5	0.01024	1.00000

14.3 Generation of Jointly Distributed Random Numbers

14.3.1 Independent random variables

Let $X_1, X_2, \ldots, X_n$ be a set of n random variables. If these variables are statistically independent, then their joint density and distribution functions can be expressed as

$$f_{X_1,\ldots,X_n}(x_1,\ldots,x_n) = \prod_{i=1}^{n} f_{X_i}(x_i) \tag{14.12}$$

$$F_{X_1,\ldots,X_n}(x_1,\ldots,x_n) = \prod_{i=1}^{n} F_{X_i}(x_i) \tag{14.13}$$

where $f_{X_i}(x_i)$ and $F_{X_i}(x_i)$ denote the marginal (individual) probability density and distribution functions of X_i. In this case the random numbers for each

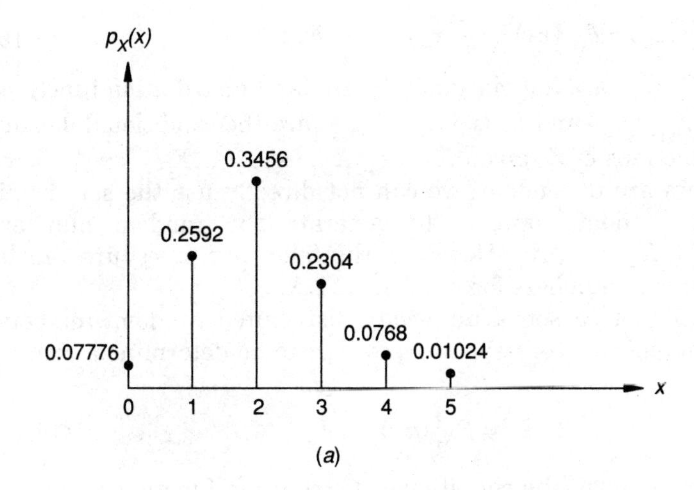

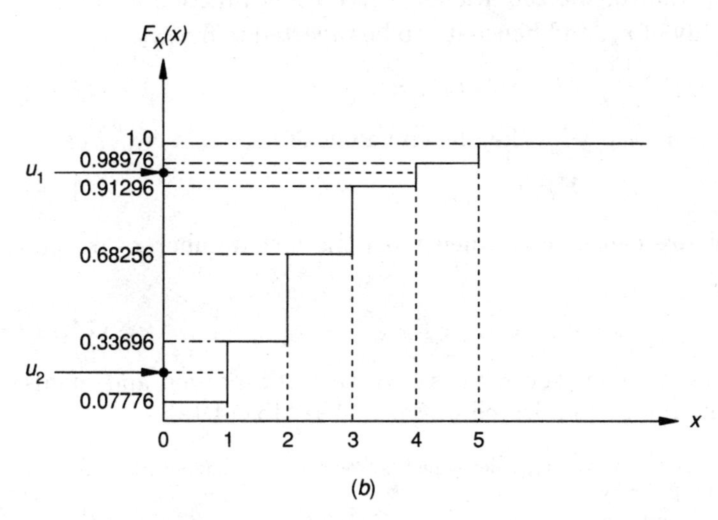

Figure 14.3

variable can be generated separately and independently of one another using the procedures described earlier.

14.3.2 Dependent random variables

For a set of n correlated (dependent) random variables $X_1, X_2, \ldots, X_n$, the joint probability density and distribution functions can be expressed as

$$f_{X_1,\ldots,X_n}(x_1,\ldots,x_n) = f_{X_1}(x_1) \cdot f_{X_2}(x_2|_{x_1}) \cdots f_{X_n}(x_n|_{x_1,\ldots,x_{n-1}}) \tag{14.14}$$

$$F_{X_1,\ldots,X_n}(x_1,\ldots,x_n) = F_{X_1}(x_1)\cdot F_{X_2}(x_2|_{x_1})\cdots F_{x_n}(X_n|_{x_1,\ldots,x_{n-1}}) \tag{14.15}$$

where $f_{X_1}(x_1)$ and $F_{X_1}(x_1)$ are the marginal density and distribution functions of X_1, and $f_{X_i}(x_i|_{x_1,x_2,\ldots,x_{i-1}})$ and $F_{X_i}(x_i|_{x_1,x_2,\ldots,x_{i-1}})$ are the conditional density and distribution functions of X_i given $X_1 = x_1, X_2 = x_2, \ldots, X_{i-1} = x_{i-1}$. Since the random numbers are dependent, we can not directly use the set of uniformly distributed random numbers to generate the random numbers corresponding to $X_1, X_2, \ldots, X_n$. However, the following procedure can be used to generate random numbers for $X_1, X_2, \ldots, X_n$.

Let $(u_1, \ldots, u_n)$ denote a set of uniformly distributed random numbers. Then the random number x_1 corresponding to X_1 can be determined from u_1 as

$$x_1 = F_{X_1}^{-1}(u_1) \tag{14.16}$$

With the value of x_1 known, the conditional distribution function $F_{X_2}(x_2|_{x_1})$ becomes a function only of x_2, and hence it can be inverted to find x_2 as

$$x_2 = F_{X_2}^{-1}(u_2|_{x_1}) \tag{14.17}$$

In general, we can generate x_i with the known values of $x_1, x_2, \ldots, x_{i-1}$ as

$$x_i = F_{X_i}^{-1}(u_i|_{x_1,x_2,\ldots,x_{i-1}}) \tag{14.18}$$

This recursive procedure can be continued until the last number x_n is generated as

$$x_n = F_{X_n}^{-1}(u_n|_{x_1,x_2,\ldots,x_{n-1}}) \tag{14.19}$$

It can be seen that the method requires the inversion of marginal and conditional distribution functions as indicated in Eqs. (14.16) to (14.19).

Example 14.8 The contact pressure, $p_{\max}$, developed between two cylinders [14.6] under a load F (see Figure 14.4) is given by

$$p_{\max} = \frac{2F}{\pi \varepsilon a} \tag{E1}$$

where 2ε denotes the width of contact after deformation:

$$\varepsilon = \sqrt{\frac{4F(1-\nu^2)\,d_1\,d_2}{\pi Ea(d_1+d_2)}} \tag{E2}$$

where F = applied load, d_1 and d_2 = diameters of cylinders, ν = Poisson's ratio of cylinders, E = Young's modulus of cylinders, and a = axial length of cylinders. Assume the load and axial length of cylinders to be independently normally distributed random variables with $(\overline{F}, \sigma_F)$ = (10,000, 2000) N, and $(\overline{a}, \sigma_a)$ = (0.2, 0.02) m. If 0.41277, 0.64903, 0.39797, 0.26741, 0.25185, 0.04836, 0.94539, 0.89573, 0.75300, and 0.22015 denote the uniformly distributed random numbers, determine the mean and standard deviation of the contact pressure developed. Assume $\nu = 0.3$, $E = 2.06\times10^{11}$ N/m², $d_1 = 0.1$ m, and $d_2 = 0.25$ m.

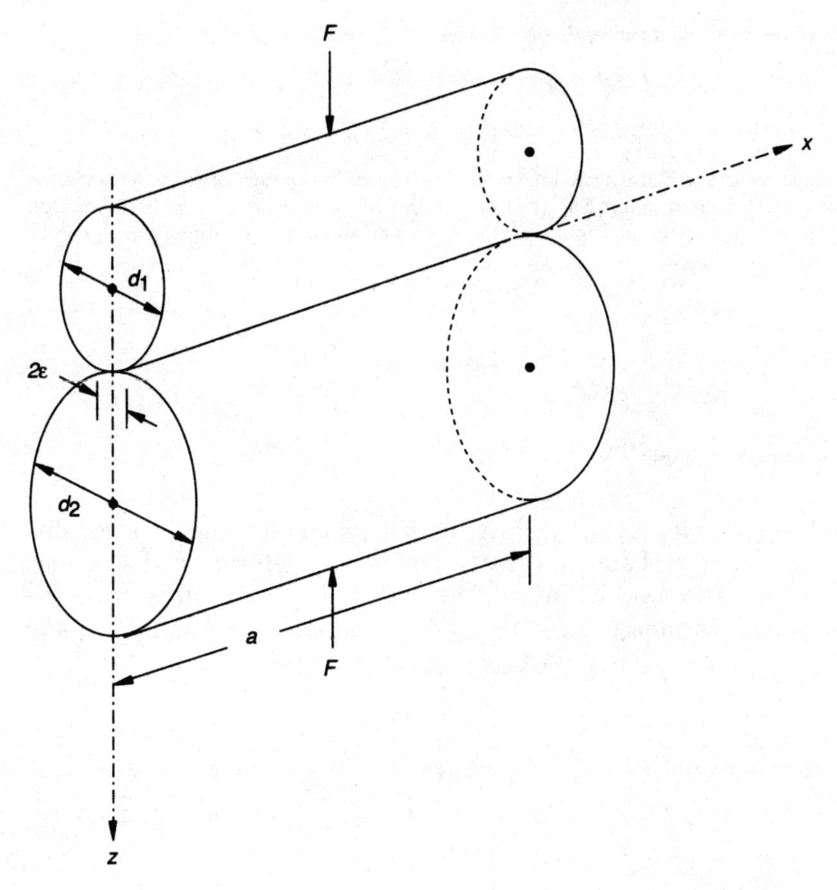

Figure 14.4

solution Since there are two random variables in the problem, we use the first five random numbers for generating the values of F and the remaining five numbers for generating the values of a. By using the definition of the standard normal distribution

$$F_Z(z) = \frac{1}{\sqrt{2\,\pi}} \int_{-\infty}^{z} e^{-\frac{1}{2}u^2} \cdot du$$

and standard normal tables, we find for the uniformly distributed number 0.41277, the corresponding standard normally distributed number x is given by

$$F_Z(z) = 0.41277 \rightarrow z = -\,0.220425$$

If z_1 and z_2 denote the standard normally distributed numbers for F and a, we have by definition,

$$z_1 = \frac{F - \bar{F}}{\sigma_F} \quad \text{and} \quad z_2 = \frac{a - \bar{a}}{\sigma_a}$$

The sample values of F and a can be determined as

$$F = \overline{F} + \sigma_F z_1 = (10000 + 2000\, z_1)\ \text{N}$$

$$a = \overline{a} + \sigma_a z_2 = (0.2 + 0.02\, z_2)\ \text{m}$$

Once the sample values of F and a are known, the corresponding contact pressure between the cylinders can be computed using Eqs. (E1) and (E2). The results are shown in Table 14.2. From the values of $p_{\max}$ given in Table 14.2, the mean and standard deviations of $p_{\max}$ can be determined as

$$\overline{p}_{\max} = 216.7737 \times 10^6\ \text{N/m}^2$$

and

$$\sigma_{p_{\max}} = 13.0799 \times 10^6\ \text{N/m}^2$$

14.3.3 Generation of correlated normal random variables

Let the random variables $X_1, X_2, \ldots, X_n$ be correlated and follow normal distribution. Their joint probability density function is defined by the means, variances, and covariances of the random variables. We assume that the mean values ($\overline{X}_i$), variances ($\sigma_i^2 = \sigma_{X_i}^2$), and covariances ($\sigma_{ij} = \sigma_{X_i X_j}$) of the variables X_i are known, that is, the vector of mean values

$$\overline{\mathbf{X}} = \begin{Bmatrix} \overline{X}_1 \\ \overline{X}_2 \\ \cdot \\ \cdot \\ \cdot \\ \overline{X}_n \end{Bmatrix} \tag{14.20}$$

TABLE 14.2

Standard uniformly distributed numbers		Corresponding standard normal numbers		Normal variates			
u_1	u_2	z_1	z_2	F (N)	a (m)	Sample value of ε(m) Eq. (E2)	Sample value of $p_{\max}$(N/m²) Eq. (E1)
0.41277	0.04836	−0.220425	−1.660967	9559.149	0.166781	151.745×10^{-6}	240.4574×10^6
0.64903	0.94539	0.382703	1.601708	10765.406	0.232034	136.526×10^{-6}	216.3419×10^6
0.39797	0.89573	−0.258605	1.257590	9482.790	0.225152	130.079×10^{-6}	206.1255×10^6
0.26741	0.75300	−0.620665	0.683961	8758.670	0.213679	128.326×10^{-6}	203.3478×10^6
0.25185	0.22015	−0.668679	−0.771687	8662.642	0.184566	137.318×10^{-6}	217.5958×10^6

and the covariance matrix

$$[V_X] = \begin{bmatrix} \sigma_1^2 & \sigma_{12} & \cdots & \sigma_{1n} \\ \sigma_{12} & \sigma_2^2 & \cdots & \sigma_{2n} \\ \cdot & \cdot & \cdots & \cdot \\ \cdot & \cdot & \cdots & \cdot \\ \cdot & \cdot & \cdots & \cdot \\ \sigma_{n1} & \sigma_{n2} & \cdots & \sigma_n^2 \end{bmatrix} \tag{14.21}$$

are known. To generate the required set of correlated normally distributed random numbers X_i, $i = 1, 2, \ldots, n$, we first generate a set of n statistically independent normally distributed random numbers W_i, $i = 1, 2, \ldots, n$ whose mean values $\overline{W}_i$ and variances σ_{Wi}^2 are unknown at this time. Express the desired correlated random variables X_i as linear functions of the independent random variables W_i [14.10] as

$$X_i = a_{i1} W_1 + a_{i2} W_2 + \cdots + a_{in} W_n$$

$$= \sum_{j=1}^{n} a_{ij} W_j; \qquad i = 1,2, \ldots, n$$

or

$$\underset{n \times 1}{\mathbf{X}} = \underset{n \times n}{[a]} \; \underset{n \times 1}{\mathbf{W}} \tag{14.22}$$

where the elements of the matrix $[a]$ are also unknown at this stage. Since the random variables X_i are expressed as linear functions of W_j, the mean values and standard deviations of X_i can be expressed as

$$\overline{\mathbf{X}} = [a]\,\overline{\mathbf{W}} \tag{14.23}$$

and

$$[V_X] = [a][V_W][a]^T \tag{14.24}$$

where $\overline{\mathbf{X}}$ and $\overline{\mathbf{W}}$ are the vectors of mean values of X_i and W_j, and $[V_X]$ and $[V_W]$ are the covariance matrices of X_i and W_j, respectively. In Eqs. (14.23) and (14.24), the vector $\overline{\mathbf{X}}$ and the matrix $[V_X]$ are known from the characteristics of the random variables to be generated, and the vector $\overline{\mathbf{W}}$ and the matrices $[a]$ and $[V_W]$ are to be determined. The Choleski decomposition method [14.7] can be used for this purpose.

In the Choleski decomposition method, a symmetric matrix $[S]$ of order $n \times n$ is decomposed into the form of a matrix product as

$$[S] = [L][D][L]^T \tag{14.25}$$

where $[L]$ is a lower triangular matrix of order $n \times n$ that contains ones on the main diagonal, and $[D]$ is a diagonal matrix of order $n \times n$. The elements of [S] and [D] [14.7, 14.10] are given by

$$D_{11} = S_{11}$$

$$L_{ii} = 1; \qquad i = 1, 2, \ldots, n$$

$$L_{j1} = \frac{S_{1j}}{D_{11}}; \qquad j \geq 2$$

$$D_{ii} = S_{ii} - \sum_{j=1}^{i-1} L_{ij}^2 D_{jj}; \qquad i \geq 2$$

$$L_{ji} = \frac{1}{D_{ii}} \left[S_{ij} - \sum_{k=1}^{i-1} L_{ik} L_{jk} D_{kk} \right]; \qquad i \geq 2; \qquad j \geq i+1 \tag{14.26}$$

By comparing Eqs. (14.24) and (14.25) we find that

$$[S] = [V_X] \tag{14.27}$$

$$[a] = [L] \tag{14.28}$$

and

$$[V_W] = [D] \tag{14.29}$$

From Eqs. (14.26) and (14.28), the elements a_{ij} and hence the linear relationships between X_i and W_j are completely known. Similarly, Eqs. (14.26) and (14.29) define the variances of W_j. Further, since the covariance matrix, $[V_W] = [D]$ is diagonal, the random variables W_j are ensured to be independent. Finally, the mean values of W_j can be determined from Eq. (14.23) as

$$\overline{\mathbf{W}} = [a]^{-1}\, \overline{\mathbf{X}} \tag{14.30}$$

It can be observed that the random variables X_i will be normally distributed since they are obtained from a linear combination of the normal random variables, W_j.

The procedure described for generating the correlated normal random numbers is indicated as a flow diagram in Fig. 14.5.

Example 14.9 The torque transmitted by a plate clutch with a single pair of friction surfaces, shown in Fig. 14.6, can be expressed as [14.6]

$$T = \frac{Ff}{4}(D + d) \text{ for uniform wear condition} \tag{E1}$$

and

$$T = \frac{Ff}{3}\left(\frac{D^3 - d^3}{D^2 - d^2}\right) \text{ for uniform pressure condition} \tag{E2}$$

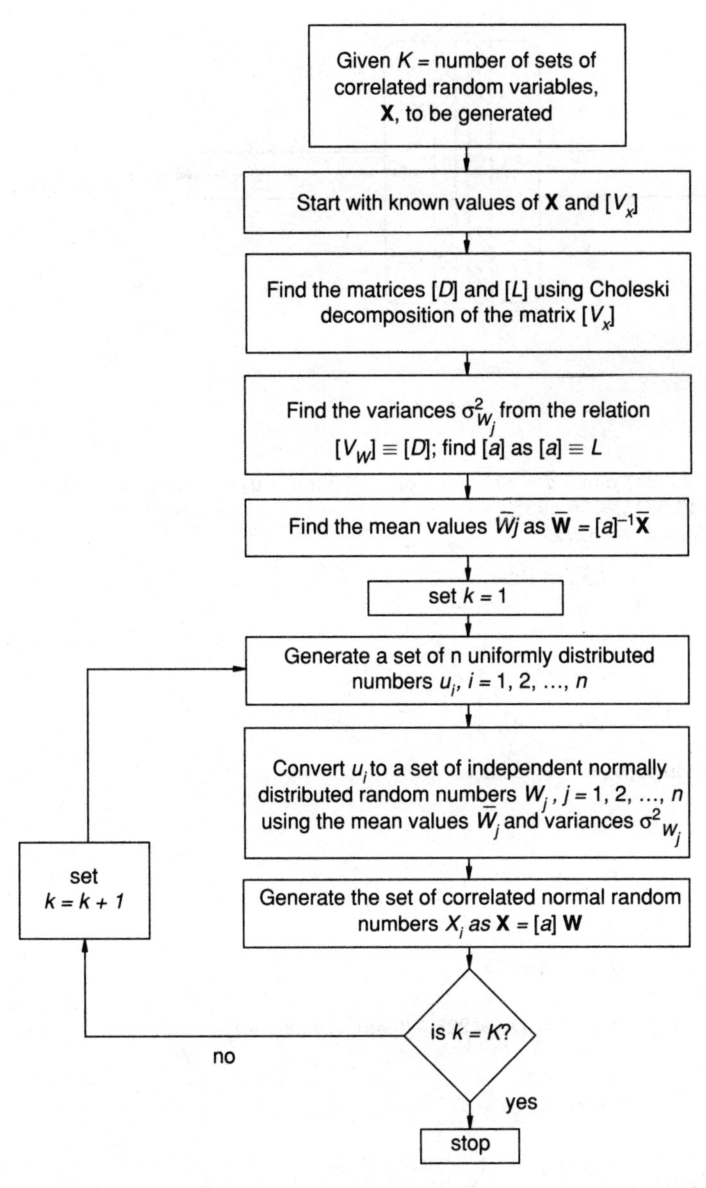

Figure 14.5

where F = axial force applied $= (\pi\, p_a\, d\,/\,2)(D - d)$, p_a = maximum pressure induced, f = coefficient of friction, D = outer diameter, and d = inner diameter. If $\bar{D} = 250$ mm, $\bar{d}$ = 200 mm, $\sigma_D = 12.5$ mm, $\sigma_d = 15$ mm, $\rho_{D,d} = 0.4$, $f = 0.25$, and $p_a = 750$ kPa, compute the torque transmitted by the clutch using the Monte Carlo simulation. Use the following uniformly distributed random numbers to generate the correlated random variables: 0.41277, 0.64903, 0.39797, 0.26741, 0.25185, 0.04836, 0.94539, 0.89573, 0.75300, 0.22015.

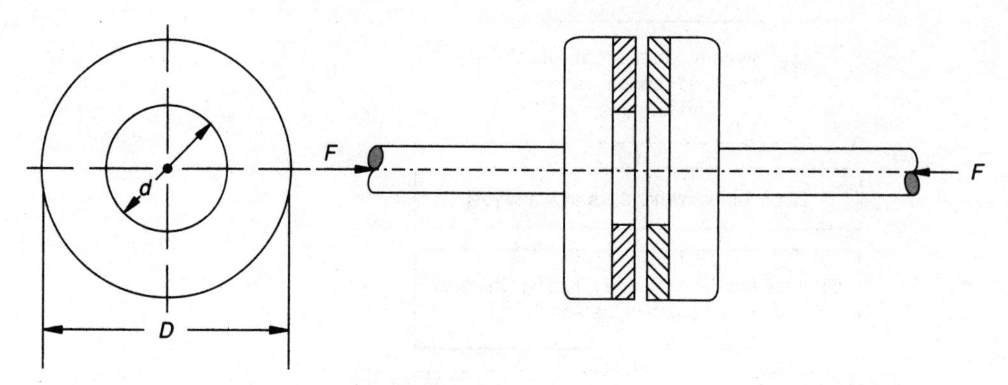

Figure 14.6

solution Given data: $\bar{D} = 250$ mm, $\bar{d} = 200$ mm, $\sigma_D = 12.5$ mm, $\sigma_d = 15$ mm, $\rho_{D,d} = 0.4 = \sigma_{D,d} / (\sigma_D \sigma_d)$, $\sigma_{D,d} = (0.4)(12.5)(15) = 75$ mm^2.

Defining $X_1 = D$ and $X_2 = d$, we have

$$\overline{\mathbf{X}} = \begin{Bmatrix} \bar{X}_1 \\ \bar{X}_2 \end{Bmatrix} = \begin{Bmatrix} 250 \\ 200 \end{Bmatrix} \text{ mm}$$

$$\left[V_x\right] = \begin{bmatrix} \sigma_{x_1}^2 & \sigma_{x_1,x_2} \\ \sigma_{x_1,x_2} & \sigma_{x_2}^2 \end{bmatrix} = \begin{bmatrix} 156.25 & 75 \\ 75 & 225 \end{bmatrix} \text{mm}^2$$

The Choleski decomposition of $[S] \equiv \left[V_X\right]$ leads to

$$D_{11} = S_{11} = 156.25$$

$$L_{11} = L_{22} = 1$$

$$L_{21} = \frac{S_{12}}{D_{11}} = \frac{75}{156.25} = 0.48$$

$$D_{22} = S_{22} - L_{21}^2\, D_{11} = 225 - (0.48)^2\,(156.25) = 189$$

$$[a] \equiv [L] = \begin{bmatrix} 1 & 0 \\ 0.48 & 1 \end{bmatrix}$$

$$\left[V_W\right] \equiv [D] = \begin{bmatrix} 156.25 & 0 \\ 0 & 189 \end{bmatrix} = \begin{bmatrix} 12.5^2 & 0 \\ 0 & 13.7477^2 \end{bmatrix}$$

The vector $\overline{\mathbf{W}}$ is determined as

$$\overline{\mathbf{W}} = [a]^{-1}\, \overline{\mathbf{X}} = \begin{bmatrix} 1 & 0 \\ -0.48 & 1 \end{bmatrix} \begin{Bmatrix} 250 \\ 200 \end{Bmatrix} = \begin{Bmatrix} 250 \\ 80 \end{Bmatrix} \text{ mm}$$

For this given uniformly distributed random number u_i, the corresponding normally distributed numbers w_i are determined. The correlated random numbers x_i are found as

TABLE 14.3

u_i	z_i	D (m)	u_i	z_i	d (m)	Torque transmitted: Uniform wear k N − m	Torque transmitted: Uniform pressure k N − m
0.41277	−0.220425	0.24724	0.04836	−1.660967	0.17584	0.39114	0.39486
0.64903	0.382703	0.25478	0.94539	1.601708	0.22432	0.24107	0.24140
0.39797	−0.258605	0.24677	0.89573	1.257590	0.21574	0.22796	0.22830
0.26741	−0.620665	0.24224	0.75300	0.683961	0.20568	0.24801	0.24857
0.25185	−0.668679	0.24164	0.22015	−0.771687	0.18538	0.32794	0.32984

$$\mathbf{x} = [a]\,\mathbf{w} = \begin{Bmatrix} w_1 \\ 0.48\, w_1 + w_2 \end{Bmatrix} \equiv \begin{Bmatrix} D \\ d \end{Bmatrix}$$

These values are shown in Table 14.3. Once the values of d and D are known, the torque transmitted by the clutch can be determined using Eqs. (E1) and (E2).

14.4 Computation of Reliability

Due to the random nature of parameters influencing the performance of an engineering system, there is a small probability of not meeting the specified performance requirements. The Monte Carlo method can be used to estimate the reliability or probability of failure of the system. In this method, a uniformly distributed random number is used to generate a sample value of each of the random parameters of the system. These parameters are then used to predict the performance of the system. The sample system's performance is compared with the specified performance requirement. The sample is considered to be a successful one if the specified performance requirement is met; otherwise it is considered to be a failed one. After generating the predetermined number of sample systems, the overall reliability of the system is computed as

$$\begin{pmatrix} \text{Reliability} \\ \text{of the system} \end{pmatrix} = \left(\frac{\text{number of successful sample systems}}{\text{total number of sample systems generated}} \right)$$

This procedure is shown as a flow diagram in Fig. 14.7.

14.4.1 Sample size and error in simulation

The random numbers generated in the Monte Carlo simulation can be used to estimate the reliability (or probability of failure) of an engineering system. It is important to know the error involved in the estimated failure probability. It is equally important to know how many simulations (i.e., sample size) are required to achieve a specified accuracy. The percent error involved with the estimated probability P_f [14.8] has been found to be

$$\%\ \text{error} = 200 \cdot \sqrt{\frac{1-P_f}{n \cdot P_f}} \tag{14.31}$$

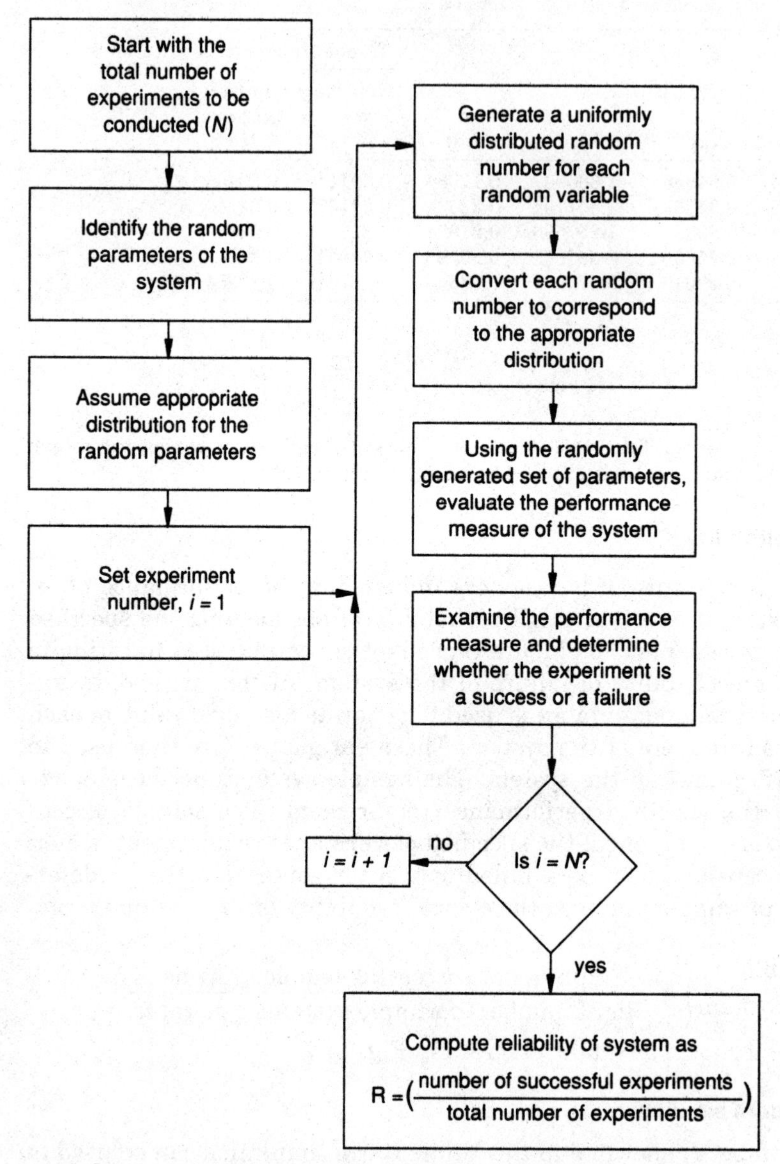

Figure 14.7 Computation of reliability of a system using Monte Carlo simulation. (Reprinted with permission from R. H. Crawford and S. S. Rao, "Probabilistic Analysis of Functioning Generating Mechanisms," *Journal of Mechanisms, Transmissions, and Automation in Design*, ASME, vol. 111, Dec. 1989.)

where n is the number of simulations (sample size) used in estimating P_f. There is a 95 percent chance that the percent error in the estimated value of P_f will be less than that given by Eq. (14.31). For example, if 1,000 simulations were performed in finding an estimated probability of failure of 0.1, Eq. (14.31) will give a 18 percent error. This means that there is a 95 percent chance that the actual failure probability will be within 0.1 ± 0.018. Equation (14.31) can also be used to find the sample size required for achieving a specified accuracy in the estimated probability of failure. For example, if an accuracy of 0.1 ± 0.01 is desired for the probability of failure, Eq. (14.31) gives the required number of simulations (n) as 3600.

14.4.2 Example: reliability analysis of a straight-line mechanism

The four-bar straight-line mechanism shown in Fig. 14.8 is considered to illustrate the Monte Carlo method for computing reliability [14.9]. The reliability of this function generating mechanism can be defined as the probability of generating the desired function with a specific accuracy. The mechanism has six design parameters, namely, r_1, r_2, r_3, r_4, r_5 and θ. The nominal values of the parameters are assumed as $r_1 = 0.963$, $r_2 = 0.764$, $r_3 = 0.528$, $r_4 = 1.815$, $r_5 = 0.778$, and $\theta = -89.65°$. These parameters are assumed to be

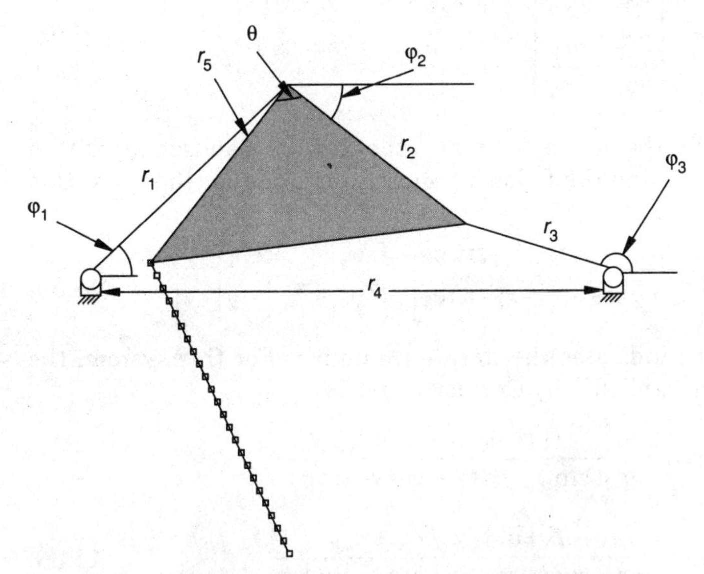

Figure 14.8 Four-bar straight line mechanism. (Reprinted with permission from R. H. Crawford and S. S. Rao, "Probabilistic Analysis of Function Generating Mechanisms," *Journal of Mechanisms, Transmissions, and Automation in Design*, ASME, vol. 111, Dec. 1989.)

random in nature, and the reliability of the mechanism is computed for normal, Weibull, and beta distributions. To calculate the reliability of the mechanism, trial values for the six design parameters are generated using random numbers. These values specify one mechanism in the sample. The input link length is r_1. Positions of the coupler point are computed by considering the vector loop equations of the mechanism

$$r_1 \cos \phi_1 + r_2 \cos \phi_2 - r_3 \cos \phi_3 - r_4 = 0$$

$$r_1 \sin \phi_1 + r_2 \sin \phi_2 - r_3 \sin \phi_3 = 0 \tag{14.32}$$

For a given mechanism design, the parameters r_1, r_2, r_3, and r_4 are known. The input angle ϕ_1 is assumed to be deterministic. For a specified value of ϕ_1, Eqs. (14.32) are solved for the values of the coupler angle, ϕ_2, and the follower angle, ϕ_3, using the Newton-Raphson method. The vector loop equations become the residual functions

$$f_1 = r_1 \cos \phi_1 + r_2 \cos \phi_2 - r_3 \cos \phi_3 - r_4$$

$$f_2 = r_1 \sin \phi_1 + r_2 \sin \phi_2 - r_3 \sin \phi_3 \tag{14.33}$$

The Jacobian matrix for these equations is given by

$$[A] = \begin{bmatrix} \dfrac{\partial f_1}{\partial \phi_2} & \dfrac{\partial f_1}{\partial \phi_3} \\ \dfrac{\partial f_2}{\partial \phi_2} & \dfrac{\partial f_2}{\partial \phi_3} \end{bmatrix} = \begin{bmatrix} -r_2 \sin \phi_2 & r_3 \sin \phi_3 \\ r_2 \cos \phi_2 & -r_3 \cos \phi_3 \end{bmatrix} \tag{14.34}$$

Given initial guesses for the angles ϕ_2 and ϕ_3, the Newton-Raphson algorithm consists of iteratively solving the following matrix equation for the correction vector $\overline{\Delta\phi}$:

$$[A^i]\,\overline{\Delta\phi} = \begin{bmatrix} -r_2 \sin \phi_2 & r_3 \sin \phi_3 \\ r_2 \cos \phi_2 & -r_3 \cos \phi_3 \end{bmatrix}^i \begin{bmatrix} \Delta\phi_2 \\ \Delta\phi_3 \end{bmatrix}^i = \begin{bmatrix} -f_1 \\ -f_2 \end{bmatrix}^i \tag{14.35}$$

where the superscript i indicates the iteration number. For this system, the solution can be written explicitly by Cramer's rule as

$$\Delta\phi_2 = \frac{f_1 \cos \phi_3 + f_2 \sin \phi_3}{r_2(\sin \phi_2 \cos \phi_3 - \cos \phi_2 \sin \phi_3)}$$

$$\Delta\phi_3 = \frac{f_2 \sin \phi_2 + f_1 \cos \phi_2}{r_3(\sin \phi_2 \cos \phi_3 - \cos \phi_2 \sin \phi_3)} \tag{14.36}$$

The updated values of the angles are then computed from

$$\phi_2^{i+1} = \phi_2^i + \Delta\phi_2$$

$$\phi_3^{i+1} = \phi_3^i + \Delta\phi_3 \tag{14.37}$$

The convergence criteria for a solution can be checked on both the residual vector and the correction vector. Once the angles are known for a given input angle, the coupler point position can be computed as

$$x_c = r_1 \cos\phi_1 + r_5 \cos(\phi_2 + \theta)$$

$$y_c = r_1 \sin\phi_1 + r_5 \sin(\phi_2 + \theta) \tag{14.38}$$

This procedure is followed over the input angle range of –14° to 42° in increments of two degrees. The resulting coupler points (x_c, y_c) are then fit to a line using least squares analysis. Finally, the perpendicular distance from the desired straight line to each of the coupler point positions is computed. If any such distance exceeds the specified error tolerance, the mechanism is declared unreliable. Also, if the Newton-Raphson iteration fails to converge after a maximum number of steps for any input angle, the mechanism is declared unreliable.

The results of reliability computations are given in Table 14.4. For each simulation, 10,000 trials were used with an error tolerance of 0.05. Plots of the density functions of three random parameters are given in Figs. 14.9–14.11. The simulation results indicate that all distributions produce reliabilities over 0.9. The beta distributions yield the highest reliability because they are bounded at both ends and have the smallest dispersion

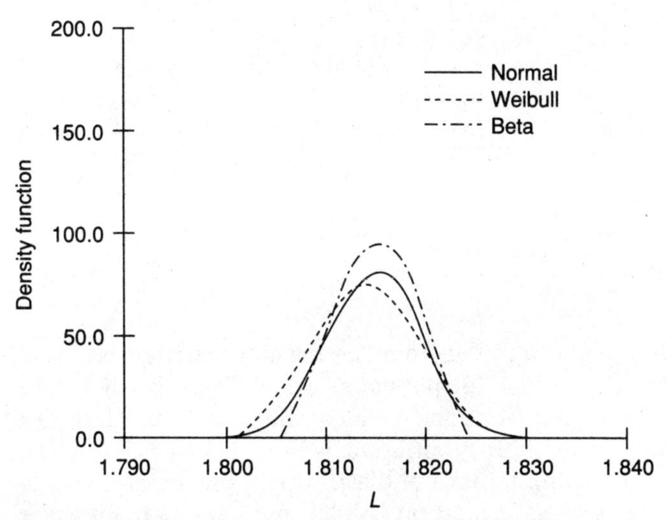

Figure 14.9 Distributions for ground link. (Reprinted with permission from R. H. Crawford and S. S. Rao, "Probabilistic Analysis of Function Generating Mechanisms," *Journal of Mechanisms, Transmissions, and Automation in Design*, ASME, vol. 111, Dec. 1989.)

TABLE 14.4 Results of reliability simulations

Normal Distribution		
Parameter	mean	standard deviation
Input link	0.963	0.005
Coupler	0.764	0.005
Follower	0.528	0.005
Ground link	1.815	0.005
Coupler radius	0.778	0.005
Coupler angle	−1.565	0.01
Reliability = 0.9514		

Weibull Distribution			
Parameter	b	h	l
Input link	3.0	0.964	0.95
Coupler	3.0	0.765	0.75
Follower	3.0	0.529	0.51
Ground link	3.0	1.816	1.80
Coupler radius	3.0	0.779	0.76
Couple angle	3.0	−1.564	−1.57
Reliability = 0.919			

Beta Distribution				
Parameter	q	r	l	u
Input link	3.0	3.0	0.953	0.973
Coupler	3.0	3.0	0.754	0.774
Follower	3.0	3.0	0.518	0.538
Ground link	3.0	3.0	1.805	1.825
Coupler radius	3.0	3.0	0.768	0.788
Coupler angle	3.0	3.0	−1.575	−1.555
Reliability = 0.9873				

SOURCE: R. H. Crawford and S. S. Rao, "Probabilistic Analysis of Function Generating Mechanisms," *Journal of Mechanisms, Transmissions, and Automation in Design*, ASME, vol. 111, Dec. 1989.

about the mean. The reliability computed from the normal distributions was somewhat lower due to the wider dispersions, which can be seen in Figs. 14.9–14.11. The Weibull distributions yielded the smallest reliability because these distributions have slightly different means. Using Eq. (14.31), the probability of failure, according to normal distribution, has been found to be 0.0486 using 10,000 simulations. Equation (14.31) gives the percent error as 8.8489. Thus, it is likely that the actual probability of failure will be within 0.0486 ± 0.0043 or the actual reliability will be within 0.9514 ± 0.0043.

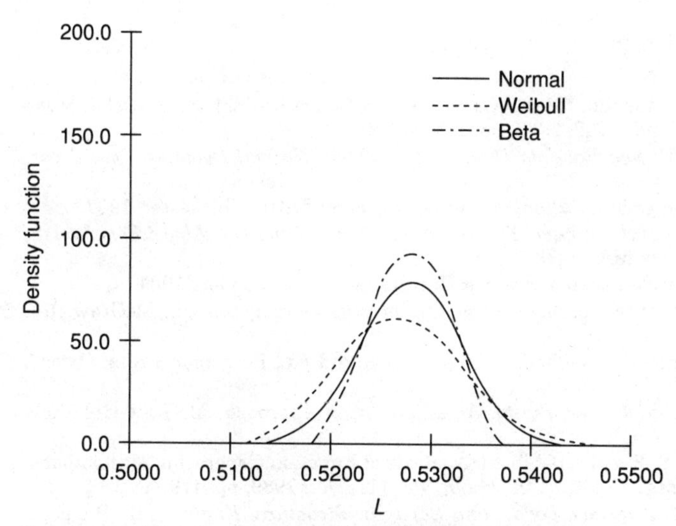

Figure 14.10 Distributions for follower. (Reprinted with permission from R. H. Crawford and S. S. Rao, "Probabilistic Analysis of Function Generating Mechanisms," *Journal of Mechanisms, Transmissions, and Automation in Design*, ASME, vol. 111, Dec. 1989.)

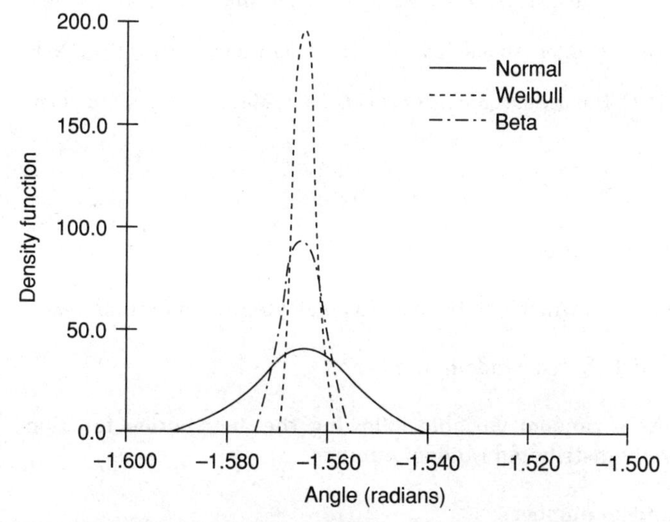

Figure 14.11 Distributions for coupler angle. (Reprinted with permission from R. H. Crawford and S. S. Rao, "Probabilistic Analysis of Function Generating Mechanisms," *Journal of Mechanisms, Transmissions, and Automation in Design*, ASME, vol. 111, Dec. 1989.)

References and Bibliography

14.1. R. F. Warner and A. P. Kabaila, "Monte Carlo Study of Structural Safety," *ASCE J. of the Structural Div.*, vol. 94, no. ST12, 1968, pp. 2847–2859.

14.2. Rand Corporation, *A Million Random Digits with 100,000 Normal Deviates*, Free Press, Glencoe, IL, 1955.

14.3. F. Neelamkavil, *Computer Simulation and Modelling*, John Wiley, Chichester, 1987.

14.4. D. E. Knuth, *The Art of Computer Programming: Seminumerical Algorithms*, vol. 2, Addison-Wesley, Reading, MA, 1969.

14.5. K. D. Tocher, *The Art of Simulation*, English Universities Press, London, 1963.

14.6. J. E. Shigley and C. R. Mischke, *Mechanical Engineering Design*, 5th ed., McGraw-Hill, New York, 1989.

14.7. S. S. Rao, *The Finite Element Method in Engineering*, 2nd ed., Pergamon Press, Oxford, England, 1989.

14.8. M. L. Shooman, *Probabilistic Reliability: An Engineering Approach*, McGraw-Hill, New York, 1968.

14.9. R. H. Crawford and S. S. Rao, Probabilistic Analysis of Function Generating Mechanisms, *ASME J. of Mech., Trans., and Auto. in Design*, vol. 111, Dec. 1989, pp. 479–481.

14.10. G. C. Hart, *Uncertainty Analysis, Loads, and Safety in Structural Engineering*, Prentice-Hall, Englewood Cliffs, NJ, 1982.

14.11. A. H. S. Ang and W. H. Tang, *Probability Concepts in Engineering Planning and Design*, vol. II: *Decision, Risk, and Reliability*, John Wiley, New York, 1984.

14.12. M. F. Spotts, "Predicting Length of Assemblies with Monte Carlo Simulation," *Machine Design*, Nov, 20, 1980, pp. 84–88.

14.13. G. S. Fishman, *Concepts and Methods in Discrete Event Digital Simulation*, John Wiley, New York, 1973.

14.14. *The Encyclopedia Americana*, international ed., vol. 19, Americana Corporation, New York, 1971.

14.15. J. M. Hammersley and D. C. Handscomb, *Monte Carlo Methods*, Methuen & Co. Ltd., London, 1964.

Review Questions

14.1 What is Monte Carlo simulation?

14.2 List five physical processes which can be used to generate random numbers?

14.3 What is a uniformly distributed random number?

14.4 How do you generate a random variable following the distribution function $F_X(x)$ using a uniformly distributed random number?

14.5 What is a pseudo-random number?

14.6 State the characteristics of a good random number generator.

14.7 How do you generate pseudo-random numbers?

14.8 How do you generate a discrete random number?

14.9 How do you find the reliability of a system using Monte Carlo simulation?

14.10 Generate the numbers given by the formula

$$X_{i+1} = (8\,X_i + 1)(\text{mod } 14)$$

with $X_1 = 2$.

Problems

14.1 The probability density function of a random variable X is given by

$$f_X(x) = \begin{cases} \dfrac{1}{x^2}; & x \ge 1 \\ 0; & x < 1 \end{cases}$$

Generate the random numbers x_i corresponding to X in terms of a set of standard uniformly distributed numbers u_i, $i = 1, 2, \ldots, N$.

14.2 The probability distribution function of a random variable X is given by

$$F_X(x) = 1 - \left(1 - \frac{x}{k}\right)^2; \qquad 0 \le x \le k$$

Generate the random numbers x_i corresponding to X in terms of a set of standard uniformly distributed numbers u_i, $i = 1, 2, \ldots, N$.

14.3 If u_i, $i = 1, 2, \ldots, N$ denote a set of standard uniformly distributed random numbers, find the corresponding numbers x_i of a random variable X which follows the following triangular distribution (see Fig. 14.12):

$$f_X(x) = \begin{cases} k\left(\dfrac{x-a}{b-a}\right); & a \le x \le b \\ k\left(\dfrac{c-x}{c-b}\right); & b \le x \le c \end{cases}$$

where $k = \left(\dfrac{2}{c-a}\right)$

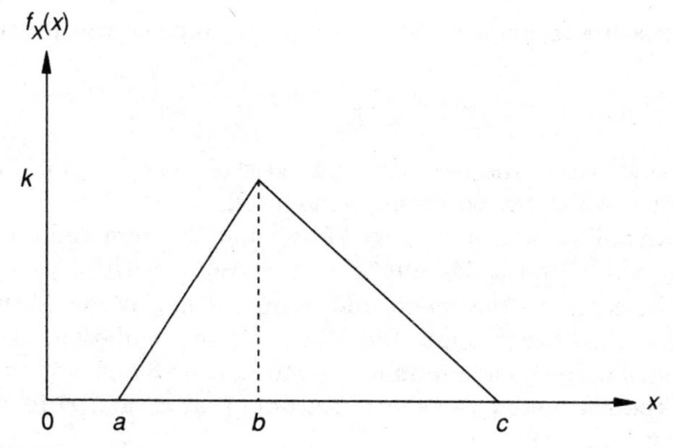

Figure 14.12

TABLE 14.5 Standard Uniformly Distributed Random Numbers

Row 1	0.52478	0.22835	0.33307	0.73842	0.67277
Row 2	0.32880	0.76457	0.94489	0.82597	0.04836
Row 3	0.58815	0.01919	0.22225	0.38562	0.45731
Row 4	0.91743	0.99315	0.70350	0.78240	0.22015
Row 5	0.66047	0.26430	0.22415	0.98215	0.10413
Row 6	0.54380	0.10492	0.59665	0.42368	0.15138
Row 7	0.71899	0.68860	0.33415	0.72559	0.19902
Row 8	0.40024	0.74215	0.93857	0.04988	0.24389
Row 9	0.22094	0.89237	0.41999	0.12790	0.87990
Row 10	0.77646	0.33177	0.62684	0.34119	0.09212

14.4 Consider a set of five standard uniformly distributed random numbers given in the first row of Table 14.5. Generate the corresponding numbers x_i of a log-normally distributed random variable X

$$f_X(x) = \frac{1}{x\,\sigma\sqrt{2\pi}} \exp\left[-\frac{(\ln x - \mu)^2}{2\sigma^2}\right]$$

with $\sigma = 1.0$ and $\mu = 2.0$.

14.5 The stress induced in a wire rope (σ) when it is bent around a sheave (Fig. 14.13) is approximately given by

$$\sigma = E\frac{d_w}{D}$$

where E = Young's modulus of the wire, d_w = diameter of the wire and D = diameter of the sheave. If d_w and D follow normal distribution with $\bar{d}_w = 2$ mm, $\bar{D} = 200$ mm, $\sigma_{d_w} = 0.2$ mm, $\sigma_D = 15$ mm, $\rho_{d_w,D} = 0.3$, and $E = 2.07 \times 10^{11}$ Pa, find the mean and standard deviations of the stress induced in the wire using the Monte Carlo simulation. Use the uniformly distributed random numbers given in the first two rows of Table 14.5 to generate the correlated random numbers. Compare the results with those given by the partial derivative rule.

14.6 The bearing pressure (p) induced by a loaded wire rope on the sheave is given by

$$p = 2\frac{F}{d_r D}$$

where F = tensile force in the rope, d_r = diameter of the rope and D = diameter of the sheave over which the wire rope is mounted.

(a). If the mean values of d_r and D are 25 mm and 200 mm and the variances are 6.25 mm^2 and 225 mm^2, respectively, with $\rho_{d_r,D} = 0.25$ and $F = 4000$ N, compute the mean and standard deviations of the bearing pressure on the sheave using the Monte Carlo simulation. Use the uniformly distributed random numbers given in rows 3 and 4 of Table 14.5 to generate the correlated variables. Assume that d_r and D follow normal distribution.

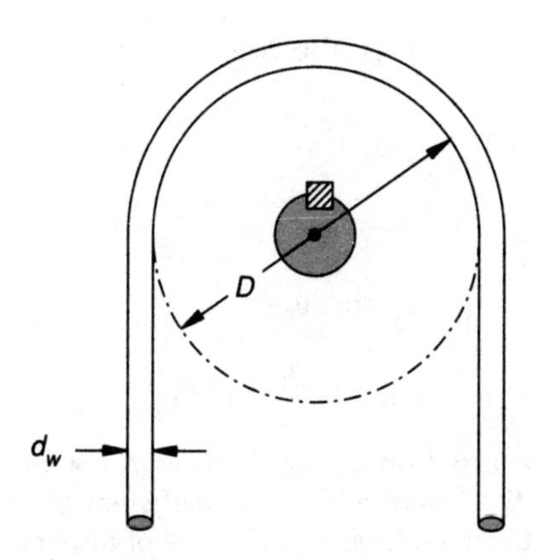

Figure 14.13

(b). Find the mean and standard deviations of p using the partial derivative rule.

14.7 The length of a roller chain (L) mounted over two sprockets (see Fig. 14.14) is approximately given by

$$L = 2C + \left(\frac{N_1 + N_2}{2}\right) p + \left(\frac{N_2 - N_1}{2\pi}\right)^2 \frac{p^2}{C}$$

where p = chain pitch, C = center distance, N_1 = number of teeth on a small sprocket, and N_2 = number of teeth on large sprocket. The following data is known for a particular application

$$N_1 = 20; N_2 = 30; \overline{C} = 30 \text{ in}; \sigma_C = 0.5 \text{ in}; \overline{p} = 0.625 \text{ in}; \sigma_p = 0.04 \text{ in}; \rho_{p,C} = 0.$$

Find the mean and standard deviations of the length of the chain using the uniformly distributed random numbers given in rows 5 and 6 of Table 14.5.

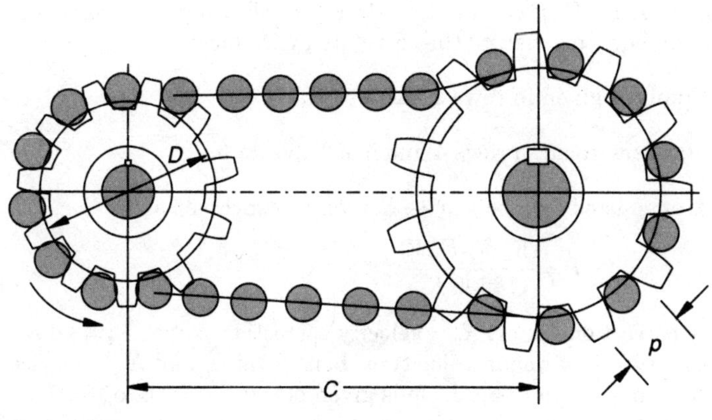

Figure 14.14

14.8 The actuating force (F) necessary to apply the brake shown in Fig. 5.12 is given by [14.6]

$$F = \frac{M_n - M_f}{a}$$

where

$$M_n = p_a trb\left(\frac{\theta_2}{2} - \frac{1}{4}\sin 2\theta_2\right)$$

$$M_f = p_a trf\left(r - r\cos\theta_2 - \frac{b}{2}\sin^2\theta_2\right)$$

where p_a = maximum pressure exerted on the brake lining, t = face width of brake shoes, r = inner radius of the brake drum, f = coefficient of friction, a = linear distance between the brake shoe-hinge and the line of application of the actuating force, b = radius of the brake shoe, and θ_2 = angular distance between the brake shoe-hinge and the point of application of the actuating force. For the brake shown in Fig. 5.12, the data is given by $\bar{b} = 120$ mm, $\sigma_b = 4$ mm, $\bar{p}_a = 10^6$ Pa, $\sigma_{p_a} = 10^5$ Pa, $\bar{t} = 30$ mm, $\sigma_t = 1$ mm, $\bar{r} = 150$ mm, $\sigma_r = 10$ mm, $\bar{f} = 0.35$, $\sigma_f = 0.035$, $\bar{a} = 200$ mm, $\sigma_a = 10$ mm, $\bar{\theta}_2 = 120°$, and $\sigma_{\theta_2} = 3°$. Assuming all parameters to be independent, estimate the mean and standard deviations of the actuating force needed to apply the brake using Monte Carlo simulation. Use five random numbers for each parameter starting from the first row in Table 14.5.

14.9 The energy absorbed by a clutch or brake is given by

$$E = \frac{J_1 J_2(\omega_1 - \omega_2)^2}{2(J_1 + J_2)}$$

where J_1 and J_2 are the mass moments of inertia and ω_1 and ω_2 are the angular velocities (see Fig. 14.15). If $\bar{J}_1 = 1.0$ kg-m^2, $\bar{J}_2 = 0.5$ kg-m^2, $\sigma_{J_1} = \sigma_{J_2} = 0.075$ kg-m^2, $\rho_{J_1,J_2} = 0.5$, $\bar{\omega}_1 = 200$ rpm, $\bar{\omega}_2 = 1000$ rpm, $\sigma_{\omega_1} = \sigma_{\omega_2} = 25$ rpm, and $\rho_{\omega_1,\omega_2} = 0.4$, estimate the mean and standard deviations of the energy absorbed by the clutch using Monte Carlo simulation. Use the following uniformly distributed random numbers to generate the correlated variables:

For J_1 and J_2: Numbers given in rows 2 and 3 of Table 14.5

For ω_1 and ω_2: Numbers given in rows 4 and 5 of Table 14.5

14.10 The horsepower transmitted by a belt drive (P) can be expressed as

$$P = \frac{K_p\, K_v\, F_a\, V}{16500\, K_s}$$

where K_p = pulley correction factor, K_v = velocity correction factor, F_a = allowable tension in belt (lb), V = linear velocity of belt (ft/min), and K_s = service factor. The data for a flat leather belt drive is given below: $\bar{K}_p = 0.6$, $\sigma_{K_p} = 0.06$, $\bar{K}_v = 0.9$, $\sigma_{K_v} = 0.09$, $\bar{V} = 3000$ ft/min, $\sigma_V = 100$ ft/min, $\bar{F}_a = 80$ lb, $\sigma_{F_a} = 4$ lb,

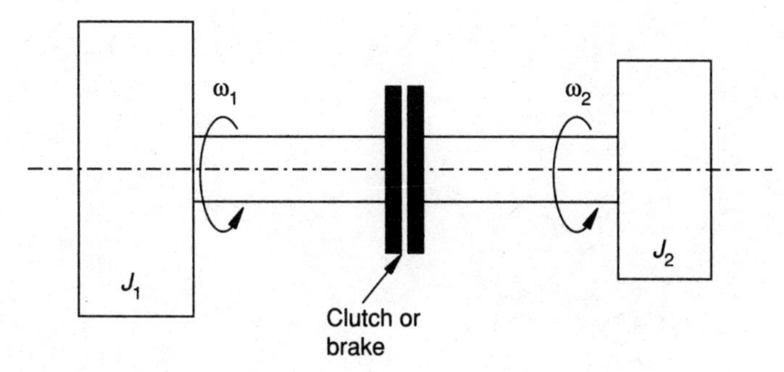

Figure 14.15

$\overline{K}_s = 1.1$, and $\sigma_{K_s} = 0.1$. Assuming that all the random variables are independent, estimate the mean and standard deviations of the horsepower transmitted by the belt drive using the Monte Carlo simulation method. Use five uniformly distributed numbers, starting from row 1 of Table 14.5, to simulate each random variable.

14.11 The contact pressure (p) developed at the interface when a steel gear with a central hole is shrink-fitted over a hollow steel shaft (Fig. 14.16) is given by

$$p = \frac{E\delta}{R}\left[\frac{(r_o^2 - R^2)\ (R^2 - r_i^2)}{2R^2(r_o^2 - r_i^2)}\right]$$

where E = Young's modulus of steel, δ = total radial interference, r_o = pitch circle radius of the gear, r_i = inner radius of the shaft, and R = interface radius. If $E = 2.07 \times 10^{11}$ N/m^2, $\delta = 1$ mm, $R = 300$ mm, $\overline{r}_o = 350$ mm, $\overline{r}_i = 250$ mm, $\sigma_{r_o} = 20$ mm, $\sigma_{r_i} = 15$ mm, $\rho_{r_o,r_i} = 0.3$, determine the mean value and the standard deviation of the contact pressure using the Monte Carlo simulation procedure. Use the uniformly distributed random numbers given in rows 4 and 5 of Table 14.5 to generate the correlated random variables.

14.12 The maximum stresses induced at the inner and outer surfaces [14.6] of a crane hook (see Fig. 14.17) are given by

$$\sigma_i = \frac{W}{A} + \frac{Mc_i}{Aer_i}; \qquad \sigma_o = \frac{W}{A} - \frac{Mc_o}{Aer_o}$$

where W = weight lifted, A = cross-sectional area, M = bending moment induced, c_i = distance from neutral axis to inner fiber, c_o = distance from neutral axis to outer fiber, e = distance from centroidal axis to neutral axis, r_i = radius of inner fiber, and r_o = radius of outer fiber, with the radius of neutral axis, for the rectangular cross-section shown in Fig. 14.17, is given by

$$r_n = \frac{w}{\ln\left(\dfrac{r_o}{r_i}\right)}$$

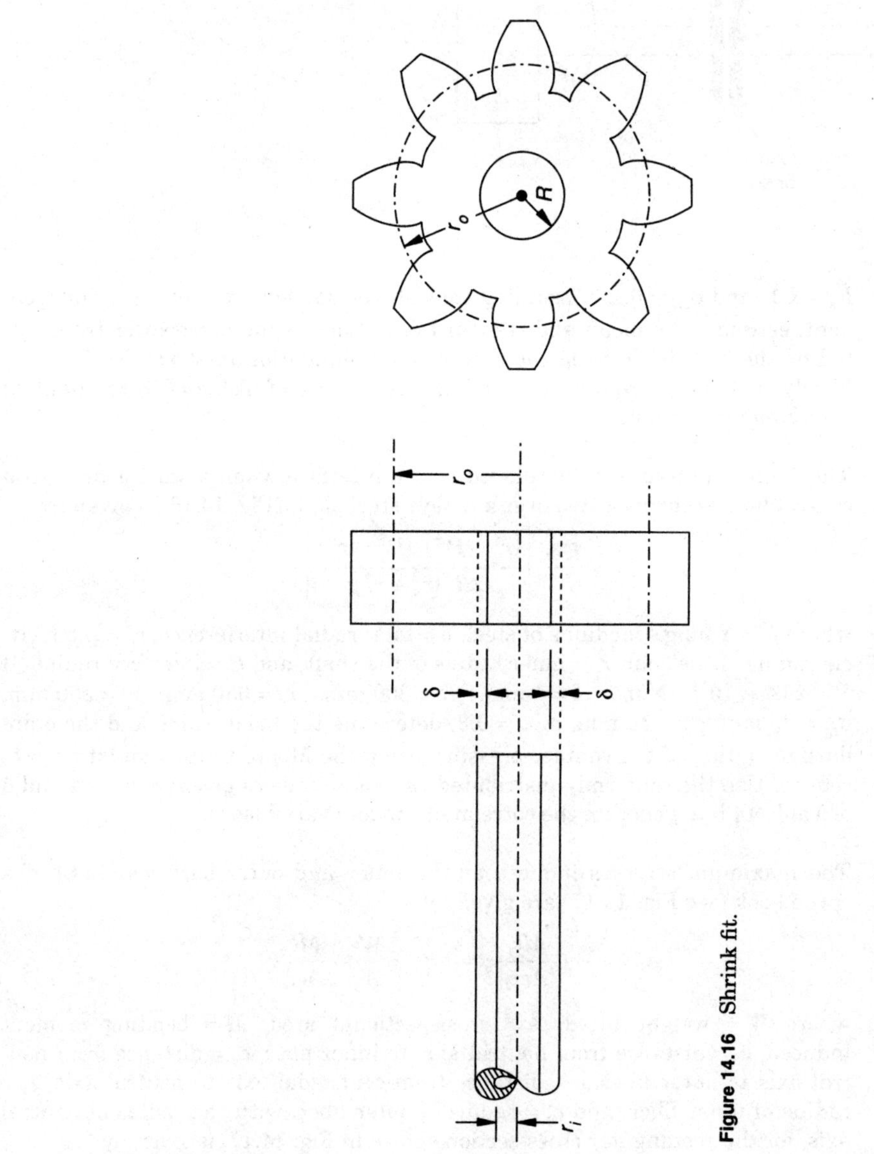

Figure 14.16 Shrink fit.

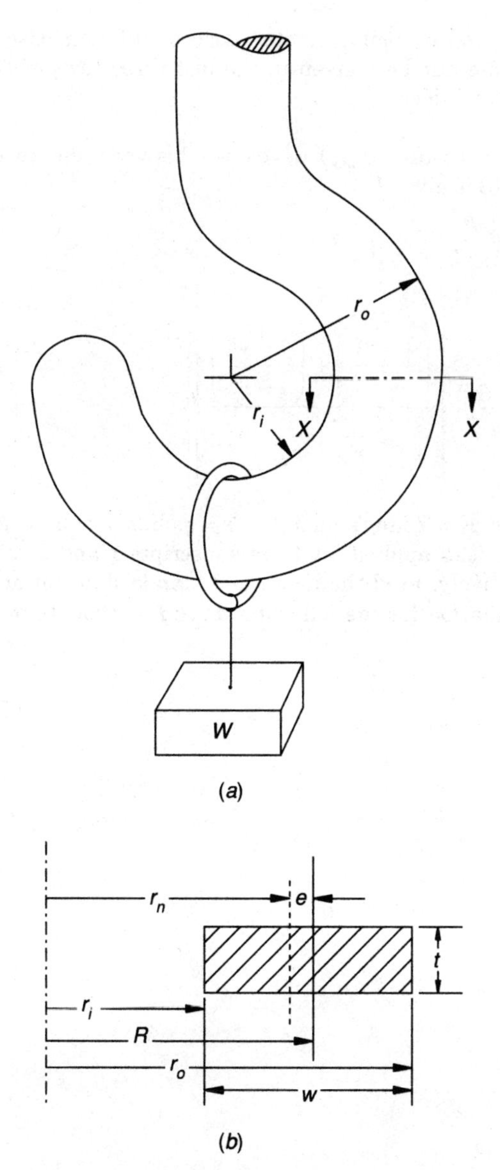

Figure 14.17 (a) Crane hook; and (b) Section X–X.

and the radius of the centroidal axis is given by $R = (r_o + r_i)/2$, so that $e = R - r_n$. The following data is applicable for a specific crane hook: $\overline{r}_i = 2$ in, $\overline{r}_o = 6$ in, $w = 4$ in, $t = 3/4$ in, $\overline{W} = 5000$ lb, $\sigma_{r_i} = 0.1$ in, $\sigma_{r_o} = 0.5$ in, $\sigma_W = 1000$ lb, and $\rho_{r_i, r_o} = 0.5$. Estimate the mean and standard deviations of the stresses induced σ_i and σ_o using Monte Carlo simulation. Use the uniformly distributed random numbers given in the last three rows of Table 14.5 to generate the random variables.

14.13 The maximum contact pressure ($p_{\max}$) developed between the two spheres [14.6] shown in Fig. 14.18 is given by

$$p_{\max} = \frac{3P}{2\pi\varepsilon^2}$$

with

$$\varepsilon = \left\{ \frac{3P}{4} \left[\frac{\left(\frac{1-\nu_1^2}{E_1}\right) + \left(\frac{1-\nu_2^2}{E_2}\right)}{\frac{1}{r_1} + \frac{1}{r_2}} \right] \right\}^{1/3}$$

where ν = Poisson's ratio, E = Young's modulus, r = radius of sphere, ε = radius of the contact area, P = load applied, and the subscripts 1 and 2 refer to the spheres 1 and 2, respectively. Find the mean and standard deviations of $p_{\max}$ using Monte Carlo simulation for the following data: $\bar{P}$ = 5000 lb, σ_P = 500 lb,

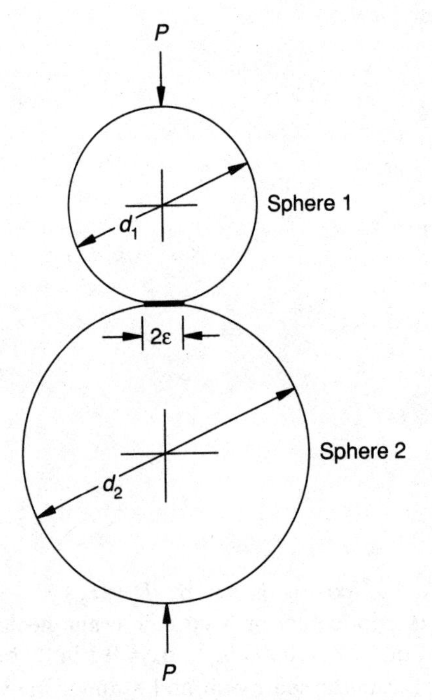

Figure 14.18

$\bar{\nu}_1 = 0.29$, $\sigma_{\nu_1} = 0.029$, $\bar{E}_1 = 30 \times 10^6$ lb/in^2, $\sigma_{E_1} = 3 \times 10^6$ lb/in^2, $\sigma_{\nu_1, E_1} = 0.3$, $\bar{\nu}_2 = 0.32$, $\sigma_{\nu_2} = 0.016$, $\bar{E}_2 = 15 \times 10^6$ lb/in^2, $\sigma_{E_2} = 0.75 \times 10^6$ lb/in^2, $\rho_{\nu_2, E_2} = 0.2$, $r_1 = 2$ in, $r_2 = 3$ in. Use five uniformly distributed random numbers starting from row 1 of Table 14.5 to generate each of the random variables.

14.14 The maximum contact pressure (p_{max}) developed between two cylinders made of different materials (see Fig. 14.4) is given by [14.6]

$$p_{max} = \frac{2F}{\pi \varepsilon a}$$

with

$$\varepsilon = \left\{ \frac{2F}{\pi a} \left[\frac{\left(\frac{1 - \nu_1^2}{E_1} \right) + \left(\frac{1 - \nu_2^2}{E_2} \right)}{\frac{1}{d_1} + \frac{1}{d_2}} \right] \right\}^{1/2}$$

where ν = Poisson's ratio, E = Young's modulus, d = diameter of cylinder, ε = semi-width of contact, a = axial length of contact, F = load applied, and the subscripts 1 and 2 refer to the cylinders 1 and 2, respectively. The data for a specific case follow:

$$\bar{F} = 5000 \text{ lb};\quad \sigma_F = 500 \text{ lb};\quad \bar{\nu}_1 = 0.29;\quad \bar{E}_1 = 30 \times 10^6 \text{ lb/in}^2;$$
$$\sigma_{\nu_1} = 0.029;\quad \sigma_{E_1} = 3 \times 10^6 \text{ lb/in}^2;\quad \rho_{\nu_1, E_1} = 0.3;\quad \bar{\nu}_2 = 0.32;$$
$$\bar{E}_2 = 15 \times 10^6 \text{ lb/in}^2;\quad \sigma_{\nu_2} = 0.016;\quad \sigma_{E_2} = 0.75 \times 10^6 \text{ lb/in}^2;$$
$$\rho_{\nu_2, E_2} = 0.2;\quad a = 20 \text{ in};\quad d_1 = 4 \text{ in};\quad d_2 = 6 \text{ in}.$$

Use five uniformly distributed random numbers starting from row 3 of Table 14.5 to generate each of the random variables and estimate the mean value and standard deviation of p_{max} using the Monte Carlo simulation method.

14.15 The torque required (T) to raise the load in a square threaded screw jack (Fig. 14.19) is given by [14.6]

$$T = \frac{F\, d_m}{2} \left(\frac{\pi \mu d_m + e}{\mu d_m - \mu e} \right)$$

where F = load being raised, d_m = mean diameter of the screw, e = lead of the threads, and μ = coefficient of friction. In a power screw, $\bar{d}_m = 30$ mm, $\sigma_{d_m} =$ 1 mm, $\bar{\mu} = 0.1$, $\sigma_\mu = 0.01$, $\bar{F} = 5000$ N, $\sigma_F = 1000$ N, and $e = 5$ mm with d_m, μ and F following uniform distributions:

a. Use the uniformly distributed random numbers given in rows 4, 5, and 6 of Table 14.5 to generate the random variables d_m, μ and F, and find the mean and standard deviations of the torque, T.

b. Find the mean value and standard deviations of the torque T using the partial derivative rule.

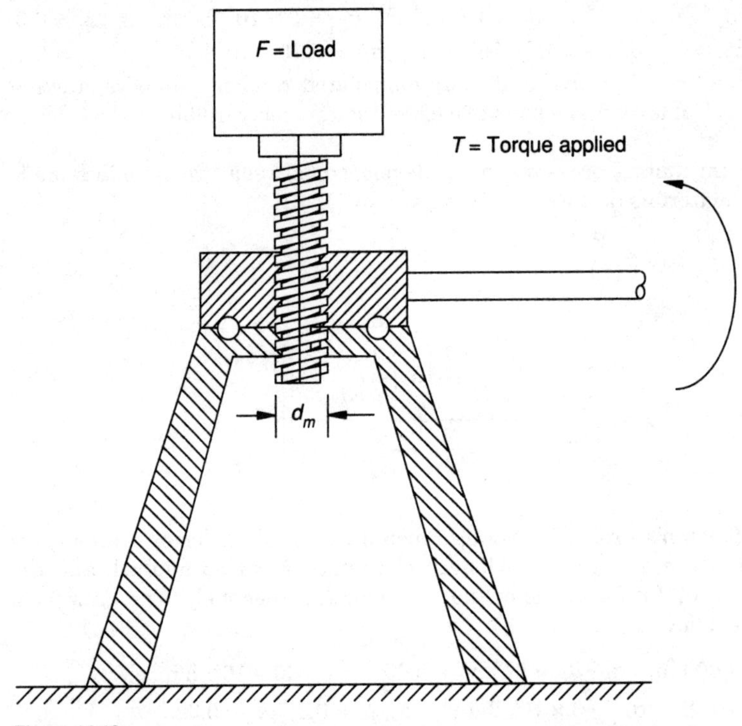

Figure 14.19

c. Discuss why the answers in parts (a) and (b) are different.

d. Assuming that instead of using just five values in part (a), we use a very large number of (infinite) values for each random variable, do you expect the results of parts (a) and (b) to be different? If so, why?

Chapter

15

Reliability Testing

Biographical Note

Waloddi Weibull

Waloddi Weibull was born on June 18, 1887, in Sweden. He joined the Royal Swedish Coast Guard in 1904 and left the military service in 1917. He studied at the Royal Institute of Technology and at Stockholm University. Later he became a consulting engineer and was the inventor of ball bearings and electric hammers. He published his first scientific paper on the propagation of explosive waves in 1914 and became a full professor at the Royal Institute of Technology in 1924. He published two important papers on the statistical distribution of material strength in the late 1930s. The probability distribution he proposed for describing material strength has become known as Weibull distribution, *and it finds application in reliability, statistics, fatigue, fracture, and many other fields. After his retirement from the Royal Institute of Technology in 1953, he became a consultant to the U.S. Air Force Materials Laboratory, published a book on* Fatigue Testing and Analysis of Results *in 1961, and accepted a position as a visiting Professor at Columbia University's Institute for the Study of Fatigue and Reliability. He received many awards, honors, and medals before he died in France on October 12, 1979 [15.13].*

15.1 Introduction

Reliability tests are intended to find whether a system can operate satisfactorily for a specified period of time under prescribed operating conditions. Different types of reliability tests are conducted at various stages of the life cycle of a system as indicated in the following list:

1. *Tests conducted during the design stage.* At the end of the design stage, prototypes are built and tested to analyze the failure modes and reliability of the system. The results of the tests are used to check whether the system is behaving as intended. The results are also used to modify or redesign the system for improved reliability.
2. *Tests conducted during the construction (or manufacturing) stage.* Before the system is put into service, qualification and acceptance tests are conducted to prove that the design standards of reliability are met. In acceptance tests, the system or component is tested to determine whether it should be accepted or rejected (on an individual unit or lot basis). Based on the results, better quality control methods can be used to reduce the defects in construction or manufacture. In qualification tests, the system is tested to determine whether it truly qualifies for its intended application. In some cases, the manufacturer has to demonstrate the reliability to fulfill the contractual requirements. The contract might include an incentive clause whereby the profit margin is readjusted either upwards or downwards depending on the value of the demonstrated reliability compared to the originally specified value.
3. *Tests conducted during the operating stage.* Tests are conducted during the operating stage of the system to find the failure rate and reliability of the system. These results are used to verify the reliability analysis conducted previously and also to find whether any modifications are needed in the design and/or operating procedures to improve the reliability. These tests are also useful to establish proper maintenance and parts replacement schedules. In addition, these tests can be used to determine reasonable warranty policies that will result in improved sales, while assuring a reasonable profit.

As stated in Chapter 6, the life of a system can be divided into three phases, based on the failure rate, as follows:

1. Infant-mortality phase (also known as burn-in phase) during which a high failure rate is observed due to manufacturing defects
2. Operating-life phase during which a constant failure rate is observed due to random failures

3. Aging or wear-out phase during which a high failure rate is observed due to mechanical wear of the components

Although some reliability tests are conducted during the burn-in phase, most of the reliability tests are conducted during the useful life of a system, since this is the period of interest to the customer.

15.1.1 Objectives of reliability tests

Depending on the objectives, reliability tests can be classified into four categories as follows:

1. *Longevity tests.* The objective is to find the length of the useful life of the system that has a constant failure rate.
2. *MTBF tests.* The objective is to determine the mean time between failures of the system.
3. *Operating life tests.* The objective is to find the ability of the system to perform without failure for a prescribed minimum time period.
4. *Reliability margin tests.* The objective is to establish the margin of safety between the extreme operating conditions of the system.

15.1.2 Details of a reliability test

In the case of consumer-oriented products and systems, the reliability requirements are related to the cost of the product/system. For military equipment and systems, the reliability requirements are usually specified in the military specifications. Before a test is conducted, the inputs and outputs of the system must be clearly defined, including their nominal values and permissible deviations or tolerances. In multicomponent systems, the failure criteria must be defined clearly. Sometimes, failures may have to be classified into major and minor ones, catastrophic and noncatastrophic ones, etc. The sample size or the number of units (components or systems) tested and the duration of the test must be determined from the required confidence level and the limitations of cost and time involved in conducting the test. Although a statistically correct sample size can be determined in all the cases, the following sample sizes are commonly used:

1. For testing individual components, 10 to several hundred units.
2. For testing subsystems, a few to 20 or 30 units.
3. For testing complete systems, 1 to 10 units. The duration of a test depends on the component or system and its failure characteristics. It may range from few hundred hours to several thousands of hours.

15.2 Analysis of Failure Time

When experiments are conducted to find the failure time of a component or system, the data may be collected in two ways; as individual data, or as grouped data. When individual data are collected, the failure times of individual components are noted. When grouped data are collected, the number of failures in different time intervals is recorded.

15.2.1 Analysis of individual failure data

To analyze the individual failure data, consider a test involving n components. Let the failure times of the components be arranged in an increasing order as $t_1 \leq t_2 \leq \cdots \leq t_i \leq \cdots \leq t_n$ where t_i denotes the failure time of component i and is also known as the ith rank statistic of the test. The cumulative distribution function, $F(t)$, by definition, represents the probability of realizing the failure time of the component less than or equal to t. Thus, $F(t_i)$ denotes the fraction of components failed in time t_i

$$F(t_i) = \frac{i}{n} \tag{15.1}$$

It has been found that $F(t_i)$ given by Eq. (15.1) provides reasonably accurate values of the distribution function when n is large. However, Eq. (15.1) does not give very accurate estimates of $F(t)$ when n is small, of the order of 10 or less. In such cases, Eq. (15.1) indicates that all the components fail for values of t larger than t_n. However, if another test is conducted with a larger number of units (of the order of $5n$), it is likely that some components may fail at times larger than t_n. Hence other formulas have been suggested for estimating the cumulative distribution function, $F(t)$, as follows:

$$F(t_i) = \frac{i}{n+1} \tag{15.2}$$

$$F(t_i) = \frac{i - 0.5}{n} \tag{15.3}$$

$$F(t_i) = \frac{(0.5)^{1/n}\,(2\,i - n - 1) + n - 1}{n - 1} \tag{15.4}$$

$$F(t_i) = \frac{i - 0.3}{n + 0.4} \tag{15.5}$$

where the right hand sides of Eqs. (15.2) and (15.5) are called the *mean rank* and the *median rank*, respectively. Equation (15.1) implies that when a large number of components are tested, equal number of components fail in each of the time intervals, $t_{i+1} - t_i$; $i = 1, 2, \ldots, n - 1$. Equation (15.2), on the other hand, denotes that the number of components failing beyond time t_n is same as the number of failures occurring in each of the time intervals, $t_{i+1} - t_i$;

$i = 1, 2, \ldots, n - 1$. Equations (15.3) to (15.5) are based on other, more advanced, statistical arguments [15.1–15.3]. If Eq. (15.1) is used, the reliability of the component at time $t = t_i$, $R(t_i)$, can be expressed as

$$R(t_i) = 1 - F(t_i) = \frac{n - i}{n} \tag{15.6}$$

This indicates that i components have failed and $(n - i)$ survived at $t = t_i$. By using a forward-finite difference formula, the probability density function of the failure time can be expressed as

$$f(t) = \frac{F(t_{i+1}) - F(t_i)}{t_{i+1} - t_i} = \frac{\frac{i+1}{n} - \frac{i}{n}}{t_{i+1} - t_i} = \frac{1}{n(t_{i+1} - t_i)};$$

$$t_i \le t \le t_{i+1}; \qquad i = 1, 2, \ldots, n-1 \tag{15.7}$$

The failure rate or hazard function of the component can be estimated as (see Eq. 6.11)

$$h(t) = \frac{f(t)}{R(t)} = \frac{1}{(n - i)(t_{i+1} - t_i)}; \qquad t_i \le t \le t_{i+1}; \qquad i = 1, 2, \ldots, n - 1 \tag{15.8}$$

Example 15.1 The failure times of ten automobile brakes are observed to be 43,500, 52,000, 63,500, 72,000, 84,500, 93,500, 101,000, 111,500, 116,000 and 123,500 miles of operation. Plot the probability density, probability distribution, reliability, and the hazard functions of the failure time of the brakes.

solution Equations (15.1), (15.7), (15.6), and (15.8) can be used to find the probability distribution, probability density, reliability, and hazard functions, respectively, of the life of brakes. The computations are indicated in Table 15.1. These results are shown graphically in Figs. 15.1a to c.

TABLE 15.1

i	t_i (miles)	$t_{i+1} - t_i$ (miles)	$F(t_i) = \frac{i}{n}$	$R(t_i) = \frac{n-i}{n}$	$f(t) = \frac{1}{n(t_{i+1} - t_i)}$ $(\times 10^{-6})$	$h(t) = \frac{f(t)}{R(t)}$ $(\times 10^{-6})$
0	0	43,500	0	1.0	2.2989	2.2989
1	43,500	8500	0.1	0.9	11.7647	13.0719
2	52,000	11,500	0.2	0.8	8.6957	10.8696
3	63,500	8500	0.3	0.7	11.7647	16.8067
4	72,000	12,500	0.4	0.6	8.0000	13.3333
5	84,500	9000	0.5	0.5	11.1111	22.2222
6	93,500	7500	0.6	0.4	13.3333	33.3332
7	101,000	10,500	0.7	0.3	9.5238	31.7460
8	111,500	4500	0.8	0.2	22.2222	111.1110
9	116,000	7500	0.9	0.1	13.3333	133.3330
10	123,500	——	1.0	0.0	——	——

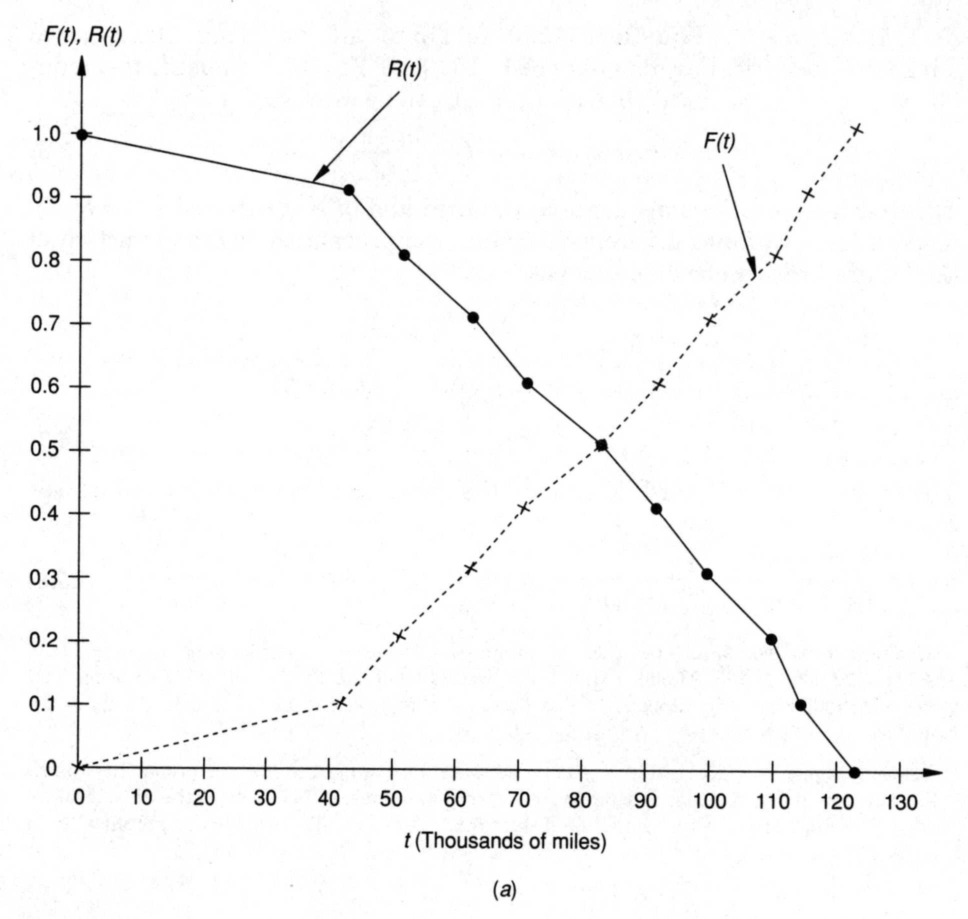

Figure 15.1*a*

15.2.2 Analysis of grouped failure data

Let a test be started with n identical components or systems. Let the number of components surviving at times $t_1, t_2, \ldots,$ and t_k be denoted as $n_1, n_2, \ldots,$ and n_k, respectively ($n_1 \geq n_2 \geq \cdots \geq n_k$). Thus, the failure times of individual components are not available; only the time intervals in which groups of components failed are available. Then the probability distribution and reliability functions at time $t = t_i$ can be expressed as

$$F(t_i) = \frac{n - n_i}{n}; \qquad i = 1, 2, \ldots, k \tag{15.9}$$

$$R(t_i) = \frac{n_i}{n}; \qquad i = 1, 2, \ldots, k \tag{15.10}$$

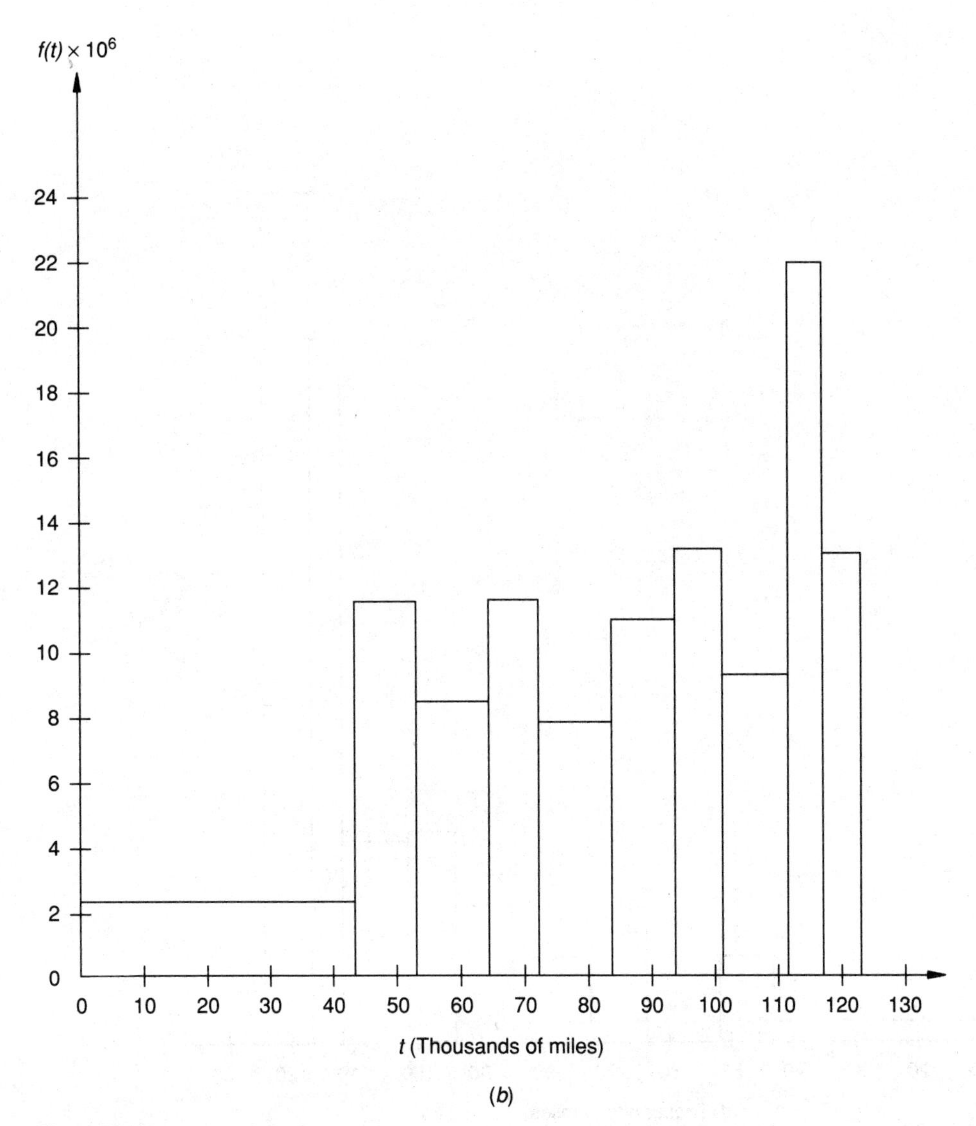

Figure 15.1*b*

The probability density function of the failure time can be determined using a forward difference formula as:

$$f(t) = \frac{F(t_{i+1}) - F(t_i)}{t_{i+1} - t_i} = \frac{(n - n_{i+1}) - (n - n_i)}{n\,(t_{i+1} - t_i)} = \frac{n_i - n_{i+1}}{n\,(t_{i+1} - t_i)};$$

$$t_i < t < t_{i+1}; \qquad i = 1, 2, \ldots, k-1 \tag{15.11}$$

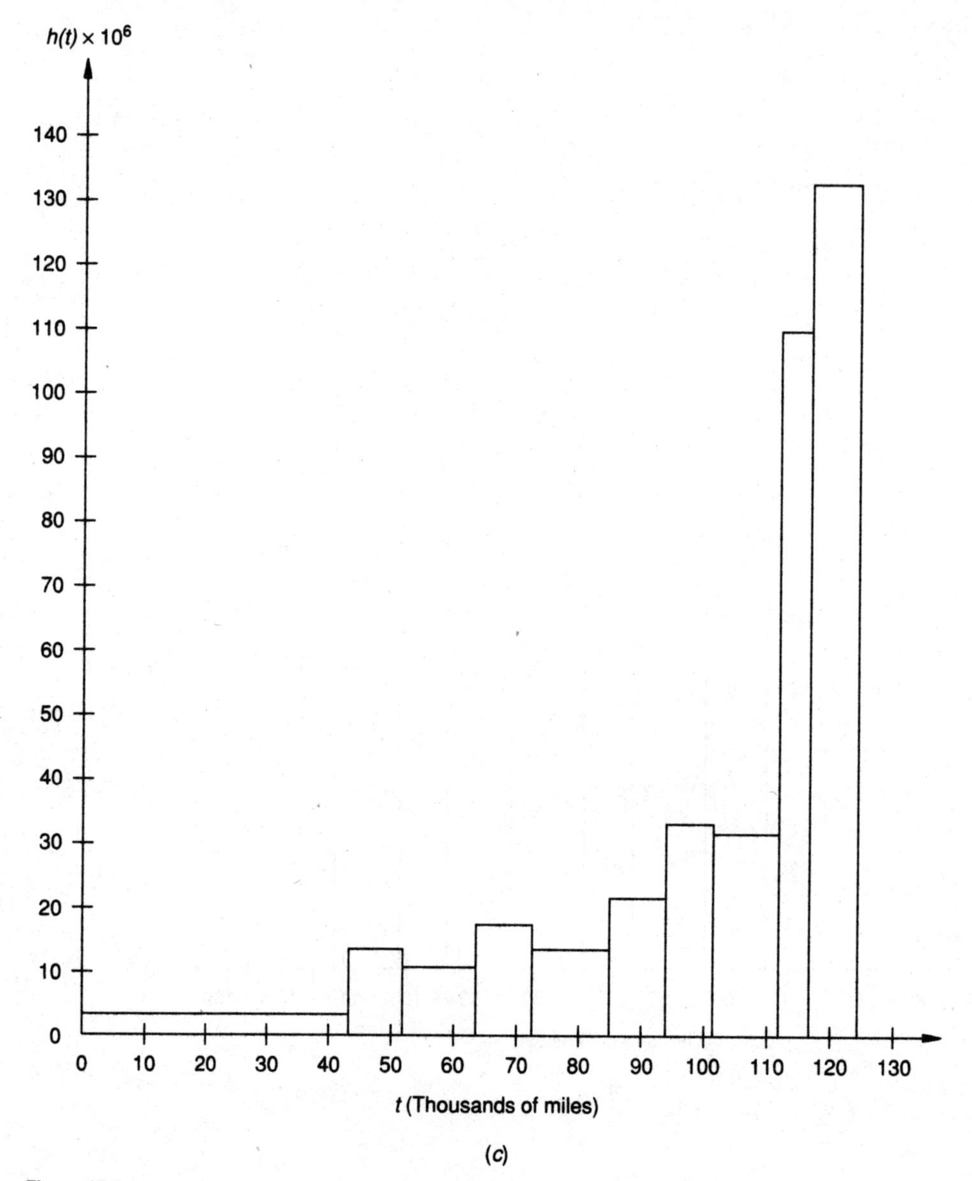

Figure 15.1c

The hazard function or the failure rate of the component can be estimated as

$$h(t) = \frac{f(t)}{R(t)} = \frac{n_i - n_{i+1}}{n_i(t_{i+1} - t_i)}; \qquad t_i < t < t_{i+1}; \qquad i = 1, 2, \ldots, k-1 \qquad (15.12)$$

15.3 Accelerated Life Testing

Usually, life testing is expensive in terms of both money and time. If a component or system is highly reliable, the testing procedure takes a long period of time before observing the failure. Hence, it is desirable to speed up the testing procedure for collecting the failure data. Accelerated testing procedures aim at reducing the time required for life testing by using one of the following strategies:

1. By conducting a test only until a part of the sample fails.
2. By using a magnified load.
3. By using sudden-death testing.

These methods are described below in detail.

15.3.1 Testing until partial failure

In this method, a large number of units are tested simultaneously until the failure of a fraction of the units is observed. Let the failure times of the units be expressed as $t_1 \le t_2 \le \cdots \le t_i \le \cdots$ where t_i denotes the failure time of the ith component. If we want to observe the failure times of m components, there are two approaches. One approach is to start with a sample of m components and conduct the test until all the components fail. The other approach is to start with a larger sample of n components ($n > m$) and continue the test until the first m components fail. If the average MTBF is to be estimated from the test results, both the approaches yield approximately the same value of MTBF. However, in general, the approach using the larger sample requires a smaller testing time. This can be seen intuitively as follows. Let there be one component, say A, with an extremely large value of failure time (life) among the m components. In the first approach, where we observe m failures out of m components, we need to conduct the test until the component A fails. However, in the second approach, we need not wait until the component A fails; some other component out of the remaining $(n - m)$ components fails before the component A does. Thus the overall testing time can be reduced by starting with a larger sample of n components.

The expected waiting time to observe the mth failure in a sample of m units [15.4, 15.5], $E(t_{m,m})$, is given by

$$E(t_{m,m}) = \theta \sum_{i=1}^{m} \frac{1}{(m - i + 1)} \tag{15.13}$$

where θ is the MTBF of the unit. Similarly the expected waiting time to observe the mth failure in a sample of n units can be expressed as

$$E(t_{m,n}) = \theta \sum_{i=1}^{m} \frac{1}{(n - i + 1)} \tag{15.14}$$

The saving in time realized in using a sample of n units instead of m units for observing m failures can be seen from the ratio

$$\text{Fractional time} = \frac{E(t_{m,n})}{E(t_{m,m})} = \frac{\sum_{i=1}^{m} \frac{1}{(n-i+1)}}{\sum_{i=1}^{m} \frac{1}{(m-i+1)}} \tag{15.15}$$

This ratio is computed for different combinations of m and n and the results are given in Table 15.2. As an example, consider a test involving the operation of five units. Let the waiting period be 100 hours to observe the failure of all the 5 units. If we use a sample of 10 units, we need to wait, on the average, only 0.2828(100) = 28.28 hours to observe the failure of the first 5 units.

If n units are tested until m failures are observed, the $(n - m)$ unfailed units may still be useful for normal application. Even if the units are not suitable for further use, the savings in time realized in testing are well worth the cost of the additional units.

15.3.2 Magnified loading

Most accelerated tests involve application of magnified stress levels (such as temperature, voltage, load, etc.) as opposed to the normal use values in order to increase the rate of failure and reduce the testing time. In many cases, the use of magnified stress levels also results in a fewer number of units required for test. When magnified loads are used, the most important consideration is to find the correlation between the two (normal and magnified) load conditions. This requires a knowledge of the failure characteristics of the unit under normal and magnified load conditions. It is to be noted that the failure mode should not be changed with the intensified stress level. Then, by ex-

TABLE 15.2 Ratio of Expected Waiting Times to Observe the *m*th Failure in Samples of Size *n* and *m*

	Value of n							
Value of m	5	10	15	20	25	30	40	50
1	0.2000	0.1000	0.0667	0.0500	0.0400	0.0333	0.0250	0.0200
2	0.3000	0.1407	0.0921	0.0684	0.0544	0.0452	0.0338	0.0269
3	0.4273	0.1833	0.1173	0.0863	0.0683	0.0565	0.0420	0.0334
4	0.6160	0.2299	0.1432	0.1042	0.0819	0.0675	0.0499	0.0396
5	1.0	0.2828	0.1705	0.1224	0.0956	0.0784	0.0577	0.0457
6	——	0.3452	0.1997	0.1413	0.1095	0.0894	0.0654	0.0516
7	——	0.4226	0.2315	0.1611	0.1237	0.1005	0.0732	0.0575
8	——	0.5258	0.2669	0.1820	0.1385	0.1119	0.0810	0.0634
9	——	0.6819	0.3069	0.2043	0.1538	0.1236	0.0888	0.0694
10	——	1.0	0.3533	0.2283	0.1699	0.1356	0.0968	0.0753

trapolating the results suitably to the "normal use" conditions, we can obtain reasonably accurate estimates of the life of the component under the "normal use" conditions.

A practical example where accelerated testing has proved to be useful is in the testing of impregnated paper-based capacitors. For these capacitors, the mean life (θ) is related to the applied DC voltage (V) as [15.5]

$$\theta = \frac{k}{V^p} \tag{15.16}$$

where k and p are constants. A value of $p = 5$ has been found to be applicable for this case. Let the average life of a capacitor be 10 years under normal use conditions in which the applied voltage is V_1. Then the nominal life of capacitors will be

$$\theta_1 = \frac{k}{V_1^5} = 10 \text{ years} \tag{15.17}$$

When the capacitors are subjected to a DC voltage that is twice the normal value, the average life will be

$$\theta_2 = \frac{k}{(2\,V_1)^5} \tag{15.18}$$

This gives

$$\frac{\theta_1}{\theta_2} = \frac{10}{\theta_2} = \frac{(2V_1)^5}{V_1^5} = 32 \tag{15.19}$$

which yields $\theta_2 = 3.75$ months, a substantial reduction of mean life from the nominal value of 10 years.

Another application of magnified loading is in fatigue testing. In *fatigue testing*, the number of cycles (N) which causes failure is determined under a series of constant stress amplitudes (S). From this data, the standard S–N curve is plotted for the material of the specimens. It is well known that in practice very few components are subjected to constant amplitude loading during their entire lives. Hence, fatigue tests involving variable amplitude of the stress are developed. If the amplitude of the stress is increased, the fatigue life of the specimen will be smaller. Thus, the results obtained from a series of fatigue tests with different stress amplitudes can be plotted as a straight line on a log-stress (S) and log-fatigue life (N) graph as shown in Fig. 15.2. For example, for the AISI 1045 steel, the fatigue strength relationship between the stress and the number of cycles is given by $SN^{0.085} = 154.0$ where S is in k(lb/in^2) and N is in cycles. In general, for most structural steels, the S–N relationship [15.6] can be expressed as

$$SN^a = b \tag{15.20}$$

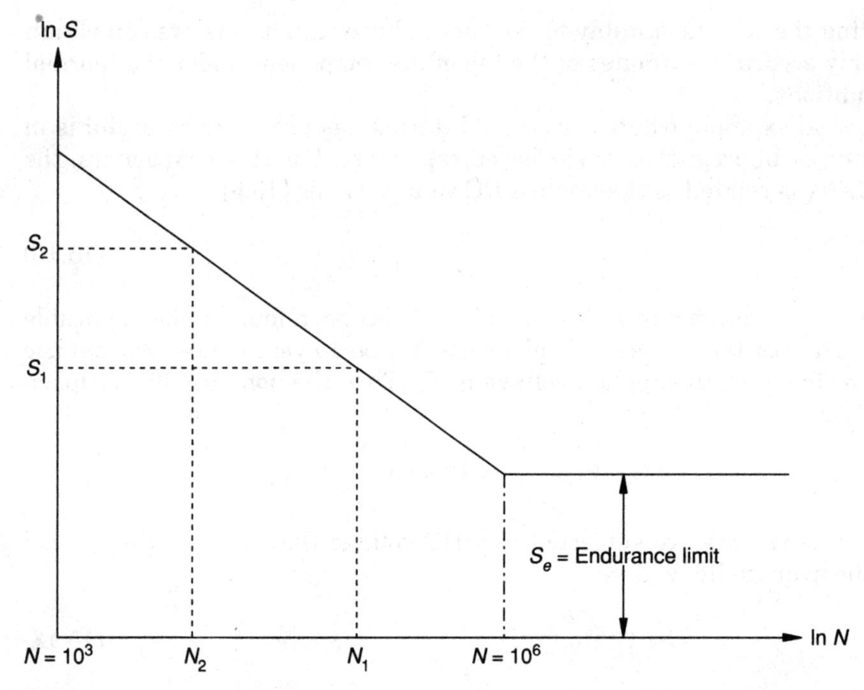

Figure 15.2

where a and b are constants. Equation (15.20) can be used to predict the fatigue life (N_2) at a stress intensity of S_2 once the fatigue life (N_1) at the stress intensity S_1 is known, as

$$S_1 N_1^a = b = S_2 N_2^a$$

or

$$N_2 = N_1 \left(\frac{S_1}{S_2} \right)^{1/a} \tag{15.21}$$

Let the stress intensity be doubled for a component made up of AISI 1045 steel for which $a = 0.085$. Then the number of cycles to failure (fatigue life) is reduced by a factor of

$$\frac{N_1}{N_2} = \left(\frac{S_2}{S_1} \right)^{1/a} = 2^{1/0.085} = 2^{11.7647} = 3480.68$$

When acceleration is to be achieved by increasing the temperature, the Arrhenius and Eyring models are commonly used.

Arrhenius model. According to the Arrhenius model, the rate of a chemical reaction or failure mechanism (process) is given by the empirical formula [15.1, 15.7]

$$\text{Rate of reaction} = ce^{-a/(kT)} \tag{15.22}$$

where a is a constant known as the activation energy, k is known as the Boltzman's constant, T is the ambient temperature (in degrees K), and c is a normalization constant. In Eq. (15.22), the value of a depends on the specific reaction or the failure model, while the value of k is given by 86.3×10^{-6} eV/°K. Usually the life of a component is assumed to be inversely proportional to the rate of the failure process, and hence

$$L = c_2 e^{a/(kT)} \tag{15.23}$$

where L represents the life of the component such as the minimum or average life and c_2 is a constant. Thus, if T_u and L_u denote the temperature and life during normal use and T_a and L_a indicate the temperature and life during the accelerated testing, Eq. (15.23) yields

$$\frac{L_a}{L_u} = e^{a/\{k(1/T_a - 1/T_u)\}} \tag{15.24}$$

The acceleration factor, A_f, achieved during the accelerated testing is given by

$$A_f = \frac{L_u}{L_a} = e^{a/\{k\,(1/T_u - 1/T_a)\}} \tag{15.25}$$

Eyring model. According to the Eyring model [15.1, 15.7], the rate of reaction is given by

$$\text{Rate of reaction} = cTe^{-a/(kT)} \tag{15.26}$$

The Eyring model is more general and can be used when stresses other than temperature are involved. While the Arrhenius model is based on an empirical equation, the Eyring model has a theoretical basis and can be derived from chemical-reaction rate theory and quantum mechanics. However, the Arrhenius model is more frequently used in the reliability testing of several types of components such as microcircuits.

15.3.3 Sudden-death testing

When each component is relatively inexpensive, the sudden-death testing can be used to reduce the overall testing time. If each component is relatively expensive, this method is not suitable. In this method, a large number of components are tested by dividing them into several groups with five or six components in each group. For example, if 60 components are available for testing, they are randomly divided into ten groups of six components each. Then the components in each group are tested simultaneously until one component fails. As soon as the first component fails, the remaining components of the group are removed or withdrawn from the test (a sudden death to these components). This procedure is continued until all the ten groups have been tested. Thus the procedure gives ten numbers, each representing the smallest value of life in a random sample size of six. If we use the median ranking procedure, an estimate of the true fraction of failures occuring in time t_i is given by (see Eq. (15.5)

$$F(t_i) = \frac{i - 0.3}{n + 0.4} \tag{15.27}$$

where i denotes the number of failures observed in a sample of size n in time t_i. Since $i = 1$ and $n = 6$ in the present case, we have $F(t_i) = (1 - 0.3)/(6 + 0.4) = 0.1094$. Thus, the 10 failure times represent the lives of the weakest 10.94 percent of the population. The Weibull distribution can be used to describe the extremal distribution corresponding to the smallest lives of the components. Hence the ten observed lives, when plotted on the Weibull graph, lie on a straight line as shown in Fig. 15.3. This straight line denotes the distribution of the lives of the weakest 10.94 percent of the population instead of that of the lives of the entire population. The median or 50 percent life corresponding to this data can be identified (shown as point A in Fig. 15.3). The vertical line drawn through the point A intersects the 10.94 percent line at point B as shown in Fig. 15.3. The straight line drawn through point B, parallel to the sudden death line, can be taken as the population line to denote the lives of the population. This assumes that the slopes of the sudden-death line and the population line are the same. The characteristics of the population can be estimated from the population line. For example, the median or 50 percent life of the population can be found from the life corresponding to point C in Fig. 15.3.

Example 15.2 Sixty switches are randomly divided into 10 equal groups and each group is subjected to sudden-death testing. As soon as the first unit in each group fails, the remaining units of the group are withdrawn from the test. The failure times observed for the ten failed units, arranged in an increasing order, are given by 1.5, 2.2, 3.5, 3.9, 5.0, 5.2, 6.4, 6.8, 8.2, and 10.1 hours. Find the median life of the population of switches.

solution Since each failure is the first failure observed in a sample of 6 units, it corresponds to a median rank of $(i - 0.3)/(n + 0.4) = (1 - 0.3)/(6 + 0.4) = 0.1094$. Hence the ten failure times observed correspond to the lives of the weakest 10.94 percent of the population of switches. These lives are plotted on a Weibull paper using the median ranks for a sample of ten ($n = 10$). The calculations are indicated in Table 15.3 and the points $(t_i, F(t_i))$; $i = 1, 2, \ldots, 10$ are plotted in Fig. 15.3. The median life of the 10 failed units, corresponding to point A, can be seen to be ≈ 5.0 hours. The point of intersection of the horizontal line at $F(t) = 0.1094$ and the vertical line through the point A lies on the population line. The population line is drawn parallel to the sudden death line. The median or 50 percent life of the population of switches can be identified from point C as ≈ 11 hours.

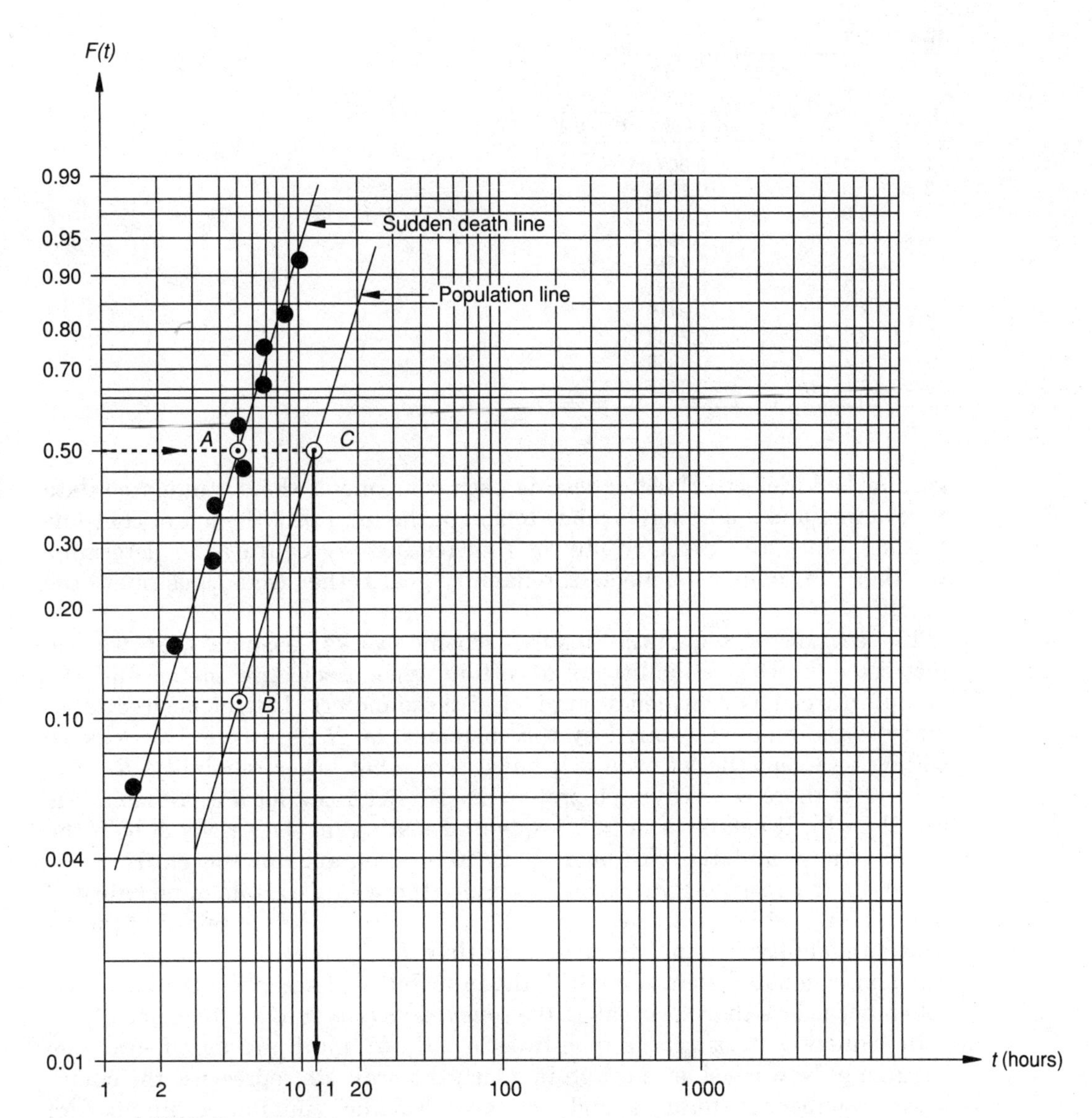

Figure 15.3 Weibull probability paper.

15.4 Sequential Life Testing

The sequential life testing procedure was developed by Abraham Wald during World War II [15.8]. It has been proved that sequential testing requires fewer units to be tested compared to the other testing methods. In sequential life testing, the sample size is not fixed. Components from a lot are examined one by one. Criteria are established so that each component may be classified as

TABLE 15.3

i	t_i (hours)	$F(t_i) = \frac{i - 0.3}{n + 0.4} = \frac{i - 0.3}{10 + 0.4}$
1	1.5	0.0673
2	2.2	0.1635
3	3.5	0.2596
4	3.9	0.3558
5	5.0	0.4519
6	5.2	0.5481
7	6.4	0.6442
8	6.8	0.7404
9	8.2	0.8365
10	10.1	0.9327

good or bad (defective). After testing each component, the accumulated data is reviewed and a decision is made to accept the lot, reject the lot, or continue testing. Thus, the primary aim of this testing procedure is to determine whether the component meets a reliability goal rather than to estimate the MTBF.

To see how a sequential testing scheme works, assume that we are interested in the reliability of a certain type of electric motor and the manufacturer has provided a large lot of these motors. Let the desired reliability of the motor (required by the customer) be R_u. However, we may be willing to accept the lot even if it has a somewhat lower reliability, R_l, provided that there is a very high probability of rejecting a lot if its reliability is less than R_l. It can be seen that sequential testing involves risks to both the manufacturer and the customer. Since the acceptance of the entire lot is based on the results of a random sample, there is a certain probability of rejecting a good lot or accepting a bad lot. The probability of rejecting the lot whose reliability is equal to or greater than R_u is known as the *producer's risk* and is denoted as α. Similarly, the probability of accepting the lot whose reliability is less than R_l is called the *consumer's risk* and is denoted as β.

In sequential testing, the quantities R_u, R_l, α, and β are specified at the beginning. Now consider a graph in which the ordinate represents the cumulative number of failures and the abscissa the cumulative number of successes as shown in Fig. 15.4. In this figure, two lines, known as the "accept" and "reject" lines, are plotted. The equations for these lines [15.5, 15.9] are given by

$$\text{Accept line:} \quad f \ln\left(\frac{1 - R_l}{1 - R_u}\right) + s \ln\left(\frac{R_l}{R_u}\right) = \ln\left(\frac{1 - \beta}{\alpha}\right) \tag{15.28}$$

$$\text{Reject line:} \quad f \ln\left(\frac{1 - R_l}{1 - R_u}\right) + s \ln\left(\frac{R_l}{R_u}\right) = \ln\left(\frac{\beta}{1 - \alpha}\right) \tag{15.29}$$

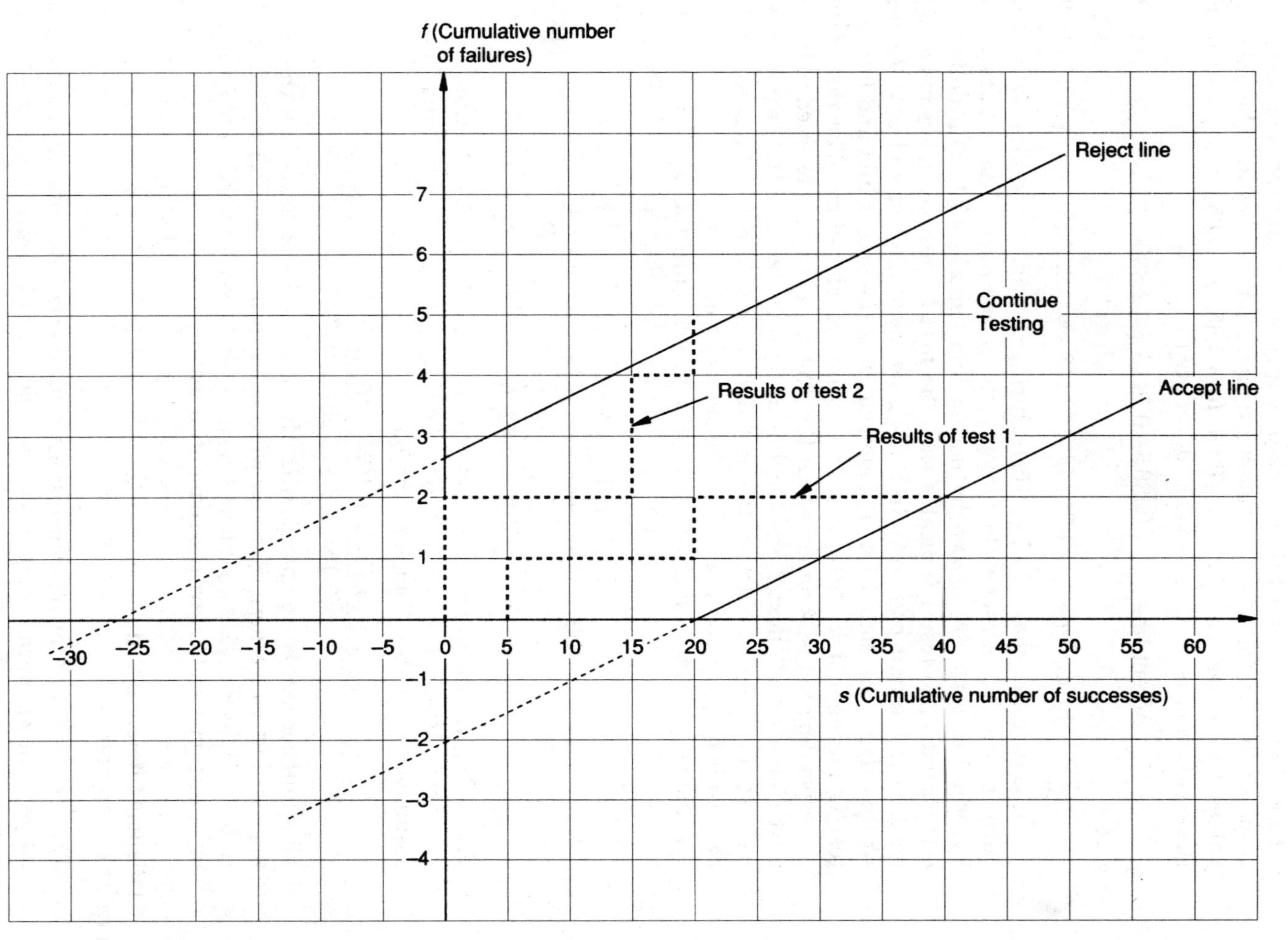
f (Cumulative number of failures)
s (Cumulative number of successes)
Reject line
Accept line
Continue Testing
Results of test 2
Results of test 1
7
6
5
4
3
2
1
−1
−2
−3
−4
−30
−25
−20
−15
−10
−5
0
5
10
15
20
25
30
35
40
45
50
55
60

Figure 15.4

where f and s denote the cumulative number of failures and successes, respectively. In the electric motor example, let us assume the following values for the parameters: $R_u = 0.95$, $R_l = 0.85$, $\alpha = 0.05$, and $\beta = 0.10$. For these values, Eqs. (15.28) and (15.29) reduce to

$$\text{Accept line:} \quad 1.0986f - 0.1113\, s = 2.8904 \tag{15.30}$$

and

$$\text{Reject line:} \quad 1.0986f - 0.1113\, s = -2.2513 \tag{15.31}$$

These lines are shown in Fig. 15.4. During testing, the results are continuously plotted in between the accept and reject lines. The results of two hypothetical tests are shown by dotted lines in Fig. 15.4. The dotted line corresponding to test 1 indicates that the first 5 units were observed to be good, next one unit bad, next 15 units good, next one bad and next 20 good. At this stage, the dotted line is found to cross the accept line and hence the test is terminated by accepting the entire lot of motors. The dotted line corresponding to test 2 denotes that the first 2 units were observed to be bad, next 15 units good, next 2 units bad, next 5 units good and the next unit was bad. At this stage, the dotted line is found to cross the reject line and hence the test is terminated by rejecting the entire lot of motors.

Although the exact sample size is not known until the test has been terminated, it is possible to estimate the average sampling number (ASN) as

$$\text{ASN} = \frac{(1-\alpha)\ln\left(\dfrac{\beta}{1-\alpha}\right) + \alpha\ln\left(\dfrac{1-\beta}{\alpha}\right)}{(1-R_u)\ln\left(\dfrac{1-R_l}{1-R_u}\right) + R_u\ln\left(\dfrac{R_l}{R_u}\right)} \tag{15.32}$$

The ASN gives the average number of units to be tested in sequential testing. For the particular values of $R_l = 0.85$, $R_u = 0.95$, $\alpha = 0.05$, and $\beta = 0.10$ assumed for the electric motor example, Eq. (15.32) gives a value of

$$\text{ASN} = \frac{(0.95)\ln 0.1053 + (0.05)\ln 18}{(0.15)\ln 3 + (0.95)\ln 0.8947} = 39.2712$$

This number can be compared with the test results indicated by the dotted lines in Fig. 15.4 where 42 units were used in test 1 and 25 in test 2. The ASN is useful for planning purposes since, obviously, certain number of units must be procured to start the sequential test.

15.5 Statistical Inference and Parameter Estimation

The procedure of drawing conclusions from the sample data, about the population from which the sample is drawn, is known as *statistical inference.* The two major types of statistical inference are called parameter estimation and hypothesis testing. The *parameter or point estimation* refers to the method of estimating a single numerical value for a parameter of the distribution from n sample values. In many practical problems, we need to determine whether a statement about some parameter is true or false. The statement is known as the hypothesis and the process of determining the truth or falseness of the hypothesis is known as *hypothesis testing.*

Reliability tests can be used to estimate not only the parameters of the distribution function but also the parameters related to reliability such as the failure rate and MTBF of the system. In each test, a number of components (or systems or units) are operated in a controlled environment under prescribed conditions until all the components have failed. This requires a clear definition of failure, especially for a multicomponent system. As soon as a failure is observed, the time to failure is recorded and the test is terminated after all the components or units have failed. In some cases, the test is terminated after the failure of a predetermined number of components or units.

15.5.1 Maximum-likelihood method

Several methods are available for estimating the parameters of a distribution. One of the popular methods is the maximum-likelihood estimation method. Consider a random variable X which follows the probability density function $f(x;\theta)$ where θ is the parameter to be estimated (such as the mean or standard deviation in normal distribution and the mean in exponential distribution). Let the sample values from the tests be given by $\{x_1, x_2, \ldots, x_n\}$. These values are assumed to be drawn from a population described by the probability density function $f(x;\theta)$. Then we will be interested in finding the probability of obtaining the specific sample $\{x_1, x_2, \ldots, x_n\}$ from all possible samples of size n from the population. In other words, among the various possible values of θ, we want to find the particular value $\hat{\theta}$ which maximizes the likelihood of obtaining the set of observations $\{x_1, x_2, \ldots, x_n\}$. The maximum likelihood method of estimation is intended to find the value of $\hat{\theta}$.

The probability of obtaining the ith sample value in the interval $x_i - dx_i/2$ and $x_i + dx_i/2$ can be assumed to be $f(x_i;\theta)\,dx_i$ (from the definition of the probability density function). Assuming the sample values to be independent, the joint probability of finding the first sample value in the range $x_1 - dx_1/2$ and $x_1 + dx_1/2$, the second sample value in the range $x_2 - dx_2/2$, and $x_2 + dx_2/2, \ldots,$ and the nth sample value in the range $x_n - dx_n/2$ and $x_n + dx_n/2$ is

$$[f(x_1;\theta)\,dx_1]\,[f(x_2;\theta)\,dx_2]\cdots[[f(x_n;\theta)\,dx_n] = \prod_{i=1}^{n} f(x_i;\theta)\,dx_i$$

$$= L(\theta)\,dx_1\,dx_2\cdots dx_n \tag{15.33}$$

where the function L given by

$$L(\theta) = \prod_{i=1}^{n} f(x_i; \theta) \tag{15.34}$$

is called the likelihood function of the sample. The maximum of the function $L(\theta)$ can be determined by setting the partial derivative of L with respect to θ equal to zero and solving the resulting equation for $\theta = \hat{\theta}$. Since most probability density functions $f(x;\theta)$ involve exponential terms, it is more convenient to maximize the natural logarithm of the likelihood function to find $\hat{\theta}$. Thus Eq. (15.34) gives

$$\ln L(\theta) = \ln \prod_{i=1}^{n} f(x_i; \theta) = \sum_{i=1}^{n} \ln f(x_i; \theta) \tag{15.35}$$

By differentiating Eq. (15.35), we obtain

$$\frac{\partial \ln L(\theta)}{\partial \theta} = 0 \tag{15.36}$$

the solution of which gives the desired maximum likelihood estimate $\hat{\theta}$.

This procedure can be used for estimating the parameters of any distribution. It can also be generalized to distributions involving two or more parameters.

Example 15.3 A reliability test is conducted within the useful life period of a system for which the failure time (T) follows the exponential distribution

$$f_T(t;\theta) = \frac{1}{\theta} e^{-t/\theta}; \qquad t > 0,\ \theta > 0 \tag{E1}$$

where θ indicates the MTBF. Find the maximum-likelihood estimate of the MTBF based on n sample values.

solution Let the sample values from the test be $t_1, t_2, \ldots, t_n$ with $t_1 \le t_2 \le \cdots \le t_n$. The corresponding likelihood function is given by

$$L(\theta) = \left(\frac{1}{\theta} e^{-t_1/\theta}\right)\left(\frac{1}{\theta} e^{-t_2/\theta}\right) \cdots \left(\frac{1}{\theta} e^{-t_n/\theta}\right) = \frac{1}{\theta^n} e^{-(1/\theta)\sum_{i=1}^{n} t_i} \tag{E2}$$

so that

$$\ln L(\theta) = -n \ln \theta - \frac{1}{\theta} \sum_{i=1}^{n} t_i \tag{E3}$$

For the maximum of $\ln L(\theta)$,

$$\frac{\partial \ln L(\theta)}{\partial \theta} = -\frac{n}{\theta} + \frac{1}{\theta^2} \sum_{i=1}^{n} t_i = 0 \tag{E4}$$

which yields the value of θ as

$$\theta = \hat{\theta} = \frac{1}{n} \sum_{i=1}^{n} t_i \tag{E5}$$

This value of $\hat{\theta}$ represents the maximum likelihood estimate of the MTBF available from the observed data.

Example 15.4 The sample values $x_1, x_2, \ldots, x_n$ have been observed for a random variable X following normal distribution with unknown mean μ and standard deviation σ. Determine the maximum likelihood estimators of μ and σ.

solution The probability density function is

$$f_X(x) = \frac{1}{\sqrt{2\pi}\,\sigma}\, e^{-1/2\,[(x-\mu)/\sigma]^2} \tag{E1}$$

where $\mu = \theta_1$ and $\sigma = \theta_2$ are the unknown parameters of the distribution. The likelihood function, $L(\theta_1, \theta_2)$, can be defined as

$$L(\theta_1, \theta_2) = \prod_{i=1}^{n} \frac{1}{\sqrt{2\,\pi}\,\sigma}\, e^{-1/2\,[(x_i-\mu)/\sigma]^2} = \frac{1}{(\sqrt{2\,\pi}\,\sigma)^n} \prod_{i=1}^{n} e^{-1/2\,[(x_i-\mu)/\sigma]^2} \tag{E2}$$

so that

$$\ln L(\theta_1, \theta_2) = -\,n\,\ln(\sqrt{2\,\pi}\,\sigma) - \frac{1}{2\,\sigma^2} \sum_{i=1}^{n} (x_i - \mu)^2 \tag{E3}$$

The conditions for maximizing $\ln L(\theta_1, \theta_2)$ are given by

$$\frac{\partial L(\theta_1, \theta_2)}{\partial\,\mu} = \frac{1}{\sigma^2} \sum_{i=1}^{n} (x_i - \mu) = 0 \tag{E4}$$

$$\frac{\partial L(\theta_1, \theta_2)}{\partial \sigma} = -\,\frac{n}{\sqrt{2\,\pi}\,\sigma}\,(\sqrt{2\,\pi}\,) - \frac{1}{2}\,\frac{(-\,2)}{\sigma^3} \sum_{i=1}^{n} (x_i - \mu)^2 = 0 \tag{E5}$$

The solution of Eqs. (E4) and (E5) gives the maximum likelihood estimators of μ and σ, namely $\hat{\mu}$ and $\hat{\sigma}$, as

$$\mu = \hat{\mu} = \frac{1}{n} \sum_{i=1}^{n} x_i \tag{E6}$$

$$\sigma^2 = \hat{\sigma}^2 = \frac{1}{n} \sum_{i=1}^{n} (x_i - \mu)^2 \tag{E7}$$

15.6 Confidence Intervals

Theoretically, an infinite number of sample values have to be collected to find the true value of a parameter such as the mean value of a random quantity. However, this is not possible in practice due to the limitations of cost and time. Hence, representative samples are tested, and the value of the parameter (such as the mean of the population or universe) is estimated from the sample values. Since only a finite sample size is used, the sample mean, for example, will not be equal to the population mean. However, the difference between the sample mean ($\overline{X}$) and the population mean (μ_X) decreases as the sample size increases. Thus, an infinite sample size is needed to obtain 100 percent assurance that the sample mean is equal to the population mean. The statistical sampling techniques can be used to find from the sample values, the range, known as the confidence interval, in which the population mean lies.

As stated earlier, the sample mean is an example of a point estimate of a population parameter. In addition to obtaining the point estimate, we would also like to find the accuracy of this estimate. For this the concept of confidence intervals is used. The *confidence interval* indicates the range of values within which the true value of the point estimate lies. If the point estimate happens to be the sample mean, the confidence interval indicates the range in which the true mean of the population lies. In fact a probability can be associated with any confidence interval. For example, we might say that "the true population mean of yield strength of steel lies between 25,000 lb/in^2 and 35,000 lb/in^2 with a probability of 99%." The lower and upper bound values of the confidence interval are known as the *confidence limits*. The probability associated with the confidence interval is called the *confidence level*. In the preceding example, the confidence limits are 25,000 lb/in^2 and 35,000 lb/in^2, and the confidence level is 99 percent. Once the confidence level is stated for a parameter such as the mean value as p, it indicates that the probability of the true population mean lying in the confidence interval is p. Thus there is a probability of $(1 - p)$ that the real mean does not lie within the given confidence interval. This quantity is known as the *level of significance* so that

$$\text{Level of significance} = 1 - \text{Confidence level} \tag{15.37}$$

In the case of the preceding example of the yield strength of steel, the level of significance is equal to $1 - 0.99 = 0.01$. Figure 15.5 illustrates the concepts of confidence interval, confidence level, and the level of significance. It can be

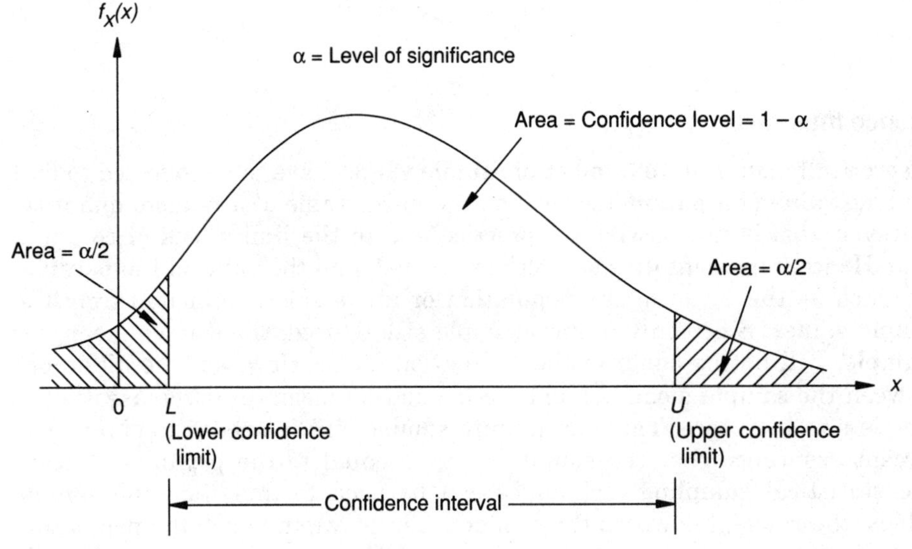

Figure 15.5 Two-sided confidence interval.

seen that the confidence level is equal to the area under the probability density function between the lower confidence limit (L) and the upper-confidence limit (U)

$$P(L \leq x \leq U) = 1 - \alpha \tag{15.38}$$

For a symmetric probability density function, we can see that

$$L = -x_{(\alpha/2)} \quad \text{and} \quad U = x_{(\alpha/2)} \tag{15.39}$$

In some applications, a one-sided confidence interval is used as indicated in Fig. 15.6. For example, in reliability testing, there is no need to specify an upper bound on the mean failure time. The larger the mean failure time, the better it will be from reliability point of view. In such a case, the confidence level denotes the area under the probability density curve from the lower limit while the level of significance represents the area under the curve up to the lower limit

$$P(x \geq L) = 1 - \alpha \tag{15.40}$$

15.6.1 Confidence interval on the mean of a normal random variable of known standard deviation

Let a normally distributed random variable X have an unknown population mean μ_X and a known standard deviation σ_X. Let $x_1, x_2, \ldots, x_n$ denote a random sample of size n, which gives the sample mean as $\overline{X}$. It has been

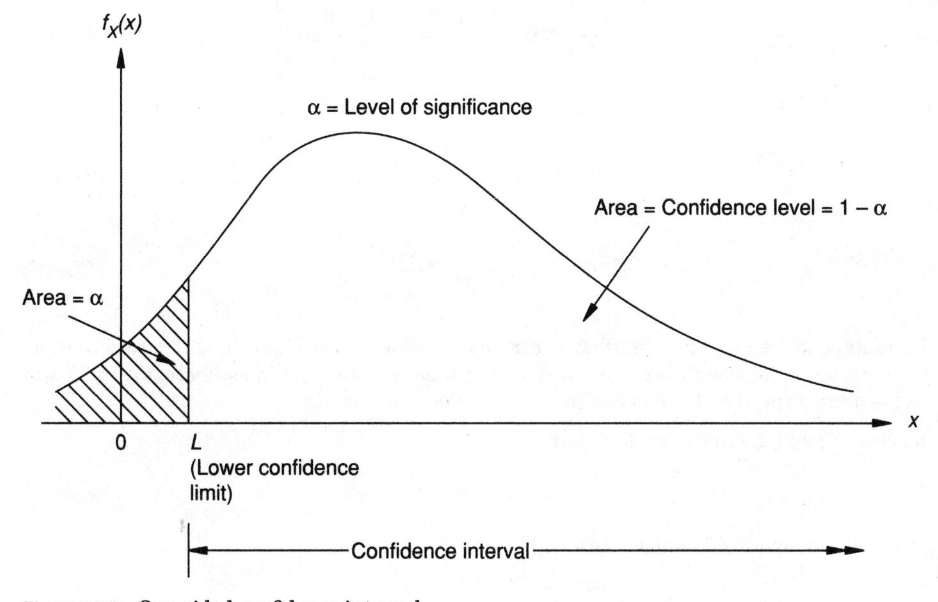

Figure 15.6 One-sided confidence interval.

proved from central-limit theorem [15.10, 15.11] that the sample mean $\overline{X}$ itself is a random variable that has an approximate normal distribution with mean μ_X (population mean) and standard deviation $\sigma_X / \sqrt{n}$ (σ_X is the population standard deviation)

$$\mu_{\overline{X}} = \mu_X \tag{15.41}$$

and

$$\sigma_{\overline{X}} = \frac{\sigma_X}{\sqrt{n}} \tag{15.42}$$

Thus, the random variable $z = (\overline{X} - \mu_X)/(\sigma_X / \sqrt{n})$ will be a standard normal variate with mean 0 and standard deviation 1. The 100 $(1 - \alpha)$ percent confidence interval on the mean μ_X is given by

$$\overline{X} \pm z_{(\alpha/2)} \frac{\sigma_X}{\sqrt{n}}$$

where $z_{(\alpha/2)}$ is the 100 $(\alpha/2)$ percentage point of the normal distribution. This implies that

$$P\left[-z_{(\alpha/2)} \le \frac{\overline{X} - \mu_X}{\left(\dfrac{\sigma_X}{\sqrt{n}}\right)} \le z_{(\alpha/2)}\right] = 1 - \alpha \tag{15.43}$$

Rearranging the terms inside the square brackets in Eq. (15.43) leads to

$$P\left[\left(\overline{X} - z_{(\alpha/2)} \frac{\sigma_X}{\sqrt{n}}\right) \le \mu_X \le \left(\overline{X} + z_{(\alpha/2)} \frac{\sigma_X}{\sqrt{n}}\right)\right] = 1 - \alpha \tag{15.44}$$

where the lower and upper confidence limits, L and U, can be identified as

$$L = \overline{X} - z_{(\alpha/2)} \frac{\sigma_X}{\sqrt{n}} \tag{15.45}$$

$$U = \overline{X} + z_{(\alpha/2)} \frac{\sigma_X}{\sqrt{n}} \tag{15.46}$$

Example 15.5 A normally distributed random variable X has a known standard deviation, σ_X. Determine the sample size n needed so that the error involved in estimating μ_X by $\overline{X}$ will be less than a specified amount δ with 100 $(1 - \alpha)$ percent confidence.

solution From Eq. (15.44), we find that

$$z_{(\alpha/2)} \frac{\sigma_X}{\sqrt{n}} = \delta$$

and hence the required sample size n will be

$$n = \left(\frac{z_{(\alpha/2)}\, \sigma_X}{\delta}\right)^2$$

15.6.2 Confidence interval on the mean of a normal random variable of unknown standard deviation

Let X be the random variable with unknown mean μ_X and unknown standard deviation σ_X. Let $x_1, x_2, \ldots, x_n$ denote a random sample of size n which gives the sample mean as $\overline{X}$ and the sample standard deviation as s_X. It has been proved that the distribution of the random variable $(\overline{X} - \mu_X)/(s_X/\sqrt{n})$ is a t-distribution with $(n - 1)$ degrees of freedom [15.11, 15.12]. Note that this random variable is similar to the variable $(\overline{X} - \mu_X)/(\sigma_X/\sqrt{n})$ defined in the preceding section. Since the population standard deviation σ_X is not known, it has been replaced by the sample standard deviation s_X in the present case.

The probability density function of a random variable T following t-distribution, also known as student's t-distribution, with n degrees of freedom [15.11, 15.12] is given by

$$f_T(t) = \frac{\Gamma\left(\frac{n+1}{2}\right)}{\Gamma\left(\frac{n}{2}\right)\sqrt{n\,\pi}} \left(1 + \frac{t^2}{n}\right)^{-(n+1)/2}; \qquad -\infty < t < \infty \tag{15.47}$$

The values of the distribution function for different values of t and n are given in Appendix B.

The 100 $(1 - \alpha)$ percent confidence interval of μ_X is given by

$$\overline{X} \pm t_{(\alpha/2;\, n-1)} \frac{s_X}{\sqrt{n}}$$

where $t_{(\alpha/2;\, n-1)}$ is the $100(\alpha/2)$ percent point of the t-distribution with $n - 1$ degrees of freedom. This implies that

$$P\left[-t_{(\alpha/2;\, n-1)} \le \frac{\overline{X} - \mu_X}{\left(\frac{s_X}{\sqrt{n}}\right)} \le t_{(\alpha/2;\, n-1)}\right] = 1 - \alpha \tag{15.48}$$

or

$$P\left[\overline{X} - t_{(\alpha/2;\, n-1)} \frac{s_X}{\sqrt{n}} \le \mu_X \le \overline{X} + t_{(\alpha/2;\, n-1)} \frac{s_X}{\sqrt{n}}\right] = 1 - \alpha \tag{15.49}$$

with the lower and upper confidence limits given by

$$L = \overline{X} - t_{(\alpha/2;\, n-1)} \frac{s_X}{\sqrt{n}} \tag{15.50}$$

and

$$U = \overline{X} + t_{(\alpha/2;\, n-1)} \frac{s_X}{\sqrt{n}} \tag{15.51}$$

Example 15.6 The internal diameters of ten internal combustion engine cylinders have been found to be 120.5, 118.4, 125.7, 116.9, 119.2, 123.3, 121.5, 119.8, 120.9, and 122.6 mm. If the diameters follow normal distribution, determine the 95 percent confidence interval for the population mean μ_X.

solution Here $\alpha = 0.05$, $n = 10$, and the observed sample mean is

$$\overline{X} = \frac{1}{10}(120.5 + 118.4 + \cdots + 122.6) = 120.88 \text{ mm}$$

and the observed sample standard deviation is

$$s_X = \left\{\frac{1}{9}[(120.5 - 120.88)^2 + (118.4 - 120.88)^2 + \cdots + (122.6 - 120.88)^2]\right\}^{1/2}$$

$$= \left(\frac{58.556}{9}\right)^{1/2} = 2.5507 \text{ mm}$$

From Appendix B, we find that $t_{(9;0.025)} = 2.262$. From Eqs. (15.50) and (15.51), we can obtain the confidence limits on the mean μ_X as

$$L = \overline{X} - t_{(\alpha/2;n-1)}\frac{s_X}{\sqrt{n}} = 120.88 - \left(\frac{2.262 \times 2.5507}{\sqrt{10}}\right) = 119.0555 \text{ mm}$$

$$U = \overline{X} + t_{(\alpha/2;n-1)}\frac{s_X}{\sqrt{n}} = 120.88 + \left(\frac{2.262 \times 2.5507}{\sqrt{10}}\right) = 122.7045 \text{ mm}$$

15.6.3 Confidence interval on the standard deviation of a normal random variable with unknown mean

Let X be a normally distributed random variable with unknown mean μ_X and unknown standard deviation σ_X. Let $\{x_1, x_2, \ldots, x_n\}$ denote a random sample of size n which gives the sample mean as $\overline{X}$ and the sample standard deviation as s_X

$$\overline{X} = \frac{1}{n}\left(\sum_{i=1}^{n} x_i\right) \tag{15.52}$$

$$s_X = \left\{\frac{1}{n-1}\sum_{i=1}^{n}\left(x_i - \overline{X}\right)^2\right\}^{1/2} \tag{15.53}$$

If we consider the random variable Y as

$$Y = \frac{(n-1)\,s_X^2}{\sigma_X^2} \tag{15.54}$$

then Y can be shown to follow a chi-square distribution with $n - 1$ degrees of freedom [15.11, 15.12]. For time-dependent processes with constant-failure rates, the chi-square distribution is used for assessing the confidence on the estimated reliability. Let x_i, $i = 1, 2, \ldots, n$ be n independent values or observations of a random variable, $N(\mu,\sigma)$. Then the variates

$$z_i = \frac{x_i - \mu}{\sigma}; \qquad i = 1, 2, \ldots, n \tag{15.55}$$

will also be normally distributed with $N(0, 1)$. Let Y (that is, χ^2) be a random variable denoting the sum of the squares of the z_i

$$Y = z_1^2 + z_2^2 + \cdots + z_n^2 \tag{15.56}$$

Then Y follows the χ^2-distribution function with n degrees of freedom given by [15.11]

$$f_{\chi^2(n)}(y) = \frac{y^{\{(n/2)-1\}} e^{-y/2}}{2^{\frac{n}{2}} \Gamma\left(\frac{n}{2}\right)}; \qquad y \geq 0 \tag{15.57}$$

The mean and variance of Y are given by

$$\mu_Y = n; \qquad \sigma_Y^2 = 2n \tag{15.58}$$

The values of the χ^2-distribution function up to 20 degrees of freedom are given in Appendix C. By letting $\chi^2_{(n,\alpha/2)}$ be the value such that

$$P\left[Y > \chi^2_{(n,\alpha/2)}\right] = \frac{\alpha}{2} \tag{15.59}$$

with n degrees of freedom, we can obtain the following relation (see Fig. 15.7)

$$P\left[\chi^2_{(n-1,1-(\alpha/2))} \leq Y \leq \chi^2_{(n-1,\alpha/2)}\right] = 1 - \alpha \tag{15.60}$$

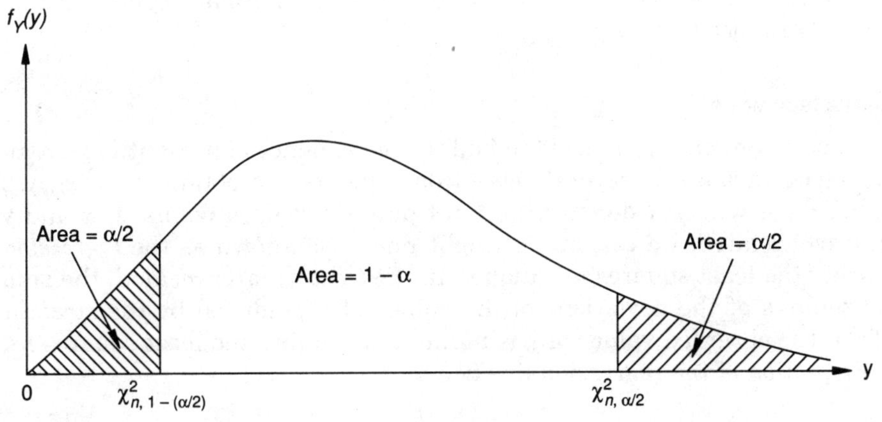

Figure 15.7 100(1 – α) percent confidence limits for Y with n degrees of freedom.

This equation can be rewritten, after substituting for Y, as

$$P\left[\frac{(n-1)\,s_X^2}{\chi^2_{(n-1,\alpha/2)}} \le \sigma_X^2 \le \frac{(n-1)\,s_X^2}{\chi^2_{(n-1,1-(\alpha/2))}}\right] = 1-\alpha \tag{15.61}$$

Example 15.7 Determine the 95 percent confidence interval for the standard deviation σ_X for the data given in Example 15.6.

solution The observed sample standard deviation is $s_X = 2.5507$ mm. The values of $\chi^2_{(9,\,0.975)}$ and $\chi^2_{(9,\,0.025)}$ can be obtained from Appendix C as

$$\chi^2_{(9,\,0.975)} = 2.700; \qquad \chi^2_{(9,\,0.025)} = 19.023$$

Equation (15.61) gives for $n = 10$ and $\alpha = 0.05$, the lower and the upper confidence limits on the standard deviation σ_X as

$$\frac{\sqrt{n-1}\; s_X}{\sqrt{\chi^2_{(n-1,\alpha/2)}}} = \frac{\sqrt{9}\;2.5507}{\sqrt{19.023}} = 1.7544 \text{ mm}$$

and

$$\frac{\sqrt{n-1}\; s_X}{\sqrt{\chi_{(n-1,1-(\alpha/2))}}} = \frac{\sqrt{9}\;2.5507}{\sqrt{2.7}} = 4.6569 \text{ mm}$$

15.7 Plotting of Reliability Data

The analytical equations for various probability distributions were given in Chapter 3. When experimental reliability data is available, it is desirable to have a simple graphical procedure to find a suitable distribution and also to estimate the parameters of the distribution. The methods presented in this section will aid in the rapid analysis of failure time data and estimation of the parameters of the appropriate probability distribution. The use of complete-, single-censored or multicensored data is discussed. The data plotting techniques are illustrated for the exponential, normal, lognormal and Weibull distributions.

15.7.1 Least squares technique

When experiments are conducted to find the dependency of a quantity y on an independent variable x, several sets of data points are obtained as (x_i, y_i), $i = 1, 2, \ldots, n$ where n denotes the total number of data points. If x and y are linearly related, we can fit a straight line, also known as the *regression line*, using the least squares technique. In the *least squares method*, the sum of the squares of the deviations of the values of y predicted by the straight line from the observed values of y is minimized. To find the least squares fit, let the equation of the regression line be

$$y = ax + b \tag{15.62}$$

where a and b are constants. By using the experimental points (x_i, y_i), $i = 1, 2, \ldots, n$, the constants a and b can be determined as (see Section 4.9.1)

$$a = \frac{n \sum_{i=1}^{n} x_i y_i - \sum_{i=1}^{n} x_i \sum_{i=1}^{n} y_i}{n \sum_{i=1}^{n} x_i^2 - \left(\sum_{i=1}^{n} x_i \right)^2} \tag{15.63}$$

and

$$b = \frac{n \sum_{i=1}^{n} x_i^2 \sum_{i=1}^{n} y_i - n \sum_{i=1}^{n} x_i \sum_{i=1}^{n} x_i y_i}{n \sum_{i=1}^{n} x_i^2 - \left(\sum_{i=1}^{n} x_i \right)^2} \tag{15.64}$$

15.7.2 Linear rectification

The least squares technique can be used to estimate the parameters of many distributions by using a procedure known as linear rectification. *Linear rectification* involves converting a given nonlinear equation into a linear form by suitably changing the variables. Let the relationship between the variables x and y be given by

$$y = \alpha x^{\beta} \tag{15.65}$$

where α and β are constants. By taking logarithms of both sides of Eq. (15.65), we get

$$\ln y = \ln \alpha + \beta \ln x \tag{15.66}$$

which represents a straight line between the variables $\ln y$ and $\ln x$. Thus the nonlinear equation (15.65) has been converted to an equivalent linear form in Eq. (15.66) using linear rectification. The concept of linear rectification is very useful because it allows us to determine in many cases the parameters of a distribution from experimental values.

15.7.3 Plotting positions

When probability distributions are to be fitted to experimental data, we need to select the plotting positions of the data points. For specificness, let the observed data represent the failure times of a system or unit. If n units are tested, the observed failure times, for convenience, can be arranged in an increasing order as

$$t_1 \le t_2 \le \cdots \le t_i \le \cdots \le t_n$$

where t_i are called the *rank statistics* of the test. The cumulative distribution function, $F(t)$, which denotes the probability of realizing the failure time of the unit less than or equal to t, can be expressed as

$$F(t_i) = \frac{i}{n} \tag{15.67}$$

To plot the test results in the form of $F(t)$ versus t, we plot the points $\{t_i, F(t_i)\}$ for $i = 1, 2, \ldots, n$. The value of $F(t_i)$ given by Eq. (15.67) thus determines the plotting (ordinate) position of the ith data point, t_i. As stated earlier (in section 15.2.1), other formulas, given by Eqs. (15.2) to (15.5), can also be used for $F(t_i)$. Equation (15.5) can be used to obtain more accurate plotting, although Eq. (15.2) is used in some cases, for simplicity, to plot the experimental data.

15.7.4 Exponential distribution

When failure time data are available, it is usual to try to fit exponential distribution first. The main reason is that exponential distribution fits best if the system has a constant failure rate. In addition, exponential distribution is one of the simplest distributions involving only one parameter. For this case, the cumulative distribution function is given by

$$F(t) = 1 - e^{-\lambda t} \tag{15.68}$$

which can be rewritten as

$$1 - F(t) = e^{-\lambda t} \tag{15.69}$$

or

$$\ln \frac{1}{1 - F(t)} = \lambda t \tag{15.70}$$

This indicates that $\ln \{1/(1 - F(t))\}$ varies linearly with time. Thus the probability paper for exponential distribution is constructed by plotting the values of $[1/(1 - f(t))]$ on a logarithmic scale against the values of t on a linear scale (semilog paper). The procedure is illustrated through Example 15.8.

Example 15.8 The failure times of 15 blenders are observed to be 30, 50, 60, 100, 130, 150, 190, 250, 270, 310, 380, 410, 530, 630, and 850 hours of operation. Plot the data on an exponential probability paper and determine the failure rate of the blenders.

solution The data are plotted twice; first by using Eq. (15.2) and then Eq. (15.5) to estimate the cumulative distribution function. The calculations are shown in Table 15.4. The resulting graphs are shown in Fig. 15.8. The solid line corresponds to Eq. (15.2) while the dotted line corresponds to Eq. (15.5). From Eq. (15.70), the failure rate, λ, can be seen to be the slope of the straight line drawn between $\ln [1/(1 - F(t))]$ and t. Thus, Fig. 15.8 gives the value of λ as $\lambda_1 = (\ln 9.2 - \ln 5.0)/(700 - 500) = 0.003049$ when $F(t_i) = \{i/(N+1)\}$ is used, and $\lambda_2 = (\ln 3.0 - \ln 1.5)/(350 - 150)$ when $F(t_i) = (i - 0.3)/(n + 0.4)$ is used.

TABLE 15.4

Data point i	Failure time t_i (hrs)	$F(t_i) = \dfrac{i}{n+1}$	$\dfrac{1}{1-F(t_i)}$	$F(t_i) = \dfrac{i-0.3}{n+0.4}$	$\dfrac{1}{1-F(t_i)}$
1	30	0.0625	1.0667	0.04545	1.04762
2	50	0.1250	1.1429	0.11039	1.12409
3	60	0.1875	1.2308	0.17532	1.21260
4	100	0.2500	1.3333	0.24026	1.31624
5	130	0.3125	1.4545	0.30519	1.43925
6	150	0.3750	1.6000	0.37013	1.58763
7	190	0.4375	1.7778	0.43506	1.77011
8	250	0.5000	2.0000	0.50000	2.00000
9	270	0.5625	2.2857	0.56493	2.29851
10	310	0.6250	2.6667	0.62987	2.70175
11	380	0.6875	3.2000	0.69481	3.27665
12	410	0.7500	4.0000	0.75974	4.16216
13	530	0.8125	5.3333	0.82468	5.70370
14	630	0.8750	8.0000	0.88961	9.05882
15	850	0.9375	16.0000	0.95454	22.00000

15.7.5 Normal distribution

When the variable t is normally distributed, we define the standard normal variate z as

$$z = \frac{t - \mu}{\sigma} \tag{15.71}$$

where μ and σ are, respectively, the mean and standard deviations of t. The cumulative distribution function is given by

$$F(t) = \Phi(z) = \Phi\left(\frac{t - \mu}{\sigma}\right) \tag{15.72}$$

and its inverse by

$$\frac{t - \mu}{\sigma} = \Phi^{-1}(F(t)) \tag{15.73}$$

or

$$t = \sigma\,\Phi^{-1}(F(t)) + \mu \tag{15.74}$$

The normal probability paper is based on Eq. (15.74) where t is plotted on the horizontal axis and $\Phi^{-1}(F(t))$ is plotted on the vertical axis. Thus, if t follows normal distribution, the data will fall on a straight line. Since $F(t) = 0.5$ and $\Phi^{-1}(0.5) = 0$ when $t = \mu$, the value of μ can be determined as the value of t for which $F(t) = 0.5$. Similarly, $F(t) = 0.8413$ and $\Phi^{-1}(0.8413) = 1.0$ when $t = \mu + \sigma$, the value of σ can be determined as the horizontal distance between the points corresponding to $F = 0.5$ and $F = 0.8413$.

Example 15.9 The strengths of 10 welded joints are found to be 35.3, 31.8, 34.0, 39.9, 35.8, 41.2, 33.5, 37.1, 36.4, and 38.6 k(lb/in^2). Plot the data on a normal probability paper and determine the parameters of the distribution from the graph.

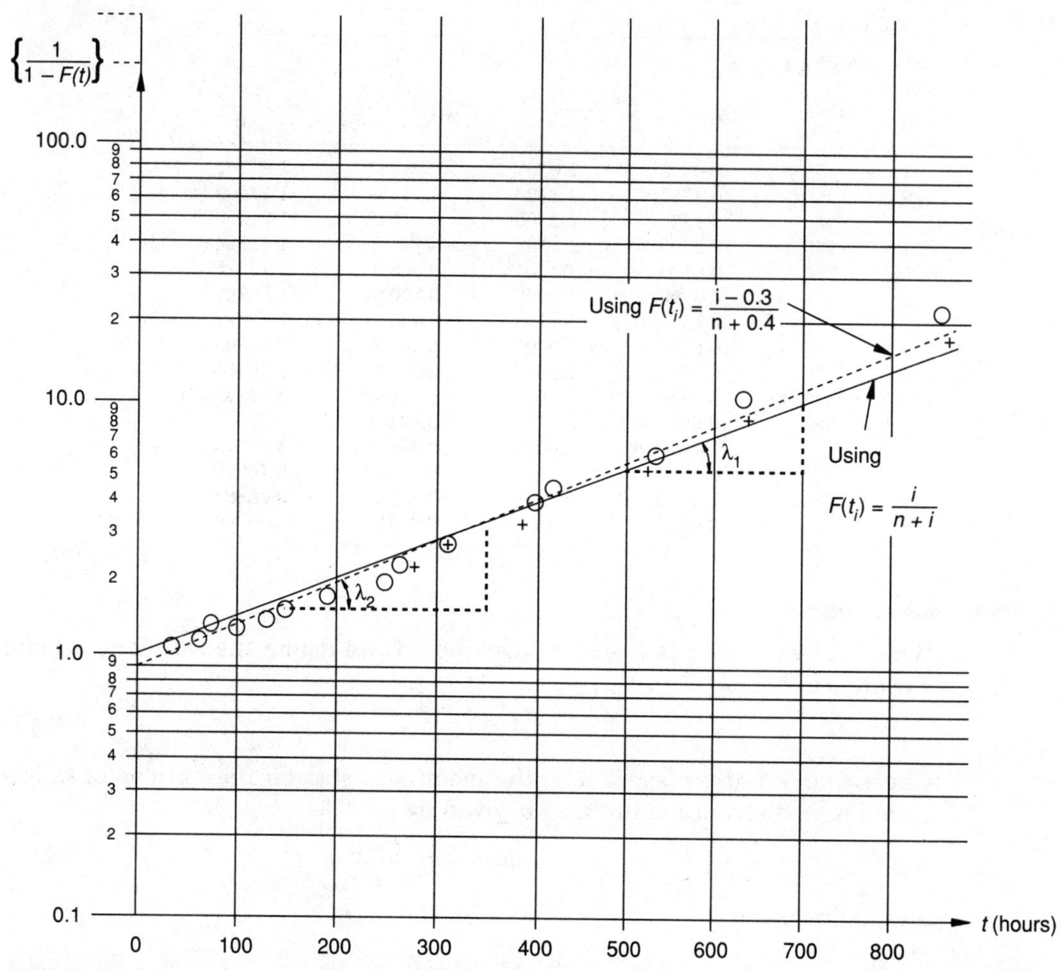

Figure 15.8

solution The strengths are arranged in an increasing order and the ith strength value x_i is plotted against the plotting position $F(x_i) = i/(n+1)$ with $n = 10$. The necessary calculations are shown in Table 15.5. The resulting graph is shown in Fig. 15.9. The mean value of the strength of the welded joint, corresponding to a value of $\Phi^{-1}(0.5) = 0.0$ or $F(x) = 0.5$, can be seen to be $\mu = 36.4$ k(lb/in^2). Similarly, the value of x corresponding to $\Phi^{-1}(0.8413) = 1.0$ or F(t) = 0.8413 gives $\mu + \sigma = 40.2$ k(lb/in^2). Since the value of μ has been found to be 36.4 k(lb/in^2), we find the value of σ as $\sigma = (\sigma + \mu) - \mu = 40.2 - 36.4 = 3.8$ k(lb/in^2). These values can be compared with the sample values of

$$\overline{X} = \frac{1}{n}\sum_{i=1}^{n} x_i = \frac{1}{10}\sum_{i=1}^{10} x_i = 36.36 \text{ k(lb/in}^2\text{)}$$

and

$$s_X = \left\{\frac{1}{n}\sum_{i=1}^{n}(\overline{X} - x_i)^2\right\}^{1/2} = \left\{\frac{1}{10}\sum_{i=1}^{10}(36.36 - x_i)^2\right\}^{1/2} = 2.7840 \text{ k(lb/in}^2\text{)}$$

TABLE 15.5

i	$F(x_i) = \dfrac{i}{n+1}$	x_i
1	0.0909	31.8
2	0.1818	33.5
3	0.2727	34.0
4	0.3636	35.3
5	0.4545	35.8
6	0.5455	36.4
7	0.6364	37.1
8	0.7273	38.6
9	0.8182	39.9
10	0.9091	41.2

15.7.6 Lognormal distribution

If the failure time, t, follows lognormal distribution, then $\ln t$ follows normal distribution and hence

$$F(\ln t) = \Phi(z) = \Phi\left(\frac{\ln t - \mu_1}{\sigma_1}\right) \tag{15.75}$$

where μ_1 and σ_1 denote, respectively, the mean value and standard deviation of $\ln t$, given by Eqs. (3.61) and (3.62)

$$\sigma_1^2 = \ln\left[\left(\frac{\sigma}{\mu}\right)^2 + 1\right] \tag{15.76}$$

$$\mu_1 = \ln \mu - \frac{1}{2}\sigma_1^2 \tag{15.77}$$

where μ and σ are, respectively, the mean and standard deviation of t. Inverting of Eq. (15.75) gives

$$\ln t = \sigma_1 \Phi^{-1}(F(\ln t)) + \mu_1 \tag{15.78}$$

Equation (15.78) denotes a linear relationship between $\ln t$ and $\Phi^{-1}(F(\ln t))$. The lognormal probability paper is constructed with $\ln t$ represented on the horizontal axis and $\Phi^{-1}(F(\ln t))$ plotted on the vertical axis. Thus if the failure time t follows lognormal distribution, the data will fall on a straight line. As in the case of normal distribution, μ_1 can be determined as the value of $\ln t$ for which $F(\ln t) = 0.5$ or $\Phi^{-1}(F(\ln t)) = 0$. Also the value of σ_1 is given by the horizontal distance between the points corresponding to $F(\ln t) = 0.5$ and $F(\ln t) = 0.8413$. Once μ_1 and σ_1 are known, the values of μ and σ can be computed using Eqs. (15.76) and (15.77).

15.7.7 Weibull distribution

The three-parameter Weibull distribution function for the failure time t is given by

$$F(t) = 1 - e^{-\{(t - t_0)/(\theta - t_0)\}^b}; \qquad t \geq t_0 \tag{15.79}$$

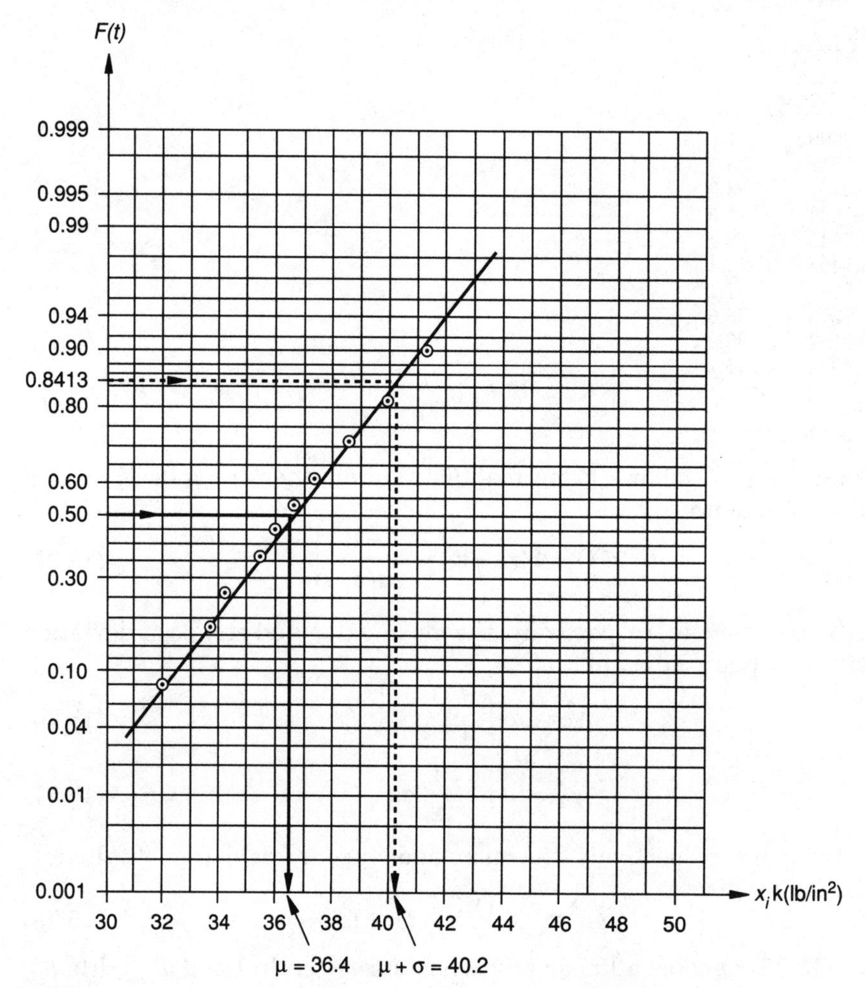

Figure 15.9 Normal probability paper.

where t_0 = minimum possible value of $t (t \geq 0)$, θ = characteristic parameter $(\theta \geq t_0)$, and b = shape parameter $(b > 0)$ denote the three parameters of the distribution. If the minimum life t_0 is taken as zero, Eq. (15.79) becomes a two-parameter Weibull distribution

$$F(t) = 1 - e^{-(t/\theta)^b} \tag{15.80}$$

Equation (15.79) can be rewritten as

$$1 - F(t) = e^{-((t - t_0)/(\theta - t_0))^b} \tag{15.81}$$

Taking logarithms on both sides of Eq. (15.81) gives

$$\ln\left(\frac{1}{1-F(t)}\right)=\left(\frac{t-t_0}{\theta-t_0}\right)^b \tag{15.82}$$

Taking logarithms again, Eq. (15.82) yields

$$\ln\ln\left(\frac{1}{1-F(t)}\right)=b\ln(t-t_0)-b\ln(\theta-t_0) \tag{15.83}$$

which can be seen to be of the form $y = cx + d$ where the ordinate (y) is $\ln\ln[1/(1-F(t))]$ and the abscissa (x) is $\ln(t-t_0)$. The resulting graph paper is called the Weibull probability paper. If the failure time t follows Weibull distribution, the data points will fall on a straight line.

To determine the parameters of the Weibull distribution, first we consider the case of a two parameter distribution for which $t_0 = 0$. For this case, Eq. (15.83) reduces to

$$\ln\ln\left(\frac{1}{1-F(t)}\right)=b\ln t-b\ln\theta \tag{15.84}$$

where b and θ are the unknown parameters. When the data is plotted as a straight line on the Weibull probability paper, the slope of the line gives the value of b. To find the value of θ, we note that when the left hand side of Eq. (15.84) is zero, we obtain $b(\ln t - \ln\theta) = 0$ or $t = \theta$. Since the left hand side of Eq. (15.84) can be seen to be zero when $F(t) = 0.632$, the value of t corresponding to $F(t) = 0.632$ in the graph is taken as the value of θ.

If the minimum failure time t_0 is not zero, we need to use the three parameter Weibull distribution. In such a case, the values of t_0, b and θ can be estimated from the experimental data by using the following two-step procedure:

Step 1. Plot the data on a Weibull paper assuming that $t_0 = 0$. Since $t_0 \neq 0$ in reality, the data points fall on a curve instead of a straight line. Draw the best curve possible, using eye judgement, through the data points as shown in Fig. 15.10. Then select three points A_1, A_2, and A_3 on the curve with coordinates (X_1, Y_1), (X_2, Y_2), and (X_3, Y_3), respectively, such that they are equally spaced along the Y-axis:

$$Y_2 - Y_1 = Y_3 - Y_2 \tag{15.85}$$

Since the ordinate of Fig. 15.10 is defined as $Y = \ln\ln[1/(1-F(t))]$, Eq. (15.85) gives

$$\ln\ln\left(\frac{1}{1-F(t_2)}\right)-\ln\ln\left(\frac{1}{1-F(t_1)}\right)$$
$$=\ln\ln\left(\frac{1}{1-F(t_3)}\right)-\ln\ln\left(\frac{1}{1-F(t_2)}\right) \tag{15.86}$$

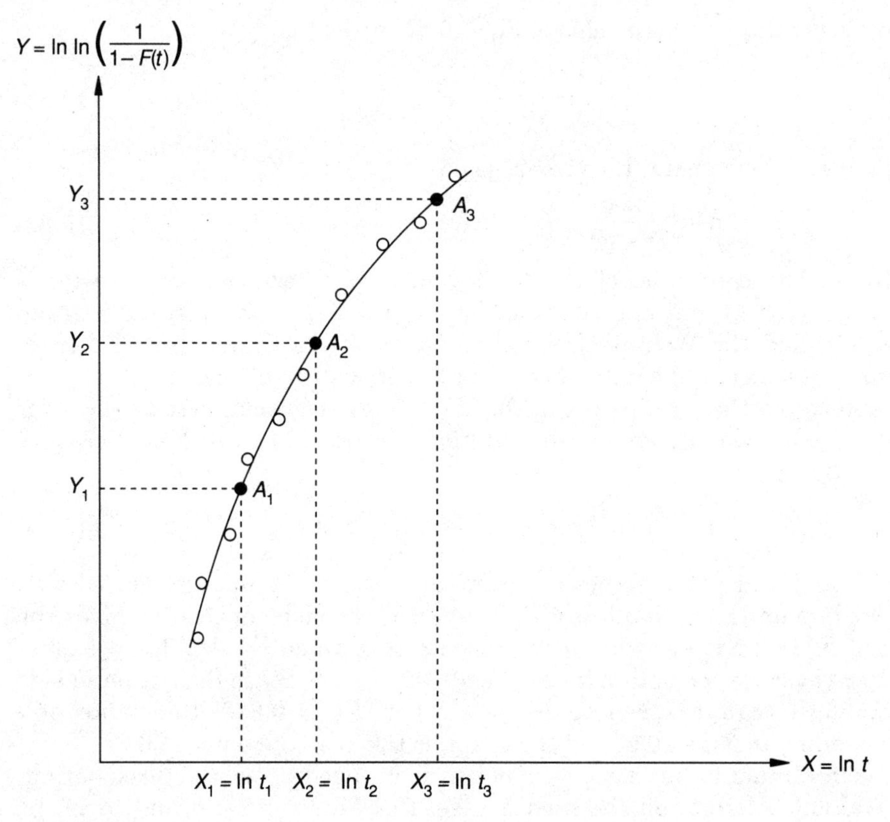

Figure 15.10

Noting that the point A_i corresponds to t_i ($i = 1, 2, 3$), Eq. (15.83) can be used to rewrite Eq. (15.86) in equivalent form as

$$[b \ln(t_2 - t_0) - b \ln(\theta - t_0)] - [b \ln(t_1 - t_0) - b \ln(\theta - t_0)]$$

$$= [b \ln(t_3 - t_0) - b \ln(\theta - t_0)] - [b \ln(t_2 - t_0) - b \ln(\theta - t_0)] \quad (15.87)$$

Equation (15.87) can be rewritten as

$$\ln\left(\frac{t_2 - t_0}{\theta - t_0}\right) - \ln\left(\frac{t_1 - t_0}{\theta - t_0}\right) = \ln\left(\frac{t_3 - t_0}{\theta - t_0}\right) - \ln\left(\frac{t_2 - t_0}{\theta - t_0}\right)$$

or

$$\ln\left(\frac{t_2 - t_0}{t_1 - t_0}\right) = \ln\left(\frac{t_3 - t_0}{t_2 - t_0}\right) \quad (15.88)$$

By taking the inverse logarithms, Eq. (15.88) gives

$$(t_2 - t_0)^2 = (t_1 - t_0)(t_3 - t_0) \tag{15.89}$$

Equation (15.89) can be solved to obtain t_0 as

$$t_0 = \left(\frac{t_1 t_3 - t_2^2}{t_1 + t_3 - 2t_2} \right) \tag{15.90}$$

Step 2. Once t_0 is known, the data are replotted with the values of $\ln(t_i - t_0)$ as abscissa and the corresponding values of $\ln \ln[1/(1 - F(t_i))]$ as ordinates for $i = 1, 2, \ldots, n$. Since the data now falls on a straight line, the parameters b and θ can be estimated from the graph as described for the case of two-parameter Weibull distribution.

Example 15.10 The failure times of eight mechanical fuses are found to be 46, 85, 115, 140, 170, 205, 270, and 360 hours. Plot the data on a two-parameter Weibull probability paper and estimate the parameters of the distribution.

solution The failure times t_i are plotted using the plotting positions $F(t_i) = (i - 0.3)/(n + 0.4)$ indicated in Table 15.6. The data points $\{t_i, F(t_i)\}$ are plotted on the Weibull probability paper as shown in Fig. 15.11. A straight line is drawn through the data points and the parameters of the distribution are computed as follows. The value of $\ln t$ (on the abscissa) corresponding to a value of $F(t) = 0.632$ gives the characteristic parameter as $\theta = \ln 202 = 5.3083$. The slope of the Weibull line gives the shape parameter b as

$$b = \frac{F(t^{(2)}) - F(t^{(1)})}{\ln t^{(2)} - \ln t^{(1)}} = \frac{0.95 - 0.07}{\ln 400 - \ln 40} = 0.3822$$

Thus the two-parameter Weibull distribution corresponding to the given failure time data is given by

$$F(t) = 1 - \exp\left\{-\left(\frac{t}{5.3083}\right)^{0.3822}\right\}$$

TABLE 15.6

Data point (i)	$F(t_i) = \dfrac{i - 0.3}{n + 0.4}$	Failure time t_i (hours)
1	0.08333	46
2	0.20238	85
3	0.32143	115
4	0.44048	140
5	0.55952	170
6	0.67857	205
7	0.79762	270
8	0.91667	360

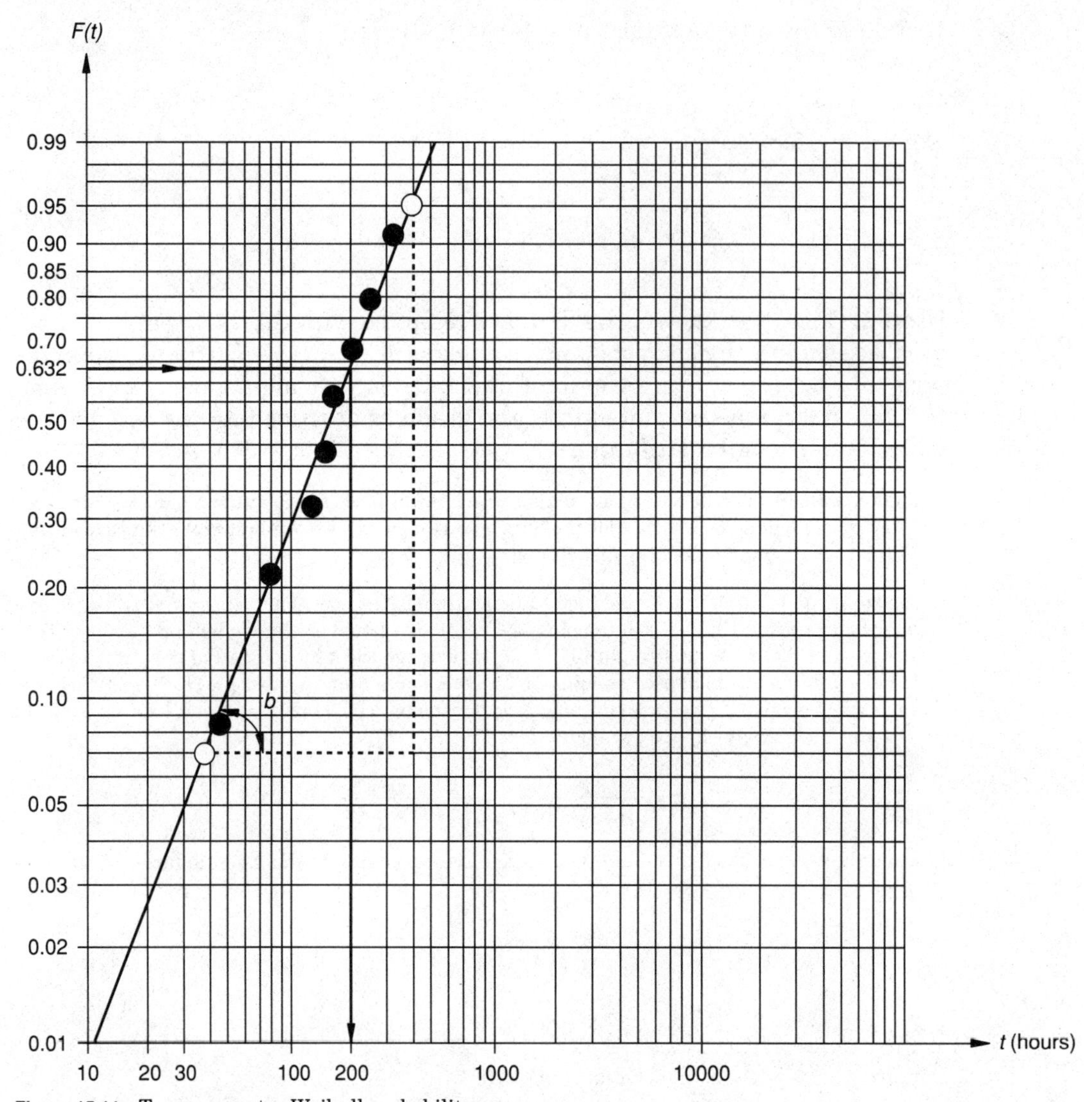

Figure 15.11 Two-parameter Weibull probability paper.

References and Bibliography

15.1. P. A. Tobias and D. C. Trindade, *Applied Reliability,* Van Nostrand Reinhold Co., New York, 1986.
15.2. E. E. Lewis, *Introduction to Reliability Engineering,* John Wiley, New York, 1987.
15.3. J. R. King, *Probability Charts for Decision Making,* Industrial Press Inc., New York, 1971.
15.4. C. O. Smith, *Introduction to Reliability in Design,* Mc Graw-Hill, New York, 1976.
15.5. N. H. Roberts, *Mathematical Methods in Reliability Engineering,* Mc Graw-Hill, New York, 1964.

15.6. J. E. Shigley and C. R. Mischke, *Mechanical Engineering Design,* 5th ed., McGraw-Hill, New York, 1989.
15.7. "Automotive Electronics Reliability Handbook," Publication no. AE-9, Society of Automotive Engineers, Inc., Warrendale, Penn., 1987.
15.8. A. Wald, *Sequential Analysis,* John Wiley, New York, 1947.
15.9. K. C. Kapur and L. R. Lamberson, *Reliability in Engineering Design,* John Wiley, New York, 1977.
15.10. E. B. Haugen, *Probabilistic Mechanical Design,* John Wiley, New York, 1980.
15.11. K. S. Trivedi, *Probability and Statistics with Reliability, Queuing and Computer Science Applications,* Prentice-Hall, Inc., Englewood Cliffs, New Jersey, 1982.
15.12. L. L. Lapin, *Probability and Statistics for Modern Engineering,* Brooks/Cole Publishing Co., Monterey, California, 1983.
15.13. R. A. Heller, "The Weibull Distribution Did Not Apply to Its Founder," in *Probabilistic Methods in the Mechanics of Solids and Structures,* (Eds.: S. Eggwertz and N. C. Lind), Springer-Verlag, Berlin, 1984, pp. xiii–xvi.

Review Questions

15.1 What type of reliability tests are conducted during the life cycle of a system?

15.2 State the objectives of a reliability test.

15.3 How many units or components are to be used in a reliability test?

15.4 Give three methods of conducting accelerated testing.

15.5 How is the "testing until partial failure" conducted?

15.6 How is acceleration achieved by using magnified loading?

15.7 What is the purpose of Arrhenius model?

15.8 What is the relationship between stress amplitude and the number of cycles in fatigue failure?

15.9 Describe the procedure used in sequential testing.

15.10 What is the significance of the average sampling number (ASN)?

15.11 What is meant by statistical inference?

15.12 What is the difference between parameter estimation and hypothesis testing?

15.13 What is the purpose of the maximum likelihood method?

15.14 Define the likelihood function.

15.15 How are the confidence level and the level of significance related?

15.16 What is confidence interval?

15.17 What is the significance of chi-square distribution?

15.18 State the principle used in the least squares method.

15.19 What is linear rectification?

15.20 What is regression analysis?

15.21 How are the plotting positions defined for observed failure data?

15.22 What is the differece between mean rank and median rank?

Problems

15.1 The following failure times were observed for fifteen hydraulic cylinders used in aircraft landing gears: 1719, 693, 1374, 2841, 1121, 1839, 947, 1746, 1972, 2477, 890, 1418, 1643, 2644, and 2165 hours. Plot the probability density, probability distribution, reliability, and hazard functions of the life of hydraulic cylinders.

15.2 The following failure data is collected for a group of 100 accelerometers:

Time interval (100 hours each)	Number of failures observed
1	21
2	13
3	9
4	7
5	6
6	6
7	5
8	7
9	11
10	15

Plot the probability density, probability distribution, reliability, and hazard functions of the failure time of the accelerometers.

15.3 Find the ratio of expected waiting times to observe the rth failure in samples of size n and r, respectively, by varying r from 1 to 20 and n from 1 to 25.

15.4 Find the magnification needed for the stress amplitude in the fatigue testing of AISI 1045 steel to reduce the testing time by a factor of 10. The S–N curve of this steel is given by $SN^{0.085} = 154.0$.

15.5 A sudden-death test is conducted by dividing randomly a group of 50 valves used in the ice-making unit of household refrigerators into groups of 5 each. The failure times observed for the first unit of each of the 10 groups are given by 192, 233, 274, 305, 359, 386, 411, 428, 457, and 585 hours. Find the median life of the population of the valves.

15.6 A lot of IC chips are subjected to a sequential test with values of $R_u = 0.95$, $R_l = 0.90$, $\alpha = 0.05$, and $\beta = 0.05$. The following observations are made sequentially about the condition of the chips: 40 good, 2 bad, 35 good, 1 bad, 30 good, and 2 bad. Plot the "accept" and "reject" lines and determine whether the lot is to be accepted or rejected. Also compare the number of chips tested with the ASN.

15.7 Find the equations of the "accept" and "reject" lines for a sequential test with $R_u = 0.99$, $R_l = 0.90$, $\alpha = 0.01$, and $\beta = 0.02$. Also estimate the number of units to be tested before making a decision about the acceptance/rejection of the lot.

15.8 The wave heights (x_i) at the site of an offshore platform are found to be 3.4, 3.7, 3.8, 4.1, 4.5, 4.8, 6.5, 5.1, 6.2, 5.3, 5.7, and 6.9 ft. Assuming Rayleigh's distribution

$$f_X(x) = \frac{x}{\alpha^2}\, e^{-1/2\,(x/\alpha)^2}; \qquad x \geq 0$$

can be used to describe the wave heights, determine the maximum likelihood estimator of the parameter α.

15.9 The sample values $x_1, x_2, \ldots, x_n$ have been observed for a random variable, X, following binomial distribution

$$p_X(x) = p^{x_i}\,(1-p)^{1-x_i}; \qquad x_i = 0, 1$$

where p is a parameter. Find the maximum likelihood estimator of p.

15.10 The value of a parameter $\hat{\theta} = \hat{\theta}(x_1, x_2, \ldots, x_n)$ estimated from the observed values $x_1, x_2, \ldots, x_n$ of the random variable X, is said to be an unbiased estimator of the parameter θ if $E[\hat{\theta}(x_1, x_2, \ldots, x_n)] = 0$. Prove that the variance of the random variable X computed using the formula

$$s_X^2 = \frac{1}{n} \sum_{i=1}^{n} (\overline{X} - x_i)^2$$

is a biased estimator of the population variance, σ^2.

15.11 Prove that the variance of the random variable X computed using the formula

$$s_X^2 = \frac{1}{n-1} \sum_{i=1}^{n} (\overline{X} - x_i)^2$$

is an unbiased estimator of the population variance σ^2.

15.12 If the value of a function $Y = Y(X_1, X_2, \ldots, X_n)$ is to be estimated from the measured values of $X_1, X_2, \ldots, X_n$, the confidence interval for Y is defined as

$$P\left[-z_{\alpha/2} < \frac{Y - \overline{Y}}{s_Y} < z_{\alpha/2}\right] = 1 - \alpha$$

where $\overline{Y} = Y(\overline{X}_1, \overline{X}_2, \ldots, \overline{X}_n)$,

$$s_Y^2 \approx \sum_{i=1}^{n} \left(\frac{\partial Y}{\partial X_i} \bigg|_{\overline{X}_1, \overline{X}_2, \ldots, \overline{X}_n} \right)^2 s_{X_i}^2$$

and $(\overline{X}_i, s_{X_i})$ are determined from the measured values of $X_i (i = 1, 2, \ldots, n)$. Thus, the $(1 - \alpha)$ confidence interval of Y can be determined as $(\overline{Y} - z_{\alpha/2}\, s_Y;\ \overline{Y} + z_{\alpha/2}\, s_Y)$. The dimensions of a casting, which is in the form of a rectangular prism, are measured independently six times and the following results are obtained:

Length (X_1): 18.2, 17.9, 18.3, 18.4, 18.1, and 18.0 in
Width (X_2): 11.9, 12.2, 12.0, 12.3, 11.8, and 12.1 in
Depth (X_3): 8.2, 8.1, 8.4, 8.0, 8.3, and 7.9 in

Determine the following:

a. The means and standard deviations of X_1, X_2, and X_3
b. The mean and standard deviation of the volume of the casting
c. The 95 percent confidence interval on the true volume of the casting

15.13 If 15 observations made for the yield strength of steel gave the sample mean as 29,800 lb/in^2 and the sample standard deviation as 1,800 lb/in^2, determine the 99 percent confidence interval for the population standard deviation of the yield strength.

15.14 The standard deviation of the yield strength of steel is known to be 1,500 lb/in^2. From a sample of 15 observations, the sample mean of the yield strength is found to be 29,800 lb/in^2. Find (a) the 95 percent confidence interval, and (b) the 99 percent confidence interval for the population mean of the yield strength of steel.

15.15 If 15 observations made for the yield strength of steel gave the sample mean as 29,800 lb/in^2 and the sample standard deviation as 1,800 lb/in^2, determine the 99 percent confidence interval for the population mean of the yield strength.

15.16 The fatigue lives of 10 helical springs are found to be 1.1, 1.2, 1.4, 1.5, 1.7, 1.9, 2.0, 2.1, 2.4, and 2.8 million cycles. Determine the following: (a) The 95 percent confidence interval on the population mean of the fatigue life; (b) The 95 percent confidence interval on the population standard deviation of the fatigue life.

15.17 The temperatures measured at various distances along the thickness of the wall of a cold storage building are given as:

Distance, x (inch)	Temperature, T (degrees F)
0	90
4	69
8	58
12	47
16	40
20	34
24	30

Determine the following relationship between the temperature and the distance using least squares technique:

1. Linear relationship: $T = a + bx$

2. Quadratic relationship: $T = a + bx + cx^2$

15.18 The average number of breakdowns experienced in different months in a machine shop due to the failure of various machines are given by 39.0, 14.5, 6.5, 75.0, 3.0, 19.5, 54.0, 31.5, 10.5, and 23.0. Plot the data on an exponential probability paper and determine the breakdown rate in the machine shop.

15.19 The tensile strengths of 12 welded joints have been observed to be 31,325, 31,556, 31,234, 31,998, 31,404, 31,263, 31,440, 31,785, 31,492, 31,357, 31,285 and 31,643 lb/in^2. Assuming that the tensile strength of welded joints follows exponential distribution, determine the parameters of the distribution graphically.

15.20 The crushing strengths of test specimens made of concrete used for a machine foundation are observed to be 18, 27, 24, 30, 22, 36, 41, 33, 29, and 34 k(lb/in^2). Plot the data by assuming normal distribution to be valid for describing the crushing strength of concrete. Also, determine the parameters of the distribution from the graph.

15.21 Solve Problem 15.20 by assuming lognormal distribution for the crushing strength of concrete.

15.22 Solve Problem 15.20 by assuming a two-parameter Weibull distribution for the crushing strength of concrete.

15.23 Tests conducted to find the ultimate tensile strength of an alloy steel gave the following results: 63.3, 61.8, 67.4, 65.3, 59.7, 64.2, 60.9, 66.1, 62.6, and 63.9 k(lb/in^2). Assuming that the ultimate tensile strength follows a two-parameter Weibull distribution, determine the parameters of the distribution.

15.24 Solve Problem 15.23 using a three-parameter Weibull distribution.

15.25 The times needed to restore power due to electrical outages by an electric power company are given by 1.0, 0.9, 2.4, 0.3, 0.7, 3.6, 2.0, 1.4, 0.4, and 1.6 hours. Plot the data on a lognormal paper and determine the mean value and standard deviation of the "restoring time" from the graph.

Appendix

A

Standard Normal Distribution Function

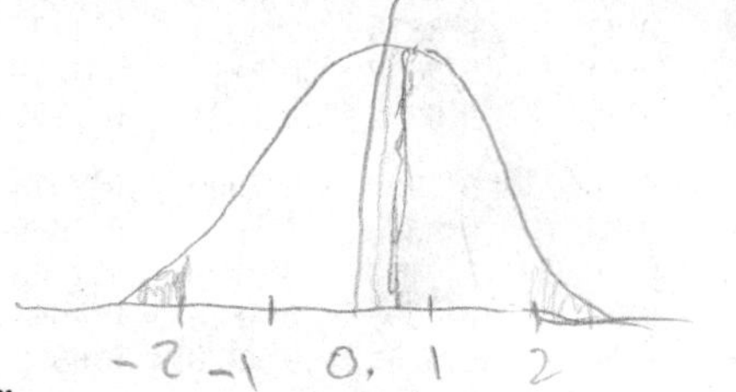

Values of $\Phi(z)$ where $\Phi(z) = \frac{1}{\sqrt{2\pi}} \int_{-\infty}^{z} e^{-x^2/2} dx$

z	.00	.01	.02	.03	.04	.05	.06	.07	.08	.09
.0	.50000	.50399	.50798	.51197	.51595	.51994	.52392	.52790	.53188	.53586
.1	.53983	.54380	.54776	.55172	.55567	.55962	.56356	.56749	.57142	.57534
.2	.57926	.58317	.58706	.59095	.59483	.59871	.60257	.60642	.61026	.61409
.3	.61791	.62172	.62552	.62930	.63307	.63683	.64058	.64431	.64803	.65173
.4	.65542	.65910	.66276	.66640	.67003	.67364	.67724	.68082	.68439	.68793
.5	.69146	.69497	.69847	.70194	.70540	.70884	.71226	.71566	.71904	.72240
.6	.72575	.72907	.73237	.73565	.73891	.74215	.74537	.74857	.75175	.75490
.7	.75804	.76115	.76424	.76730	.77035	.77337	.77637	.77935	.78230	.78524
.8	.78814	.79103	.79389	.79673	.79955	.80234	.80510	.80785	.81057	.81327
.9	.81594	.81859	.82121	.82381	.82639	.82894	.83147	.83398	.83646	.83891
1.0	.84134	.84375	.84614	.84849	.85083	.85314	.85543	.85769	.85992	.86214
1.1	.86433	.86650	.86864	.87076	.87286	.87493	.87698	.87900	.88100	.88298
1.2	.88493	.88686	.88877	.89065	.89251	.89435	.89616	.89796	.89973	.90147
1.3	.90320	.90490	.90658	.90824	.90988	.91149	.91309	.91466	.91621	.91774
1.4	.91924	.92073	.92220	.92364	.92507	.92647	.92785	.92922	.93056	.93189
1.5	.93319	.93448	.93574	.93699	.93822	.93943	.94062	.94179	.94295	.94408
1.6	.94520	.94630	.94738	.94845	.94950	.95053	.95154	.95254	.95352	.95449
1.7	.95543	.95637	.95728	.95818	.95907	.95994	.96080	.96164	.96246	.96327
1.8	.96407	.96485	.96562	.96638	.96712	.96784	.96856	.96926	.96995	.97062
1.9	.97128	.97193	.97257	.97320	.97381	.97441	.97500	.97558	.97615	.97670
2.0	.97725	.97778	.97831	.97882	.97932	.97982	.98030	.98077	.98124	.98169
2.1	.98214	.98257	.98300	.98341	.98382	.98422	.98461	.98500	.98537	.98574
2.2	.98610	.98645	.98679	.98713	.98745	.98778	.98809	.98840	.98870	.98899
2.3	.98928	.98956	.98983	$.9^2 0097$	$.9^2 0358$	$.9^2 0613$	$.9^2 0863$	$.9^2 1106$	$.9^2 1344$	$.9^2 1576$
2.4	$.9^2 1802$	$.9^2 2024$	$.9^2 2240$	$.9^2 2451$	$.9^2 2656$	$.9^2 2857$	$.9^2 3053$	$.9^2 3244$	$.9^2 3431$	$.9^2 3613$

Values of $\Phi(z)$ where $\Phi(z) = \frac{1}{\sqrt{2\pi}} \int_{-\infty}^{z} e^{-x^2/2} dx$ (Continued)

z	.00	.01	.02	.03	.04	.05	.06	.07	.08	.09
2.5	$.9^{2}3790$	$.9^{2}3963$	$.9^{2}4132$	$.9^{2}4297$	$.9^{2}4457$	$.9^{2}4614$	$.9^{2}4766$	$.9^{2}4915$	$.9^{2}5060$	$.9^{2}5201$
2.6	$.9^{2}5339$	$.9^{2}5473$	$.9^{2}5604$	$.9^{2}5731$	$.9^{2}5855$	$.9^{2}5975$	$.9^{2}6093$	$.9^{2}6207$	$.9^{2}6319$	$.9^{2}6427$
2.7	$.9^{2}6533$	$.9^{2}6636$	$.9^{2}6736$	$.9^{2}6833$	$.9^{2}6928$	$.9^{2}7020$	$.9^{2}7110$	$.9^{2}7197$	$.9^{2}7282$	$.9^{2}7365$
2.8	$.9^{2}7445$	$.9^{2}7523$	$.9^{2}7599$	$.9^{2}7673$	$.9^{2}7744$	$.9^{2}7814$	$.9^{2}7882$	$.9^{2}7948$	$.9^{2}8012$	$.9^{2}8074$
2.9	$.9^{2}8134$	$.9^{2}8193$	$.9^{2}8250$	$.9^{2}8305$	$.9^{2}8359$	$.9^{2}8411$	$.9^{2}8462$	$.9^{2}8511$	$.9^{2}8559$	$.9^{2}8605$
3.0	$.9^{2}8650$	$.9^{2}8694$	$.9^{2}8736$	$.9^{2}8777$	$.9^{2}8817$	$.9^{2}8856$	$.9^{2}8893$	$.9^{2}8930$	$.9^{2}8965$	$.9^{2}8999$
3.1	$.9^{3}0324$	$.9^{3}0646$	$.9^{3}0957$	$.9^{3}1260$	$.9^{3}1553$	$.9^{3}1836$	$.9^{3}2112$	$.9^{3}2378$	$.9^{3}2636$	$.9^{3}2886$
3.2	$.9^{3}3129$	$.9^{3}3363$	$.9^{3}3590$	$.9^{3}3810$	$.9^{3}4024$	$.9^{3}4230$	$.9^{3}4429$	$.6^{3}4623$	$.9^{3}4810$	$.9^{3}4991$
3.3	$.9^{3}5166$	$.9^{3}5335$	$.9^{3}5499$	$.9^{3}5658$	$.9^{3}5811$	$.9^{3}5959$	$.9^{3}6103$	$.9^{3}6242$	$.9^{3}6376$	$.9^{3}6505$
3.4	$.9^{3}6631$	$.9^{3}6752$	$.9^{3}6869$	$.9^{3}6982$	$.9^{3}7091$	$.9^{3}7197$	$.9^{3}7299$	$.9^{3}7398$	$.9^{3}7493$	$.9^{3}7585$
3.5	$.9^{3}7674$	$.9^{3}7759$	$.9^{3}7842$	$.9^{3}7922$	$.9^{3}7999$	$.9^{3}8074$	$.9^{3}8146$	$.9^{3}8215$	$.9^{3}8282$	$.9^{3}8347$
3.6	$.9^{3}8409$	$.9^{3}8469$	$.9^{3}8527$	$.9^{3}8583$	$.9^{3}8637$	$.9^{3}8689$	$.9^{3}8739$	$.9^{3}8787$	$.9^{3}8834$	$.9^{3}8879$
3.7	$.9^{3}8922$	$.9^{3}8964$	$.9^{4}0039$	$.9^{4}0426$	$.9^{4}0799$	$.9^{4}1158$	$.9^{4}1504$	$.9^{4}1838$	$.9^{4}2159$	$.9^{4}2568$
3.8	$.9^{4}2765$	$.9^{4}3052$	$.9^{4}3327$	$.9^{4}3593$	$.9^{4}3848$	$.9^{4}4094$	$.9^{4}4331$	$.9^{4}4558$	$.9^{4}4777$	$.9^{4}4988$
3.9	$.9^{4}5190$	$.9^{4}5385$	$.9^{4}5573$	$.9^{4}5753$	$.9^{4}5926$	$.9^{4}6092$	$.9^{4}6253$	$.9^{4}6406$	$.9^{4}6554$	$.9^{4}6696$
4.0	$.9^{4}6833$	$.9^{4}6964$	$.9^{4}7090$	$.9^{4}7211$	$.9^{4}7327$	$.9^{4}7439$	$.9^{4}7546$	$.9^{4}7649$	$.9^{4}7748$	$.9^{4}7843$
4.1	$.9^{4}7934$	$.9^{4}8022$	$.9^{4}8106$	$.9^{4}8186$	$.9^{4}8263$	$.9^{4}8338$	$.9^{4}8409$	$.9^{4}8477$	$.9^{4}8542$	$.9^{4}8605$
4.2	$.9^{4}8665$	$.9^{4}8723$	$.9^{4}8778$	$.9^{4}8832$	$.9^{4}8882$	$.9^{4}8931$	$.9^{4}8978$	$.9^{5}0226$	$.9^{5}0655$	$.9^{5}1066$
4.3	$.9^{5}1460$	$.9^{5}1837$	$.9^{5}2199$	$.9^{5}2545$	$.9^{5}2876$	$.9^{5}3193$	$.9^{5}3497$	$.9^{5}3788$	$.9^{5}4066$	$.9^{5}4332$
4.4	$.9^{5}4587$	$.9^{5}4831$	$.9^{5}5065$	$.9^{5}5288$	$.9^{5}5502$	$.9^{5}5706$	$.9^{5}5902$	$.9^{5}6089$	$.9^{5}6268$	$.9^{5}6439$
4.5	$.9^{5}6602$	$.9^{5}6759$	$.9^{5}6908$	$.9^{5}7051$	$.9^{5}7187$	$.9^{5}7318$	$.9^{5}7442$	$.9^{5}7561$	$.9^{5}7675$	$.9^{5}7784$
4.6	$.9^{5}7888$	$.9^{5}7987$	$.9^{5}8081$	$.9^{5}8172$	$.9^{5}8258$	$.9^{5}8340$	$.9^{5}8419$	$.9^{5}8494$	$.9^{5}8566$	$.9^{5}8634$
4.7	$.9^{5}8699$	$.9^{5}8761$	$.9^{5}8821$	$.9^{5}8877$	$.9^{5}8931$	$.9^{5}8983$	$.9^{6}0320$	$.9^{6}0789$	$.9^{6}1235$	$.9^{6}1661$
4.8	$.9^{6}2067$	$.9^{6}2453$	$.9^{6}2822$	$.9^{6}3173$	$.9^{6}3508$	$.9^{6}3827$	$.9^{6}4131$	$.9^{6}4420$	$.9^{6}4696$	$.9^{6}4958$
4.9	$.9^{6}5208$	$.9^{6}5446$	$.9^{6}5673$	$.9^{6}5889$	$.9^{6}6094$	$.9^{6}6289$	$.9^{6}6475$	$.9^{6}6652$	$.9^{6}6821$	$.9^{6}6981$

$\Phi(z)$	z
0.60	0.253
0.70	0.524
0.80	0.842
0.90	1.282
0.95	1.645
0.99	2.326
0.999	3.090
0.9999	3.719
0.99999	4.27
0.999999	4.75
0.9999999	5.20
0.99999999	5.61
0.999999999	6.00
0.9999999999	6.36

NOTE: $\Phi(-z) = \Phi(z)$; $\Phi(3) = 0.9^{2}8650 =$ 0.998650, etc.

Appendix

B

Critical Values of *t*-Distribution

Values of $t_{(n, \alpha/2)}$

ν	$\alpha/2 = 0.4$ $\alpha = 0.8$	0.25 0.5	0.1 0.2	0.05 0.1	0.025 0.05	0.01 0.02	0.005 0.01	0.0025 0.005	0.001 0.002	0.0005 0.001
1	0.325	1.000	3.078	6.314	12.706	31.821	63.657	127.32	318.31	636.62
2	.289	0.816	1.886	2.920	4.303	6.965	9.925	14.089	22.327	31.598
3	.277	.765	1.638	2.353	3.182	4.541	5.841	7.453	10.214	12.924
4	.271	.741	1.533	2.132	2.776	3.747	4.604	5.598	7.173	8.610
5	0.267	0.727	1.476	2.015	2.571	3.365	4.032	4.773	5.893	6.869
6	.265	.718	1.440	1.943	2.447	3.143	3.707	4.317	5.208	5.959
7	.263	.711	1.415	1.895	2.365	2.998	3.499	4.029	4.785	5.408
8	.262	.706	1.397	1.860	2.306	2.896	3.355	3.833	4.501	5.041
9	.261	.703	1.383	1.833	2.262	2.821	3.250	3.690	4.297	4.781
10	0.260	0.700	1.372	1.812	2.228	2.764	3.169	3.581	4.144	4.587
11	.260	.697	1.363	1.796	2.201	2.718	3.106	3.497	4.025	4.437
12	.259	.695	1.356	1.782	2.179	2.681	3.055	3.428	3.930	4.318
13	.259	.694	1.350	1.771	2.160	2.650	3.012	3.372	3.852	4.221
14	.258	.692	1.345	1.761	2.145	2.624	2.977	3.326	3.787	4.140
15	0.258	0.691	1.341	1.753	2.131	2.602	2.947	3.286	3.733	4.073
16	.258	.690	1.337	1.746	2.120	2.583	2.921	3.252	3.686	4.015
17	.257	.689	1.333	1.740	2.110	2.567	2.898	3.222	3.646	3.965
18	.257	.688	1.330	1.734	2.101	2.552	2.878	3.197	3.610	3.922
19	.257	.688	1.328	1.729	2.093	2.539	2.861	3.174	3.579	3.883

Values of $t_{(n, \alpha/2)}$ (Continued)

ν	$\alpha/2 = 0.4$ $\alpha = 0.8$	0.25 0.5	0.1 0.2	0.05 0.1	0.025 0.05	0.01 0.02	0.005 0.01	0.0025 0.005	0.001 0.002	0.0005 0.001
20	0.257	0.687	1.325	1.725	2.086	2.528	2.845	3.153	3.552	3.850
21	.257	.686	1.323	1.721	2.080	2.518	2.831	3.135	3.527	3.819
22	.256	.686	1.321	1.717	2.074	2.508	2.819	3.119	3.505	3.792
23	.256	.685	1.319	1.714	2.069	2.500	2.807	3.104	3.485	3.767
24	.256	.685	1.318	1.711	2.064	2.492	2.797	3.091	3.467	3.745
25	0.256	0.684	1.316	1.708	2.060	2.485	2.787	3.078	3.450	3.725
26	.256	.684	1.315	1.706	2.056	2.479	2.779	3.067	3.435	3.707
27	.256	.684	1.314	1.703	2.052	2.473	2.771	3.057	3.421	3.690
28	.256	.683	1.313	1.701	2.048	2.467	2.763	3.047	3.408	3.674
29	.256	.683	1.311	1.699	2.045	2.462	2.756	3.038	3.396	3.659
30	0.256	0.683	1.310	1.697	2.042	2.457	2.750	3.030	3.385	3.646
40	.255	.681	1.303	1.684	2.021	2.423	2.704	2.971	3.307	3.551
60	.254	.679	1.296	1.671	2.000	2.390	2.660	2.915	3.232	3.460
120	.254	.677	1.289	1.658	1.980	2.358	2.617	2.860	3.160	3.373
∞	.253	.674	1.282	1.645	1.960	2.326	2.576	2.807	3.090	3.291

SOURCE: Reprinted with permission of Biometrika Trustees, from E. S. Pearson and H. O. Hartley (eds.), *Biometrika Tables for Statisticians*, vol. I, 3d ed., Biometrika Trust, London, 1979.

Appendix

C

Critical Values of χ^2-Distribution

Values of $\chi^2 \equiv \chi^2_{(n,\alpha)}$ **given by** $\alpha = 1 - F_{\chi^2} = 2^{-n/2} \dfrac{1}{\Gamma\left(\frac{n}{2}\right)} \int_{\chi^2}^{\infty} e^{-x/2}\, x^{(n/2)-1}\, dx$

n \ α	0.995	0.990	0.975	0.950	0.900	0.750	0.500
1	392704.10^{-10}	157088.10^{-9}	982069.10^{-9}	393214.10^{-8}	0.0157908	0.1015308	0.454936
2	0.0100251	0.0201007	0.0506356	0.102587	0.210721	0.575364	1.38629
3	0.0717218	0.114832	0.215795	0.351846	0.584374	1.212534	2.36597
4	0.206989	0.297109	0.484419	0.710723	1.063623	1.92256	3.35669
5	0.411742	0.554298	0.831212	1.145476	1.61031	2.67460	4.35146
6	0.675727	0.872090	1.23734	1.63538	2.20413	3.45460	5.34812
7	0.989256	1.239043	1.68987	2.16735	2.83311	4.25485	6.34581
8	1.34441	1.64650	2.17973	2.73264	3.48954	5.07064	7.34412
9	1.73493	2.08790	2.70039	3.32511	4.16816	5.89883	8.34283
10	2.15586	2.55821	3.24697	3.94030	4.86518	6.73720	9.34182
11	2.60322	3.05348	3.81575	4.57481	5.57778	7.58414	10.3410
12	3.07382	3.57057	4.40379	5.22603	6.30380	8.43842	11.3403
13	3.56503	4.10692	5.00875	5.89186	7.04150	9.29907	12.3398
14	4.07467	4.66043	5.62873	6.57063	7.78953	10.1653	13.3393
15	4.60092	5.22935	6.26214	7.26094	8.54676	11.0365	14.3389
16	5.14221	5.81221	6.09766	7.96165	9.31224	11.9122	15.3385
17	5.69722	6.40776	7.56419	8.67176	10.0852	12.7919	16.3382
18	6.26480	7.01491	8.23075	9.39046	10.8649	13.6753	17.3379
19	6.84397	7.63273	8.90652	10.1170	11.6509	14.5620	18.3377

Values of $\chi^2 \equiv \chi^2_{(n,\,\alpha)}$ given by $\alpha = 1 - F_{\chi^2} = 2^{-n/2} \dfrac{1}{\Gamma\left(\dfrac{n}{2}\right)} \displaystyle\int_{\chi^2}^{\infty} e^{-x/2}\, x^{(n/2)-1}\, dx$ (Continued)

n \ α	0.995	0.990	0.975	0.950	0.900	0.750	0.500
20	7.43384	8.26040	9.59078	10.8508	12.4426	15.4518	19.3374
21	8.03365	8.89720	10.28293	11.5913	13.2396	16.3444	20.3372
22	8.64272	9.54249	10.9823	12.3380	14.0415	17.2396	21.3370
23	9.26043	10.19567	11.6886	13.0905	14.8480	18.1373	22.3369
24	9.88623	10.8564	12.4012	13.8484	15.6587	19.0373	23.3367
25	10.5197	11.5240	13.1197	14.6114	16.4734	19.9393	24.3366
26	11.1602	12.1981	13.8439	15.3792	17.2919	20.8434	25.3365
27	11.8076	12.8785	14.5734	16.1514	18.1139	21.7494	26.3363
28	12.4613	13.5647	15.3079	16.9279	18.9392	22.6572	27.3362
29	13.1211	14.2565	16.0471	17.7084	19.7677	23.5666	28.3361
30	13.7867	14.9535	16.7908	18.4927	20.5992	24.4776	29.3360
40	20.7065	22.1643	24.4330	26.5093	29.0505	33.6603	39.3353
50	27.9907	29.7067	32.3574	34.7643	37.6886	42.9421	49.3349
60	35.5345	37.4849	40.4817	43.1880	46.4589	52.2938	59.3347
70	43.2752	45.4417	48.7576	51.7393	55.3289	61.6983	69.3345
80	51.1719	53.5401	57.1532	60.3915	64.2778	71.1445	79.3343
90	59.1963	61.7541	65.6466	69.1260	73.2911	80.6247	89.3342
100	67.3276	70.0649	74.2219	77.9295	82.3581	90.1332	99.3341

n \ α	0.250	0.100	0.050	0.025	0.010	0.005	0.001
1	1.32330	2.70554	3.84146	5.02389	6.63490	7.87944	10.828
2	2.77259	4.60517	5.99146	7.37776	9.21034	10.5966	13.816
3	4.10834	6.25139	7.81473	9.34840	11.3449	12.8382	16.266
4	5.38527	7.77944	9.48773	11.1433	13.2767	14.8603	18.467
5	6.62568	9.23636	11.0705	12.8325	15.0863	16.7496	20.515
6	7.84080	10.6446	12.5916	14.4494	16.8119	18.5476	22.458
7	9.03715	12.0170	14.0671	16.0128	18.4753	20.2777	24.322
8	10.2189	13.3616	15.5073	17.5345	20.0902	21.9550	26.125
9	11.3888	14.6837	16.9190	19.0228	21.6660	23.5894	27.877
10	12.5489	15.9872	18.3070	20.4832	23.2093	25.1882	29.588
11	13.7007	17.2750	19.6751	21.9200	24.7250	26.7568	31.264
12	14.8454	18.5493	21.0261	23.3367	26.2170	28.2995	32.909
13	15.9839	19.8119	22.3620	24.7356	27.6882	29.8195	34.528
14	17.1169	21.0641	23.6848	26.1189	29.1412	31.3194	36.123
15	18.2451	22.3071	24.9958	27.4884	30.5779	32.8013	37.697
16	19.3689	23.5418	26.2962	28.8454	31.9999	34.2672	39.252
17	20.4887	24.7690	27.5871	30.1910	33.4087	35.7185	40.790
18	21.6049	25.9894	28.8693	31.5264	34.8053	37.1565	42.312
19	22.7178	27.2036	30.1435	32.8523	36.1909	38.5823	43.820

Values of $\chi^2 \equiv \chi^2_{(n,\alpha)}$ given by $\alpha = 1 - F_{\chi^2} = 2^{-n/2} \frac{1}{\Gamma\left(\frac{n}{2}\right)} \int_{\chi^2}^{\infty} e^{-x/2} x^{(n/2)-1}\, dx$ (Continued)

n \ α	0.250	0.100	0.050	0.025	0.010	0.005	0.001
20	23.8277	28.4120	31.4104	34.1696	37.5662	39.9968	45.315
21	24.9348	29.6151	32.6706	35.4789	38.9322	41.4011	46.797
22	26.0393	30.8133	33.9244	36.7807	40.2894	42.7957	48.268
23	27.1413	32.0069	35.1725	38.0756	41.6384	44.1813	49.728
24	28.2412	33.1962	36.4150	39.3641	42.9798	45.5585	51.179
25	29.3389	34.3816	37.6525	40.6465	44.3141	46.9279	52.618
26	30.4346	35.5632	38.8851	41.9232	45.6417	48.2899	54.052
27	31.5284	36.7412	40.1133	43.1945	46.9629	49.6449	55.476
28	32.6205	37.9159	41.3371	44.4608	48.2782	50.9934	56.892
29	33.7109	39.0875	42.5570	45.7223	49.5879	52.3356	58.301
30	34.7997	40.2560	43.7730	46.9792	50.8922	53.6720	59.703
40	45.6160	51.8051	55.7585	59.3417	63.6907	66.7660	73.402
50	56.3336	63.1671	67.5048	71.4202	76.1539	79.4900	86.661
60	66.9815	74.3970	79.0819	83.2977	88.3794	91.9517	99.607
70	77.5767	85.5270	90.5312	95.0232	100.425	104.215	112.317
80	88.1303	96.5782	101.879	106.629	112.329	116.321	124.839
90	98.6499	107.565	113.145	118.136	124.116	128.299	137.208
100	109.141	118.498	124.342	129.561	135.807	140.169	149.449

SOURCE: Reprinted with permission of Biometrika Trustees, from E. S. Pearson and H. O. Hartley (eds.), *Biometrika Tables for Statisticians*, vol. I, 3d ed., Biometrica Trust, London, 1979.

Appendix

D

Product Liability

D.1 Basic Concept

In legal terms, product liability describes an action (such as a lawsuit) in which the plaintiff (the injured party) seeks to recover damages for personal injury or loss of property from the defendent (the seller or the manufacturer) when it is alleged that the damage was caused by a defective product [D.1–D.4]. A product defect may be either a manufacturing (or production) defect or a design defect. A manufacturing defect may arise due to an error in the production process. In such a case, the product may not meet the manufacturer's own standards and may escape from detection due to the use of inadequate quality-control procedures . A classic example is the soda pop bottle that exploded, either due to an imperfection in the glass structure of the bottle or due to an inadvertent overcarbonization. The design defect may arise when inferior standards are used by the manufacturer in producing the product. Thus, the manufacturer and the designer become responsible for the reliable and safe performance of a product.

D.2 Definitions

Some of the basic terminology employed in product liability are indicated in the following list:

1. *Tort.* Any wrongful act, damage, or injury for which the aggrieved can seek redress by legal action.
2. *Privity.* A direct contractual relationship between two persons or parties.

3. *Plaintiff.* The person (organization) who (which) initiates legal action for redress of a loss.
4. *Negligence.* The failure to take reasonable amount of care which results in an injury or property damage to another person or party.
5. *Warranty.* An assurance by the seller to the purchaser that the product will be as promised.
6. *Implied warranty.* An automatic warranty, implied by law, that a product is suitable for a specific purpose and is reasonably safe for use.
7. *Express warranty.* A statement made by a seller, either orally or in writing, that the product sold is suitable for a specific use.

D.3 Theories of Product Liability

The following three legal principles form a basis in settling most product liability suits in favor of the consumer:

1. Negligence
2. Breach of express warranty
3. Breach of implied warranty

Negligence. *Negligence* can be described as the "conduct which falls below the standard established by law for the protection of others against unreasonably great risk of harm." This principle became the keystone of tort law. It can be seen that this law is flexible in that the "standard" would always be current, appropriate to the situation at hand, and whatever the court declares it to be. Hence, the manufacturer should produce the product based on the state-of-the-art technology known to them at the time of manufacture so that they will not be "negligent" in taking "reasonable care" in producing a safe product. The state-of-the-art technology may be accomplished through the following means:

In house training and seminars
Professional society meetings
Continuing education courses
Scientific and business periodicals

A proof of such endeavors by a company might prove its progressive attitude in any future product liability litigation involving its products.

Breach of express warranty. The *breach of express warranty* occurs when the claims made by the manufacturer about his product are not met, and this results in an injury or damage. The claims made by the manufacturer need not be in writing. Some examples of express warranties according to the Uniform Commercial Code, adopted by most states, are as follows:

1. Catalog descriptions of products offered for sale by mail-order businesses
2. Photographs, display material, and manuals accompanying a product
3. Any model or sample that is made part of the basis of bargain
4. Any affirmation of fact or promise made by the seller to the buyer, which then becomes part of the basis of bargain.

Breach of implied warranty. The *implied warranty* is one which is implied by the law rather than given explicitly by the seller. The Uniform Commercial Code, which has been adopted by all the states except Lousiana, forms a basis for liability suits under breach of implied warranty. The implied warranty may be in regard to merchantability or fitness of a product. The *merchantability* indicates that the mere act of offering a product for sale implies that the product is safe and has at least the minimum qualities one would expect of such a product. In this case, the plaintiff need not prove that the entire population of products is unsafe; he needs to prove that only the particular one, which may be called *the worst-case product*, was unsafe. The *fitness-type* of implied warranty arises when a manufacturer or his representative makes a wrong recommendation about the suitability of a product for a particular use. An example is that of a paint store clerk recommending to the customer a particular type of lacquer thinner to remove paint from a floor. When a customer died due to the occurrence of an explosion during the use of the thinner, the store had to pay damages on the basis of breach of warranty of fitness.

D.4 Prevention of Product Liability

In the last two decades, the number of product liability suits increased dramatically to the extent of causing what some people label it as a crisis or an epidemic. Under the docrine of strict liability, the manufacturers have become increasingly responsible for their products and designs; the reason is that the producer cannot assume that his product will be used correctly, safely, and solely for the purpose for which it is intended. For example, a person used a rotary lawn mower to trim the hedges, got injured, filed a lawsuit against the manufacturer and recovered the damages. To prevent product liability suits, the manufacturer should follow the old saying, "an ounce of prevention is worth a pound of cure," and follow the guidelines which are given.

Guidelines to prevent design defects

1. Evaluate every design from safety point of view.
2. Carefully scrutinize the functioning of the product during prototype development and initial testing phases.

3. Many problems and/or defects will be discovered by the consumers. Correct these problems and/or defects by modifying the design in a suitable manner.
4. Evaluate the risks involved and balance the design between product utility (benefits) and potential risks.
5. Conduct a thorough mathematical analysis, whcih should include kinematic, kinetic, stress, weight, and reliability analyses, during the design stage to insure the safety of the product.
6. Use proper safety and reliability factors during design.
7. Use materials with known properties.
8. Design the product to be safe against all foreseeable misuses and abuses.
9. Use well-established standards such as those recommended by ANSI, ASTM, ASME, SAE and Military Specifications in the design.

Guidelines to prevent manufacturing defects

1. Use proper process control, quality control, and inspection techniques to reduce manufacturing defects.
2. Use statistical sampling techniques to evaluate the adherence of production employees to design and manufacturing specifications.
3. If the potential risk of the product in causing injuries is high, consider using 100 percent inspection instead of statistical sampling.
4. Document all inspection, quality control, and testing activities and report the results to the product design and development department. This not only aids in making the product safer and cheaper but also in the manufacturer's defense in the event of a lawsuit.
5. Use proper labels and warnings about the use of the product.
6. Build all safety features and devices as part of the basic product instead of making them available as optional equipment.

References and Bibliography

D.1. V. J. Colangelo and P. A. Thornton, *Engineering Aspects of Product Liability*, American Society of Metals, Metals Park, Ohio, 1981.

D.2. C. E. Witherell, *How to Avoid Products Liability Lawsuits and Damages*, Noyes Publications, Park Ridge, New Jersey, 1985.

D.3. A. S. Weinstein, A. D. Twerski, H. R. Piehler, and W. A. Donaher, *Products Liability and the Reasonably Safe Product*, John Wiley, New York, 1978.

D.4. J. F. Thorpe and W. H. Middendorf, *What Every Engineer Should Know About Product Liability*, Marcel Dekker, Inc., New York, 1979.

Index